Genomics of Foodborne Bacterial Pathogens

Food Microbiology and Food Safety Series

Food Microbiology and Food Safety publishes valuable, practical, and timely resources for professionals and researchers working on microbiological topics associated with foods, as well as food safety issues and problems.

Series Editor

Michael P. Doyle, *Regents Professor and Director of the Center for Food Safety, University of Georgia, Griffith, GA, USA*

Editorial Board

For other titles published in this series, go to
www.springer.com/series/7131

Genomics of Foodborne Bacterial Pathogens

Editors

Martin Wiedmann
Department of Food Science, Cornell University, Stocking Hall 116, Ithaca, NY
14853, USA

Wei Zhang
Illinois Institute of Technology, National Center for Food Safety & Technology,
S. Archer Rd. 6502, Summit, IL 60501, USA

Editors

Martin Wiedmann
Department of Food Science
Cornell University
Stocking Hall 116
Ithaca, NY 14853, USA
mw16@cornell.edu

Wei Zhang
Illinois Institute of Technology
National Center for Food Safety
   and Technology
S. Archer Rd. 6502
Summit, IL 60501, USA
zhangw@iit.edu

ISBN 978-1-4419-7685-7   e-ISBN 978-1-4419-7686-4
DOI 10.1007/978-1-4419-7686-4
Springer New York Dordrecht Heidelberg London

Printed on acid-free paper

Springer is part of Springer Science+Business Media (www.springer.com)

Preface

Bacterial genomics is a rapidly expanding and fast moving research area. Considerable new knowledge has been gained by studying bacterial genomes and their functions. Genomics has now become an integral part of bacterial studies which provide new insights into the bacterial physiology, evolution, ecology, epidemiology, and pathogenesis.

A number of excellent review articles on recent genomic findings in various bacterial genera and species exist; however, it is difficult to find a comprehensive source of up-to-date information on the genomics of foodborne bacterial pathogens. For this reason, we formulated the idea of a book that provides in-depth reviews on the genomics of food-associated bacterial pathogens, as well as broad discussions on current applications of genomic, transcriptomic, and proteomic technologies in studying the bacterial physiology, epidemiology, and evolution, including the application of genomic tools to practical food safety problems. We hope that this book will be a useful reference for researchers interested in bacterial genomics and the use of genomic tools to investigate foodborne bacterial pathogens. As newer generations of sequencing and other genomic technologies, as well as analytical tools emerge at an unprecedented rate, we anticipate that the information provided in this book will present an introduction into the rapidly changing world of foodborne pathogen genomics.

We are most grateful for our contributors for their willingness, dedication, time, and efforts to make this book a reality. We feel humbled to co-write this book with a stellar group of internationally recognized experts in the cross-disciplinary fields of food microbiology and bacterial genomics. We specially acknowledge Susan Safren, who invited us to contribute to the *Food Microbiology & Food Safety* book series and provided us tremendous help along the journey of publishing this book. We are also thankful for Rita Beck for her kind assistance in the production of this book.

Ithaca, NY   Martin Wiedmann
Summit, IL   Wei Zhang

Contents

Contributors

Galeb S. Abu-Ali Microbial Evolution Laboratory, National Food Safety and Center, Michigan State University, East Lansing, MI, USA, abualiga@msu.edu

Helene Andrews-Polymenis Department of Microbial and Molecular Pathogenesis, College of Medicine, Texas A&M University System Health Science Center, College Station, TX 77843-11114, USA, handrews@medicine.tamhsc.edu

Eric Brown Center for Food Safety and Applied Nutrition, US Food and Drug Administration, College Park, MD, USA, eric.brown@fda.hhs.gov

Didier Cabanes Group of Molecular Microbiology, Institute for Molecular and Cell Biology, Porto, Portugal, didier@ibmc.up.pt

Rocio Canals Department of Pathology and Laboratory Medicine, Vaccine Research Institute of San Diego, University of California, Irvine, CA 92697, USA, rcanals@sdibr.org

Yi Chen Center for Food Safety and Applied Nutrition, US Food and Drug Administration, College Park, MD, USA, yi.chen@fda.hhs.gov

Pascale Cossart Unit of Bacteria Cell Interactions, Department of Cell Biology and Infection, Institute of Pasteur, Paris, France, pcossart@pasteur.fr

Edward G. Dudley Department of Food Science, Pennsylvania State University, University Park, PA, USA, egd100@psu.edu

Michelle Dziejman Department of Microbiology and Immunology, University of Rochester School of Medicine and Dentistry, Rochester, NY 14625, USA, michelle_dziejman@urmc.rochester.edu

Clifton K. Fagerquist Produce Safety and Microbiology Research Unit, US Department of Agriculture, Western Regional Research Center, Agricultural Research Service, Albany, CA 94710, USA, clifton.fagerquist@ars.usda.gov

Ozan Gundogdu Department of Infectious and Tropical Diseases, London School of Hygiene and Tropical Medicine, London, UK, ozan.gundogdu@lshtm.ac.uk

Karin Hoelzer Department of Food Science, Cornell University, Ithaca, NY 14853, USA, kh294@cornell.edu

Mark Joseph Jacobson Department of Bacteriology, University of Wisconsin, Madison, WI 53706, USA, mjjacobson@wisc.edu

Eric A. Johnson Department of Bacteriology, University of Wisconsin, Madison, WI 53706, USA, eajohnso@wisc.edu

Emily J. Kay Department of Infectious and Tropical Diseases, London School of Hygiene and Tropical Medicine, London WC1E 7HT, UK; Centre for Integrative Systems Biology, Imperial College London, London SW7 2AZ, UK, emily.kay@lshtm.ac.uk

Stephen J. Knabel Department of Food Science, Pennsylvania State University, University Park, PA, USA, sjk9@psu.edu

Anne-Brit Kolstø Department of Pharmaceutical Biosciences, University of Oslo, 0316 Oslo, Norway, a.b.kolsto@farmasi.uio.no

Jodi A. Lindsay Department of Cellular & Molecular Medicine, St George's University of London, London SW17 0RE, UK, jlindsay@sgul.ac.uk

Shannon D. Manning Microbiology and Molecular Genetics, Michigan State University, 194 Food Safety and Toxicology Building, E. Lansing, MI 48824, USA, shannon.manning@ht.msu.edu

Michael McClelland Department of Pathology and Laboratory Medicine, Vaccine Research Institute of San Diego, University of California, Irvine, CA 92697, USA, mmcclelland@sdibr.org

Sara R. Milillo Department of Food Science, Cornell University, Ithaca, NY 14853, USA, srm43@cornell.edu

Ole Andreas Økstad Department of Pharmaceutical Biosciences, University of Oslo, 0316 Oslo, Norway, aloechen@farmasi.uio.no

Nicole T. Perna Department of Genetics and the Genome Center of Wisconsin, University of Wisconsin, Madison, WI 53706, USA, ntperna@wisc.edu

Carlos A. Santiviago Departamento de Bioquímica y Biología Molecular, Facultad de Ciencias Químicas y Farmacéuticas, Universidad de Chile, Santiago, Chile, csantiviago@ciq.uchile.cl

Sandra Sousa Group of Molecular Microbiology, Institute for Molecular and Cell Biology, Porto, Portugal, srsousa@ibmc.up.pt

Joseph T. Wade New York State Department of Health, Wadsworth Center, Albany, NY, USA, jwade@wadsworth.org

Martin Wiedmann Department of Food Science, Cornell University, Ithaca, NY 14853, USA, mw16@cornell.edu

Brendan Wren Department of Infectious and Tropical Diseases, London School of Hygiene and Tropical Medicine, London, UK; Centre for Integrative Systems Biology, Imperial College London, London, UK, brendan.wren@lshtm.ac.uk

Fitnat H. Yildiz Department of Microbiology Environmental Toxicology, University of California, Santa Cruz, CA, USA, fyildiz@ucsc.edu

Wei Zhang National Center for Food Safety and Technology, Illinois Institute of Technology, Summit, IL 60501, USA, zhangw@iit.edu

Chapter 1
Overview: The Impact of Microbial Genomics on Food Safety

Sara R. Milillo, Martin Wiedmann, and Karin Hoelzer

1.1 Introduction

The first use of the term "genome" is attributed to Hans Winkler in his 1920 publication *Verbeitung und Ursache der Parthenogenesis im Pflanzen und Tierreiche* (Winkler, 1920). However, it was not until 1986 that the study of genomic concepts coalesced with the creation of a new journal by the same name (McKusick, 1997). The study of genomics was initially defined as the use or the application of "informatic tools" to study features of a sequenced genome (Strauss and Falkow, 1997). Today the field of genomics is typically considered to encompass efforts to determine the nucleic acid DNA sequence of an organism as well as the expression of genetic information using high-throughput, genome-wide methods, including transcriptomic, proteomic, and metabolomic analyses. As such, *functional genomics* describes efforts focused on understanding gene function and, increasingly, the function of other genomic elements. In contrast, *comparative genomics*, a field that has only recently become feasible due to advances in high-throughput genome analysis technologies, uses comparisons between genome sequences of organisms on a strain, a species, or a genus level to gain insights into organism biology and evolution. Unlocking the information hidden in the genomes of human pathogens, and foodborne pathogens in particular, will directly benefit society by improving public health and reducing productivity losses.

1.1.1 The Importance of Foodborne Pathogens

Illness caused by foodborne pathogens represents an important economic and public health burden worldwide. An estimated 76 million foodborne illness cases occur in the USA each year, including a considerable fraction with unknown causative agents or food sources (Mead et al., 1999). Based on this yearly case estimate and recent US census data, an estimated 25% of the US population is affected by

K. Hoelzer (✉)
Department of Food Science, Cornell University, Ithaca, NY 14853, USA
e-mail: kh294@cornell.edu

M. Wiedmann, W. Zhang (eds.), *Genomics of Foodborne Bacterial Pathogens*,
Food Microbiology and Food Safety, DOI 10.1007/978-1-4419-7686-4_1,
© Springer Science+Business Media, LLC 2011

foodborne illness each year. Viral foodborne pathogens, mainly members of the *Norovirus, Rotavirus, Astrovirus, Hepatovirus,* or *Enterovirus* genus, are thought to contribute approximately 50% of all foodborne illness cases in the USA (Mead et al., 1999; Widdowson et al., 2005). *Salmonella* is the leading bacterial foodborne pathogen in the USA, causing an estimated 14.8 foodborne illness cases/100,000 population/year, followed by *Campylobacter* (12.7 cases/100,000 population/year) and *Shigella* (6.1 cases/100,000 population/year) (CDC, 2007). Rates of foodborne illness as measured by cases of acute gastroenteritis are similar in many developed countries, for instance, England, Ireland, Canada, and Australia (Flint et al., 2005). While exact numbers are difficult to estimate, it is thought that the burden of foodborne illness in developing countries is even higher than in the developed world (WHO, 2007). Genomic research may be able to significantly improve food safety and ultimately help alleviate the burden of foodborne illness. Functional and comparative genomics are already contributing to the study of foodborne pathogens, and are exerting a measurable impact on food safety, as discussed below.

1.1.2 Genome Sequencing Technology

The number of completely sequenced and draft genomes has increased considerably over the past 5 years. The number of completed eukaryotic, bacterial, and archaeal genomes available in the Genomes Online Database (http://www.genomesonline. org/; Liolios et al., 2008) has risen from around 250 in 2005 to 1,154 in January 2010, and the trend is likely to continue. As case in point, in January 2010, 117 archaeal, 3,730 bacterial, and 1,250 eukaryotic organisms were listed in the Genomes Online database as being part of ongoing sequencing projects. As sequencing methods become more automated, faster, and less cost restrictive, genome information is expected to increase even more rapidly (Ronaghi et al., 1996). Full-genome sequence information is already available for a great variety of pathogens. As of January 2010, far in excess of 100 foodborne pathogen genomes are available in the Genome Database maintained by the National Center for Biotechnology Information (Benson et al., 2006; http://www.ncbi.nlm.nih.gov/ sites/entrez?db=genome), representing bacterial, protozoal, fungal, and viral genomes. Table 1.1 lists the number of completed and draft genome sequences available for some of the most common foodborne pathogens, including those associated with high mortality rates or high disease incidence.

Sanger sequencing. The first organism to have its entire genome published was bacteriophage MS2, an RNA virus (Fiers et al., 1976). The first DNA-based genome, of bacteriophage ΦX174, was published shortly thereafter, in 1977 (Sanger et al., 1977a). Later that same year, Sanger et al. (1977b) published an article detailing a chain termination sequencing technique often referred to as Sanger sequencing method. Briefly, Sanger sequencing reactions contain the desired DNA template in a single-stranded form, as well as labeled nucleotides, a DNA primer, and a DNA polymerase. One of four labeled chain-terminating dideoxynucleotide triphosphates, ddATP, ddTTP, ddCTP, or ddGTP, is subsequently added to the sequencing

Table 1.1   Examples of completed foodborne pathogen genome sequences[a]

Foodborne pathogen	Date of first genome release	No. of currently complete (no. of draft) sequences
Bacterial[b]		
L. monocytogenes	11/2001	4(24)
S. enterica subsp. enterica	10/2001	17(40)
E. coli	9/2001	41(73)
C. jejuni	9/2001	6(14)
Shigella flexneri	4/2003	3(3)
Protozoan[b]		
Cryptosporidium	5/2005	0(1)
Giardia	10/2007	0(1)
Viral[c]		
Norwalk virus	2/2003	1(1)
Sapovirus	7/2004	4(4)
Hepatitis A virus	11/2002	1(1)
Rotavirus	11/2005	2(2)

[a]Data collected as of 01/08/2010.
[b]Data available in Genomes Online Database.
[c]Data available in GenBank.

reaction. Once incorporated, the chain-terminating dideoxynucleotide triphosphate prevents elongation of the nascent DNA strand since it lacks the 3'-hydroxy group required to establish phosphodiester bonds between nucleotides. The newly synthesized, labeled DNA products are separated by size with a one-nucleotide resolution; each band represents a DNA fragment whose synthesis was terminated by incorporation of one of the four labeled dideoxynucleotides. The DNA sequence can then be inferred from the labeled DNA fragments, starting with the smallest fragment.

For more than 20 years, Sanger sequencing was the only sequencing method available (Schuster, 2008) and, surprisingly, its technology evolved little over three decades, despite some advances in automation (Smith et al., 1986). Full-genome sequencing using the Sanger method was most often approached by the so-called shotgun sequencing (Anderson, 1981). Shotgun sequencing relies on random breakdown of genomic DNA, followed by cloning of the fragments into carrier vectors to create a clone library. Vector inserts are subsequently sequenced, and the overlapping fragments are assembled to reflect contiguous portions of the genome referred to as "contigs." Small gaps remaining in the inferred sequence can then be directly sequenced using a variety of different methods, most commonly based on polymerase chain reaction (PCR) technology. The *Haemophilus influenzae* genome was the first bacterial genome sequenced using shotgun sequencing and was published in 1995 (Fleischmann et al., 1995). However, this approach is time consuming and expensive. The average cost of genome sequencing including annotation has been estimated to equal approximately $0.10 per nucleotide such that sequencing of one 3.5 Mb genome would infer costs of approximately $350,000 (Read et al.,

2002). In addition to the high cost, shotgun Sanger sequencing is limited by low throughput.

Sequencing by synthesis. In 1996, publication of an article entitled "Real-Time DNA Sequencing Using Detection of Pyrophosphate Release" signaled the advent of a new course for sequencing technology (Ronaghi et al., 1996). The manuscript detailed a novel DNA sequencing approach without the need for size separation, based on sequencing by synthesis and referred to as pyrosequencing. The procedure relies on the combination of three enzymes: a DNA polymerase, an ATP sulfurylase, and a luciferase. Pyrophosphate is released every time the DNA polymerase incorporates one or more nucleotides into the nascent DNA strand, one pyrophosphate per nucleotide incorporated. The sulfurylase quantitatively converts this pyrophosphate into adenosine triphosphate (ATP), which in turn fuels a luciferase-mediated reaction that emits visible light. The produced light is proportional to the amount of ATP available, itself a function of the available pyrophosphate. In order to allow reliable interpretation of the emitted light signal, only a single nucleotide type (i.e., adenosine, guanosine, thymidine, or cytidine) can be available for incorporation into the DNA at any given time, and large-scale use of this technology therefore requires sophisticated automation. In 2005, researchers from 454 Life Sciences (now partnered with Roche Applied Biosystems) and from the laboratory of George Church at Harvard Medical School independently published two innovative, high-throughput sequencing-by-synthesis methods (Margulies et al., 2005; Shendure et al., 2005).

In the *454 sequencing method*, genomic DNA is sheared to create random fragment libraries that are subsequently labeled with a common tag or adaptor. Then, the tagged fragments are diluted and individual fragments are captured onto beads, which are labeled with a complimentary tag. The bead-captured fragments are amplified by emulsion PCR. Then the DNA-coupled beads are deposited into wells of so-called PicoTiterPlates, one bead per well, where sequencing by synthesis proceeds as detailed above. Reagents for the sequencing reaction are provided by so-called flows, with "wash" steps in-between to ensure accuracy, such that light produced from a well is only due to the currently available nucleotide (see Rothberg and Leamon, 2008 for a recent review of the topic). This method developed by 454 Life Sciences is now available commercially through Roche, in the GS-FLX Genome Analyzer (Schuster, 2008).

Applied Biosystems released the *SOLiD sequencing system* (for *s*equencing by *o*ligonucleotide *li*gation and *d*etection), a four-color sequencing-by-ligation system that produces highly accurate sequence data based on the sequencing-by-synthesis technology developed by the Church lab (Shendure et al., 2005; Voelkerding et al., 2009). Fragment libraries for SOLiD sequencing are predominantly constructed using mate-pair technology (Church, 2006; Toner, 2005; Wilkinson, 2007). As in the 454 sequencing technique, template DNA is randomly fragmented, but then the fragments are separated by size, tagged, circularized, and digested by a type II restriction endonuclease, which liberates the fragment containing the newly joined ends. This brings sequences located at a specified distance apart in the fragment edges into proximity, allowing the sequences of both ends of each fragment to be determined sequentially, hence generating mate-paired sequences spaced at a known distance apart (Shendure et al., 2005). During SOLiD sequencing, fragments are immobilized on beads, amplified by emulsion PCR, and the sequence is determined through sequence-specific, sequential ligation of fluorescently labeled probes. The fluorescent signal of the ligated probe is detected, recorded, and the probe is cleaved, allowing another round of ligation, detection, and cleavage to proceed (see Rothberg and Leamon, 2008 for a recent review).

Solexa sequencing technology is offered commercially by Illumina, Inc. (Schuster, 2008). Solexa sequencing begins in a similar way as the method developed by 454 Life Sciences. Single-stranded genomic DNA is prepared, randomly fragmented, and two different adaptors are added, one for each fragment end. The DNA fragments are then added to flow cells, in which each channel is studded with complimentary adaptors to capture the fragmented DNA. One end of the DNA fragment will bind irreversibly to one flow channel adaptor, and the other end will bind reversibly to a different adaptor, forming a horseshoe shape. Free, unlabeled DNA polymerase and nucleotides are subsequently added to the flow cell, and DNA amplification proceeds for each bound fragment. After one round of amplification, two copies of the DNA fragment will result, each bound by one end to the flow cell channel. After several rounds of amplification, there will be tightly grouped clusters of each DNA fragment. Next, fluorescently labeled, reversibly chain-terminating nucleotides, DNA polymerase, and primers are added to the flow cell. The first nucleotide will be incorporated (after which amplification ceases) and then the first base in the sequence can be read using a laser to determine which labeled base was incorporated. The blocked 3' terminus of the nucleotide is removed and another round of amplification proceeds, followed by another round of laser excitation and determination of the second base (see Rothberg and Leamon, 2008 for a recent review).

VisiGen and *Helicos* are currently developing third-generation sequencing systems, which no longer require amplification of the DNA template prior to sequencing (Schuster, 2008). VisiGen technology relies on Förster resonance energy transfer (FRET) and uses fluorescently modified nucleotides as well as a fluorescently tagged DNA polymerase. The polymerase is immobilized and functions as fluorescent donor during DNA extension, thereby stimulating emission of a base-specific incorporation signature that can be detected nearly instantaneously. Helicos technology, on the contrary, relies only on fluorescently labeled nucleotides. Sample DNA is captured within flow cells, fluorescent nucleotides are incorporated into the nascent DNA strand, and all unbound nucleotides are removed during a washing step. The fluorescently labeled nucleotides incorporated in the nascent DNA strand are read by lasers, and their position is recorded. Subsequent cleavage of the fluorescent group allows the addition of further nucleotides (see Rothberg and Leamon, 2008 for a recent review).

The development of new technologies with even greater improvements in cost reduction will likely continue in the foreseeable future, due to the enormous demand for genome sequence data. As a case in point, a recent review by Genomes Online (Liolios et al., 2008) listed approximately 3,500 current and ongoing sequencing projects. The second- and third-generation sequencing techniques described above are still limited by their short read length. For instance, read lengths for SOLiD and Solexa sequencing are currently below 40 nucleotides (nts) but are likely to improve in the near future (Voelkerding et al., 2009). Pyrosequencing-based sequencing-by-synthesis methods were first reported to have a maximum read length of only approximately 100 nts, but innovative modifications have considerably increased read length already, allowing read lengths of approximately 500 nts (see Rothberg and Leamon, 2008 for an excellent review). In fact, the Sanger method had an approximate read length of 80 nts when it was first described, and as for Sanger sequencing, the read length of modern sequencing-by-synthesis methods is expected to improve considerably with time (Schuster, 2008). Sequencing-by-synthesis methods free researchers from relying on carrier vectors and clone libraries, both requirements for shotgun sequencing. Moreover, these new techniques greatly improve on the low degree of multiplicity in Sanger sequencing, already greatly reducing the cost of genome sequencing. Novel sequencing-by-synthesis methods may reduce the cost of genome sequencing per kilobase down to approx. one-tenth of that for Sanger sequencing (Shendure et al., 2005).

1.2 Comparative Genomics in Food Safety

Comparative genomics, the study of genome structure and function across organisms, includes direct comparisons of genomic sequences as well as gene expression mapping. As such, comparative genomics approaches can be classified into sequence-based approaches and those based on microarray technology.

1.2.1 Nucleotide Sequencing-Based Applications

The availability and use of genome sequences for microbial foodborne pathogens has and will continue to have major impacts on food safety. As detailed below, the increase in available genome information has improved genome annotation, allows increasingly powerful genome comparisons, supports outbreak investigations and disease surveillance efforts through novel subtyping techniques, has aided the detection of new pathogens, and provides new clues that link foodborne illness to chronic disease.

The role of genome annotation in food safety. Once genome sequencing is completed, the data must be transformed from a string of nucleotides into a more meaningful and useful collection of genes and other genetic elements. This process is called genome annotation and can be carried out in a variety of ways (Frishman, 2007). Predicted genes can be inferred using software such as GLIMMER (Delcher et al., 1999), and subsequently genes can be assigned putative functions based on their homology to previously annotated genes (e.g., using BLAST; Altschul et al., 1997). Genes can be classified into clusters of orthologous groups (e.g., with COG; Tatusov et al., 2000), characterized by their association with a known pathway (e.g., via KEGG; Kanehisa et al., 2004), or their putative translated sequences can be compared to conserved families of proteins (e.g., using Pfam; Finn et al., 2006). Automated gene prediction software is widely used for sequence analysis, but the output generally has to be revised manually (Stothard and Wishart, 2006). Three recently developed genome annotation software packages are PUMA2, ASAP, and MaGe (Glasner et al., 2006; Maltsev et al., 2006; Vallenet et al., 2006). Annotation software is steadily improving, and sequence data are often reanalyzed when new annotation software is released. Therefore several discrepant genome annotation versions may exist for some foodborne pathogens. Published genome sequences are made publicly available through a number of Web-based databases [e.g., GenBank (Benson et al., 2006), the European Molecular Biology Laboratory (EMBL; Cochrane et al., 2006), or the DNA Data Bank of Japan (Okubo et al., 2006)]. Most scientific journals require sequence deposition in one of these repositories as prerequisite for publication.

The power of genome annotation lies in the ability to facilitate genome comparisons and thus generate research hypotheses. As such, genome annotation may reveal homologies to known virulence factors or genes involved in stress survival (Raskin et al., 2006). For instance, annotation of the extraintestinal, uropathogenic *Escherichia coli* strain CFT073 genome revealed that it encodes many genes involved in fimbrial adhesion and phase switching, two characteristics associated with colonization and persistence in the intestinal lumen (Welch et al., 2002). Genome annotation also revealed that *Mycobacterium tuberculosis* possesses two copies of DNA polymerase *dnaE*. The second copy of *dnaE* (*dnaE2*) has been shown to be critical for virulence in vivo and it has been hypothesize that *dnaE2* may contribute to inducible mutagenesis, as it is upregulated following exposure to a variety of DNA-damaging treatments (Boshoff et al., 2003). *Cryptosporidium parvum* was found to encode only one pathway for the synthesis of guanine nucleotides, involving an inosine-5′-monophosphate dehydrogenase (IMPDH) encoding gene that appears to have been acquired via lateral transfer from a bacterial host (Umejiego et al., 2008). IMPDH-specific inhibitors were subsequently developed and proved more potent antiparasitic agents than the standard treatments (Umejiego et al., 2008). These and other studies continue to expand our understanding of foodborne bacterial pathogens. They provide tools to elucidate virulence and stress response mechanisms, and allow inference of evolutionary relationships between foodborne pathogens, thereby contributing to the development of novel treatment and prevention strategies.

The role of genome comparisons in food safety. Direct genome comparisons can be performed using sequence data and computer algorithms designed to search sequences and identify matching or similar sequences from other organisms (e.g., BLAST; Altschul et al., 1997). Nelson et al. (2004), for instance, used this approach to compare genome sequences of *Listeria monocytogenes* serotype 4b and 1/2a strains associated with human illness. Their analysis demonstrated that the genome sequences of *L. monocytogenes* strains associated with human disease are highly similar but identified some strain- and serotype-specific differences, including 83 serotype 1/2a- and 51 serotype 4b-specific genes. *Bacillus cereus* toxin cereulide, known to induce vomiting, has been shown to be encoded by a non-ribosomal peptide synthetase (NRPS) (Toh et al., 2004). This discovery was made possible by designing PCR primers recognizing orthologous genes in cyanobacterium *Microcystis*, and comparative sequence analysis using BLAST later showed that orthologous NRPS genes are present in the genomes of other *Bacillus* species but absent from the genomes of non-emetic *B. cereus* strains (Toh et al., 2004).

The impact of pathogen subtyping on outbreak investigation and disease surveillance. Outbreak investigations of foodborne infections are critical for determining outbreak sources and help identify potential pathogen control points along the food production chain. Increased pressure by consumers to deliver fresh, minimally processed foods (e.g., vegetables and leafy greens) has increased the risk of foodborne outbreaks on a national scale. Subtyping methods can characterize organisms isolated from foods, production environments, and human illness cases, and facilitate conclusions about contamination patterns, transmission routes, and outbreak sources.

The first subtyping methods were based on *phenotypic characteristics* of an organism, including biotyping, serotyping, and phage typing (for review, see Hyytia-Trees et al., 2007). However, these methods were laborious and often expensive, required specific reagents such as diagnostic antisera, and were limited in discriminatory power. *Molecular* subtyping methods were later introduced, fueled by a need for improved discriminatory power [e.g., pulsed field gel electrophoresis (PFGE)] and portability [e.g., multilocus sequence typing (MLST)]. More recently, the availability of genome sequences stimulated the development of new sequence-based methods for subtyping, e.g., multiple-locus variable-number tandem repeat analysis (MLVA), single-nucleotide polymorphism (SNP)-based subtyping, and whole genome sequencing (Cebula et al., 2005; Hoffmaster et al., 2002).

MLVA is increasingly used to subtype foodborne pathogens and for foodborne disease surveillance. The MLVA technique relies on the presence of variable numbers of tandem repeats (VNTR) in the genomic DNA. While short sequence DNA repeats, and specifically VNTR, have been well documented in bacterial genomes (van Belkum et al., 1998), development of MLVA-based subtyping typically requires the availability of at least one genome sequence for the target organism. The overall strategy for MLVA development is simple: design a multiplex PCR system to simultaneously amplify multiple regions of tandem repeats known to vary in length and separate the PCR products by size using a resolution of up to one base pair (Fig. 1.1). Product sizes correspond to the number of repeats and, thus, the allele present at that location. A recent study using *Bacillus anthracis*, a highly clonal organism, showed that MLVA was able to further discriminate isolates that were indistinguishable by two-enzyme PFGE (Hoffmaster et al., 2002). Recently, MLVA has also been applied to foodborne disease surveillance and outbreak detection, for example, involving *Salmonella enterica* subsp. *enterica* serotype Typhimurium (Torpdahl et al., 2006) or *E. coli* O157:H7 (Noller et al., 2003). In Denmark, MLVA helped identify an outbreak of foodborne illness caused by *Salmonella* Typhimurium and linked the outbreak to an imported cured sausage (Nygard et al., 2007).

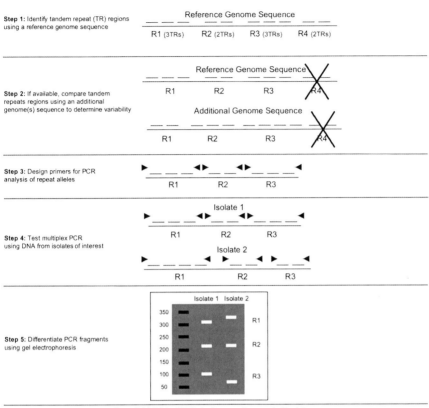

Fig. 1.1 Development of an MLVA-based subtyping method

A *SNP* (single-nucleotide polymorphism) is the result of a single-nucleotide change in an inter- or an intra-genic region of DNA of one organism relative to another, and detection of SNPs has become easier as the number of available genomes has increased (Hyytia-Trees et al., 2007). SNPs in an organism's genome can be detected using a variety of methods, including PCR-based methods, sequencing of small SNP-containing regions, and hybridization studies using microarrays (Cebula et al., 2005). SNP-based subtyping can be particularly useful for discriminating between organisms that appear to primarily evolve clonally. Table 1.2 lists SNP-based subtyping assays that have been developed for foodborne pathogens and published in peer-reviewed journals. Various academic research groups as well as a number of commercial companies have developed assays for SNP-based subtyping. For example, Applied Biosystems offers TaqMan® SNP Genotyping (RT-PCR-based assay) and the SNPlex™ Genotyping System (selective hybridization-based assay), Agencourt offers SNP discovery and re-sequencing services (using PCR-based technology), and Perkin–Elmer uses the FP-TDI system (PCR-based assay

Table 1.2 Examples of SNP-based subtyping assays designed for foodborne pathogens

Foodborne pathogen	Assay type	References
B. anthracis	Sequencing	Read et al. (2002)
L. monocytogenes	PCR w/fluorescent detection	Ducey et al. (2007)
	Real-time PCR	Rudi and Holck (2003)
	Allele-specific PCR	Moorhead et al. (2003)
S. enterica	Sequencing	Octavia and Lan (2007)
	Real-time PCR	Esaki et al. (2004)
C. jejuni	Real-time PCR	Best et al. (2007)
E. coli	Sequencing	Tartof et al. (2007)
	Sequencing	Hommais et al. (2005)

with fluorescent detection). Both Agencourt and Applied Biosystems also offer custom assay design services.

SNP-based subtyping can distinguish between very similar bacterial isolates, as demonstrated by Read and colleagues (2002) for *B. anthracis* and by Ducey et al. (2007) for *L. monocytogenes* lineage I. Read et al. (2002) analyzed an isolate of *B. anthracis* associated with the 2001 anthrax cases related to terrorist attacks in the USA and found that several SNPs in the isolate's genome sequence distinguished this isolate from a previously sequenced *B. anthracis* reference strain. This observation provided the basis for a powerful new tool that allows molecular characterization of *B. anthracis* isolates. Ducey et al. (2007) identified 413 SNPs that differentiated the 65 *L. monocytogenes* strains included in their study into 35 sequence types. This data, in combination with information on genomic insertions and deletions, allowed the creation of a new tool for foodborne outbreak detection and epidemiological investigations. In addition, some SNPs in the *Salmonella* Typhimurium genome show potential for use in SNP-based subtyping strategies (Hu et al., 2006). SNP-based subtyping is expected to increase in popularity as more genome sequences become available for SNP screening.

Whole genome sequencing is the most thorough and discriminatory method for subtyping microorganisms and relies on whole genome alignments, obtained, for example, through software such as MUMmer (Delcher et al., 1999) or MAUVE (Darling et al., 2004). Whole genome sequencing allows inference of clonal groups and phylogenetic relationships, with a high degree of certainty. Whole genome sequencing is not yet a feasible subtyping method for many bacterial, archaeal, and eukaryotic organisms but is increasingly used for viral subtyping. Currently, approx. 2,400 complete viral genomes are available in the Genome database (http://www.ncbi.nlm.nih.gov/genomes/GenomesGroup.cgi?taxid=10239&opt=Virus).

The currently completed viral genomes differ considerable in size and include small satellite virus genomes of approx. 200 nts as well as the genomes of significantly larger viruses, for instance poxviruses with genome sizes of over 350,000 nts. The widespread use of whole genome sequencing for important viral foodborne pathogens in the near future appears likely (Djikeng et al., 2008; Hardy, 2005).

Identification of unknown pathogens. Despite extensive efforts in the industrialized and developing world, the burden of foodborne illnesses on society has remained high, with an estimated 76 million foodborne illness cases occurring in the USA each year (Mead et al., 1999). In as many as two-thirds of foodborne outbreaks, no causative agent can be identified, indicating that a considerable number of foodborne pathogens may have yet to be described (Olsen et al., 2000). In fact, it has been estimated that once in every 2 years a previously undescribed – or in rare cases newly emerging – foodborne pathogen becomes widely recognized (Tauxe, 2002). The dramatic increase in available genome sequences has certainly contributed to the detection of previously unknown pathogens. Genome sequencing can therefore assist in the recognition and study of new foodborne pathogens. The primary means for defining new bacterial taxonomic groups is based on16S rRNA gene sequencing, and the increase in available sequence data greatly facilitates phylogenetic comparisons (Jay, 2003). Some examples of newly recognized bacterial foodborne pathogens include *Campylobacter concisus*, *Enterobacter sakazakii*, and enteroaggregative *E. coli* (Drudy et al., 2006; Fouts et al., 2005; Newell, 2005).

The identification of unknown microorganisms is complicated by the ability to distinguish the new species from related species, uncertainty of the genetic diversity encompassed by the new species, the complexities of establishing causal relationships between infection and human disease, and uncertainty of the underlying virulence mechanisms. Comparative genomics can provide very useful tools to address and overcome these issues. For instance, the availability of *Campylobacter* genome sequences has provided new grounds for developing sequence-based methods that allow distinction between *Campylobacter concisus*, *Campylobacter jejuni*, and *Campylobacter coli* (Fouts et al., 2005; Newell, 2005). A recent study based on 16S rRNA gene sequences showed that *E. sakazakii* is a genetically diverse species meriting further study (Lehner et al., 2004). Enteroaggregative *E. coli* is characterized by a unique and clearly distinct virulence-associated phenotype, but the determining virulence factors have so far remained unclear (Harrington et al., 2006). Genome sequencing efforts will enable the development of better techniques for discrimination among and between pathogens and facilitate comparative and functional genomics studies that will help elucidate virulence factors.

Metagenomics is defined as the genomic analysis of microbial communities using culture-independent methods (Schloss and Handelsman, 2003). Since metagenomics relies on genomic sequence data and is culture independent, this approach can be used to analyze the totality of microorganisms present in a particular environment without biasing against unculturable organisms and organisms that are difficult to culture using standard methodology. The first publication of comprehensive metagenomic research analyzed the microbial communities of soil environments (Rondon et al., 2000). In this study, DNA sequencing of all 16S rRNA genes present in a sample allowed approximation of the microbial diversity in the environment from which the sample was taken. Metagenomic principles have also been used to estimate the diversity of microorganisms in other natural or production environments, and metagenomic analysis has recently been used to study the microbial community of the human gut (Gill et al., 2006). Metagenomic approaches have also been applied to investigate unexplained epidemics. For instance, honey bee colonies affected by colony collapse disorder (CCD), a condition threatening to destroy the US honey bee population, were subjected to metagenomic analysis, and the results were compared to those obtained from unaffected controls (Cox-Foster et al., 2007). Several microorganisms were found to be associated with CCD, with moderate to high epidemiological support. Similarly, metagenomic analysis has recently been applied to study novel causative agents of human urinary tract infections (UTIs). Metagenomic analysis revealed associations between certain microorganisms and UTIs which could not be detected using standard culture methods (Imirzalioglu et al., 2008). Surprisingly, obligate anaerobic bacteria were detected in urine samples from patients who were suspected of UTIs but whose urine specimens were culture negative. In the future, metagenomic analyses may also contribute significantly to the diagnosis and detection of unculturable pathogens. For instance, Nakamura et al. (2008) report successfully using metagenomic approaches to diagnose *C. jejuni* infection in a culture-negative patient, and an increasing role of metagenomics for diagnostic applications in the future appears likely.

Emerging or re-emerging pathogens. Some well-recognized protozoan and viral pathogens have raised increasing concern in recent years. The underlying mechanisms are still subject to debate, but increasing commodity trade may be – at least partially – responsible. For instance, the increasing prevalence of human disease associated with *Cyclospora cayetanensis*, a protozoan parasite, has been linked to globalization of the food industry, and outbreaks have been linked to imported contaminated fruits from South America (Skovgaard, 2007). Hepatitis E virus is an important and well-recognized cause of viral hepatitis in developing countries (Smith, 2001). However, sporadic cases are increasingly being reported in the industrialized world, and contaminated pork products have been implicated as possible outbreak sources (Widdowson et al., 2003). An increased understanding of genome sequences, sequence diversity, and virulence determinants will allow for better detection and surveillance of these and other emerging or re-emerging pathogens, thereby benefitting public health and protecting international trade relations.

Foodborne pathogens and chronic disease. The importance of foodborne pathogens as underlying causes of chronic diseases is gradually being recognized. Associations between infections with a variety of foodborne pathogens and severe chronic and autoimmune disorders have been established. It has been estimated that 2–3% of foodborne illnesses result in chronic disease; however, the effects on economy and public health may be greater than those by all acute foodborne diseases (Archer and Kvenberg, 1985). Several chronic or autoimmune diseases have been associated with prior bacterial or viral infection, including reactive arthritis, Kawasaki disease, Guillain–Barré syndrome, and diarrhea-positive hemolytic uremic syndrome (HUS) and related renal failure (Bunning et al., 1997). Two of these diseases in particular have been linked to important foodborne pathogens: Guillain–Barré syndrome (GBS) to *C. jejuni* and HUS to *E. coli* O157:H7 (Razzaq, 2006; Tsang, 2002). GBS is a disorder of the nervous system, resulting in acute paralysis; HUS is a combination of three major symptoms: hemolytic anemia, thrombopenia, and acute renal failure (Razzaq, 2006; Tsang, 2002). GBS mortality, caused primarily by acute respiratory failure, can vary from 4 to 20%, and an estimated 12% of HUS patients die as a result of renal failure (Garg et al., 2003; Mehta, 2006). However, the underlying molecular mechanisms are not fully understood. Molecular mimicry has been detected between *C. jejuni* surface molecules and certain auto-reactive antibodies found in GBS patients (Tsang, 2002). Shiga toxins produced by *E. coli* O157:H7 can bind to Gb3 receptors on kidney cells, possibly stimulating the severe kidney damage associated with HUS (Razzaq, 2006). Increased understanding of the genome sequences will help reveal the interactions between pathogens and the human body, and provide clues as to how these pathogens elicit chronic illnesses. Increased availability of genome sequences will also help reveal other associations between chronic diseases and causative agents, and elucidate interactions between pathogen virulence factors and host receptors.

1.2.2 Microarray-Based Analyses

Microarrays evolved from Southern blotting, a technique where DNA fragments are immobilized on a substrate such as a nylon membrane and subsequently detected by hybridization with a known probe. Microarrays offer greater precision than blotting techniques and allow for higher throughput due to their multiplexing capacity (Stears et al., 2003). Traditional solid-phase microarrays consist of probe DNA immobilized on a solid surface, most commonly a glass slide, where several thousand different probes can be combined on a single chip. Bead arrays represent an alternative to solid-phase microarrays in that the DNA probes are instead spotted onto identifiable polystyrene beads. Probes are often oligonucleotides of 15–120 nts, which represent part, most, or all of a genetic region of interest (Schoolnik, 2002; Stears et al., 2003). However, PCR products, cDNA clones, or bacterial artificial chromosome (BAC) clones, ranging in length from 100 nts to 100 kb, have also been used as probes, especially in cases where the available genomic sequence information is limited (Gresham et al., 2008; Stears et al., 2003). Probes are spotted onto microarrays or in situ synthesized. In the former case, a priori-generated nucleotide fragments are printed onto the array surface using a specialized dispenser, for instance so-called inkjets or pins (Stears et al., 2003). Alternatively, short or longer oligonucleotides can be in situ synthesized directly on the chip (see Stears et al., 2003 for a recent review of the topic). Spotting strategies allow for greater assay versatility and are widely used for custom array manufacture, particularly in research settings, while synthesis technologies are used only for commercial manufacture (Stears et al., 2003).

Two-color microarrays are widely used for comparative genomics and gene expression profiling, by either using genomic DNA or, in the case of the latter, RNA, which is first converted to cDNA. In the former application, target DNA is collected from different organisms or cell types, while in the latter, RNA expression is compared under different experimental conditions. In both cases, the DNA is differentially labeled, most commonly using fluorescent dyes with non-overlapping fluorescence spectra, and then equal amounts of both target DNAs are hybridized to the array (see Gresham et al., 2008 for a recent review). After this, the hybridization reaction is incubated for a sufficient amount of time to reach equilibrium. The tightness of biding between probe and each target DNA molecule is correlated with the degree of sequence homology. The tighter the binding between the two DNA molecules is, the less likely the target DNA molecule will dissociate and be removed during subsequent rigorous washes, resulting in a more intense fluorescent signal. The microarray can be read using a variety of detection technologies such as confocal optics or photomultiplier tubes (Stears et al., 2003). After reading the array for each probe separately, the results for each well are overlaid. The use of two distinct labels for two different samples hybridized simultaneously to the same array allows for absolute as well as relative quantification. Depending on whether genomic DNA is isolated for gene comparison studies or mRNA is used to measure gene expression, these results provide either inference of the degree of sequence homology between related genomes or measurement of relative gene transcription under different experimental conditions. One-color microarrays are similar to two-color arrays but allow only analysis of one sample per time and therefore only generate absolute quantification results.

In recent years, a variety of new microarray applications have been developed. Genome tiling arrays rely on large sets of overlapping probes that cover a gene or whole genome in an unbiased fashion (Liu, 2007). Tiling arrays have, for instance, been used for transcriptome mapping, to characterize protein binding sites in the genome (i.e., ChIP-chip arrays), to map genome-wide methylation sites (i.e., MEDIP-chip or DNase-chip arrays), to compare DNA copy number differences (i.e., ARRAYCGH-chip array), and to describe nucleosome localization (Liu, 2007). Microarrays have also been used successfully for single-nucleotide polymorphism typing, to confirm the presence of pathogens in foods, animal feed or clinical diagnostics, or to test pathogens for the presence of specific antimicrobial resistance or virulence genes (LaFramboise, 2009; Rasooly and Herold, 2008).

Genome comparison using microarrays. Microarrays can be used for direct genomic comparisons. For instance, one study compared *L. monocytogenes* genomic DNA from 113 different strains and detected specific patterns of gene presence or absence among *L. monocytogenes* strains from different serotypes and lineages (Doumith et al., 2004). These results provided potential genetic targets for more efficient disease surveillance and outbreak tracking. Reen et al. (2005) used a similar approach to study genomic diversity among *S. enterica* isolates. They determined that the five recognized *Salmonella* pathogenicity islands were present in all isolates examined but identified more than 30 genomic regions that varied in presence among the tested isolates. Comparably little genomic diversity seemed to exist among the *Salmonella* study isolates within particular serovars, despite their wide geographic and temporal distribution.

Gene expression microarrays. Gene expression mapping is extensively used to compare levels of gene transcription under a set of a priori-specified experimental conditions. Comparative microarray data for gene expression mapping of *E. coli* O157:H7 grown in minimal media identified a dramatic effect of growth phase on gene transcription, which affected many genes thought to be involved in virulence

(Bergholz et al., 2007). For instance, Shiga toxin gene transcription increased significantly as *E. coli* O157:H7 transitioned into the stationary phase of growth. In another study, gene expression mapping was used to monitor host immune responses to *Salmonella* Enteritidis (Lillehoj et al., 2007). In this study, researchers followed changes in host cell gene transcription over 24 h. They noted that many host cell genes associated with cell adhesion and proliferation were downregulated following infection. Microarray analysis also revealed that although *Brucella* spp. lipopolysaccharide (LPS) induces a less dramatic immune response than does *E. coli* LPS, host cell defense-related genes such as those encoding proinflammatory cytokines and chemokines are upregulated following infection (Eskra et al., 2003; Rajashekara et al., 2006).

Comparative genomic studies such as the ones described above are extremely valuable and greatly contribute to food safety research. Studies such as these allow the identification of genomic regions responsible for niche adaptation, can reveal novel virulence genes, and may help explain the acquisition of antimicrobial resistance. The information gained from these studies can be used to control foodborne illness by allowing for the development of more specific and effective preventive measures and can help envision innovative treatment approaches.

1.3 Functional Genomics in Food Safety

The rise in the number of sequenced genomes has given birth to a new field of study, functional genomics. Functional genomics describes a collection of techniques that use genome sequences to derive likely functional attributes and interactions of the genes encoded therein (Suen et al., 2007). Interpreting the functions of genes and their possible contributions to virulence and stress response is critical for improving overall food safety. DNA sequence comparisons, microarray technology, RNA-Seq, protein expression analyses, and metabolic analyses are examples of experimental techniques that can be applied to functional genomics. Several of these methods are described in great detail in later chapters. The following section provides only a brief overview, and the reader is encouraged to refer to the respective chapters for more details.

1.3.1 DNA Sequence Comparisons

Numerous sequence-based techniques have been designed to facilitate exploration of probable gene functions and protein interactions. Several publically available databases compile gene homology and provide protein prediction services for a single or a range of organisms [e.g., LEGER for *Listeria* spp. (Dieterich et al., 2006), Prolinks (Bowers et al., 2004), PLEX (Date and Marcotte, 2005), and STRING (von Mering et al., 2005) for several organisms]. Based on the assumption that orthologous genes are conserved across bacteria from different genera and that these are likely to be linked functionally, a gene's phylogenetic profile can offer clues about likely functions (Suen et al., 2007). For example, the *Vibrio cholerae* genome was compared to 57 other bacteria, enabling reconstruction of the SoxR oxidative stress response pathway in *V. cholerae* by using sequence homology and knowledge of the pathway in the other related organisms (Date and Marcotte, 2003).

1.3.2 Transcriptome Profiling Using RNA-Seq

RNA-Seq is a recently developed transcriptomics approach that depends on high-throughput sequencing technology. In short, RNA of interest in enriched, converted into a cDNA library, adaptors are attached to each end of the cDNA fragments, and each molecule is subsequently sequenced using high-throughput DNA sequencing technology (see Wang et al., 2009 for a recent excellent review of the topic). RNA-Seq offers several advantages over more traditional transcriptomics approaches, including high versatility and independence from genome annotation, precision in detecting transcription boundaries, and inference regarding sequence diversity in the transcribed region. RNA-Seq is still in its infancy but has already provided valuable insights to foodborne pathogen research. For instance, Oliver et al. (2009) found that 83% of all *L. monocytogenes* genes were transcribed in stationary phase cells, as well as 67 noncoding RNA molecules (ncRNAs). They further demonstrated that RNA-Seq provides a valuable tool for quantitative and comparative transcriptomics, particularly when combined with appropriate bioinformatic tools. Similarly, RNA-Seq analysis of *V. cholerae* identified a large number of putative ncRNAs and a novel sRNA regulator of the carbon metabolism (Liu et al., 2009). As high-throughput sequencing technology becomes cheaper and more easily accessible, the importance of RNA-Seq for functional genomics will likely increase dramatically.

1.3.3 Microarray Technology

Microarrays can be used for functional genomic analyses. The function of a gene of interest can be inferred when microarrays are used to compare gene expression profiles for a wild-type (wt) strain versus an isogenic null mutant. For instance, Manganelli et al. (2001) studied the role of extracytoplasmic function (ECF) alternative sigma factor E (σ^E) in *M. tuberculosis* using microarrays and compared gene transcription of a *sigE* null mutant with wt *M. tuberculosis*. After exposure to the detergent sodium dodecyl sulfate, 23 genes were selectively upregulated in the presence of σ^E but not upregulated in the absence of σ^E. *Mycobacterium tuberculosis* σ^E therefore seems to play a functional role in regulating genes induced in response to cell wall damage. A study of the foodborne pathogen *L. monocytogenes* also used microarrays to interpret the function of a different alternative sigma factor, σ^B (Kazmierczak et al., 2003). The study showed that in addition to regulating stress-responsive systems, σ^B has a role in regulating virulence factors internalin A and internalin B. This has since been confirmed by several other studies, and subsequent microarray studies suggest that the transcriptional regulator σ^B

is involved in complex regulatory networks that regulate virulence gene expression as well as stress response (Ollinger et al., 2009; Raengpradub et al., 2008; Toledo-Arana et al., 2009). Microarray techniques were also used to monitor host responses in mice orally infected with *L. monocytogenes* or its non-pathogenic relative *Listeria innocua* (Lecuit et al., 2007). In this study, ileal RNA was collected from mice infected with both species, and then the host response profiles were compared, revealing a striking difference in the response to the two organisms despite their high degree of genetic similarity. Seven hundred and seventy-three genes were differentially regulated, 614 upregulated, and 159 downregulated, following infection with *L. monocytogenes* compared to *L. innocua* inoculated or the negative control mice. Further analyses comparing infections with *L. monocytogenes* wt to those with *inlA* or *hly* knockout strains suggested that the virulence factor Hly (listeriolysin) may be a more important determinant of host cell response compared to internalin A.

1.3.4 Protein Expression Analysis

Gene function studies can also directly compare and contrast protein expression under different conditions. This approach often provides important information as it has been noted repeatedly that cellular mRNA levels, the readouts of all transcriptome studies, which are generally more convenient and simpler to measure, frequently do not correspond well with protein levels (Greenbaum et al., 2003; Gygi et al., 1999; Holt et al., 2000). The study of gene function in terms of the proteins produced is called proteomics. The word "proteome" was first used in 1995, referring to all proteins encoded by an organism's genome (Blackstock and Weir, 1999).

Classical proteomics. Classical proteomic approaches rely on protein separation by two-dimensional gel electrophoresis (2-DE) and subsequent peptide mass mapping for protein identification (Cash, 2003). For example, the proteome of *E. sakazakii* was recently analyzed under different osmotic stresses using such a combination of 2-DE and matrix-assisted laser desorption ionization time-of-flight mass spectrometry (MALDI-TOF-MS) (Riedel and Lehner, 2007). Proteins were isolated and separated using 2-DE. Protein spots of interest were then excised from the gel and partially enzymatically digested. The digested protein fragments were again separated, and each fragments' mass was measured. Finally, the resulting mass profile was used to search fractionation databases for protein matches. This study revealed that the two different osmotic stresses resulted in distinct but overlapping protein expression profiles. Burns-Keliher et al. (1997) selectively labeled proteins produced by *Salmonella* Typhimurium during infection of epithelial cells and, using a similar proteomic analysis, determined that 34 Typhimurium proteins were uniquely expressed in the intracellular environment. A subsequent study found that protein expression by *S.* Typhimurium changes in different host cell types (Burns-Keliher et al., 1998). Fifty-eight proteins were detected which were uniquely expressed during growth in one of the five different host cell types studied. This demonstrated that *Salmonella* serotype Typhimurium tailors its response to varying host environments, which may be of paramount importance when designing new *Salmonella* vaccines.

Protein arrays. Protein arrays, also referred to as protein chips or protein-binding microarrays, offer an alternative to classical proteomics approaches. Some of the earliest methods have been reviewed by Nagayama (1997), and the reader is referred to this publication for a comprehensive review. Currently, multiple slightly different protein arrays are offered commercially, which are based on different support systems such as glass slide or nanowells. Each of the formats offers a high-throughput option for analyzing protein–protein, protein–DNA, protein–RNA, protein–ligand, or protein–substrate interactions (Zhu and Snyder, 2001). Together with protein mass-spectrophotometric identification methods, protein arrays have incredible potential for identifying and characterizing protein interactions and gene function (Walter et al., 2000).

Multidimensional protein identification technology (MudPIT). MudPIT, also referred to as multidimensional liquid chromatography and mass spectrometry (MDLC/MS/MS), represents another novel proteomics approach. MudPIT relies on a combination of two-dimensional capillary chromatography using strong cation exchangers and reverse phase chromatography, paired with mass spectrometry. Peptides are separated by charge and hydrophobicity, and sequenced by tandem mass spectrometry. Modern protein labeling reagents such as iTRAQ (sold commercially by Applied Biosystems) can be used to differentially label protein samples, and therefore allow for easy multiplexing and the generation of relative as well as absolute protein expression data. MudPIT has recently been used to compare the proteomes of *L. monocytogenes* lineage I and II strains during early stationary phase growth (Donaldson et al., 2009). The strains differed significantly in the expression of proteins involved in flagellar biosynthesis, DNA repair, stress response, and cell wall physiology, indicating a potential role of some or all of these proteins in determining the pathogenicity differences between the two lineages.

1.3.5 Metabolomic Studies

Metabolomics is another avenue with which to explore functional genomics. Metabolomics is the analysis, including identification and measurement, of all metabolites within a biochemical system (Oldiges et al., 2007). The field of metabolomic research has been expanding considerably in recent years, and the rate of metabolomic research publication has increased rapidly (Oldiges et al., 2007). For instance, a metabolomic study of *L. monocytogenes* membranes revealed a change in membrane lipid composition that is associated with growth at low temperatures (Mastronicolis et al., 2006). The total lipid anteiso-15:0/anteiso-17:0 fatty acid ratio and the neutral lipid membrane content increased markedly at 5°C, with a 10-fold increase observed for the former. These two changes are thought to reflect the mechanistic response of *L. monocytogenes* to low temperatures.

1.4 Impacts of Comparative and Functional Genome Analyses on Food Safety

The use of genome sequence data has clearly impacted food safety, including the development of new technologies and assays that are routinely used to detect and characterize foodborne pathogens. Functional genomics, on the contrary, mainly focused on improving our understanding of pathogen biology in varying environments.

1.4.1 Impacts of Comparative Genomics – Identification of Virulence Factors and Targets for Growth Inhibition

Comparative genomics had an important impact on the study of *L. monocytogenes* virulence when whole genome comparisons between *L. monocytogenes* and the closely related, non-pathogen species *L. innocua* identified a virulence factor specific to *L. monocytogenes* (Dussurget et al., 2002; Glaser et al., 2001; Vazquez-Boland et al., 2001). Dussurget and colleagues (2002) located a bile salt hydrolase gene specific to the *L. monocytogenes* genome and subsequently determined that this gene contributed to bile resistance, intestinal colonization, and systemic infection of the liver. The identification of new, previously uncharacterized virulence factors is one strength of comparative genomics and has lasting implications for understanding foodborne pathogen virulence and the treatment of foodborne illness. Another example of a recent study that used comparative genomics to further food safety was performed by Chan et al. (2007). In this study, the researchers described a microarray-based analysis of genes expressed by *L. monocytogenes* during growth at refrigeration temperatures. They found that a number of genes required for cell proliferation (e.g., RNA helicases) were upregulated under these conditions, while virulence factors were downregulated. This study identified potential targets to inhibit *L. monocytogenes* growth at low temperatures, thereby leading to more effective ways to control *L. monocytogenes* growth during food storage prior to consumption. Multilocus sequence typing and full-genome sequencing recently enabled Kingsley et al. (2009) to characterize a multidrug-resistant epidemic clone of *Salmonella* Typhimurium predominantly associated with invasive disease in sub-Saharan Africa and to identify the genetic basis for antimicrobial resistance and virulence. The authors showed that MDR-encoding genes were located on a virulence-associated plasmid and that the bacterial genome showed marked signs of degradation, providing evidence for niche adaptation and the co-selection of multidrug resistance and virulence. In combination with other research approaches and similar studies of other foodborne pathogens, comparative genomics research as outlined above will allow new strategies to control pathogen growth and improve food safety.

1.4.2 Impacts of Functional Genomics – Development of New Therapeutics and Vaccines

Functional genomics can be applied in several ways that impact vaccine development and the discovery of new therapeutics. An array constructed of 198 small molecules with putative antibacterial effects was used recently to probe for novel antibacterial agents against *Staphylococcus aureus* (Bowman et al., 2007). While further studies are required to determine the safety and applicability of the antibacterial agents identified in this study, the results are promising and may revolutionize the treatment of *Staphylococcal* infections. Similar approaches have also recently been used to identify small molecules that inhibit virulence of *V. cholerae*. Virstatin, an inhibitor of transcriptional regulator ToxT, has been shown to inhibit expression of cholera toxin and the toxin co-regulated pilus in vitro, and studies in mice have proven promising (Hung et al., 2005). Studies such as these provide innovative alternatives to classical antimicrobial drugs and may open the way for similar methods to be developed for other pathogens, increasing the hope that novel therapeutics may soon be available, particularly for multidrug-resistant foodborne pathogens such as non-typhoidal *Salmonella* (Alcaine et al., 2007). A protein-based study to identify novel vaccine targets recently used in vitro transcription and translation to study 197 *B. anthracis* proteins and their immunoreactivity (Gat et al., 2006). This study identified 30 novel, immunoreactive proteins related to virulence, representing numerous targets for development of new *B. anthracis* vaccines. As functional genomics techniques continue to be refined, they will thus, undoubtedly, lead to the identification of new drug targets, which will ultimately lead to the development of novel therapeutics and vaccines for various foodborne pathogens, thus improving the ability to prevent and treat foodborne illnesses.

1.5 Conclusions

The field of genomics has clearly already had a considerable impact on microbial food safety and will surely continue to do so in the future. Genomics has advanced food safety in numerous ways, and the applications are likely to diversify even more in the future. Genomics not only has considerably advanced our understanding of pathogen genetics but has also provided novel insights into virulence factors and the control of pathogen transmission, undoubtedly important contributions to food safety. Genomics will increase our understanding of how foodborne pathogens interact with natural and host environments, and provides clues that allow for translation of this knowledge into practical applications. As such, genomics will provide novel treatment strategies and help lessen the incidence and severity of foodborne illness worldwide.

References

Alcaine SD, Warnick LD, Wiedmann M (2007) Antimicrobial resistance in nontyphoidal *Salmonella*. J Food Prot 70:780–790

Altschul SF, Madden TL, Schaffer AA, Zhang J, Zhang Z, Miller W, Lipman DJ (1997) Gapped blast and psi-blast: a new generation of protein database search programs. Nucleic Acids Res 25:3389–3402

Anderson S (1981) Shotgun DNA sequencing using cloned DNAse i-generated fragments. Nucleic Acids Res 9:3015–3027

Archer DL, Kvenberg JE (1985) Incidence and cost of foodborne diarrhoeal disease in the United States. J Food Prot 48:887–894

Benson DA, Karsch-Mizrachi I, Lipman DJ, Ostell J, Wheeler DL (2006) Genbank. Nucleic Acids Res 34:D16–D20

Bergholz TM, Wick LM, Qi W, Riordan JT, Ouellette LM, Whittam TS (2007) Global transcriptional response of *Escherichia coli* O157:H7 to growth transitions in glucose minimal medium. BMC Microbiol 7:97

Best EL, Fox AJ, Owen RJ, Cheesbrough J, Bolton FJ (2007) Specific detection of *Campylobacter jejuni* from faeces using single nucleotide polymorphisms. Epidemiol Infect 135:839–846

Blackstock WP, Weir MP (1999) Proteomics: quantitative and physical mapping of cellular proteins. Trends Biotechnol 17:121–127

Boshoff HI, Reed MB, Barry CE 3rd, Mizrahi V (2003) Dnae2 polymerase contributes to in vivo survival and the emergence of drug resistance in *Mycobacterium tuberculosis*. Cell 113: 183–193

Bowers PM, Pellegrini M, Thompson MJ, Fierro J, Yeates TO, Eisenberg D (2004) Prolinks: a database of protein functional linkages derived from coevolution. Genome Biol 5:R35

Bowman MD, O'Neill JC, Stringer JR, Blackwell HE (2007) Rapid identification of antibacterial agents effective against *Staphylococcus aureus* using small-molecule macroarrays. Chem Biol 14:351–357

Bunning VK, Lindsay JA, Archer DL (1997) Chronic health effects of microbial foodborne disease. World Health Stat Q 50:51–56

Burns-Keliher L, Nickerson CA, Morrow BJ, Curtiss R 3rd (1998) Cell-specific proteins synthesized by *Salmonella* Typhimurium. Infect Immun 66:856–861

Burns-Keliher LL, Portteus A, Curtiss R 3rd (1997) Specific detection of *Salmonella* Typhimurium proteins synthesized intracellularly. J Bacteriol 179:3604–3612

Cash P (2003) Proteomics of bacterial pathogens. Adv Biochem Eng Biotechnol 83:93–115

CDC (Centers for Disease Control and Prevention) (2007) Preliminary FoodNet data on the incidence of infection with pathogens transmitted commonly through food – 10 states, 2006. MMWR Morb Mortal Wkly Rep 56:336–339

Cebula TA, Jackson SA, Brown EW, Goswami B, LeClerc JE (2005) Chips and SNPs, bugs and thugs: a molecular sleuthing perspective. J Food Prot 68:1271–1284

Chan YC, Raengpradub S, Boor KJ, Wiedmann M (2007) Microarray-based characterization of the *Listeria monocytogenes* cold regulon in log- and stationary-phase cells. Appl Environ Microbiol 73:6484–6498

Church GM (2006) Technology transfer and commercial scientific advisory roles (last revised 11/8/2006). http://arep.med.harvard.edu/gmc/tech.html

Cochrane G, Aldebert P, Althorpe N, Andersson M, Baker W, Baldwin A, Bates K, Bhattacharyya S, Browne P, van den Broek A, Castro M, Duggan K, Eberhardt R, Faruque N, Gamble J, Kanz C, Kulikova T, Lee C, Leinonen R, Lin Q, Lombard V, Lopez R, McHale M, McWilliam H, Mukherjee G, Nardone F, Pastor MP, Sobhany S, Stoehr P, Tzouvara K, Vaughan R, Wu D, Zhu W, Apweiler R (2006) EMBL nucleotide sequence database: developments in 2005. Nucleic Acids Res 34:D10–D15

Cox-Foster DL, Conlan S, Holmes EC, Palacios G, Evans JD, Moran NA, Quan PL, Briese T, Hornig M, Geiser DM, Martinson V, vanEngelsdorp D, Kalkstein AL, Drysdale A, Hui J, Zhai J, Cui L, Hutchison SK, Simons JF, Egholm M, Petti JS, Lipkin WI (2007) A metagenomic survey of microbes in honey bee colony collapse disorder. Science 318:283–287

Darling AC, Mau B, Blattner FR, Perna NT (2004) Mauve: multiple alignment of conserved genomic sequence with rearrangements. Genome Res 14:1394–1403

Date SV, Marcotte EM (2003) Discovery of uncharacterized cellular systems by genome wide analysis of functional linkages. Nat Biotech 21:1055–1062

Date SV, Marcotte EM (2005) Protein function prediction using the protein link explorer (PLEX). Bioinformatics 21:2558–2559

Delcher AL, Harmon D, Kasif S, White O, Salzberg SL (1999) Improved microbial gene identification with GLIMMER. Nucleic Acids Res 27:4636–4641

Dieterich G, Karst U, Fischer E, Wehland J, Jansch L (2006) Leger: knowledge database and visualization tool for comparative genomics of pathogenic and non-pathogenic *Listeria* species. Nucleic Acids Res 34:D402–D406

Djikeng A, Halpin R, Kuzmickas R, Depasse J, Feldblyum J, Sengamalay N, Afonso C, Zhang X, Anderson NG, Ghedin E, Spiro DJ (2008) Viral genome sequencing by random priming methods. BMC Genomics 9:5

Donaldson JR, Nanduri B, Burgess SC, Lawrence ML (2009) Comparative proteomic analysis of *Listeria monocytogenes* strains F2365 and EGD. Appl Environ Microbiol 75:366–373

Doumith M, Cazalet C, Simoes N, Frangeul L, Jacquet C, Kunst F, Martin P, Cossart P, Glaser P, Buchrieser C (2004) New aspects regarding evolution and virulence of *Listeria monocytogenes* revealed by comparative genomics and DNA arrays. Infect Immun 72:1072–1083

Drudy D, Mullane NR, Quinn T, Wall PG, Fanning S (2006) *Enterobacter sakazakii*: an emerging pathogen in powdered infant formula. Clin Infect Disease 42:996–1002

Ducey TF, Page B, Usgaard T, Borucki MK, Pupedis K, Ward TJ (2007) A single nucleotide-polymorphism-based multilocus genotyping assay for subtyping lineage I isolates of *Listeria monocytogenes*. Appl Environ Microbiol 73:133–147

Dussurget O, Cabanes D, Dehoux P, Lecuit M, Buchrieser C, Glaser P, Cossart P (2002) *Listeria monocytogenes* bile salt hydrolase is a PrfA-regulated virulence factor involved in the intestinal and hepatic phases of listeriosis. Mol Microbiol 45:1095–1106

Esaki H, Noda K, Otsuki N, Kojima A, Asai T, Tamura Y, Takahashi T (2004) Rapid detection of quinolone-resistant *Salmonella* by real time SNP genotyping. J Microbiol Methods 58:131–134

Eskra L, Mathison A, Splitter G (2003) Microarray analysis of mRNA levels from RAW264.7 macrophages infected with *Brucella abortus*. Infect Immun 71:1125–1133

Fiers W, Contreras R, Duerinck F, Haegeman G, Iserentant D, Merregaert J, Min Jou W, Molemans F, Raeymaekers A, Van den Berghe A, Volckaert G, Ysebaert M (1976) Complete nucleotide sequence of bacteriophage ms2 RNA: primary and secondary structure of the replicase gene. Nature 260:500–507

Finn RD, Mistry J, Schuster-Bockler B, Griffiths-Jones S, Hollich V, Lassmann T, Moxon S, Marshall M, Khanna A, Durbin R, Eddy SR, Sonnhammer EL, Bateman A (2006) Pfam: clans, web tools and services. Nucleic Acids Res 34:D247–D251

Fleischmann RD, Adams MD, White O, Clayton RA, Kirkness EF, Kerlavage AR, Bult CJ, Tomb JF, Dougherty BA, Merrick JM et al (1995) Whole-genome random sequencing and assembly of *Haemophilus influenzae* Rd. Science 269:496–512

Flint J, Duynhoven YV, Angulo FJ, DeLong SM, Braun P, Kirk M, Scallan E, Fitzgerald M, Adak GK, Sockett P, Ellis A, Hall G, Gargouri N, Walke H, Braam P (2005) Estimating the burden of acute gastroenteritis, foodborne disease, and pathogens commonly transmitted by food: an international review. Clin Infect Dis 41:698–704

Fouts DE, Mongodin EF, Mandrell RE, Miller WG, Rasko DA, Ravel J, Brinkac LM, DeBoy RT, Parker CT, Daugherty SC, Dodson RJ, Durkin AS, Madupu R, Sullivan SA, Shetty JU, Ayodeji MA, Shvartsbeyn A, Schatz MC, Badger JH, Fraser CM, Nelson KE (2005) Major structural differences and novel potential virulence mechanisms from the genomes of multiple *Campylobacter* species. PLoS Biol 3:e15

Frishman D (2007) Protein annotation at genomic scale: the current status. Chem Rev 107:3448–3466

Garg AX, Suri RS, Barrowman N, Rehman F, Matsell D, Rosas-Arellano MP, Salvadori M, Haynes RB, Clark WF (2003) Long-term renal prognosis of diarrhea-associated hemolytic uremic syndrome: a systematic review, meta-analysis, and meta-regression. JAMA 290:1360–1370

Gat O, Grosfeld H, Ariel N, Inbar I, Zaide G, Broder Y, Zvi A, Chitlaru T, Altboum Z, Stein D, Cohen S, Shafferman A (2006) Search for *Bacillus anthracis* potential vaccine candidates by a functional genomic—serologic screen. Infect Immun 74:3987–4001

Gill SR, Pop M, DeBoy RT, Eckburg PB, Turnbaugh PJ, Samuel BS, Gordon JI, Relman DA, Fraser-Liggett CM, Nelson KE (2006) Metagenomic analysis of the human distal gut microbiome. Science 312:1355–1359

Glaser P, Frangeul L, Buchrieser C, Rusniok C, Amend A, Baquero F, Berche P, Bloecker H, Brandt P, Chakraborty T, Charbit A, Chetouani F, Couve E, de Daruvar A, Dehoux P, Domann E, Dominguez-Bernal G, Duchaud E, Durant L, Dussurget O, Entian KD, Fsihi H, Garcia-del Portillo F, Garrido P, Gautier L, Goebel W, Gomez-Lopez N, Hain T, Hauf J, Jackson D, Jones LM, Kaerst U, Kreft J, Kuhn M, Kunst F, Kurapkat G, Madueno E, Maitournam A, Vicente JM, Ng E, Nedjari H, Nordsiek G, Novella S, de Pablos B, Perez-Diaz JC, Purcell R, Remmel B, Rose M, Schlueter T, Simoes N, Tierrez A, Vazquez-Boland JA, Voss H, Wehland J, Cossart P (2001) Comparative genomics of *Listeria* species. Science 294:849–852

Glasner JD, Rusch M, Liss P, Plunkett G 3rd, Cabot EL, Darling A, Anderson BD, Infield-Harm P, Gilson MC, Perna NT (2006) Asap: a resource for annotating, curating, comparing, and disseminating genomic data. Nucleic Acids Res 34:D41–D45

Greenbaum D, Colangelo C, Williams K, Gerstein M (2003) Comparing protein abundance and mRNA expression levels on a genomic scale. Genome Biol 4:117

Gresham D, Dunham MJ, Botstein D (2008) Comparing whole genomes using DNA microarrays. Nat Rev Genet 9:291–302

Gygi SP, Rochon Y, Franza BR, Aebersold R (1999) Correlation between protein and mRNA abundance in yeast. Mol Cell Biol 19:1720–1730

Hardy ME (2005) Norovirus protein structure and function. FEMS Microbiol Lett 253:1–8

Harrington SM, Dudley EG, Nataro JP (2006) Pathogenesis of enteroaggregative *Escherichia coli* infection. FEMS Microbiol Lett 254:12–18

Hoffmaster AR, Fitzgerald CC, Ribot E, Mayer LW, Popovic T (2002) Molecular subtyping of *Bacillus anthracis* and the 2001 bioterrorism-associated anthrax outbreak, United States. Emerg Infect Dis 8:1111–1116

Holt LJ, Enever C, de Wildt RM, Tomlinson IM (2000) The use of recombinant antibodies in proteomics. Curr Opin Biotechnol 11:445–449

Hommais F, Pereira S, Acquaviva C, Escobar-Paramo P, Denamur E (2005) Single nucleotide polymorphism phylotyping of *Escherichia coli*. Appl Environ Microbiol 71:4784–4792

Hu H, Lan R, Reeves PR (2006) Adaptation of multilocus sequencing for studying variation within a major clone: evolutionary relationships of *Salmonella enterica* serovar Typhimurium. Genetics 172:743–750

Hung DT, Shakhnovich EA, Pierson E, Mekalanos JJ (2005) Small-molecule inhibitor of *Vibrio cholerae* virulence and intestinal colonization. Science 310:670–674

Hyytia-Trees EK, Cooper K, Ribot EM, Gerner-Smidt P (2007) Recent developments and future prospects in subtyping of foodborne bacterial pathogens. Future Microbiol 2:175–185

Imirzalioglu C, Hain T, Chakraborty T, Domann E (2008) Hidden pathogens uncovered: metagenomic analysis of urinary tract infections. Andrologia 40:66–71

Jay JM (2003) A review of recent taxonomic changes in seven genera of bacteria commonly found in foods. J Food Prot 66:1304–1309

Kanehisa M, Goto S, Kawashima S, Okuno Y, Hattori M (2004) The KEGG resource for deciphering the genome. Nucleic Acids Res 32:D277–D280

Kazmierczak MJ, Mithoe SC, Boor KJ, Wiedmann M (2003) *Listeria monocytogenes* sigma B regulates stress response and virulence functions. J Bacteriol 185:5722–5734

Kingsley RA, Msefula CL, Thomson NR, Kariuki S, Holt KE, Gordon MA, Harris D, Clarke L, Whitehead S, Sangal V, Marsh K, Achtman M, Molyneux ME, Cormican M, Parkhill J, MacLennan CA, Heyderman RS, Dougan G (2009) Epidemic multiple drug resistant *Salmonella* Typhimurium causing invasive disease in sub-Saharan Africa have a distinct genotype. Genome Res 19:2279–2287

LaFramboise T (2009) Single nucleotide polymorphism arrays: a decade of biological, computational and technological advances. Nucleic Acids Res 37:4181–4193

Lecuit M, Sonnenburg JL, Cossart P, Gordon JI (2007) Functional genomic studies of the intestinal response to a foodborne enteropathogen in a humanized gnotobiotic mouse model. J Biol Chem 282:15065–15072

Lehner A, Tasara T, Stephan R (2004) 16S rRNA gene based analysis of *Enterobacter sakazakii* strains from different sources and development of a PCR assay for identification. BMC Microbiol 4:43

Lillehoj HS, Kim CH, Keeler CL Jr, Zhang S (2007) Immunogenomic approaches to study host immunity to enteric pathogens. Poult Sci 86:1491–1500

Liolios K, Mavromatis K, Tavernarakis N, Kyrpides NC (2008) The genomes on line database (GOLD) in 2007: status of genomic and metagenomic projects and their associated metadata. Nucleic Acids Res 36:D475–D479

Liu XS (2007) Getting started in tiling microarray analysis. PLoS Comput Biol 3:1842–1844

Liu JM, Livny J, Lawrence MS, Kimball MD, Waldor MK, Camilli A (2009) Experimental discovery of sRNAs in *Vibrio cholerae* by direct cloning, 5S/tRNA depletion and parallel sequencing. Nucleic Acids Res 37. doi:10.1093/nar/gkp080

Maltsev N, Glass E, Sulakhe D, Rodriguez A, Syed MH, Bompada T, Zhang Y, D'Souza M (2006) PUMA2 – grid-based high-throughput analysis of genomes and metabolic pathways. Nucleic Acids Res 34:D369–D372

Manganelli R, Voskuil MI, Schoolnik GK, Smith I (2001) The *Mycobacterium tuberculosis* ECF sigma factor sigmaE: role in global gene expression and survival in macrophages. Mol Microbiol 41:423–437

Margulies M, Egholm M, Altman WE, Attiya S, Bader JS, Bemben LA, Berka J, Braverman MS, Chen Y-J, Chen Z, Dewell SB, Du L, Fierro JM, Gomes XV, Godwin BC, He W, Helgesen S, Ho CH, Irzyk GP, Jando SC, Alenquer MLI, Jarvie TP, Jirage KB, Kim J-B, Knight JR, Lanza JR, Leamon JH, Lefkowitz SM, Lei M, Li J, Lohman KL, Lu H, Makhijani VB, McDade KE, McKenna MP, Myers EW, Nickerson E, Nobile JR, Plant R, Puc BP, Ronan MT, Roth GT, Sarkis GJ, Simons JF, Simpson JW, Srinivasan M, Tartaro KR, Tomasz A, Vogt KA, Volkmer GA, Wang SH, Wang Y, Weiner MP, Yu P, Begley RF, Rothberg JM (2005) Genome sequencing in microfabricated high-density picolitre reactors. Nature 437:376–380

Mastronicolis SK, Boura A, Karaliota A, Magiatis P, Arvanitis N, Litos C, Tsakirakis A, Paraskevas P, Moustaka H, Heropoulos G (2006) Effect of cold temperature on the composition of different lipid classes of the foodborne pathogen *Listeria monocytogenes*: focus on neutral lipids. Food Microbiol 23:184–194

McKusick VA (1997) Genomics: structural and functional studies of genomes. Genomics 45: 244–249

Mead PS, Slutsker L, Dietz V, McCaig LF, Bresee JS, Shapiro C, Griffin PM, Tauxe RV (1999) Food-related illness and death in the United States. Emerg Infect Dis 5:607–625

Mehta S (2006) Neuromuscular disease causing acute respiratory failure. Respir Care 51: 1016–1021

Moorhead SM, Dykes GA, Cursons RT (2003) An SNP-based PCR assay to differentiate between *Listeria monocytogenes* lineages derived from phylogenetic analysis of the sigB gene. J Microbiol Methods 55:425–432

Nagayama K (1997) Protein arrays: concepts and subjects. Adv Biophys 34:3–23

Nakamura, Maeda, Miron, Yoh, Izutsu, Kataoka, Honda, Yasunaga, Nakaya, Kawai, Hayashizaki, Horii, Iida, 2008] Nakamura S, Maeda N, Miron IM, Yoh M, Izutsu K, Kataoka C, Honda T, Yasunaga T, Nakaya T, Kawai J, Hayashizaki Y, Horii T, Iida T (2008) Metagenomic diagnosis of bacterial infections. Emerg Infect Dis 14:1784–1786

Nelson KE, Fouts DE, Mongodin EF, Ravel J, DeBoy RT, Kolonay JF, Rasko DA, Angiuoli SV, Gill SR, Paulsen IT, Peterson J, White O, Nelson WC, Nierman W, Beanan MJ, Brinkac LM, Daugherty SC, Dodson RJ, Durkin AS, Madupu R, Haft DH, Selengut J, Van Aken S, Khouri H, Fedorova N, Forberger H, Tran B, Kathariou S, Wonderling LD, Uhlich GA, Bayles DO,

Luchansky JB, Fraser CM (2004) Whole genome comparisons of serotype 4b and 1/2a strains of the food-borne pathogen *Listeria monocytogenes* reveal new insights into the core genome components of this species. Nucleic Acids Res 32:2386–2395

Newell DG (2005) *Campylobacter concisus*: an emerging pathogen? Eur J Gastroenterol Hepatol 17:1013–1014

Noller AC, McEllistrem MC, Pacheco AG, Boxrud DJ, Harrison LH (2003) Multilocus variable-number tandem repeat analysis distinguishes outbreak and sporadic *Escherichia coli* O157:H7 isolates. J Clin Microbiol 41:5389–5397

Nygard K, Lindstedt BA, Wahl W, Jensvoll L, Kjelso C, Molbak K, Torpdahl M, Kapperud G (2007) Outbreak of *Salmonella* Typhimurium infection traced to imported cured sausage using MLVA-subtyping. Euro Surveill 12:E0703155

Octavia S, Lan R (2007) Single-nucleotide-polymorphism typing and genetic relationships of *Salmonella enterica* serovar Typhi isolates. J Clin Microbiol 45:3795–3801

Okubo K, Sugawara H, Gojobori T, Tateno Y (2006) DDBJ in preparation for overview of research activities behind data submissions. Nucleic Acids Res 34:D6–D9

Oldiges M, Lutz S, Pflug S, Schroer K, Stein N, Wiendahl C (2007) Metabolomics: current state and evolving methodologies and tools. Appl Microbiol Biotechnol 76:495–511

Oliver HF, Orsi RH, Ponnala L, Keich U, Wang W, Sun Q, Cartinhour SW, Filiatrault MJ, Wiedmann M, Boor KJ (2009) Deep RNA sequencing of *L. monocytogenes* reveals overlapping and extensive stationary phase and sigma B-dependent transcriptomes, including multiple highly transcribed noncoding RNAs. BMC Genomics 10:641

Ollinger J, Bowen B, Wiedmann M, Boor KJ, Bergholz TM (2009) *Listeria monocytogenes* {sigma}B modulates PrfA-mediated virulence factor expression. Infect Immun 77:2113–2124

Olsen SJ, MacKinnon LC, Goulding JS, Bean NH, Slutsker L (2000) Surveillance for foodborne-disease outbreaks – United States, 1993–1997. MMWR CDC Surveill Summ 49:1–62

Raengpradub S, Wiedmann M, Boor KJ (2008) Comparative analysis of the {sigma}B-dependent stress responses in *Listeria monocytogenes* and *Listeria innocua* strains exposed to selected stress conditions. Appl Envir Microbiol 74:158–171

Rajashekara G, Eskra L, Mathison A, Petersen E, Yu Q, Harms J, Splitter G (2006) *Brucella*: functional genomics and host–pathogen interactions. Anim Health Res Rev 7:1–11

Raskin DM, Seshadri R, Pukatzki SU, Mekalanos JJ (2006) Bacterial genomics and pathogen evolution. Cell 124:703–714

Rasooly A, Herold KE (2008) Food microbial pathogen detection and analysis using DNA microarray technologies. Foodborne Pathog Dis 5:531–550

Razzaq S (2006) Hemolytic uremic syndrome: an emerging health risk. Am Fam Physician 74:991–996

Read TD, Salzberg SL, Pop M, Shumway M, Umayam L, Jiang L, Holtzapple E, Busch JD, Smith KL, Schupp JM, Solomon D, Keim P, Fraser CM (2002) Comparative genome sequencing for discovery of novel polymorphisms in *Bacillus anthracis*. Science 296:2028–2033

Reen FJ, Boyd EF, Porwollik S, Murphy BP, Gilroy D, Fanning S, McClelland M (2005) Genomic comparisons of *Salmonella enterica* serovar Dublin, Agona, and Typhimurium strains recently isolated from milk filters and bovine samples from Ireland, using a *Salmonella* microarray. Appl Environ Microbiol 71:1616–1625

Riedel K, Lehner A (2007) Identification of proteins involved in osmotic stress response in *Enterobacter sakazakii* by proteomics. Proteomics 7:1217–1231

Ronaghi M, Karamohamed S, Pettersson B, Uhlen M, Nyren P (1996) Real-time DNA sequencing using detection of pyrophosphate release. Anal Biochem 242:84–89

Rondon MR, August PR, Bettermann AD, Brady SF, Grossman TH, Liles MR, Loiacono KA, Lynch BA, MacNeil IA, Minor C, Tiong CL, Gilman M, Osburne MS, Clardy J, Handelsman J, Goodman RM (2000) Cloning the soil metagenome: a strategy for accessing the genetic and functional diversity of uncultured microorganisms. Appl Environ Microbiol 66:2541–2547

Rothberg JM, Leamon JH (2008) The development and impact of 454 sequencing. Nat Biotechnol 26:1117–1124

Rudi K, Holck AL (2003) Real-time closed tube single nucleotide polymorphism (SNP) quantification in pooled samples by quencher extension (qext). Nucleic Acids Res 31:e117

Sanger F, Air GM, Barrell BG, Brown NL, Coulson AR, Fiddes CA, Hutchison CA, Slocombe PM, Smith M (1977a) Nucleotide sequence of bacteriophage phi x174 DNA. Nature 265:687–695

Sanger F, Nicklen S, Coulson AR (1977b) DNA sequencing with chain-terminating inhibitors. Proc Natl Acad Sci 74:5463–5467

Schloss PD, Handelsman J (2003) Biotechnological prospects from metagenomics. Curr Opin Biotechnol 14:303–310

Schoolnik GK (2002) Functional and comparative genomics of pathogenic bacteria. Curr Opin Microbiol 5:20–26

Schuster SC (2008) Next-generation sequencing transforms today's biology. Nat Methods 5:16–18

Shendure J, Porreca GJ, Reppas NB, Lin X, McCutcheon JP, Rosenbaum AM, Wang MD, Zhang K, Mitra RD, Church GM (2005) Accurate multiplex polony sequencing of an evolved bacterial genome. Science 309:1728–1732

Skovgaard N (2007) New trends in emerging pathogens. Int J Food Microbiol 120:217–224

Smith JL (2001) A review of hepatitis E virus. J Food Prot 64:572–586

Smith LM, Sanders JZ, Kaiser RJ, Hughes P, Dodd C, Connell CR, Heiner C, Kent SB, Hood LE (1986) Fluorescence detection in automated DNA sequence analysis. Nature 321:674–679

Stears RL, Martinsky T, Schena M (2003) Trends in microarray analysis. Nat Med 9:140–145

Stothard P, Wishart DS (2006) Automated bacterial genome analysis and annotation. Curr Opin Microbiol 9:505–510

Strauss EJ, Falkow S (1997) Microbial pathogenesis: genomics and beyond. Science 276:707–712

Suen G, Arshinoff BI, Taylor RG, Welch RD (2007) Practical applications of bacterial functional genomics. Biotechnol Genet Eng Rev 24:213–242

Tartof SY, Solberg OD, Riley LW (2007) Genotypic analyses of uropathogenic *Escherichia coli* based on fimH single nucleotide polymorphisms (SNPs). J Med Microbiol 56:1363–1369

Tatusov RL, Galperin MY, Natale DA, Koonin EV (2000) The COG database: a tool for genome-scale analysis of protein functions and evolution. Nucleic Acids Res 28:33–36

Tauxe RV (2002) Emerging foodborne pathogens. Int J Food Microbiol 78:31–41

Toh M, Moffitt MC, Henrichsen L, Raftery M, Barrow K, Cox JM, Marquis CP, Neilan BA (2004) Cereulide, the emetic toxin of *Bacillus cereus*, is putatively a product of nonribosomal peptide synthesis. J Appl Microbiol 97:992–1000

Toledo-Arana A, Dussurget D, Nikitas G, Sesto N, Guet-Revillet H, Balestrino D, Loh E, Gripenland J, Tiensuu T, Vaitkevicius K, Barthelemy M, Vergassola M, Nahori MA, Soubigou G, Régnault B, Coppée JY, Lecuit M, Johansson J, Cossart P (2009) The *Listeria* transcriptional landscape from saprophytism to virulence. Nature 459:950–956

Toner B (2005) Harvard' s church calls for open source, non-anonymous personal genome project. Genome Web Daily News, 14 Nov. http://www.genomeweb.com/issues/news/125128-1.html

Torpdahl M, Sorensen G, Ethelberg S, Sando G, Gammelgard K, Porsbo LJ (2006) A regional outbreak of *S.* Typhimurium in Denmark and identification of the source using MLVA typing. Euro Surveill 11:134–136

Tsang RS (2002) The relationship of *Campylobacter jejuni* infection and the development of Guillain–Barre syndrome. Curr Opin Infect Dis 15:221–228

Umejiego NN, Gollapalli D, Sharling L, Volftsun A, Lu J, Benjamin NN, Stroupe AH, Riera TV, Striepen B, Hedstrom L (2008) Targeting a prokaryotic protein in a eukaryotic pathogen: identification of lead compounds against cryptosporidiosis. Chem Biol 15:70–77

Vallenet D, Labarre L, Rouy Z, Barbe V, Bocs S, Cruveiller S, Lajus A, Pascal G, Scarpelli C, Medigue C (2006) MAGE: a microbial genome annotation system supported by synteny results. Nucleic Acids Res 34:53–65

Vazquez-Boland JA, Dominguez-Bernal G, Gonzalez-Zorn B, Kreft J, Goebel W (2001) Pathogenicity islands and virulence evolution in *Listeria*. Microbes Infect 3:571–584

Voelkerding KV, Dames SA, Durtschi JD (2009) Next-generation sequencing: from basic research to diagnostics. Clin Chem 55:641–650

Walter G, Bussow K, Cahill D, Lueking A, Lehrach H (2000) Protein arrays for gene expression and molecular interaction screening. Curr Opin Microbiol 3:298–302

Wang Z, Gerstein M, Snyder M (2009) RNA-Seq: a revolutionary tool for transcriptomics. Nat Rev Genet 10:57–63

Welch RA, Burland V, Plunkett G 3rd, Redford P, Roesch P, Rasko D, Buckles EL, Liou SR, Boutin A, Hackett J, Stroud D, Mayhew GF, Rose DJ, Zhou S, Schwartz DC, Perna NT, Mobley HL, Donnenberg MS, Blattner FR (2002) Extensive mosaic structure revealed by the complete genome sequence of uropathogenic *Escherichia coli*. Proc Natl Acad Sci 99:17020–17024

Widdowson MA, Sulka A, Bulens SN, Beard RS, Chaves SS, Hammond R, Salehi ED, Swanson E, Totaro J, Woron R, Mead PS, Bresee JS, Monroe SS, Glass RI (2005) Norovirus and foodborne disease, United States, 1991–2000. Emerg Infect Dis 11:95–102

Wilkinson M (2007) ABI launch SOLiD gene sequencer. Labtechnologist.com (7/11/2007). http://www.labtechnologist.com/news/ng.asp?id=77192-applied-biosystems-abi-dna-sequencing-solid

Widdowson MA, Jaspers WJ, van der Poel WH, Verschoor F, de Roda Husman AM, Winter HL, Zaaijer HL, Koopmans M (2003) Cluster of cases of acute hepatitis associated with hepatitis E virus infection acquired in the Netherlands. Clin Infect Dis 36:29–33

Winkler H (1920) Verbeitung und Ursache der Parthenogenesis im Pflanzen und Tierreiche. Springer, Jena

WHO (World Health Organization) (2007) Food safety and foodborne illness. Fact Sheet No. 327

van Belkum A, Scherer S, van Alphen L, Verbrugh H (1998) Short-sequence DNA repeats in prokaryotic genomes. Microbiol Mol Biol Rev 62:275–293

von Mering C, Jensen LJ, Snel B, Hooper SD, Krupp M, Foglierini M, Jouffre N, Huynen MA, Bork P (2005) String: known and predicted protein–protein associations, integrated and transferred across organisms. Nucleic Acids Res 33:D433–D437

Zhu H, Snyder M (2001) Protein arrays and microarrays. Curr Opin Chem Biol 5:40–45

Chapter 2
Genomics of *Bacillus* Species

Ole Andreas Økstad and Anne-Brit Kolstø

2.1 The Genus *Bacillus*

Members of the genus *Bacillus* are rod-shaped spore-forming bacteria belonging to the Firmicutes, the low G+C gram-positive bacteria. The *Bacillus* genus was first described and classified by Ferdinand Cohn in Cohn (1872), and *Bacillus subtilis* was defined as the type species (Soule, 1932). The genus is large, encompassing more than 60 species with a great genetic diversity (Priest, 1993) (Fig. 2.1), most of which are considered non-pathogenic. *Bacillus* species may be divided into five or six groups (groups I–VI), based on 16S rRNA phylogeny or phenotypic features, respectively (Priest, 1993).

The *Bacillus* genus includes a range of species of human interest. This is mostly due to either (1) the use of the bacteria in industrial applications, such as for example in the making of biotechnological products (insect toxins, peptide antibiotics, enzymes for detergents, etc.) (Priest, 1993); (2) the employment of the spore as a model system for studying bacterial cellular differentiation, and its resistance to decontaminating agents or treatments; or (3) the role of certain *Bacillus* species in causing human disease. The latter interest can be followed back to the late nineteenth century and the studies of Louis Pasteur, using heat-attenuated cells of *Bacillus anthracis* as the first anti-bacterial vaccine, and Robert Koch, elucidating the role of a specific microorganism (*B. anthracis*) in causing a specific disease (anthrax).

2.2 Pathogenicity of *Bacillus* Species

Several Bacilli may be linked to opportunistic infections, e.g. in post-surgical wounds, cancer patients, or immunocompromised individuals. Pathogenicity among *Bacillus* spp. is however mainly a feature of organisms belonging to the *B. cereus* group, a subgroup of the *B. subtilis* group (group II) within the *Bacillus* genus

O.A. Økstad, A.-B. Kolstø (✉)
Department of Pharmaceutical Biosciences, University of Oslo, 0316 Oslo, Norway
e-mail: aloechen@farmasi.uio.no; a.b.kolsto@farmasi.uio.no

M. Wiedmann, W. Zhang (eds.), *Genomics of Foodborne Bacterial Pathogens*,
Food Microbiology and Food Safety, DOI 10.1007/978-1-4419-7686-4_2,
© Springer Science+Business Media, LLC 2011

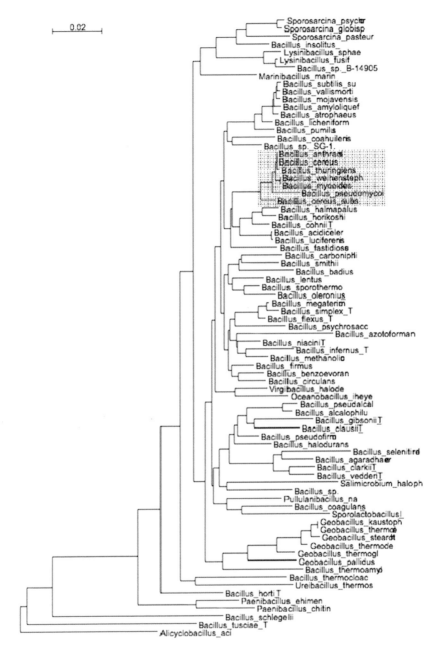

Fig. 2.1 Phylogeny of species within the *Bacillus* genus, based on 16S rRNA sequence data. The *B. cereus* group members are *boxed* in *pink*. The *horizontal bar* indicates a genetic distance of 0.02

(Fig. 2.1), and which are commonly found in the environment (reviewed by Drobniewski, 1993). In line with this, although *Bacillus licheniformis, Bacillus pumilus,* and *B. mojavensis* have all been implicated in food poisoning incidents (Salkinoja-Salonen et al., 1999; Nieminen et al., 2007; Apetroaie-Constantin et al.,

Fig. 2.2 Scanning electron micrograph (Hitachi HHS/2R) of *B. cereus* ATCC 10987

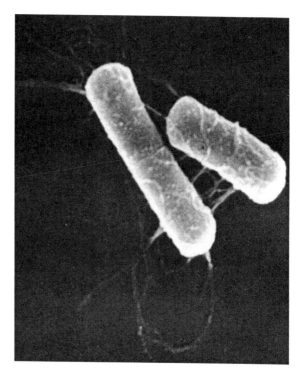

2009), the majority of reported cases of *Bacillus* food poisoning are caused by *B. cereus* and its close relatives (Fig. 2.2). In this chapter, we will therefore discuss the genomics of *Bacillus cereus* group bacteria in relation to their roles as etiological agents of two food poisoning syndromes.

2.3 The *Bacillus cereus* Group

The *B. cereus* group (*B. cereus sensu lato*) includes six approved species: *B. anthracis*, *B. cereus* (*sensu stricto*), *B. thuringiensis*, *B. mycoides*, *B. pseudomycoides*, and *B. weihenstephanensis*. In addition, a remote cluster of three thermophilic strains has been identified within the group (Lund et al., 2000; Fagerlund et al., 2007; Auger et al., 2008). This cluster has been suggested as a new species: *B. cytotoxicus* (or *B. cytoxis*) (Fig. 2.3) (Lapidus et al., 2008). The phylogeny of the *B. cereus* group has been mapped extensively by various methods, including multilocus enzyme electrophoresis (MLEE; Helgason et al., 1998, 2000a, c), amplified fragment length polymorphism (AFLP; Keim et al., 1997a; Jackson et al., 1999; Ticknor et al., 2001; Hill et al., 2004; Mignot et al., 2004), and by what is currently considered the gold standard for such studies, multilocus sequence typing (MLST; Helgason et al., 2004; Barker et al., 2005; Tourasse et al., 2006). Altogether five MLST schemes exist for the group, based largely on non-overlapping genes and strain sets. Strains from all five schemes have however been integrated into one phylogeny using supertree methodology (Tourasse and

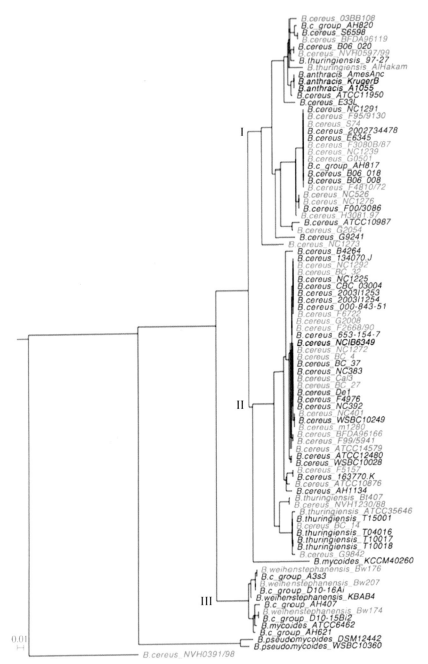

Fig. 2.3 Molecular phylogeny of *Bacillus cereus* group bacteria as analyzed by multilocus sequence typing (MLST), employing supertree technology. The species displayed form a subtree of 91 isolates extracted from the supertree of 1,400 isolates available in the SuperCAT database provided at the University of Oslo typing website (http://mlstoslo.uio.no/). The supertree is based on a

Kolstø, 2008), resulting in the hitherto most comprehensive view of the *B. cereus* group population, encompassing more than 1,400 isolates (SuperCAT database; http://mlstoslo.uio.no). By MLST analysis, the *B. cereus* group population can be grouped into at least three main clusters of isolates (Fig. 2.3); cluster I contains *B. anthracis* and related *B. cereus* and *B. thuringiensis* strains and carries mostly isolates of clinical origin; cluster II harbours *B. cereus* and *B. thuringiensis* strains from a variety of sources including food poisoning events; while cluster III contains *B. weihenstephanensis* and *B. mycoides* isolates.

2.3.1 B. cereus, B. anthracis, *and* B. thuringiensis

The three most frequently studied species within the *B. cereus* group are *B. anthracis, B. thuringiensis,* and *B. cereus. B. cereus* is a common bacterium of the soil and can colonize invertebrate guts as a symbiont, an environment which has been suggested to be its natural habitat (Margulis et al., 1998; Jensen et al., 2003). It is however a frequent cause of human food poisoning, as well as various opportunistic and nosocomial infections, e.g. in the immunocompromised or following trauma to the eye (reviewed by Drobniewski, 1993; Kotiranta et al., 2000; Bottone, 2010). *B. cereus* can cause two types of food poisoning syndromes, namely, the emetic syndrome, due to the synthesis of cereulide, a small, heat-stable non-ribosomally synthesized dodecadepsipeptide, and the diarrhoeal syndrome, caused by enterotoxins (Drobniewski, 1993; Stenfors Arnesen et al., 2008). Other potential virulence factors include secreted phospholipases, haemolysins, proteases, and other degradative enzymes. These proteins are extracellular virulence factors, and their expression is under the control of the global pleiotropic transcriptional regulator PlcR (Agaisse et al., 1999; Økstad et al., 1999b). Genes encoding the proteinaceous virulence factors, enterotoxins included, are located on the chromosome, while the genes responsible for synthesis of the emetic toxin are located on a large (270 kb) plasmid, pCER270 (Hoton et al., 2005; Ehling-Schulz et al., 2006).

B. thuringiensis is also frequent in soil, is an entomopathogenic bacterium, and is the most commonly used commercial biopesticide worldwide (Soberon et al., 2007). Its identification and classification are based on the production of insecticidal proteinaceous toxin crystals (Cry and Cyt proteins) during sporulation (Schnepf et al., 1998; Aronson, 2002), a feature recognized by microscopy. The Cry and Cyt toxins are of different classes and exhibit variable specificities towards the larvae of different classes of insects (reviewed in Whiteley and Schnepf, 1986; Schnepf et al., 1998).

Fig. 2.3 (continued) combination of sequence data from all five published MLST schemes available for the *B. cereus* group (Tourasse and Kolstø, 2008). Three main phylogenetic clusters of the *B. cereus* group population are indicated by *Roman numerals*, and strains are *coloured* by source of isolation, following their representation in SuperCAT (*red*, human; *purple*, animal; *dark brown*, soil; *orange*, food; *blue*, dairy; *grey*, other)

B. *anthracis* is a highly monomorphic species within the *B. cereus* group, showing very little genetic variation (Fig. 2.3; Keim et al., 1997b; Van Ert et al., 2007). In the environment, the bacterium primarily exists as a highly stable, dormant spore in the soil. Nevertheless, it has been claimed that the organism can grow and persist outside the host, in the rhizosphere of plants (Saile and Koehler, 2006). *B. anthracis* is the cause of anthrax, primarily a disease of herbivores, but may also cause isolated cases of infections in man. In several regions of the world, including parts of Africa and Asia, *B. anthracis* is endemic or hyperendemic, while being sporadic in Australia and the United States (http://www.vetmed.lsu.edu/whocc/mp_world.htm). Anthrax takes three forms: cutaneous, gastro-intestinal, or inhalational (reviewed in Mock and Fouet, 2001). While the cutaneous form is easily treatable with antimicrobials, the gastrointestinal and inhalational forms of the disease are more severe, as has been demonstrated by human deaths following the ingestion of meat from animals that died from anthrax disease (reviewed by Beatty et al., 2003), as well as in the bioterror attacks against the United States in fall of 2001, when letters containing *B. anthracis* spores were distributed through the US Postal Service, killing five people (Jernigan et al., 2001). Anthrax disease is primarily caused by two virulence traits: (1) the synthesis of two A–B type toxins from three toxin components, namely, lethal toxin (LT) being composed of protective antigen (PA) and lethal factor (LF), and edema toxin (ET) which is composed of PA and edema factor (EF) (Mock and Fouet, 2001; Mock and Mignot, 2003; Young and Collier, 2007), and (2) the presence of a poly-γ-D-glutamic acid (polyglutamate) capsule which is important for *B. anthracis* survival in a host, as it helps the bacterium evade the host immune system by protecting vegetative cells from phagocytotic killing during infection (Preisz, 1909; Drysdale et al., 2005; Candela and Fouet, 2006; Richter et al., 2009).

Although the *B. cereus* group in general is phylogenetically heterogeneous, strains of the same species, as well as of different species, may be very closely related and phylogenetically intermixed, when employing genetic markers at the chromosomal level (Fig. 2.3; a more complete representation of the *B. cereus* group phylogeny can be found at http://mlstoslo.uio.no). Bacteria belonging to the *B. cereus* group generally exhibit complex genomes; different strains may carry plasmids in variable numbers (1–>12) and sizes (2–600 kb; frequently >80 kb) (reviewed in Kolstø et al., 2009), some of which are conjugative or mobilizable and can host a number of different IS elements. Typically, strains also contain bacteriophages which may be integrated in the chromosome as prophages or which may replicate as independent linear elements (Carlson et al., 1994a; Rasko et al., 2005; Verheust et al., 2005; Sozhamannan et al., 2006; Lapidus et al., 2008). The traditional species distinctions of *B. anthracis* and *B. thuringiensis* were largely based on their different pathogenic specificities towards vertebrates and insect larvae, respectively. Interestingly, however, the genes coding for the typical species-specific virulence properties of both *B. anthracis* and *B. thuringiensis* are plasmid-borne, a fact unknown at the time of species designation; in *B. thuringiensis* the crystal protein toxins (Cry or Cyt) causing pathogenicity to insects are almost exclusively encoded by genes present on plasmids of various sizes and often associated with IS elements (Schnepf et al., 1998). Similarly, in *B. anthracis* two large plasmids, pXO1 (182 kb) and pXO2 (95 kb), are necessary for full *B. anthracis* virulence (Mock and Fouet, 2001; Passalacqua and Bergman, 2006). pXO1 carries the genes coding for the anthrax toxin components (*pag, lef,* and *cya*), all located within a 44.8 kb pathogenicity island (PAI; Okinaka et al., 1999b) which also encodes the transcriptional activator AtxA and the repressor PagR that regulate the expression of the toxin genes (Uchida et al., 1993; Dai et al., 1995; Mignot et al., 2003; Fouet and Mock, 2006). pXO2 encodes the other major *B. anthracis* virulence factor, the poly-γ-D-glutamic acid (polyglutamate) capsule, in a five-gene operon (*capBCADE*). Capsule expression is activated by the transcriptional regulators AcpA and AcpB, and the capsule operon and its regulator genes are all located in a 35 kb PAI (Pannucci et al., 2002b; Van der Auwera et al., 2005). Expression of *acpA* and *acpB* (located on pXO2) is under the control of AtxA (encoded on pXO1) (Fouet and Mock, 2006; Perego and Hoch, 2008); thus cross-talk occurs between the two virulence plasmids.

The phylogenetic intermixing of strains of different species, in particular *B. cereus* and *B. thuringiensis* isolates (Fig. 2.3; http://mlstoslo.uio.no), and the fact that the main phenotypical traits classically used to define each of the *B. cereus* group species are carried by plasmids, including the insect toxicity of *B. thuringiensis* and human pathogenicity of *B. anthracis*, have led to disputes regarding the species definitions within the group (Helgason et al., 2000b; Rasko et al., 2005). Hitherto, no species-specific property outside the plasmid-borne *cry* and *cyt* genes, has been identified for *B. thuringiensis*, and *B. thuringiensis* strains can carry the same chromosomally encoded virulence genes as *B. cereus*, including genes for enterotoxins, phospholipases, haemolysins, and proteases (Han et al., 2006; Scarano et al., 2009). These genes may in fact be important for the virulence of the bacterium, following its entry into the insect larvae haemocol (Fedhila et al., 2002, 2003, 2004, 2006). Thus, a *B. thuringiensis* strain that has lost the *cry*- or *cyt*-containing plasmids will be indistinguishable from *B. cereus* and will be identified as such. Therefore, although the current nomenclature is kept, largely based on the well-established differences observed in pathogenicity profiles of the two species towards insects, the two bacterial species are indistinguishable when a chromosomal phylogeny is reconstructed based on a sufficient number of isolates (http://mlstoslo.uio.no). *B. cereus* and *B. thuringiensis* have therefore been suggested to constitute one species in genetic terms (Helgason et al., 2000c).

Members of the *B. cereus* group are found in various habitats in the environment, including different types of soils, plant leaves, the rhizosphere, the intestinal tract of soil invertebrates, as well as man-made settings such as food production factories and hospital environments (Jensen et al., 2003), where they may constitute reservoirs for disease. The ubiquitous presence of several *B. cereus* group members in a great variety of natural habitats, combined with the ability to survive in nutrient-poor and otherwise hostile environments due to spore-forming abilities, contributes to their role as common polluter organisms (Drobniewski, 1993). The presence of *B. cereus* in these habitats allows the organism to easily spread to different foods, including milk and milk products (via the udders of grazing cows), as well as rice and other carbohydrate-rich foodstuffs. From these locations it may cause gastro-intestinal disease.

2.3.2 Bacillus mycoides, Bacillus pseudomycoides, B. weihenstephanensis, *and* "Bacillus cytotoxicus"

The three remaining species of the *B. cereus* group, *Bacillus mycoides, Bacillus pseudomycoides,* and *B. weihenstephanensis* have been studied less extensively. *B. mycoides* and *B. pseudomycoides* are characterized based on one morphological property – rhizoidal growth on solid medium. Phylogenetically, *B. mycoides* strains are widely distributed within the *B. cereus* group population, while *B. pseudomycoides* may appear to be limited to a remote cluster (Fig. 2.3). Only a very low number of *B. pseudomycoides* strains have been identified, however, and there is a need for more strains to be mapped by molecular phylogeny before firm conclusions about a possible clonal distribution pattern for this species can be reached. Although both *B. mycoides and B. pseudomycoides* are generally regarded as nonpathogenic, certain strains of *B. mycoides* have been linked to cases of food-borne disease (McIntyre et al., 2008).

B. weihenstephanensis is a psychrotolerant species in the *B. cereus* group and carries specific 16S rRNA and *cspA* (major cold-shock protein) signatures (Thorsen et al., 2006). Psychrotolerant *B. cereus* group members however do not necessarily belong to *B. weihenstephanensis,* as psychrotolerant strains of *B. cereus* also exist (Stenfors and Granum, 2001). *B. weihenstephanensis* may produce enterotoxins even at refrigerator temperatures (Baron et al., 2007), and some *B. weihenstephanensis* strains can produce emetic toxin (cereulide) (Thorsen et al., 2006), further emphasizing the food-borne intoxication potential of this species.

The food poisoning strain *B. cereus* NVH391-98 was originally isolated from cases of severe gastroenteritis linked to an outbreak in an elderly home in France, in which three people were killed (Lund et al., 2000). Molecular typing identified this strain as a remote member of the *B. cereus* group (Fagerlund et al., 2007), and following the discovery of two novel *B. cereus* group strains forming a phylogenetic cluster together with the NVH391-98 strain (Auger et al., 2008), the new species name, *B. cytotoxicus,* was proposed (Lapidus et al., 2008). The species designation however remains to be formally approved. These isolates are all thermotolerant, but the degree of CytK and Nhe enterotoxin production, as well as cytotoxicity, was variable between the strains in the cluster; while *B. cereus* NVH391-98 produced high levels of the CytK cytotoxin, *B. cereus* 883/00 produced little or no CytK and Nhe enterotoxin, and was not cytotoxic to Vero cells (Fagerlund et al., 2007).

2.4 Genome Sequencing of the *Bacillus cereus* Group

The first *B. cereus* group genome project, that of the *B. anthracis* Ames Porton model strain, was initiated in 2000 at the Institute for Genomic Research (TIGR) and published in 2003 (Read et al., 2003a). The bioterror events during the fall of 2001, involving the distribution of letters containing *B. anthracis* spores via the US Postal System, have however spurred the sequencing of a multitude of *B. cereus* group strains during the past decade. Following the sequencing of the bioterror attack strain *B. anthracis* Ames Florida (Read et al., 2002) and the first *B. cereus* strain (ATCC 14579, type strain; Ivanova et al., 2003), 105 additional genomes have been or are being sequenced (Entrez genome project: http://www.ncbi.nlm.nih.gov/genomeprj; Genomes OnLine database: www.genomesonline.org), providing an unprecedented reservoir for doing comparative genomics. The genome sequences cover all six approved species within the group, in addition to the *B. cereus* NVH391-98 strain from the remote "*Bacillus cytotoxicus*" cluster. In one of the latest genome sequencing projects involving *B. cereus* group isolates, more than 50 strains have been sequenced to draft stage using next-generation sequencing technology. Strains were selected so as to provide the best possible coverage of the group phylogeny (Timothy D. Read, personal communication), as mapped by MLST (http://mlstoslo.uio.no). Among the 108 strains that have been or are being sequenced (per 20 May 2010; http://www.ncbi.nlm.nih.gov/genomeprj; www.genomesonline.org) (Kolstø et al., 2009), there are eight isolates that were linked to cases of food-borne disease or were isolated from foodstuffs (Table 2.1).

Table 2.1 Sequenced *B. cereus* group isolates linked to food-borne disease or isolated from contaminated food. Genome project and source information were obtained from the NCBI Entrez Genome Project database (http://www.ncbi.nlm.nih.gov/genomeprj)

Isolate/strain	Source	Genome publication
B. cereus ATCC 10987	Spoiled cheese, Canada, 1930	Rasko et al. (2004)
B. cereus NVH 0597/99	Isolated from spice mix, believed to be the cause of a food poisoning outbreak in Norway, 1999	Not published
B. cereus m1293	Isolated from cream cheese	Not published
B. cereus m1550	Isolated from uncooked chicken	Not published
B. cereus MM3	Isolated from food	Not published
B. cereus F4810/72 (AH187)	Emetic food poisoning outbreak, 1972 (emetic type strain)	Not published
B. cereus H3081.97	Emetic strain	Not published
B. cereus NVH391/98 ("*B. cytotoxicus*")	Outbreak of severe gastroenteritis, elderly home, France, 1998	Lapidus et al. (2008)

2.4.1 The **B. cereus** *Group from a Genomic Perspective*

Prior to 2000, only sporadic genome sequence data had been produced from *B. cereus* group organisms. The *B. anthracis* virulence plasmids, pXO1 and pXO2, were sequenced in 1999 (Okinaka et al., 1999a, b), while more than 100 kb of random genome sequence data was produced from *B. cereus* ATCC 10987 and ATCC 14579 (type strain) in the late 1990s (Økstad et al., 1999a; b). Today, the *B. cereus* group is one of the groups of closely related bacteria with the highest number of sequenced genomes. The first genomes sequenced from the group, those of *B. anthracis* Ames and *B. cereus* strains ATCC 14579 and ATCC 10987, were all closed and serve as reference genomes for later projects. Although the current trend of sequencing genomes to high-quality draft format also applies to the current *B. cereus* group projects, altogether 17 genomes have now been closed, showing that chromosome sizes for *B. cereus* group members are in the range 5.2–5.4 Mb (with the exception of *B. cereus* NVH391-98 which is 4.1 Mb), confirming previous pulsed-field gel electrophoresis-based estimations (Carlson et al., 1992; Carlson and Kolstø, 1993; Carlson et al., 1996; Lovgren et al., 2002). Chromosomes have a GC content of around 35.3–35.4%, and strains typically carry a large number of rRNA operons (11 and 13 for *B. anthracis* Ames Porton and *B. cereus* ATCC 14579, respectively). A comparison of *B. anthracis* Ames to 19 *B. cereus* and *B. thuringiensis* isolates by comparative genome hybridization (CGH) revealed that the 2-Mb region surrounding the chromosomal replication terminus contained a significantly higher proportion of strain-specific genes and was clearly of higher plasticity than the rest of the chromosome (Read et al., 2003a), suggesting that gene mobility events are probably more frequent in the terminus region. This is in line with studies in other bacterial species, describing the region surrounding the replication terminus as a high-plasticity region (Suyama and Bork, 2001).

Prior to the sequencing of the first representatives of the *B. cereus* group members, many scientists in the field had viewed *B. anthracis* as more fundamentally different from the other species in the group, based on several phenotypic character differences such as β-lactamase sensitivity, lack of haemolytic activity, and sensitivity to γ-phage. Strikingly, however, comparison of *B. cereus* ATCC 14579 to the draft sequence of *B. anthracis* Florida (A2012), and of *B. anthracis* Ames Porton to the 19 *B. cereus* and *B. thuringiensis* strains by comparative genome hybridization (CGH), revealed that only four regions of the chromosome were unique to *B. anthracis* and that these regions corresponded to four prophages (Ivanova et al., 2003; Read et al., 2003a). Thus, no candidates for unique chromosomal genes of importance to *B. anthracis* virulence could be identified. Most of the toxicity specifically linked to anthrax disease thus seems to be linked to the pXO1 and pXO2 plasmids, although mechanisms of cross-talk between the two plasmids and between plasmids and chromosome seem to be of importance (Uchida et al., 1997; Mignot et al., 2003; Chitlaru et al., 2006).

In addition to the four prophages, one other principle difference between *B. anthracis* and the other *B. cereus* group members is the characteristic nonsense mutation in the global pleiotropic transcriptional regulator of extracellular virulence factors, PlcR, which renders the protein non-functional (Agaisse et al., 1999; Kolstø et al., 2009). In most *B. cereus* and *B. thuringiensis* strains, PlcR is active and important for expression of a range of chromosomally encoded virulence factors, including phospholipases, proteases, haemolysins, and enterotoxin. PlcR is however clearly not necessary for *B. anthracis* virulence (reviewed in Kolstø et al., 2009).

Given the fact that plasmid-borne factors are key elements for the toxicity of *B. thuringiensis* towards insect larvae and for the virulence of *B. anthracis* towards vertebrates, plasmid content and distribution has been a topic of considerable interest to the scientific community studying *B. cereus* group organisms. In general, *B. cereus* group strains are well known for having the potential for harbouring one or more, often several, plasmids in the cell at the same time (reviewed in Kolstø et al., 2009). The discovery of pXO1 plasmid-like sequences in a large proportion of 19 *B. cereus* and *B. thuringiensis* strains tested for pXO1 gene markers by CGH (Read et al., 2003a) was at the time striking and indicated that plasmids with similarity to pXO1 could potentially exist in species other than *B. anthracis*. This was later confirmed by sequencing of *B. cereus* ATCC 10987 (Rasko et al., 2004) and by Panucci and co-workers, who screened a large number of *B. cereus* group strains for pXO1 gene markers by PCR (Pannucci et al., 2002a). Orthologs to the genes making up the pXO1 PAI containing the anthrax toxin genes and their regulators, as well as the *gerX* locus essential for germination within host macrophages, were however generally not present in the non-*B. anthracis* strains. pXO2-like sequences seemed to be occurring less frequently in *B. cereus* group species other than *B. anthracis* (Pannucci et al., 2002b; Read et al., 2003a).

Many chromosomal features have been identified as being common among *B. cereus* group isolates, unifying the various *B. cereus* group bacteria and emphasizing their relatedness; chromosomes are generally in the same size range for sequenced isolates and are to a large extent syntenic, with common orthologous genes being organized in a conserved order (Ivanova et al., 2003; Rasko et al., 2004). Also, a core gene set has been identified, counting $3,000 \pm 200$ genes (Lapidus et al., 2008) out of the in excess of 5,000 genes that make up a typical *B. cereus* group chromosome (excluding *B. cereus* NVH391-98, which has an unusually small genome; Lapidus et al., 2008). Furthermore, each strain characteristically has in the order of 400–800 strain-specific genes (Lapidus et al., 2008), which may potentially be involved in niche adaptation processes, and which are contributing to a fairly large *B. cereus* group pan-genome (20–25,000 genes estimated in Lapidus et al., 2008). Another unifying feature for the *B. cereus* group bacteria is the presence of several ubiquitous interspersed repeat elements in the size range 100–400 bp, many of which seem to be unique to the group (Tourasse et al., 2006). These repeats, named *bcr1–bcr18*, are non-protein coding, but some seem to be expressed at the RNA level and are predicted to constitute non-autonomous mobile elements belonging to the class 'miniature inverted-repeat transposable elements' (MITEs) (Økstad et al., 2004; Tourasse et al., 2006; Klevan et al., 2007). Although no specific function has yet been assigned to any of the *bcr* repeats, MITEs are known in other bacteria to be involved in a variety of processes, including regulation of transcription and mRNA degradation, DNA methylation, integration host factor (IHF) binding, and creation of novel gene loci, to mention some (Delihas, 2008). Phylogenetic studies of the *B. cereus* group had already prior to the genomics era shown that closely related strains of *B. cereus* and *B. anthracis* exist, as do close strains of *B. cereus* and *B. thuringiensis*, and that the two latter species are intermixed phylogenetically based on chromosomal markers (Carlson et al., 1994b; Helgason et al., 2000c), leading to the suggestion that *B. cereus* and *B. thuringiensis* belong to the same species in purely genetic terms (Helgason et al., 2000c). Also, the idea has been presented that *B. anthracis* is in reality an over-sampled *B. cereus* (Rasko et al., 2005), which is highly monomorphic and genetically constrained. For the moment, however, species designations are kept, given the fundamental differences in pathogenicity profiles between the different species in the group.

2.4.2 Phages of the B. cereus *Group*

For several important pathogenic bacteria, including the ethiological agents of cholera, diphtheria, and enterohaemorrhagic diarrhoeas, bacteriophages are important vectors for the transfer of virulence factors. In general, phages constitute important sources of gene flow in bacteria. *B. cereus* group bacteria can host a range of phages; however, carriage of specific toxin genes has not yet been observed in these elements; while *B. anthracis* ubiquitously carries the four unique prophages (lambda01–lambda04) in its chromosome, *B. cereus* ATCC 10987 carries three chromosomal prophages, while the *B. cereus* type strain, ATCC 14579, has three prophages integrated in its chromosome and in addition carries a linear extrachromosomal phage-like element, pBClin15 (Carlson et al., 1992; Ivanova et al., 2003). Similarly, *B. thuringiensis* 97–27 and *B. cereus* E33L (formerly known as zebra killer, ZK) carry a variety of phages in their chromosome and plasmid(s) (Han et al., 2006). The phages generally exhibit a host range which is limited to the species from which they have been isolated. However, *B. cereus* strains that are very similar to *B. anthracis*, such as *B. cereus* ATCC 4342, may be susceptible to *B. anthracis* phages, such as *Gamma* and *Cherry* (reviewed in Rasko et al., 2005).

2.5 Genome Dynamics Related to Food-Borne Disease

B. cereus can cause food-borne disease by two mechanisms – either by intoxication, following the ingestion of foodstuffs containing pre-formed emetic toxin (cereulide) produced by an emetic strain contaminating a food matrix environment, or by gastro-intestinal infection of vegetative *B. cereus* strains thought to form one or more enterotoxins in the intestine, following the ingestion of spores or viable cells with food or milk products (reviewed in Stenfors Arnesen et al., 2008).

2.5.1 Enterotoxins

The identity of the proteins conferring the enterotoxic activity inherent to *B. cereus* is still controversial. The three cytotoxins haemolysin BL (Hbl, three-component toxin), non-haemolytic enterotoxin (Nhe, three-component toxin), and cytotoxin K (CytK, single-component toxin of the β-barrel pore-forming toxin family) are however generally considered to be the causes of the *B. cereus* diarrhoeal syndrome (Beecher and Macmillan, 1991; Lund and Granum, 1996; Lund et al., 2000) and are specified by genes carried on the chromosome. In addition to the well-established Hbl, Nhe, and CytK cytotoxins, several additional candidate proteins have been suggested as potential contributors to the enterotoxic activity of *B. cereus*, including haemolysin II and haemolysin III (Baida and Kuzmin, 1995; Baida et al., 1999), EntFM (Shinagawa et al., 1991; Asano et al., 1997; Tran et al., 2010), phospholipases C (Kuppe et al., 1989), cereolysin O (Kreft et al., 1983), and InhA2 (Fedhila et al., 2003). Again, these proteins are chromosomally encoded, and the synthesis of several of them is, like Hbl, Nhe, and CytK, subject to regulation by PlcR (Gohar et al., 2008). Most likely, the toxins can act synergistically to cause *B. cereus* food-borne disease (Stenfors Arnesen et al., 2008).

Nhe, originally characterized following a large food poisoning outbreak in Norway in 1995 (Lund and Granum, 1996), is the most commonly found enterotoxin gene complex and is probably ubiquitous in *B. cereus* group bacteria (reviewed in Stenfors Arnesen et al., 2008). The complex, encoded by the *nheA*, *nheB*, and *nheC* genes (Granum et al., 1999), is, with one recently discovered exception, encoded chromosomally. In *B. weihenstephanensis* KBAB4, however, a second copy of the *nhe* locus is hosted on a 400 kb plasmid, pBWB401 (Lapidus et al., 2008). The identity of the *nhe* locus between strains is, with the exception of the NVH391-98 strain which is phylogenetically remote, generally around 90%, but is approaching 100% within clonal clusters such as the *B. anthracis* cluster and the emetic clusters. The plasmid-borne *nhe* locus in *B. weihenstephanensis* KBAB4 is however only around 58% identical to the chromosomal *nhe* copies. Whether this locus gives rise to a functional enterotoxin is at present unknown. The maximal cytotoxic activity of Nhe towards Vero cells was obtained when the ratio between the Nhe components was 10:10:1 (NheA, NheB, NheC; Lindback et al., 2004), and the enterotoxic activity of Nhe has recently been explained (Fagerlund et al., 2008); exposure of plasma membrane to Nhe leads to rapid membrane lysis, and Nhe has been shown to form pores in lipid bilayers, leading to colloidosmotic lysis. Indeed, Nhe has more recently been shown, in spite of its name, to exhibit haemolytic activity towards erythrocytes from several mammalian species (Fagerlund et al., 2008).

The Hbl complex constitutes another pore-forming enterotoxin in the *B. cereus* group and exhibits haemolytic activity towards erythrocytes from several animal species. It is encoded by the *hblC*, *hblD*, and *hblA* genes, encoding components L2, L1, and B, respectively. In addition to the three structural genes, the most common variant of the *hbl* locus carries a fourth gene, *hblB*, downstream of *hblCDA*. *hblB*, which has probably originated by duplication of a large part of *hblA* and fusion to an ORF in the 3′ end (Økstad et al., 1999b), is however most probably a pseudogene, since it has not been shown to be transcribed to a detectable level and the *hblCDA* transcript seems to terminate within *hblB* (Lindback et al., 1999). Other variants of the *hbl* locus do however exist: in *B. cereus* 03BB108 and *B. weihenstephanensis* KBAB4 an *hbl* locus exists which consists of the *hblCDA* genes only. Like *B. cereus* MGBC145 (Beecher and Wong, 2000), the 03BB108 strain also carries a second *hbl* locus, which in the latter strain is of the *hblCDA* type. The *hbl* locus is less frequently present in *B. cereus* group strains compared to *nhe* and was identified in approximately 60% of strains in a PCR screening procedure (Pruss et al., 1999).

The *hblCDAB* operon is located on the chromosome, in a conserved location between strains, and is part of a 17.7 kb genome insertion bordered on one side of a degenerate IS*Rso11* transposase fragment which has been suggested to have been acquired as a mobile element (Han et al., 2006). The insertion also contains genes encoding germination proteins and a transcriptional regulator (*trrA*) and putative histidine kinase that could have the potential for forming a two-component system (Økstad et al., 1999b). In contrast, the *hblCDA* operon seems to exhibit a less conserved genomic localization between strains; in *B. weihenstephanensis* KBAB4 the locus is chromosomally encoded and is flanked by genes encoding a putative β-lactamase and an S-layer domain protein, respectively. In the *B. cereus* 03BB108 draft genome sequence (GenBank entry: ABDM00000000), however, the corresponding *hblCDA* genes are flanked by sequences with similarity to the *B. anthracis* pXO1 plasmid, possibly indicating that the 03BB108 *hblCDA* genes are plasmid-borne (reviewed in Stenfors Arnesen et al., 2008).

The Hbl and Nhe proteins are related in sequence and have probably arisen by several gene duplication events from a common ancestor locus. The crystal structure of HblB has been solved and shows high structural similarity to ClyA (also known as HlyE or SheA), a pore-forming cytolysin from *Escherichia coli, Salmonella enterica* serovars Typhi and Paratyphi, and *Shigella flexneri* (Oscarsson et al., 1996; Wallace et al., 2000; Oscarsson et al., 2002). The HblB structure consists of five α-helix bundles wrapped around each other in a left-handed super-coil and a hydrophobic β-hairpin flanked by two short α-helices (PDB entry 2nrj; Fagerlund et al., 2008). Interestingly, NheB and NheC exhibit sufficient sequence similarity to HblB to allow modelling of their structures based on the HblB crystal structure, and although the toxin components exhibit a very limited primary sequence identity, their conserved structure suggests that the Hbl/Nhe family and the ClyA family constitute a new superfamily of toxins (Fagerlund et al., 2008).

Although much knowledge regarding the structure, function, and genetic organization of the Nhe and Hbl enterotoxins has accumulated over the past two decades, still much remains to be discovered. Any potential effect of the heterogeneity in genome organization and duplication of the *nhe* and *hbl* loci in different *B. cereus* group strains remains to be solved. Furthermore, no host cell receptor has yet been identified neither for Hbl nor for Nhe; while all three Hbl components can bind to the surface of erythrocytes (Beecher and Wong, 1997), out of the three Nhe components only NheB and NheC has been shown to exhibit surface binding to Vero cells (Lindback et al., 2004; Lindbäck et al., 2010). Even though both Nhe and Hbl appear to constitute tri-partite pore-forming toxins, details of the molecular interaction of the various protein components of each complex are yet to be resolved, as is the degree to which oligomerization occurs during formation of the transmembrane pore in the plasma membrane of target cells (reviewed in Stenfors Arnesen et al., 2008).

CytK is a 34 kDa single-component protein toxin of the β-barrel pore-forming toxin family. It was first identified following a gastroenteritis outbreak in a French nursing home in 1998, in which *B. cereus* NVH391-98 ("*B. cytotoxicus*") was, as already mentioned, identified as the agent causing the disease. In the course of the outbreak, several patients presented with bloody diarrhoea, and three elderly people died (Lund et al., 2000). CytK was originally identified as the prime virulence factor, given its necrotic, haemolytic, and enterotoxic effects, and since no Hbl or Nhe was apparently present in the NVH391-98 strain (Lund et al., 2000). Later, however, the *nhe* genes have been identified in the *B. cereus* NVH391-98 strain, although with a lower sequence identity to Nhe from other *B. cereus* group strains (Fagerlund et al., 2007). CytK was originally described to be divided into two gene families in different *B. cereus* group isolates, CytK-1 and CytK-2, where CytK-1 corresponded to the variant identified in *B. cereus* NVH391-98, which was thought to represent a particularly toxic variant of the protein. Later the sequence divergence of the CytK protein (CytK-1) from this strain has, as is the case for Nhe, instead been shown to reflect the general phylogenetic divergence of this strain from the rest of the *B. cereus* group. The potent cytotoxicity of the *B. cereus* NVH391-98 strain is probably rather due to an exceptionally high CytK expression level in this isolate (Brillard and Lereclus, 2004).

It is important to note that the genes coding for the enterotoxin components and degradative enzymes are often found in strains of several species within the *B. cereus* group, *B. thuringiensis* and *B. anthracis* included, suggesting that *B. cereus* group organisms in general may have the potential to be pathogenic (Read et al., 2003b; Han et al., 2006; Hendriksen et al., 2006; McIntyre et al., 2008; Stenfors Arnesen et al., 2008). Using the term "non-pathogenic strain" for environmental isolates of *B. cereus* that have not been linked to disease therefore does not seem appropriate, as it is impossible to know whether these isolates could cause infection in man, given the right dose and setting. *B. anthracis* is however different from the other species in the group, in carrying a nonsense mutation in PlcR which makes the protein non-functional (Agaisse et al., 1999). Therefore, *B. anthracis* encodes a very limited extracellular proteome compared to the other *B. cereus* group species (Gohar et al., 2005), and the chromosomal virulence factors belonging to the PlcR regulon (Gohar et al., 2008) are not synthesized, enterotoxins included. *B. anthracis* virulence was not increased in a mouse infection model (intranasal) by transfer of a functional *plcR* gene back into host cells, indicating that the pathogenicity of *B. anthracis* is not dependent on the chromosomal virulence factors expressed as part of the PlcR regulon in other *B. cereus* group bacteria.

2.5.2 Emetic toxin

The emetic toxin, cereulide, is a small (1.2. kDa) non-ribosomally synthesized dodecadepsipeptide, produced by a non-ribosomal peptide synthetase system (*ces*) found in emetic *B. cereus* strains (Ehling-Schulz et al., 2005a). The toxin is heat-stable, acid and protease resistant and can be pre-formed in foodstuffs contaminated with an emetic strain of *B. cereus*, leading to emesis within 0.5–6 h post ingestion (reviewed in Stenfors Arnesen et al., 2008), and occasionally more severe intoxications such as liver failure and death (Mahler et al., 1997; Dierick et al., 2005). Expression of the toxin is known to be induced towards the end of logarithmic growth, reaching a maximum during the early stationary growth phase (reviewed in Stenfors Arnesen et al., 2008). Although the mechanisms regulating emetic toxin production are not well characterized, its expression is affected by factors like oxygen, pH, and temperature and is regulated by the transitional state regulator AbrB, but not by PlcR (Lucking et al., 2009). Among the sequenced strains from the *B. cereus* group are two emetic isolates, including the reference emetic strain *B. cereus* F4810/72, isolated in 1972 from human vomit following an emetic food poisoning outbreak.

The peptide synthetase responsible for cereulide synthesis is encoded by a 24 kb gene cluster (*ces*) on a 208 kb plasmid, pCER270, with similarity to *B. anthracis* pXO1 and other pXO1-like plasmids (Hoton et al., 2005; Ehling-Schulz et al., 2006; Rasko et al., 2007). The cluster is comprised of seven genes, including typical non-ribosomal peptide synthetase (NRPS) genes such as a phosphopanteth-einyl transferase and genes encoding modules for the activation and incorporation of monomers in the growing peptide chain. Also, a potential hydrolase and an ABC transporter are encoded by the *ces* cluster. Interestingly, the pPER270 plas-mid is similar to pBC10987 from *B. cereus* ATCC 10987, isolated from spoiled cheese in Canada in 1930 (Herron, 1930; Rasko et al., 2004), and pPER272, a plasmid hosted by *B. cereus* strains AH820 and AH818 isolated from the peri-odontal pocket and root canal, respectively, in patients with periodontal disease (Helgason et al., 2000a; Rasko et al., 2007). The pXO1 pathogenicity island encod-ing the anthrax toxin genes and associated regulators is however missing both in pCER270 and in pPER272 and is replaced by a 77 kb insertion in pCER270 which is bordered by transposase and resolvase genes that could potentially have been involved in the insertion of the region (Helgason et al., 2000a; Rasko et al., 2007).

Cereulide production has been mapped to two phylogenetically separated clonal clusters within the *B. cereus* group population, mostly consisting of *B. cereus* isolates (Ehling-Schulz et al., 2005b; Vassileva et al., 2007). However, specific *B. weihenstephanensis* strains have also been found that are capable of forming emetic toxin, even at temperatures as low as 8°C (Thorsen et al., 2006). Given that the genetic determinants of emetic toxin production are plasmid-borne, this implies that acquisition of a plasmid encoding emetic toxin production has proba-bly occurred more than once during evolution and that pCER270 may be subject to lateral transfer between strains.

2.6 Potential for Causing Food-Borne Disease – A General Feature of the *B. cereus* Group?

B. thuringiensis is an insect pathogen; however, isolates may also have the potential to act as opportunistic pathogens in humans and animals, possibly causing tissue necrosis, pulmonary infections, or food poisoning (Hernandez et al., 1999; Ghelardi et al., 2007; McIntyre et al., 2008). In line with this, the sequencing of *B. thuringien-sis* strains Al-Hakam, 97–27 (*var.* konkukian) and ATCC 35646 (*var.* israelensis), has shown that *B. thuringiensis* strains can carry the same chromosomal virulence factors that are typical to *B. cereus* (i.e. enterotoxins, haemolysins, phospholipases, proteases (Han et al., 2006), as can other species in the *B. cereus* group. With what frequency human infections by *B. thuringiensis* actually occur is however at present unknown, since *B. cereus* group strains isolated from human infections (food poisoning included) are most often not tested for the presence of crystal toxin genes or for the production of such toxins. Also, it is plausible that *B. thuringiensis*, during an infection in man, may lose the plasmid encoding its entomopathogenic properties, as these plasmids are often less stably maintained at 37°C than at lower growth temperatures. This would make the isolate practically indistinguishable from *B. cereus*. In fact, the ability to cause the *B. cereus* diarrhoeal syndrome may be an inherent feature of all *B. cereus* group species carrying a functional PlcR regulator gene. Indeed, *B. thuringiensis* was identified in food or clinical samples from four outbreaks, and *B. mycoides* was identified in one outbreak, sampled in British Columbia, Canada, in the period 1991–2005 (McIntyre et al., 2008). Seemingly, contrary to the emetic isolates which group into two clusters phylogenetically, isolates causing the diarrhoeal syndrome are located throughout the *B. cereus* group phylogeny and may frequently share identical molecular typing data (based on chromosomal markers) with environmental strains isolated from soil or plants (Fig. 2.3; Tourasse et al., in press; http://mlstoslo.uio.no). Held together with the fact that genes encoding the non-haemolytic enterotoxin complex (Nhe) are found ubiquitously among *B. cereus* group organisms and that *B. cereus* group species other than *B. cereus sensu stricto* have been characterized as the cause of gastroenteritis (McIntyre et al., 2008), this emphasizes the opportunistic nature of *B. cereus* group bacteria and the potential for other *B. cereus* group species to cause food-borne disease. Also, notably, it is not merely the presence or the absence of toxin and other virulence genes that determines toxicity. Gene expression levels can clearly be highly variable between isolates, exemplified by *B. cereus* NVH391-98, which as mentioned is highly toxic, and has been shown to synthesize higher levels of the *cytK* mRNA compared to other *B. cereus* group strains (Brillard and Lereclus, 2004).

2.7 Future Perspectives – Importance of Plasmids to the Biology of *B. cereus* Group Bacteria

Plasmids are key elements in the encoding of several of the phenotypes characteristic of each species in the *B. cereus* group, including *B. thuringiensis* entomopathogenicity and *B. anthracis* virulence. With the discovery of *B. cereus* strains which are able to cause severe disease symptoms resembling those of anthrax and encode the anthrax toxins from variants of the pXO1 plasmid as well as producing a capsule, the principle differences separating these species are getting increasingly blurry. Such strains include *B. cereus* G9241, which was isolated from the sputum and blood of a patient with life-threatening pneumonia, and carries a 191 kb plasmid (pBCX01) with 99.6% identity to pXO1 in regions shared between the two plasmids, as well as a second 218 kb plasmid, pBC218, which has a gene cluster encoding a polysaccharide capsule (Hoffmaster et al., 2004). Perhaps even more striking, during recent years *B. cereus* strains causing anthrax-like disease in great apes have been isolated in Côte d'Ivoire and Cameroon (Leendertz et al., 2004; Klee et al., 2006; Leendertz et al., 2006). These strains (*B. cereus* CI and CA, respectively) carry the *B. anthracis* toxin and capsule genes on plasmids of sizes corresponding to pXO1 and pXO2 and have a frameshift mutation in the *plcR* gene, however in a different position than the nonsense mutation universally found in *B. anthracis* strains (Klee et al., 2006). The mutation would produce a PlcR protein with a modified C-terminus, a part of the protein known to be involved in the specific interaction with PapR, its cognate peptide pheromone, which is necessary for activation of the PlcR regulon (Slamti and Lereclus, 2002, 2005; Bouillaut et al., 2008). It is therefore conceivable that as is the case in *B. anthracis*, expression of the PlcR regulon could be abolished in the CI and CA strains (Klee et al., 2006), and functional experiments performed in the CI strain point in this direction (Klee et al., 2010). Further studies to investigate whether other *B. cereus* strains capable of synthesizing anthrax toxins and capsule exist and may occur more frequently than previously known would seem justified.

By comparative analysis to *B. cereus* strains, it is known that very few genes are specific to *B. anthracis* (Read et al., 2003a). Taken together, it is thus apparent that what principally separates *B. anthracis* from *B. cereus* is the following: (1) being located to the *B. anthracis* cluster phylogenetically, (2) the presence of the four unique prophages (lambda01–lambda04) in the *B. anthracis* chromosome, and (3) the unique nonsense mutation in *plcR*, which is found only in *B. anthracis* (reviewed in Kolstø et al., 2009). *B. cereus* strains may harbour a range of plasmids of various sizes and families, many of which are poorly characterized, as well as the pXO1- and pXO2-like plasmids carrying anthrax toxin and capsule synthesis genes mentioned above, which was previously thought to be a specific and unique feature to *B. anthracis*. There is to date, a lack of knowledge of what features *B. cereus* and *B. thuringiensis* plasmids encode, such as novel putative virulence genes or genes potentially involved in the adaptation to specific niches. *B. cereus* plasmids can be mobile, or capable of mobilizing other plasmids in the group, e.g. pXO14 which is efficient in mobilizing pXO1 and pXO2 (Reddy et al., 1987). It is however not known to what extent horizontal transfer of pXO1 and pXO2 plasmids may occur in the *B. cereus* population. Given the fact that newly emerging pathogens (CI/CA strains) seem to arise from transfer of such plasmids, that *B. cereus* strains encoding alternative capsules (G9241) and possibly other virulence traits from plasmid elements exist, and that knowledge of plasmid diversity in the group is rather limiting, a systematic sequencing approach targeting *B. cereus* group plasmids seems warranted. Finally, it should be kept in mind that a considerable fraction of the annotated genes in *B. cereus* group genomes are still categorized as hypothetical or conserved hypothetical genes of unknown function and that these genes are represented both on the chromosome and on the plasmids. What contributions proteins encoded by these genes make to *B. cereus* group biology is still enigmatic.

Acknowledgements The authors wish to thank Dr. Nicolas Tourasse for performing the phylogenetic analyses for Figs. 2.1 and 2.3.

References

Agaisse H, Gominet M, Økstad OA, Kolstø AB, Lereclus D (1999) PlcR is a pleiotropic regulator of extracellular virulence factor gene expression in Bacillus thuringiensis. Mol Microbiol 32:1043–1053

Apetroaie-Constantin C, Mikkola R, Andersson MA, Teplova V, Suominen I, Johansson T, Salkinoja-Salonen M (2009) Bacillus subtilis and B. mojavensis strains connected to food poisoning produce the heat stable toxin amylosin. J Appl Microbiol 106:1976–1985

Aronson A (2002) Sporulation and delta-endotoxin synthesis by *Bacillus thuringiensis*. Cell Mol Life Sci 59:417–425

Asano SI, Nukumizu Y, Bando H, Iizuka T, Yamamoto T (1997) Cloning of novel enterotoxin genes from Bacillus cereus and Bacillus thuringiensis. Appl Environ Microbiol 63:1054–1057

Auger S, Galleron N, Bidnenko E, Ehrlich SD, Lapidus A, Sorokin A (2008) The genetically remote pathogenic strain NVH391-98 of the Bacillus cereus group is representative of a cluster of thermophilic strains. Appl Environ Microbiol 74:1276–1280

Baida G, Budarina ZI, Kuzmin NP, Solonin AS (1999) Complete nucleotide sequence and molecular characterization of hemolysin II gene from Bacillus cereus. FEMS Microbiol Lett 180:7–14

Baida GE, Kuzmin NP (1995) Cloning and primary structure of a new hemolysin gene from Bacillus cereus. Biochim Biophys Acta 1264:151–154

Barker M, Thakker B, Priest FG (2005) Multilocus sequence typing reveals that Bacillus cereus strains isolated from clinical infections have distinct phylogenetic origins. FEMS Microbiol Lett 245:179–184

Baron F, Cochet MF, Grosset N, Madec MN, Briandet R, Dessaigne S et al (2007) Isolation and characterization of a psychrotolerant toxin producer, Bacillus weihenstephanensis, in liquid egg products. J Food Prot 70:2782–2791

Beatty ME, Ashford DA, Griffin PM, Tauxe RV, Sobel J (2003) Gastrointestinal anthrax: review of the literature. Arch Intern Med 163:2527–2531

Beecher DJ, Macmillan JD (1991) Characterization of the components of hemolysin BL from Bacillus cereus. Infect Immun 59:1778–1784

Beecher DJ, Wong AC (1997) Tripartite hemolysin BL from Bacillus cereus. Hemolytic analysis of component interactions and a model for its characteristic paradoxical zone phenomenon. J Biol Chem 272:233–239

Beecher DJ, Wong AC (2000) Tripartite haemolysin BL: isolation and characterization of two distinct homologous sets of components from a single Bacillus cereus isolate. Microbiology 146(Pt 6):1371–1380

Bottone EJ (2010) Bacillus cereus, a volatile human pathogen. Clin Microbiol Rev 23: 382–398

Bouillaut L, Perchat S, Arold S, Zorrilla S, Slamti L, Henry C et al (2008) Molecular basis for group-specific activation of the virulence regulator PlcR by PapR heptapeptides. Nucleic Acids Res 36:3791–3801

Brillard J, Lereclus D (2004) Comparison of cytotoxin cytK promoters from Bacillus cereus strain ATCC 14579 and from a B. cereus food-poisoning strain. Microbiology 150:2699–2705

Candela T, Fouet A (2006) Poly-gamma-glutamate in bacteria. Mol Microbiol 60:1091–1098

Carlson CR, Caugant DA, Kolstø AB (1994a) Genotypic diversity among *Bacillus cereus* and *Bacillus thuringiensis* strains. Appl Environ Microbiol 60:1719–1725

Carlson CR, Caugant DA, Kolstø AB (1994b) Genotypic diversity among Bacillus cereus and Bacillus thuringiensis strains. Appl Environ Microbiol 60:1719–1725

Carlson CR, Gronstad A, Kolstø AB (1992) Physical maps of the genomes of three Bacillus cereus strains. J Bacteriol 174:3750–3756

Carlson CR, Johansen T, Kolstø AB (1996) The chromosome map of Bacillus thuringiensis subsp. canadensis HD224 is highly similar to that of the Bacillus cereus type strain ATCC 14579. FEMS Microbiol Lett 141:163–167

Carlson CR, Kolstø AB (1993) A complete physical map of a Bacillus thuringiensis chromosome. J Bacteriol 175:1053–1060

Chitlaru T, Gat O, Gozlan Y, Ariel N, Shafferman A (2006) Differential proteomic analysis of the Bacillus anthracis secretome: distinct plasmid and chromosome CO_2-dependent cross talk mechanisms modulate extracellular proteolytic activities. J Bacteriol 188:3551–3571

Cohn F (1872) Untersuchungen über Bacterien. Beitrage zur Biologie der Pflanzen 1:127–244

Dai Z, Sirard JC, Mock M, Koehler TM (1995) The atxA gene product activates transcription of the anthrax toxin genes and is essential for virulence. Mol Microbiol 16:1171–1181

Delihas N (2008) Small mobile sequences in bacteria display diverse structure/function motifs. Mol Microbiol 67:475–481

Dierick K, Van Coillie E, Swiecicka I, Meyfroidt G, Devlieger H, Meulemans A et al (2005) Fatal family outbreak of Bacillus cereus-associated food poisoning. J Clin Microbiol 43:4277–4279

Drobniewski FA (1993) Bacillus cereus and related species. Clin Microbiol Rev 6:324–338

Drysdale M, Heninger S, Hutt J, Chen Y, Lyons CR, Koehler TM (2005) Capsule synthesis by *Bacillus anthracis* is required for dissemination in murine inhalation anthrax. EMBO J 24: 221–227

Ehling-Schulz M, Fricker M, Grallert H, Rieck P, Wagner M, Scherer S (2006) Cereulide synthetase gene cluster from emetic *Bacillus cereus*: structure and location on a mega virulence plasmid related to *Bacillus anthracis* toxin plasmid pXO1. BMC Microbiol 6:20

Ehling-Schulz M, Vukov N, Schulz A, Shaheen R, Andersson M, Martlbauer E, Scherer S (2005a) Identification and partial characterization of the nonribosomal peptide synthetase gene responsible for cereulide production in emetic Bacillus cereus. Appl Environ Microbiol 71:105–113

Ehling-Schulz M, Svensson B, Guinebretiere MH, Lindback T, Andersson M, Schulz A et al (2005b) Emetic toxin formation of Bacillus cereus is restricted to a single evolutionary lineage of closely related strains. Microbiology 151:183–197

Fagerlund A, Brillard J, Furst R, Guinebretiere MH, Granum PE (2007) Toxin production in a rare and genetically remote cluster of strains of the Bacillus cereus group. BMC Microbiol 7:43

Fagerlund A, Lindback T, Storset AK, Granum PE, Hardy SP (2008) Bacillus cereus Nhe is a pore-forming toxin with structural and functional properties similar to the ClyA (HlyE, SheA) family of haemolysins, able to induce osmotic lysis in epithelia. Microbiology 154:693–704

Fedhila S, Daou N, Lereclus D, Nielsen-LeRoux C (2006) Identification of Bacillus cereus internalin and other candidate virulence genes specifically induced during oral infection in insects. Mol Microbiol 62:339–355

Fedhila S, Gohar M, Slamti L, Nel P, Lereclus D (2003) The Bacillus thuringiensis PlcR-regulated gene inhA2 is necessary, but not sufficient, for virulence. J Bacteriol 185:2820–2825

Fedhila S, Guillemet E, Nel P, Lereclus D (2004) Characterization of two Bacillus thuringiensis genes identified by in vivo screening of virulence factors. Appl Environ Microbiol 70: 4784–4791

Fedhila S, Nel P, Lereclus D (2002) The InhA2 metalloprotease of Bacillus thuringiensis strain 407 is required for pathogenicity in insects infected via the oral route. J Bacteriol 184:3296–3304

Fouet A, Mock M (2006) Regulatory networks for virulence and persistence of *Bacillus anthracis*. Curr Opin Microbiol 9:160–166

Ghelardi E, Celandroni F, Salvetti S, Fiscarelli E, Senesi S (2007) *Bacillus thuringiensis* pulmonary infection: critical role for bacterial membrane-damaging toxins and host neutrophils. Microbes Infect 9:591–598

Gohar M, Faegri K, Perchat S, Ravnum S, Økstad OA, Gominet M et al (2008) The PlcR virulence regulon of Bacillus cereus. PLoS One 3:e2793

Gohar M, Gilois N, Graveline R, Garreau C, Sanchis V, Lereclus D (2005) A comparative study of Bacillus cereus, Bacillus thuringiensis and Bacillus anthracis extracellular proteomes. Proteomics 5:3696–3711

Granum PE, O'Sullivan K, Lund T (1999) The sequence of the non-haemolytic enterotoxin operon from Bacillus cereus. FEMS Microbiol Lett 177:225–229

Han CS, Xie G, Challacombe JF, Altherr MR, Bhotika SS, Brown N et al (2006) Pathogenomic sequence analysis of *Bacillus cereus* and *Bacillus thuringiensis* isolates closely related to *Bacillus anthracis*. J Bacteriol 188:3382–3390

Helgason E, Caugant DA, Lecadet MM, Chen Y, Mahillon J, Lovgren A et al (1998) Genetic diversity of Bacillus cereus/B. thuringiensis isolates from natural sources. Curr Microbiol 37:80–87

Helgason E, Caugant DA, Olsen I, Kolstø AB (2000a) Genetic structure of population of Bacillus cereus and B. thuringiensis isolates associated with periodontitis and other human infections. J Clin Microbiol 38:1615–1622

Helgason E, Tourasse NJ, Meisal R, Caugant DA, Kolstø AB (2004) Multilocus sequence typing scheme for bacteria of the Bacillus cereus group. Appl Environ Microbiol 70: 191–201

Helgason E, Økstad OA, Caugant DA, Johansen HA, Fouet A, Mock M et al (2000b) *Bacillus anthracis, Bacillus cereus,* and *Bacillus thuringiensis* – one species on the basis of genetic evidence. Appl Environ Microbiol 66:2627–2630

Helgason E, Økstad OA, Caugant DA, Johansen HA, Fouet A, Mock M et al (2000c) *Bacillus anthracis, Bacillus cereus,* and *Bacillus thuringiensis* – one species on the basis of genetic evidence. Appl Environ Microbiol 66:2627–2630

Hendriksen NB, Hansen BM, Johansen JE (2006) Occurrence and pathogenic potential of *Bacillus cereus* group bacteria in a sandy loam. Antonie Van Leeuwenhoek 89:239–249

Hernandez E, Ramisse F, Cruel T, le Vagueresse R, Cavallo JD (1999) *Bacillus thuringiensis* serotype H34 isolated from human and insecticidal strains serotypes 3a3b and H14 can lead to death of immunocompetent mice after pulmonary infection. FEMS Immunol Med Microbiol 24:43–47

Herron WM (1930) Rancidity in cheddar cheese. Queen's University, Kingston, ON, Canada

Hill KK, Ticknor LO, Okinaka RT, Asay M, Blair H, Bliss KA et al (2004) Fluorescent amplified fragment length polymorphism analysis of Bacillus anthracis, Bacillus cereus, and Bacillus thuringiensis isolates. Appl Environ Microbiol 70:1068–1080

Hoffmaster AR, Ravel J, Rasko DA, Chapman GD, Chute MD, Marston CK et al (2004) Identification of anthrax toxin genes in a Bacillus cereus associated with an illness resembling inhalation anthrax. Proc Natl Acad Sci USA 101:8449–8454

Hoton FM, Andrup L, Swiecicka I, Mahillon J (2005) The cereulide genetic determinants of emetic Bacillus cereus are plasmid-borne. Microbiology 151:2121–2124

Ivanova N, Sorokin A, Anderson I, Galleron N, Candelon B, Kapatral V et al (2003) Genome sequence of Bacillus cereus and comparative analysis with Bacillus anthracis. Nature 423:87–91

Jackson PJ, Hill KK, Laker MT, Ticknor LO, Keim P (1999) Genetic comparison of Bacillus anthracis and its close relatives using amplified fragment length polymorphism and polymerase chain reaction analysis. J Appl Microbiol 87:263–269

Jensen GB, Hansen BM, Eilenberg J, Mahillon J (2003) The hidden lifestyles of *Bacillus cereus* and relatives. Environ Microbiol 5:631–640

Jernigan JA, Stephens DS, Ashford DA, Omenaca C, Topiel MS, Galbraith M et al (2001) Bioterrorism-related inhalational anthrax: the first 10 cases reported in the United States. Emerg Infect Dis 7:933–944

Keim P, Kalif A, Schupp J, Hill K, Travis SE, Richmond K et al (1997a) Molecular evolution and diversity in Bacillus anthracis as detected by amplified fragment length polymorphism markers. J Bacteriol 179:818–824

Keim P, Kalif A, Schupp J, Hill K, Travis SE, Richmond K et al (1997b) Molecular evolution and diversity in *Bacillus anthracis* as detected by amplified fragment length polymorphism markers. J Bacteriol 179:818–824

Klee SR, Brzuszkiewicz EB, Nattermann H, Brüggemann H, Dupke S, Wollherr A, Franz T, Pauli G, Appel B, Liebl W, Couacy-Hymann E, Boesch C, Meyer FD, Leendertz FH, Ellerbrok H, Gottschalk G, Grunow R, Liesegang H (2010) The genome of a Bacillus isolate causing anthrax in chimpanzees combines chromosomal properties of B. cereus with B. anthracis virulence plasmids. PLoS One 5(7):e10986.

Klee SR, Ozel M, Appel B, Boesch C, Ellerbrok H, Jacob D et al (2006) Characterization of Bacillus anthracis-like bacteria isolated from wild great apes from Cote d'Ivoire and Cameroon. J Bacteriol 188:5333–5344

Klevan A, Tourasse NJ, Stabell FB, Kolstø AB, Økstad OA (2007) Exploring the evolution of the Bacillus cereus group repeat element bcr1 by comparative genome analysis of closely related strains. Microbiology 153:3894–3908

Kolstø AB, Tourasse NJ, Økstad OA (2009) What sets Bacillus anthracis apart from other Bacillus species? Annu Rev Microbiol 63:451–476

Kotiranta A, Lounatmaa K, Haapasalo M (2000) Epidemiology and pathogenesis of Bacillus cereus infections. Microbes Infect 2:189–198

Kreft J, Berger H, Hartlein M, Muller B, Weidinger G, Goebel W (1983) Cloning and expression in Escherichia coli and Bacillus subtilis of the hemolysin (cereolysin) determinant from Bacillus cereus. J Bacteriol 155:681–689

Kuppe A, Evans LM, McMillen DA, Griffith OH (1989) Phosphatidylinositol-specific phospholipase C of Bacillus cereus: cloning, sequencing, and relationship to other phospholipases. J Bacteriol 171:6077–6083

Lapidus A, Goltsman E, Auger S, Galleron N, Segurens B, Dossat C et al (2008) Extending the Bacillus cereus group genomics to putative food-borne pathogens of different toxicity. Chem Biol Interact 171:236–249

Leendertz FH, Ellerbrok H, Boesch C, Couacy-Hymann E, Matz-Rensing K, Hakenbeck R et al (2004) Anthrax kills wild chimpanzees in a tropical rainforest. Nature 430:451–452

Leendertz FH, Lankester F, Guislain P, Neel C, Drori O, Dupain J et al (2006) Anthrax in Western and Central African great apes. Am J Primatol 68:928–933

Lindback T, Fagerlund A, Rodland MS, Granum PE (2004) Characterization of the Bacillus cereus Nhe enterotoxin. Microbiology 150:3959–3967

Lindbäck T, Hardy SP, Dietrich R, Sødring M, Didier A, Moravek M, Fagerlund A, Bock S, Nielsen C, Casteel M, Granum PE, Märtlbauer E (2010) Cytotoxicity of the Bacillus cereus Nhe enterotoxin requires specific binding order of its three exoprotein components. Infect Immun. 78(9):3813–3821

Lindback T, Økstad OA, Rishovd AL, Kolstø AB (1999) Insertional inactivation of hblC encoding the L2 component of Bacillus cereus ATCC 14579 haemolysin BL strongly reduces enterotoxigenic activity, but not the haemolytic activity against human erythrocytes. Microbiology 145(Pt 11):3139–3146

Lovgren A, Carlson CR, Kang D, Eskils K, Kolstø AB (2002) Physical mapping of the Bacillus thuringiensis subsp. kurstaki and alesti chromosomes. Curr Microbiol 44:81–87

Lucking G, Dommel MK, Scherer S, Fouet A, Ehling-Schulz M (2009) Cereulide synthesis in emetic Bacillus cereus is controlled by the transition state regulator AbrB, but not by the virulence regulator PlcR. Microbiology 155:922–931

Lund T, De Buyser ML, Granum PE (2000) A new cytotoxin from Bacillus cereus that may cause necrotic enteritis. Mol Microbiol 38:254–261

Lund T, Granum PE (1996) Characterisation of a non-haemolytic enterotoxin complex from Bacillus cereus isolated after a foodborne outbreak. FEMS Microbiol Lett 141:151–156

Mahler H, Pasi A, Kramer JM, Schulte P, Scoging AC, Bar W, Krahenbuhl S (1997) Fulminant liver failure in association with the emetic toxin of Bacillus cereus. N Engl J Med 336: 1142–1148

Margulis L, Jorgensen JZ, Dolan S, Kolchinsky R, Rainey FA, Lo SC (1998) The Arthromitus stage of Bacillus cereus: intestinal symbionts of animals. Proc Natl Acad Sci USA 95: 1236–1241

McIntyre L, Bernard K, Beniac D, Isaac-Renton JL, Naseby DC (2008) Identification of Bacillus cereus group species associated with food poisoning outbreaks in British Columbia, Canada. Appl Environ Microbiol 74:7451–7453

Mignot T, Couture-Tosi E, Mesnage S, Mock M, Fouet A (2004) In vivo Bacillus anthracis gene expression requires PagR as an intermediate effector of the AtxA signalling cascade. Int J Med Microbiol 293:619–624

Mignot T, Mock M, Fouet A (2003) A plasmid-encoded regulator couples the synthesis of toxins and surface structures in Bacillus anthracis. Mol Microbiol 47:917–927

Mock M, Fouet A (2001) Anthrax. Annu Rev Microbiol 55:647–671

Mock M, Mignot T (2003) Anthrax toxins and the host: a story of intimacy. Cell Microbiol 5:15–23

Nieminen T, Rintaluoma N, Andersson M, Taimisto AM, Ali-Vehmas T, Seppala A et al (2007) Toxinogenic Bacillus pumilus and Bacillus licheniformis from mastitic milk. Vet Microbiol 124:329–339

Okinaka R, Cloud K, Hampton O, Hoffmaster A, Hill K, Keim P et al (1999a) Sequence, assembly and analysis of pX01 and pX02. J Appl Microbiol 87:261–262

Okinaka RT, Cloud K, Hampton O, Hoffmaster AR, Hill KK, Keim P et al (1999b) Sequence and organization of pXO1, the large Bacillus anthracis plasmid harboring the anthrax toxin genes. J Bacteriol 181:6509–6515

Økstad OA, Hegna I, Lindback T, Rishovd AL, Kolstø AB (1999a) Genome organization is not conserved between Bacillus cereus and Bacillus subtilis. Microbiology 145(Pt 3):621–631

Økstad OA, Gominet M, Purnelle B, Rose M, Lereclus D, Kolstø AB (1999b) Sequence analysis of three Bacillus cereus loci carrying PlcR-regulated genes encoding degradative enzymes and enterotoxin. Microbiology 145(Pt 11):3129–3138

Økstad OA, Tourasse NJ, Stabell FB, Sundfaer CK, Egge-Jacobsen W, Risoen PA et al (2004) The bcr1 DNA repeat element is specific to the Bacillus cereus group and exhibits mobile element characteristics. J Bacteriol 186:7714–7725

Oscarsson J, Mizunoe Y, Uhlin BE, Haydon DJ (1996) Induction of haemolytic activity in Escherichia coli by the slyA gene product. Mol Microbiol 20:191–199

Oscarsson J, Westermark M, Lofdahl S, Olsen B, Palmgren H, Mizunoe Y et al (2002) Characterization of a pore-forming cytotoxin expressed by Salmonella enterica serovars typhi and paratyphi A. Infect Immun 70:5759–5769

Pannucci J, Okinaka RT, Sabin R, Kuske CR (2002a) Bacillus anthracis pXO1 plasmid sequence conservation among closely related bacterial species. J Bacteriol 184:134–141

Pannucci J, Okinaka RT, Williams E, Sabin R, Ticknor LO, Kuske CR (2002b) DNA sequence conservation between the Bacillus anthracis pXO2 plasmid and genomic sequence from closely related bacteria. BMC Genomics 3:34

Passalacqua KD, Bergman NH (2006) *Bacillus anthracis*: interactions with the host and establishment of inhalational anthrax. Future Microbiol 1:397–415

Perego M, Hoch JA (2008) Commingling regulatory systems following acquisition of virulence plasmids by *Bacillus anthracis*. Trends Microbiol 16:215–221

Preisz H (1909) Experimentelle studien über virulenz, empfänglichkeit und immunität beim milzbrand. Zeitschr Immunität-Forsch 5:341–452

Priest FG (1993) Systematics and ecology of Bacillus. In: Bacillus subtilis and other Gram-positive bacteria - Biochemistry, physiology, and molecular genetics. In: Sonenshein AL, Hoch JA, Losick R (eds.) ASM press, American Society for Microbiology, Washington, D.C. ISBN 1-55581-053-5.

Pruss BM, Dietrich R, Nibler B, Martlbauer E, Scherer S (1999) The hemolytic enterotoxin HBL is broadly distributed among species of the Bacillus cereus group. Appl Environ Microbiol 65:5436–5442

Rasko DA, Altherr MR, Han CS, Ravel J (2005) Genomics of the Bacillus cereus group of organisms. FEMS Microbiol Rev 29:303–329

Rasko DA, Ravel J, Økstad OA, Helgason E, Cer RZ, Jiang L et al (2004) The genome sequence of Bacillus cereus ATCC 10987 reveals metabolic adaptations and a large plasmid related to Bacillus anthracis pXO1. Nucleic Acids Res 32:977–988

Rasko DA, Rosovitz MJ, Økstad OA, Fouts DE, Jiang L, Cer RZ et al (2007) Complete sequence analysis of novel plasmids from emetic and periodontal Bacillus cereus isolates reveals a common evolutionary history among the B. cereus-group plasmids, including Bacillus anthracis pXO1. J Bacteriol 189:52–64

Read TD, Peterson SN, Tourasse N, Baillie LW, Paulsen IT, Nelson KE et al (2003a) The genome sequence of Bacillus anthracis Ames and comparison to closely related bacteria. Nature 423:81–86

Read TD, Peterson SN, Tourasse NJ, Baillie LW, Paulsen IT, Nelson KE et al (2003b) The genome sequence of *Bacillus anthracis* Ames and comparison to closely related bacteria. Nature 423:81–86

Read TD, Salzberg SL, Pop M, Shumway M, Umayam L, Jiang L et al (2002) Comparative genome sequencing for discovery of novel polymorphisms in Bacillus anthracis. Science 296: 2028–2033

Reddy A, Battisti L, Thorne CB (1987) Identification of self-transmissible plasmids in four Bacillus thuringiensis subspecies. J Bacteriol 169:5263–5270

Richter S, Anderson VJ, Garufi G, Lu L, Budzik JM, Joachimiak A et al (2009) Capsule anchoring in *Bacillus anthracis* occurs by a transpeptidation reaction that is inhibited by capsidin. Mol Microbiol 71:404–420

Saile E, Koehler TM (2006) *Bacillus anthracis* multiplication, persistence, and genetic exchange in the rhizosphere of grass plants. Appl Environ Microbiol 72:3168–3174

Salkinoja-Salonen MS, Vuorio R, Andersson MA, Kampfer P, Andersson MC, Honkanen-Buzalski T, Scoging AC (1999) Toxigenic strains of Bacillus licheniformis related to food poisoning. Appl Environ Microbiol 65:4637–4645

Scarano C, Virdis S, Cossu F, Frongia R, De Santis EP, Cosseddu AM (2009) The pattern of toxin genes and expression of diarrheal enterotoxins in Bacillus thuringiensis strains isolated from commercial bioinsecticides. Vet Res Commun 33(Suppl 1):257–260

Schnepf E, Crickmore N, Van Rie J, Lereclus D, Baum J, Feitelson J et al (1998) *Bacillus thuringiensis* and its pesticidal crystal proteins. Microbiol Mol Biol Rev 62:775–806

Shinagawa K, Sugiyama J, Terada T, Matsusaka N, Sugii S (1991) Improved methods for purification of an enterotoxin produced by Bacillus cereus. FEMS Microbiol Lett 64:1–5

Slamti L, Lereclus D (2002) A cell-cell signaling peptide activates the PlcR virulence regulon in bacteria of the Bacillus cereus group. EMBO J 21:4550–4559

Slamti L, Lereclus D (2005) Specificity and polymorphism of the PlcR-PapR quorum-sensing system in the Bacillus cereus group. J Bacteriol 187:1182–1187

Soberon M, Pardo-Lopez L, Lopez I, Gomez I, Tabashnik BE, Bravo A (2007) Engineering modified Bt toxins to counter insect resistance. Science 318:1640–1642

Soule M (1932) Identity of Bacillus subtilis, Cohn 1872. J Infect Dis 51:191–215

Sozhamannan S, Chute MD, McAfee FD, Fouts DE, Akmal A, Galloway DR et al (2006) The *Bacillus anthracis* chromosome contains four conserved, excision-proficient, putative prophages. BMC Microbiol 6:34

Stenfors Arnesen LP, Fagerlund A, Granum PE (2008) From soil to gut: Bacillus cereus and its food poisoning toxins. FEMS Microbiol Rev 32:579–606

Stenfors LP, Granum PE (2001) Psychrotolerant species from the Bacillus cereus group are not necessarily Bacillus weihenstephanensis. FEMS Microbiol Lett 197:223–228

Suyama M, Bork P (2001) Evolution of prokaryotic gene order: genome rearrangements in closely related species. Trends Genet 17:10–13

Thorsen L, Hansen BM, Nielsen KF, Hendriksen NB, Phipps RK, Budde BB (2006) Characterization of emetic Bacillus weihenstephanensis, a new cereulide-producing bacterium. Appl Environ Microbiol 72:5118–5121

Ticknor LO, Kolstø AB, Hill KK, Keim P, Laker MT, Tonks M, Jackson PJ (2001) Fluorescent amplified fragment length polymorphism analysis of Norwegian Bacillus cereus and Bacillus thuringiensis soil isolates. Appl Environ Microbiol 67:4863–4873

Tourasse NJ, Helgason E, Klevan A, Sylvestre P, Moya M, Haustant M, Økstad OA, Fouet A, Mock M, Kolstø AB. Extended and global phylogenetic view of the Bacillus cereus group population by combination of MLST, AFLP, and MLEE genotyping data. Food Microbiology. In Press.

Tourasse NJ, Helgason E, Økstad OA, Hegna IK, Kolstø AB (2006) The Bacillus cereus group: novel aspects of population structure and genome dynamics. J Appl Microbiol 101:579–593

Tourasse NJ, Kolstø AB (2008) SuperCAT: a supertree database for combined and integrative multilocus sequence typing analysis of the Bacillus cereus group of bacteria (including B. cereus, B. anthracis and B. thuringiensis). Nucleic Acids Res 36:D461–D468

Tran SL, Guillemet E, Gohar M, Lereclus D, Ramarao N (2010) CwpFM (EntFM) is a Bacillus cereus potential cell wall peptidase implicated in adhesion, biofilm formation, and virulence. J Bacteriol 192:2638–2642

Uchida I, Hornung JM, Thorne CB, Klimpel KR, Leppla SH (1993) Cloning and characterization of a gene whose product is a trans-activator of anthrax toxin synthesis. J Bacteriol 175: 5329–5338

Uchida I, Makino S, Sekizaki T, Terakado N (1997) Cross-talk to the genes for Bacillus anthracis capsule synthesis by atxA, the gene encoding the trans-activator of anthrax toxin synthesis. Mol Microbiol 23:1229–1240

Van der Auwera GA, Andrup L, Mahillon J (2005) Conjugative plasmid pAW63 brings new insights into the genesis of the *Bacillus anthracis* virulence plasmid pXO2 and of the *Bacillus thuringiensis* plasmid pBT9727. BMC Genomics 6:103

Van Ert MN, Easterday WR, Huynh LY, Okinaka RT, Hugh-Jones ME, Ravel J et al (2007) Global genetic population structure of *Bacillus anthracis*. PLoS ONE 2:e461

Vassileva M, Torii K, Oshimoto M, Okamoto A, Agata N, Yamada K et al (2007) A new phylogenetic cluster of cereulide-producing Bacillus cereus strains. J Clin Microbiol 45:1274–1277

Verheust C, Fornelos N, Mahillon J (2005) GIL16, a new gram-positive tectiviral phage related to the *Bacillus thuringiensis* GIL01 and the *Bacillus cereus* pBClin15 elements. J Bacteriol 187:1966–1973

Wallace AJ, Stillman TJ, Atkins A, Jamieson SJ, Bullough PA, Green J, Artymiuk PJ (2000) E. coli hemolysin E (HlyE, ClyA, SheA): X-ray crystal structure of the toxin and observation of membrane pores by electron microscopy. Cell 100:265–276

Whiteley HR, Schnepf HE (1986) The molecular biology of parasporal crystal body formation in Bacillus thuringiensis. Annu Rev Microbiol 40:549–576

Young JA, Collier RJ (2007) Anthrax toxin: receptor binding, internalization, pore formation, and translocation. Annu Rev Biochem 76:243–265

Chapter 3
Post-genome Analysis of the Foodborne Pathogen *Campylobacter jejuni*

Emily J. Kay, Ozan Gundogdu, and Brendan Wren

3.1 *Campylobacter jejuni* – The Organism and Disease

The human pathogen *Campylobacter jejuni* is part of the genus *Campylobacter* that lies within the epsilon proteobacteria subclass of bacteria. The nearest family in phylogenetic terms is the *Helicobacteraceae* which includes the *Helicobacter* and *Wolinella* genuses. *Campylobacter* species are Gram-negative, curved rod shaped or spiral and are motile (via polar flagella). They range from 0.2 to 0.8 μm wide and 0.5 to 5 μm long (Debruyne et al., 2008) (Fig. 3.1). The optimum growth temperature for *C. jejuni* is 42°C in microaerobic conditions (85% nitrogen, 10% carbon dioxide and 5% oxygen). This reflects the fact that *C. jejuni* is a frequent commensal microorganism of the gastrointestinal tract of birds (Altekruse et al., 1999).

There are currently 18 recognized *Campylobacter* species (Debruyne et al., 2008). A dendrogram of *Campylobacter* species is presented in Fig. 3.2, based on 16S rRNA gene sequences (Fagerquist et al., 2007). Several of the species are pathogenic to animals, and nine have been associated with disease in humans (Table 3.1). *Campylobacter* were previously thought to be purely animal pathogens until the 1970s when it was discovered that *C. jejuni* and *Campylobacter coli* caused diarrhoea in man (Skirrow, 1977). Since then, as identification and isolation methods have improved, a role for other *Campylobacter* species in human disease has been identified, although, in the case of *Campylobacter rectus*, *Campylobacter gracilis* and *Campylobacter concisus*, the link to disease is not clearcut as bacteria have been isolated from both healthy and symptomatic individuals (Debruyne et al., 2008).

C. coli and *C. jejuni* cause most human diseases, responsible for 5 and 90% of reported campylobacteriosis cases, respectively (Fitzgerald et al., 2008), although many cases of campylobacteriosis are not distinguished to species level. In the USA *Campylobacter* is responsible for an estimated 2.5 million cases of foodborne illness

E.J. Kay (✉)
Department of Infectious and Tropical Diseases, London School of Hygiene and Tropical Medicine, London WC1E 7HT, UK; Centre for Integrative Systems Biology, Imperial College London, London SW7 2AZ, UK
e-mail: emily.kay@lshtm.ac.uk

M. Wiedmann, W. Zhang (eds.), *Genomics of Foodborne Bacterial Pathogens*, Food Microbiology and Food Safety, DOI 10.1007/978-1-4419-7686-4_3, © Springer Science+Business Media, LLC 2011

Fig. 3.1 Electron micrograph
of *Campylobacter jejuni*
(courtesy of Abdi Elmi)

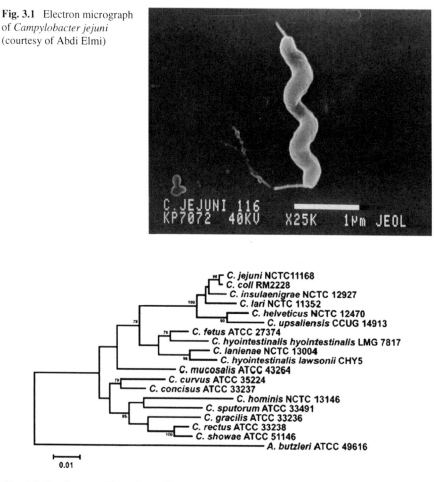

Fig. 3.2 Dendrogram illustrating 16S rRNA comparison between *Campylobacter* species. The dendrogram was constructed using the neighbouring-joining algorithm and the Kimura two-parameter distance estimation method. Bootstrap values of >75%, generated from 500 replicates, are shown at the nodes. The scale bar represents substitutions per site. The 16S rRNA sequence of the type strain of *Arcobacter butzleri* was included to root the tree (Fagerquist et al., 2007) – Reproduced with permission of the Royal Society of Chemistry

per year, 17% of hospitalizations resulting from foodborne infections and an estimated 5% of foodborne-related deaths (Mead et al., 1999). In contrast *Salmonella*, another enteric pathogen, is estimated to cause 1.4 million cases of foodborne illness per year, although hospitalization and mortality rates are higher for this pathogen (Mead et al., 1999). Recent figures suggest that rates of infection with *Campylobacter*, after an initial decrease, have remained relatively constant since 2001 (Fig. 3.3). Even though cases of gastroenteritis caused by *Campylobacter* outnumber those caused by *Salmonella*, outbreaks of campylobacteriosis are more rare with 61 *Campylobacter* outbreaks, affecting 1,440 people, reported to the Centers for Disease Control and Prevention (CDC) between 1998 and 2002 compared with

Table 3.1 Pathogenic *Campylobacter* and their reservoirs

Species	Clinical presentation	Reservoirs
Campylobacter jejuni	Enteritis, systemic illness, Guillain–Barré syndrome	Birds, cattle, sheep, pigs, domestic pets
Campylobacter coli	Enteritis	Pigs, birds
Campylobacter lari	Enteritis, bacteraemia	Seagulls, other animals, shellfish
Campylobacter fetus	Enteritis, systemic illness in immunodeficient patients	Sheep, cattle
Campylobacter upsaliensis	Enteritis	Domestic pets
Campylobacter hyointestinalis	Enteritis, in rare cases bacteraemia	Pigs, cattle
Campylobacter rectus	Putative periodontal pathogen	?
Campylobacter gracilis	Associated with serious deep tissue infection	?
Campylobacter concisus	Isolated from cases of enteritis and periodontal illness but role in pathogenesis unclear	?

Compiled from Debruyne et al. (2008)

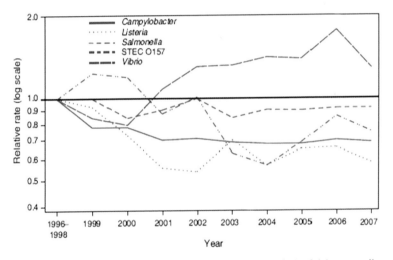

Fig. 3.3 Relative rates compared with 1996–1998 baseline period of laboratory-diagnosed cases of infection with *Campylobacter* and other pathogens by year – Foodborne Diseases Active Surveillance Network, United States, 1996–2007. STEC O157 – Shiga toxin-producing *Escherichia coli* (CDC, 2008)

585 for *Salmonella*, affecting 16,821 (CDC, 2006). This highlights the fact that many cases of campylobacteriosis are sporadic rather than outbreak associated.

C. *jejuni* has been isolated from the gastrointestinal tract of birds (particularly poultry), cattle and domestic pets, where it rarely causes disease. Transmission to humans has been reported from a variety of sources including raw or undercooked meat, especially poultry (Tam, 2001). As many as 75% of retail chickens are contaminated by *Campylobacter* in the USA (Cui et al., 2005; Zhao et al., 2001). Other sources of infection include unpasteurized milk, bird-pecked milk on doorsteps and untreated water: *Campylobacter* may be shed into surface water by birds and can survive for many weeks at low temperatures. However, most infections remain unexplained by recognized risk factors (Tam, 2001).

C. jejuni can cause a spectrum of disease ranging from mild watery diarrhoea to severe inflammatory diarrhoea and has also been associated with extraintestinal disease. In its uncomplicated form the symptoms of *C. jejuni* infection include a high fever, nausea and abdominal cramps, followed by profuse diarrhoea which commonly lasts 2–3 days. In severe cases illness can last for more than a week (Blaser and Engberg, 2008). Antibiotics are not usually administered as *Campylobacter* infections are self-limiting; however, for severe cases drugs such as ciprofloxacin and erythromycin are administered (Altekruse et al., 1999).

There are marked differences in disease outcomes in developed and developing countries. Patients in the developed world often have inflammatory bloody diarrhoea, whereas those in the developing world have non-inflammatory watery diarrhoea (Coker et al., 2002). In developing countries disease is rarely associated with adults and primarily affects young children, <2 years old (Coker et al., 2002).

Infants, the elderly and immunocompromised patients who are infected with *C. jejuni* are also known to have further complications (Crushell et al., 2004). Complications of *C. jejuni* infection include endocarditis, meningitis and neurological sequelae such as Guillain–Barré syndrome (GBS), which causes neuromuscular paralysis (Nachamkin et al., 1998), and Miller Fisher syndrome (MFS), considered to be a variant of GBS which causes paralysis of the ocular muscles. Post-infection complications of *C. jejuni* infection are rare with about 1% developing reactive arthritis and 0.1% developing GBS (Leach, 1997).

How *C. jejuni* is able to cause disease remains somewhat of a paradox as the microorganism is not thought to multiply in foods and has been shown to be susceptible to a wide range of food processing treatments, along with having fastidious growth requirements for laboratory culture (Solomon and Hoover, 1999). However, only a low dose is required to cause disease: 500 cells, if not lower, depending on the infecting strain (Black et al., 1988; Robinson, 1981). The lack of information on this problematic pathogen was the driving force for initiating the original *C. jejuni* genome project on strain NCTC 11168 published in 2000 (Parkhill et al., 2000). Subsequent *C. jejuni* genome projects have revealed considerable genomic diversity between strains (Fouts et al., 2005; Hofreuter et al., 2006; Pearson et al., 2007; Poly et al., 2007).

3.2 Genome Sequencing

3.2.1 Campylobacter jejuni

To date there have been five *C. jejuni* isolates sequenced summarized in Table 3.2, and other strains are in the process of being sequenced. The NCTC 11168 strain was isolated in 1977 from a patient with diarrhoea in the UK (Skirrow, 1977), and

Table 3.2 *Campylobacter jejuni* sequencing projects at May 2008 with sequence characteristics

Campylobacter jejuni strain	Source[a]	State of genome sequence project	Genome size (Mb)[b]	GC content (%)	Number of plasmids	GenBank accession number	Sequencing centre[c]	Publication
Campylobacter jejuni subsp. *jejuni* NCTC 11168	Clinical isolate	Completed	1.60	30.6	0	AL111168.1	Sanger	Parkhill et al. (2000)
Campylobacter jejuni subsp. *jejuni* RM1221	Retail chicken	Completed	1.80	30.3	0	CP000025.1	JCVI	Fouts et al. (2005)
Campylobacter jejuni subsp. *jejuni* 81116 (NCTC 11828)	Clinical isolate	Completed	1.60	30.3	0	CP000814.1	BBSRC IFR	Pearson et al. (2007)
Campylobacter jejuni subsp. *jejuni* 81-176	Clinical isolate	Completed	1.68	30.5	2	CP000538.1	JCVI	Unpublished
Campylobacter jejuni subsp. *jejuni* 81-176	Clinical isolate	In progress	**	30.6	**	AASL00000000	Yale University	Hofreuter et al. (2006)
Campylobacter jejuni subsp. *jejuni* CG8486	Clinical isolate	In progress	**	30.4	**	AASY00000000	NMRC	Poly et al. (2007)
Campylobacter jejuni subsp. *jejuni* 260.94	Clinical isolate (GBS associated)	In progress	**	30.5	**	AANK00000000	JCVI	Unpublished
Campylobacter jejuni subsp. *jejuni* 84-25	Human cerebrospinal fluid	In progress	**	30.4	**	AANT00000000	JCVI	Unpublished
Campylobacter jejuni subsp. *jejuni* CF93-6	Clinical isolate (MFS)	In progress	**	30.5	**	AANJ00000000	JCVI	Unpublished
Campylobacter jejuni subsp. *jejuni* HB93-13	Human faeces (GBS)	In progress	**	30.6	**	AANQ00000000	JCVI	Unpublished

Data obtained from NCBI website (http://www.ncbi.nlm.nih.gov/)

** Denotes information not yet available

[a] Source abbreviations: GBS, Guillain–Barré syndrome; MFS, Miller Fisher syndrome

[b] Denotes genome is estimated

[c] Institute abbreviations: JCVI, J. Craig Venter Institute; BBSRC IFR, BBSRC Institute of Food Research; NMRC, Naval Medical Research Centre

its genome sequence provided the first real detailed insight into this pathogen. Since then, further strains with varying characteristics have been sequenced, e.g. *C. jejuni* RM1221 isolated in 2000 from the skin of a chicken isolated in the USA (Fouts et al., 2005; Miller et al., 2000). The virulent strain, 81-176, isolated in 1981 from a US patient, where an outbreak from raw milk was implicated as the source, was sequenced in 2006 (Hofreuter et al., 2006). More recently *C. jejuni* strain CG8486, a clinical isolate from Thailand (Poly et al., 2007), and strain 81116 (NCTC 11828) (Pearson et al., 2007) have been published.

Key features identified in the original *C. jejuni* NCTC 11168 genome anno-tation were the four surface-located glycan structures including a novel capsule, lipo-oligosaccharide (LOS), an N-linked glycosylation system and an O-linked gly-cosylation system that decorates the flagellum (see Section 3.3.1). There was a notable lack of insertion or phage-associated sequences and very few repeat gene sequences within the genome. Hypervariable sequences allowing phase variation were identified in genes mainly encoding proteins involved in the biosynthesis or modification of surface structures (Parkhill et al., 2000). Known and potential pathogenicity factors were identified including genes that encode the cytolethal dis-tending toxin (CDT), haemolysin-like toxins, several putative type II export genes and chemotaxis genes. The NCTC 11168 genome was shown to be one of the most compact genomes sequenced with 1,654 coding sequences (CDSs) representing 94.3% of the genome, with a relatively low G/C percentage at 30.55% (Parkhill et al., 2000). The NCTC 11168 genome was re-annotated in 2007 (Gundogdu et al., 2007) to incorporate new information from the literature and databases and enhanced annotation techniques that were unavailable when the genome was pub-lished in 2000 (Parkhill et al., 2000). Importantly 18.2% of product functions have been revised and 90% of the CDSs have additional information which highlights the importance of re-annotation to curate new information into existing genome projects.

The genome of *C. jejuni* strain RM1221 is larger than NCTC 11168 at 1.78 Mb and predicted to encode 1,835 proteins. The genomes of NCTC 11168 and RM1221 were shown to be syntenic with the exception of four genomic islands in RM1221. Three of the islands were phage derived and the fourth likely to be an integrated plasmid (Fouts et al., 2005). The LOS and capsular polysaccharide loci contained many different genes. Other interesting observations were the identification of a number of homopolymeric tracts which can lead to phase variation and thus expres-sion of CDSs. Phase variation is an important tool employed by pathogenic bacteria to adapt to host immune surveillance (Hallet, 2001), influencing expression of vir-ulence phenotypes and creating phenotypic diversity in clonal populations (Gogol et al., 2007; Salaun et al., 2003; van der Woude and Baumler, 2004). In RM1221 although there are many poly G tracts, only eight could be seen to vary between sequence reads in the genome assembly. A summary of genome traits for the five *C. jejuni* sequences is presented in Table 3.3.

The more recent *C. jejuni* sequences 81116, 81-176 and CG8486 show simi-lar features in that they all contain homopolymeric tracts, the glycostructure loci are variable and there is variation in pseudogenes, with some being intact in some

Table 3.3 Genome features of five *Campylobacter jejuni* genomes

Category	Trait	Strain				
		C. jejuni	*C. jejuni*	*C. jejuni*	*C. jejuni*	*C. jejuni*
General	Strain	NCTC 11168	RM1221	81-176	81116	CG8486
	Serotype	HS:2	HS:53	HS:23/36	HS:6	HS:4
	MLST[a]	ST-21 (43)	ST-354 (354)	ST-42 (913)	ST-283 (267)	Unknown
	Origin	Clinical	Chicken	Clinical	Clinical	Clinical
Genome properties	Genome size (Mb)[b]	1.64	1.77	1.62	1.63	1.60
	GC content (%)	30.55	30.31	30.62	30.54	30.43
	Predicted CDS numbers	1654(1643)[e]	1835	1568	1626	1588
	Pseudogenes	20(19)[e]	47	0[c]	1[c]	3[c]
	Poly G/C tracts[d]	29(22)	25(8)	19	17	23
	Plasmids	0	0	2	0	0

Data obtained from Fouts et al. (2005), Parkhill et al. (2000), Gundogdu et al. (2007), Hofreuter et al. (2006), Poly et al. (2007) and Pearson et al. (2007)

[a] ST represents clonal complex; ( ) indicates sequence type

[b] Indicates genome size made by approximation; CG8486 genome sequence is currently in 19 contigs

[c] Indicates approximate number of pseudogenes

[d] Poly G/C tracts represent total found; ( ) indicate tracts greater than 7 or more nucleotides in length and have been shown to vary during sequencing project

[e] ( ) indicates number after NCTC11168 re-annotation

strains and not in others. *C. jejuni* strain 81116 is unusual in that it lacks CfrA and Cj0178, outer membrane iron-uptake receptors, as well as only encoding one copy of TonB, an inner membrane iron transporter (Pearson et al., 2007). Strain 81-176 also lacks CfrA and Cj0178 and contains putative components of novel electron transport, energy metabolism and respiratory pathways compared to NCTC 11168. These include a putative DMSO reductase system and additional cytochrome *c* (Hofreuter et al., 2006), which also appear to be present in 81116. Strain 81-176 contains two plasmids (which will be discussed in Section 3.3.6). Strain CG8486 has arsenate resistance genes as in RM1221 and also a 45 kb plasmid with a high degree of similarity to pTet from 81-176, including Tet(O), tetracycline resistance determinant.

3.2.2 Campylobacter *Species Other Than* C. jejuni

In addition to *C. jejuni* genomes a number of other *Campylobacter* species have been sequenced or are in the process of being sequenced (see Table 3.4). Many cause enteritis in humans, although *Campylobacter hominis* has also been sequenced which is not known to cause disease in man, which will provide a useful reference for pathogenic *Campylobacter* species. In 2005, Fouts et al. published the sequences of *Campylobacter lari* RM2100, *Campylobacter upsaliensis* RM3195 and *C. coli* RM2228. *C. lari* can cause enteritis and is often found in cats, dogs, chickens and seagulls where soiled water is a potential transmission route (Lastovica and Skirrow, 2000). The sequenced strain *C. lari* RM2100 was originally a human isolate. *C. upsaliensis* is mainly isolated from domestic pets but can cause enteritis in humans. The sequenced strain RM3195 was isolated from a 4-year-old boy with Guillain–Barré syndrome. *C. coli* is the second most common *Campylobacter* to cause enteritis after *C. jejuni* and has been isolated from pigs and poultry. The sequenced strain (RM2228) was isolated from a chicken carcass and is known to be multi-drug resistant (Fouts et al., 2005).

A total of 1,084 CDSs were shared between the three other *Campylobacter* species and *C. jejuni* RM1221. *C. coli* shares a considerable amount of synteny with RM1221 whereas *C. upsaliensis* and *C. lari* show little if any synteny. Many genes involved in host colonization, including *cadF*, *cdt*, *ciaB* and flagellin genes, are conserved across the species, but variations that appear to be species specific are evident for the LOS and capsule locus. *C. upsaliensis* has three clusters of capsule genes, instead of one flanked by *kps* genes, as is found in *C. jejuni*, *C. lari* and *C. coli*. A detailed comparison is shown in Table 3.5. *C. upsaliensis* may contain a *lic*ABCD locus (Fouts et al., 2005) which in commensal *Neisseria* species has been shown to modify LOS with phosphorylcholine aiding attachment to host cells (Serino and Virji, 2002). *C. upsaliensis* RM3195, *C. lari* RM2100 and *C. coli* RM2228 all have plasmids which will be discussed in Section 3.3.6. Interestingly, *C. coli* RM2228 and *C. lari* RM2100 have relatively few varying poly G/C tracts, 1 and 3, respectively, whereas *C. upsaliensis* has 22, which is comparable to the number found in *C. jejuni* NCTC 11168 (Fouts et al., 2005).

Table 3.4 *Campylobacter* species other than *Campylobacter jejuni* sequencing projects at May 2008 with sequence characteristics

Campylobacter species/strain	Source[a]	State of genome sequence project	Genome size (Mb)[b]	GC content (%)	Number of plasmids	GenBank accession number	Sequencing centre[c]	Publication
Campylobacter concisus 13826	Human faeces	Completed	2.15	39.3	2	CP000792.1	JCVI	Unpublished
Campylobacter curvus 525.92	Human faeces	Completed	2.00	44.5	0	CP000767.1	JCVI	Unpublished
Campylobacter fetus subsp. Fetus 82-40	Human blood	Completed	1.80	33.3	0	CP000487.1	JCVI	Unpublished
Campylobacter hominis ATCC BAA-381	Human faeces	Completed	1.70	32.0	1	CP000776.1	JCVI	Unpublished
Campylobacter coli RM2228	Chicken carcass	In progress	1.86	31.1	**	AAFL00000000	JCVI	Fouts et al. (2005)
Campylobacter lari RM2100	Clinical isolate	In progress	1.56	29.6	**	AAFK00000000	JCVI	Fouts et al. (2005)
Campylobacter upsaliensis RM3195	Human faeces (GBS)	In progress	1.77	34.3	**	AAFJ00000000	JCVI	Fouts et al. (2005)

Data obtained from NCBI website (http://www.ncbi.nlm.nih.gov/)

** Denotes information not yet available

[a] Source abbreviations: GBS, Guillain–Barré syndrome

[b] Denotes genome is estimated

[c] Institute abbreviations: JCVI, J. Craig Venter Institute

Table 3.5 Genome features of five *Campylobacter* genomes

Category	Trait	Strain				
		Campylobacter jejuni	*C. jejuni*	*Campylobacter coli*	*Campylobacter lari*	*Campylobacter upsaliensis*
General	Strain	NCTC 11168	RM1221	RM2228	RM2100	RM3195
	Origin	Clinical	Chicken	Chicken	Clinical	Clinical
Genome properties	Genome size (Mb)[b]	1.64	1.77	1.68	1.5	1.66
	GC content (%)	30.55	30.31	31.37	29.64	34.54
	ORF numbers (less pseudogenes)	1654(1643)[d]	1835	1764	1554	1782
	Pseudogenes	20(19)[d]	47	7[c]	4[c]	11[c]
	Assigned function	1286(1408)[d]	1124	1304	1130	1203
	Phage/genomic island regions	0	4	0	1	1
	Plasmids	0	0	1	1	2
	CRISPR structures	Yes	Yes	No	No	No
Virulence	Bacterial adherence – CadF	1	1	1	1	1
	Bacterial adherence – PEB1(four genes)	4	4	4	4	4
	Bacterial adherence – JlpA	1	1	1	1	1
	Bacterial adherence – 43 kDa MOMP	1	1	1	1	1
	Motility	63	66	58	56	55
	Two-component systems[a]	15	15	15	13	11
	Response regulator	9	9	9	8	7
	Sensor histidine kinase	6	6	6	5	4
	Toxin production and resistance	19	20	18	16	15

Data obtained from Fouts et al. (2005), Parkhill et al. (2000) and Gundogdu et al. (2007)

[a] Indicates data originally obtained from TIGR role category
[b] Indicates genome size made by approximation
[c] Indicates approximate number of pseudogenes
[d] () indicates number after NCTC11168 re-annotation

An interesting variation between *Campylobacter* genome sequences is the presence of clustered regularly interspaced short palindromic repeats (CRISPR) (Godde and Bickerton, 2006; Jansen et al., 2002; Schouls et al., 2003). CRISPR regions are thought to be mobile elements; however, their exact biological role is under much discussion and currently only speculated (Bolotin et al., 2005). In conjunction with this, in *C. jejuni* NCTC 11168, the three CDSs adjacent to the CRISPR repeats were identified as CRISPR-associated proteins (Gundogdu et al., 2007). Interestingly, *C. coli*, *C. upsaliensis* and *C. lari* lack CRISPR repeats (Fouts et al., 2005).

In summary, *Campylobacter* genomes show variation at the genomic level between species and strains. This no doubt has implications for the survival of these bacteria and also for the range of severity of disease in humans. Here we analyse the salient features of these unique bacteria, with emphasis on *C. jejuni* as the most important human pathogen among the *campylobacters*.

3.3 Genome Sequence Features

3.3.1 Surface Structures

A high proportion of the *C. jejuni* NCTC 11168 genome was found to be devoted to the biosynthesis of glycostructures. These structures include LOS, capsule polysaccharide and glycosylation systems (Fig. 3.4). Gene clusters corresponding to the N-linked glycosylation system have been identified in other *Campylobacter* species

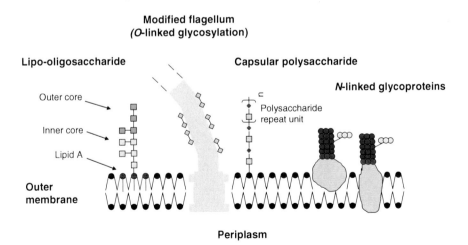

Fig. 3.4 Illustration of four main glyco-surface structures in *Campylobacter jejuni*. Capsular polysaccharide, lipo-oligosaccharide and O-linked glycosylation all vary with strain. N-linked glycosylation is conserved: *red circles* represent Di-*N*-acetyl bacillosamine; *blue circles* represent *N*-acetyl-galactosamine and *yellow circles* represent glucose. Figure modified from Karlyshev et al. (2005b)

(Fouts et al., 2005; Szymanski and Wren, 2005). The O-linked glycosylation system by contrast is highly variable and is responsible for decorating the flagella with various branching sugars. Surface structures have been shown to play an important role in pathogenesis and survival.

3.3.1.1 Lipo-oligosaccharide

One of the main glycostructures in *C. jejuni* is the LOS, a lipid A molecule joined to a core oligosaccharide but lacking O-antigen repeats. In NCTC 11168 the major disaccharide in lipid A is GlcN3N-GlcN (2,3-diamino-2,3-dideoxy-D-glucose and D-glucosamine) phosphorylated with *PP*Etn (pyrophosphorylethanolamine), and this is acylated with palmitic or lauric acid (Szymanski et al., 2003b). However, this disaccharide varies as do the phosphorylation patterns. In the genome sequence of NCTC 11168 the lipid A biosynthesis genes *lpxA*, *lpxC*, *lpxD*, *lpxB* and *lpxK* were found along with the Kdo transferase, *kdtA*, and the late acetylase gene, *htrB* (Karlyshev et al., 2005b), although not in a single locus. Conversely, genes for biosynthesis of the core oligosaccharide of LOS are co-located, adjacent to the N-linked glycosylation locus. The LOS gene cluster (cj1133–cj1152) is one of the most highly studied regions within this bacterium, with structural studies and enzymes characterized by x-ray crystallography (Chiu et al., 2004).

LOS contains common sugars such as glucose, galactose and *N*-acetylgalactosamine. The size of the oligosaccharide locus ranges from 8 to 15 kb (Parker et al., 2005). DNA sequences from the LOS region of 11 *C. jejuni* strains have been compared and assigned to one of three classes (Gilbert et al., 2002). This has subsequently been subdivided (Gilbert et al., 2004) and new classes proposed (Parker et al., 2005).

Not only is LOS important in maintaining structural integrity of the bacterium, it is also involved in adherence and can vary to avoid the host immune system. It has been shown in *C. jejuni* strain 81-176 that deep core mutants, such as *waaC* (Kanipes et al., 2006) and *waaF* (Kanipes et al., 2004), that produce a severely truncated LOS structure, are impaired in their ability to adhere to and invade tissue culture cells. In contrast, more distal mutants such as *igtF* and *galT*, which demonstrate truncated LOS and a loss of distal residues, respectively, are able to invade at levels comparable to wild type (Kanipes et al., 2008). LOS structure can be varied through several mechanisms including alterations in gene content, recombination between and within genes and homopolymeric tract variation (phase variation), all of which act at different rates. LOS diversity may be important for the ability to colonize a wide variety of hosts and intestinal niches, as well as for survival in non-intestinal environments.

The main region of variation is *htrB* to *waaV*. Even small differences in sequence can lead to marked differences in functionality, e.g. *cst* gene for sialyltransferase where different amino acid substitutions are associated with α-2,3 and α-2,8 activity or different acceptors (Gilbert et al., 2002). In addition at least nine genes in the LOS locus have homopolymeric tracts. In NCTC 11168 *wlaN* tract length variation has been linked with variation in terminal sugar structures (Linton et al., 2000). A number of these tract length variations in LOS biosynthesis genes have been linked to alterations in ganglioside mimics.

The development of GBS syndrome has been associated with ganglioside mimics and occurs in roughly 1:1,000 cases of campylobacteriosis. Sialic acid is thought to be the major component of molecular mimicry which may lead to autoimmune diseases such as GBS and Miller Fisher syndromes (MFS). Three genes, *cgtA/neuA1*, *neuB1* and *neuC1*, are required for sialic acid biosynthesis. In many strains *cgtA* and *neuA1* are separate CDSs, instead of the single, merged *cgtA/neuA1* CDS of NCTC 11168 which produces a bi-functional protein (Gilbert et al., 2002). Sialic acid is important for immune avoidance as a mutant lacking LOS sialic acid residues showed greater immunoreactivity and decreased serum resistance (Guerry et al., 2000). Mass spectrometry and sequence information have been used to predict the structure of LOS from different strains and link them to observed pathology (Godschalk et al., 2007). This method showed that 73% of GBS-associated strains expressed LOS with ganglioside mimics. GM1a was shown to be the most prevalent ganglioside mimic, often being found as a GM1a/GD1a mixture, suggesting the formation of ganglioside complexes.

There also seems to be a degree of cross-talk between the different glycostructure biosynthesis and modification pathways. For example, a *waaC* mutant was shown to lack a 3-*O*-methyl group in the capsule of strain 81-176 (Kanipes et al., 2006). However, the degree of cross-talk between pathways appears to vary between strains. For example, mutation of *galE* affects LOS, capsule and N-linked glycans in strain NCTC 11168 (Bernatchez et al., 2005), whereas mutation of *galE* in strain 81116 resulted in a truncated core polysaccharide but had no affect on capsule (Fry et al., 2000).

3.3.1.2 Capsule

The genome sequence of NCTC 11168 revealed the presence of a locus of genes with similarity to type II/III capsular polysaccharide-related genes (Parkhill et al., 2000). This was later experimentally verified as capsular polysaccharide and shown to be the major determinant of the Penner serotyping scheme (Karlyshev and Wren, 2001; Karlyshev et al., 2000, 2001).

From a number of strains the capsule has been shown to range in size from 15 to 34 kb, containing many horizontally acquired genes due to organization of the locus. The central biosynthetic genes are flanked by highly conserved *kps* genes, which are responsible for assembly and translocation of the capsular polysaccharide (CPS) to the cell surface and provide a favourable area for recombination (Karlyshev et al., 2005a).

The structure of several CPS from different strains has been determined including NCTC 11168 (St. Michael et al., 2002), 81116 (Kilcoyne et al., 2006), RM1221 (Gilbert et al., 2007), HS:1 (McNally et al., 2005), HS:19 (McNally et al., 2006b) and CG8486 (Chen et al., 2008). The CPS of NCTC 11168 was shown to be composed of beta-d-Ribp, beta-d-GalfNAc, alpha-d-GlcpA6(NGro), a uronic acid

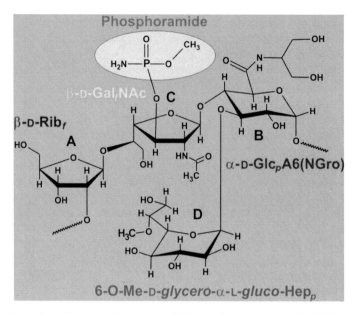

Fig. 3.5 Illustration of the capsule structure of *Campylobacter jejuni* strain NCTC 11168. The phosphoramide group is a phase-variable modification

amidated with 2-amino-2-deoxyglycerol at C-6 and 6-*O*-methyl-d-glycero-alpha-l-gluco-heptopyranose as a side-branch (St. Michael et al., 2002). In addition the CPS from NCTC 11168 contains an unusual phosphoramide side-branch modification (Szymanski et al., 2003b) (Fig. 3.5). A partial metabolomics approach has recently been used to identify the gene clusters responsible for the side-branch structures in this strain (McNally et al., 2007). The CDSs responsible for the phosphoramide side-branch are known to be present in the capsule locus of other strains, including NCTC 12517 (HS:19), G1 and 81-176 (Karlyshev et al., 2005a). Some other strains, including 81-176 and other HS:23 and HS:36 strains, also probably contain heptosyl side-branches, encoded by *hddC*, *gmhA2* and *hddA* (Karlyshev et al., 2005a).

Capsular polysaccharide contains negatively charged molecules which increases resistance to phagocytosis. Hydration of these molecules may also protect the bacteria from desiccation outside the avian or human host (as has been shown in other bacteria) (Gibson et al., 2006), and indeed capsule mutants have increased hydrophobicity (Bacon et al., 2001). CPS has also been shown to be important for adherence, invasion and resistance to human serum. However, the contribution of CPS to colonization/adherence may differ depending on host. For example, a *kpsE* mutant in 81116 was shown to be defective in adherence and invasion of human embryonic epithelial cells, but only a slight defect was seen in chick colonization (Bachtiar et al., 2007). Mutation in capsule has also been shown to alter sensitivity to phage infection (Coward et al., 2006).

So far, between two and six genes with a potential for phase variation have been found in different strains (Karlyshev et al., 2005a). Phase variability leads to variation in silver staining, antibody reactivity and side-branches between examined single colonies (Szymanski et al., 2003b). This may indicate population heterogeneity, indicating that the capsule is a dynamic structure in *Campylobacter*.

3.3.1.3 Flagella

The initial stage of infection or colonization involves *Campylobacter* moving towards cell surfaces. *Campylobacter* displays a distinctive corkscrew-like movement, thought to be an adaptation allowing penetration of the mucus overlying the intestinal epithelium. There are two genes encoding flagellin in *C. jejuni*: *flaA* and *flaB*, and these genes are arranged in tandem on the chromosome and show a high degree of sequence identity to each other. Indeed, intragenomic and intergenomic recombination between *flaA* and *flaB* genes of *C. jejuni* has been demonstrated to generate antigenic diversity (Harrington et al., 1997).

The flagella are constructed from multimers of flagellin; FlaA flagellin protein is the major component with a small amount of FlaB flagellin protein. These flagellin proteins are attached by a hook protein to a basal structure which is embedded in the membrane and, along with the stator units (MotA and MotB plus FliMNG), acts as a motor for rotation (Garza et al., 1995). When *flaA* is mutated a truncated flagellar filament is formed composed only of FlaB resulting in a reduced motility, whereas when *flaB* is mutated a full-length flagellar filament is formed with no apparent structural alteration or reduction in motility (Guerry et al., 1991).

The flagella are post-translationally modified by phosphorylation and glycosylation (which will be discussed further in Section 3.3.2.1) (Szymanski et al., 2003a), and they exhibit phase and antigenic variation. The *flaA* and *flaB* genes are independently transcribed by different types of promoter, σ^{28} and σ^{54} dependent, respectively. The expression of *flaB* seems to be environmentally regulated by temperature and pH (Alm et al., 1993).

There is a role for flagellin in colonization that is distinct from motility. For example, the flagella export apparatus can also be used to secrete non-flagellar proteins (see Section 3.3.5). Thus the flagella appear to have a dual function of being required for motility and as a channel for secretion of selected proteins. The production of flagella is necessary for virulence as determined by genetically matched isolates showing differences in invasion and secretion of Cia proteins as the result of the downregulation of flagellar class II genes (Malik-Kale et al., 2007).

3.3.2 Glycosylation

3.3.2.1 O-Linked Glycosylation

The O-linked glycosylation locus is located adjacent to the genes encoding the flagellin structural proteins: FlaA and FlaB. There are about 50 genes in the locus of strain NCTC 11168 including seven *maf* (motility accessory family) genes involved in variation of motility by a slipped-strand mispairing mechanism (Karlyshev et al., 2002), resulting in on/off switching of flagella. The *ptm* locus (*cj1314–cj1338*) is a highly variable region within the glycosylation locus (Szymanski et al., 2003a). Certain strains have been shown to be missing stretches of genes within the glycosylation locus. For example, in strain 81-176 orthologues of *cj1318–cj1332* are absent, as are orthologues of *cj1335* and *cj1336* (Thibault et al., 2001). Genes *cj1321– cj1326* have been shown to be associated with the animal clade cluster (see Section 3.4.1) (Champion et al., 2005), suggesting a possible role for glycosylation in host specificity.

It was originally thought that flagellins were modified with sialic acid (Neu5Ac) as they bound to a sialic acid-specific lectin (Doig et al., 1996). However, recent studies have shown that flagellin is actually modified with monosaccharides of the related sugar, pseudaminic acid (Pse5Ac7Ac), in strain 81-176 (Thibault et al., 2001). Pseudaminic acid residues may be further modified, e.g. with the acetamidino-substituted analogue (PseAm). Glycosylation in this strain occurs at 19 serine and threonine residues in the central, surface-exposed, domain of the flagellin protein (Thibault et al., 2001). However, the site of attachment does not seem to be linked to a consensus sequence (Szymanski et al., 2003a), and there is also no candidate glycosyltransferase in the same locus, making the mechanism of attachment unclear. A function for many of the genes in the glycosylation locus has yet to be elucidated. In addition, the glycosylation in different strains is likely to vary. Recent analysis of the *C. coli* strain VC167 has shown the flagellin to be modified with legionaminic acid (McNally et al., 2007). As yet there is no full structural data for the NCTC 11168 flagellin glycosylation, and a function for the extra genes present in the locus of this strain has yet to be analysed.

The function of flagellar glycosylation in *Campylobacter* is unclear. However, some flagellar glycosylation mutants are non-motile, but flagellar subunits accumulate intracellularly, suggesting that glycosylation has a role in assembly/secretion (Goon et al., 2003). The flagella of the closely related species *Helicobacter pylori* are also glycosylated with pseudaminic acid, which has similarly been shown to be necessary for motility; so perhaps glycosylation plays a key role in polar flagellar function and assembly (Schirm et al., 2003). In addition, as flagella are major antigens, glycosylation may present an immune avoidance strategy, or because many *C. jejuni* phages are flagellatropic, glyco-modifications may be important in preventing phage attachment and subsequent phage lysis (Coward et al., 2006).

3.3.2.2 N-Linked Glycosylation

In addition to an O-linked glycosylation, *C. jejuni* also contains an N-linked glycosylation locus, which by contrast is highly conserved between strains. The locus has also been found in *C. coli*, *C. lari* and *C. upsaliensis*. Before the NCTC 11168 genome sequence was published (Parkhill et al., 2000), the N-linked glycosylation locus of 81116 (*wlaB-wlaM*) was cloned and sequenced (Fry et al., 1998), although at the time it was proposed to function in LOS and O-antigen biosynthesis. In NCTC 11168 the locus comprises the genes *cj1119c–cj1131c*, adjacent to the LOS biosynthesis locus, but is transcribed in the opposite direction. A function has been attributed to each of the 14 genes in this locus (Linton et al., 2005), with the exception of *wlaJ*, which is present in some strains. The key enzyme oligosaccharyl transferase PglB is the only protein necessary for the transfer of glycan to the common N-linked sequon. The relaxed specificity of PglB has been exploited to create novel *N*-glycan structures, which could be used for in vivo production of glycan polysaccharide–protein conjugate vaccines (Feldman et al., 2005). The *pgl* gene cluster has been successfully transferred to *Escherichia coli* and a co-expressed protein, AcrA, glycosylated to produce for the first time recombinant glycoproteins (Wacker et al., 2002).

The structure of the N-linked glycan was shown by NMR and mass spectrometry to be the heptasaccharide GalNAc-α1,4-GalNAc-α1,4-[Glcβ1,3]GalNAc-α1,4-GalNAc-α1,4-GalNAc-α1,3-Bac-β1,*N*-Asn, where Bac is bacillosamine, 2,4-diacetamido-2,4,6-trideoxyglucopyranose, and GalNAc is *N*-acetyl-galactosamine (Young et al., 2002). This bacillosamine-containing heptasaccharide is attached to proteins at the asparagine in the motif Asn-Xaa-Ser/Thr (Wacker et al., 2002). It has been proposed that in *C. jejuni*, nucleotide-activated sugars are sequentially assembled on a lipid carrier, before being added to the protein (Szymanski et al., 2003a).

Over 30 potential glycoproteins have been identified, most of which are designated as unknown function (Young et al., 2002). It is difficult to determine the function of N-linked protein glycosylation as disrupting the pathway removes the glycan from all glycoproteins, although it is known to be important phenotypically. For example, the *pglH* mutant showed a reduced ability to adhere to and invade Caco-2 cells; in addition, a reduced ability to colonize chicks was also seen (Karlyshev et al., 2004). N-linked glycosylation is also required for competence in some strains (Larsen et al., 2004), and there may be other important functions analogous to those in eukaryotes. Although the structure of the glycan and the function of each of the genes in the pathway have been resolved there are still unanswered questions. For example, it is still unknown why all potential glycosylation sites are not modified or how glycosylation pathways are regulated. There also appears to be a degree of cross-talk between the O-linked and the N-linked glycosylation pathways; PseB is able to perform a secondary reaction producing the same product as PglF, as demonstrated by the return of N-linked glycan in a double *pglF/pseC* mutant (Guerry et al., 2007).

3.3.3 Adhesion and Invasion

C. jejuni adheres and invades human epithelial cells as part of its colonization process. Typically, bacteria would often have surface appendages such as pili to aid this process; however, the *Campylobacter* genome projects have not identified any such potential CDSs (Fouts et al., 2005; Parkhill et al., 2000). In *C. jejuni* a number of CDSs similar to type II-like secretion systems were identified. However, an actual pilus-like structure was not identified (Wiesner et al., 2003; Young et al., 2007). Adhesion in *C. jejuni* takes place with the aid of a number of proteins (adherence factors) produced by the bacterium. Here we briefly describe some of the common adhesins.

CadF (*Campylobacter* adhesion to Fibronectin) is a 37-kDa outer membrane protein that allows the binding of the bacterium to the extracellular matrix component fibronectin (Konkel et al., 1997; Monteville et al., 2003). Research on CadF mutants show reduced binding and invasion in a chick colonization model (Monteville et al., 2003; Ziprin et al., 1999). The exact mechanism of activity has not yet been elucidated.

JlpA (Jejuni lipoprotein A) is another *Campylobacter* adherence factor that has been shown to be a surface-exposed lipoprotein involved in the process of adherence in *C. jejuni* to epithelial cells (Jin et al., 2001). This is believed to occur by interacting with cell surface heat shock protein (Hsp)90 alpha, allowing initiation of signalling pathways leading to activation of pro-inflammatory responses by activating NF-κB and p38 MAP kinase (Jin et al., 2003).

CapA (*Campylobacter* adhesion protein A) is another putative adhesin that is an autotransporter homologous to other autotransporter adhesin proteins (Ashgar et al., 2007). Research has shown that CapA-deficient mutants show decreased adherence to Caco-2 cells and also in chick models (Ashgar et al., 2007).

Peb1 is a putative adhesin that is unusual in that it is located in the periplasm and has been shown to be vital for adherence to HeLa cells (Kervella et al., 1993; Pei et al., 1998). Similarity searches show that Peb1 shares amino acid similarity with ATP-binding cassette (ABC) transporters which are periplasmic-binding proteins (Pei et al., 1998). It has been shown that Peb1 binds to aspartate and glutamate with high affinity and that Peb1-deficient mutants do not grow without these two amino acids as the major carbon source (Leon-Kempis et al., 2006). Again, the exact method of action of Peb1 is still unclear.

Other important adhesion factors include MOMPs (major outer membrane proteins) that include a putative porin and a multifunction surface protein (Moser et al., 1997). LOS, which is a key component of the outer surface membrane of *C. jejuni*, is often involved in primary contact with external surfaces. LOS diversity is an important feature in allowing *Campylobacter* to colonize different hosts (Guerry et al., 2002; Linton et al., 2000). Finally, the *C. jejuni* capsule is most likely involved in bacterial survival and also has contact with external surfaces. LOS and the capsule are discussed in previous sections.

In terms of invasion, it has been shown that increased adherence does not lead to increased invasion (Hu and Kopecko, 1999). This indicates that other factors must exist for invasion to proceed. One such example is CiaB (*Campylobacter* invasion antigen), an invasion factor involved after adhesion has occurred allowing internalization to take place. CiaB is a 73-kDa protein secreted by the flagellin export apparatus and has been shown to be required for efficient invasion of *C. jejuni* into intestinal epithelial cells (Konkel et al., 1999b; Rivera-Amill and Konkel, 1999; Young et al., 1999). Research indicates that *C. jejuni*-secreted proteins may play a role in using host-cell structures, e.g. microtubules and microfilaments, to allow bacterial entry into the cell (Konkel et al., 1999a, 2004).

C. jejuni has a number of mechanisms which most likely occur concurrently to allow adherence to occur on intestinal epithelial cells. These factors are clearly effective as *C. jejuni* must initially adhere and then invade to cause disease in humans.

3.3.4 Cytolethal Distending Toxin

Cytolethal distending toxin (CDT) is part of the bacterial armoury involved in host colonization. CDT CDSs have been identified in *C. coli*, *Campylobacter fetus*, *C. upsaliensis* and *C. lari* (Asakura et al., 2007; Fouts et al., 2005). Initially described by Johnson and Lior (1988), research has steadily increased our under-standing of this potentially important factor in *C. jejuni* (Pickett et al., 1996; Purdy et al., 2000; Whitehouse et al., 1998). CDT acts by causing eukaryotic cells to arrest in either G_1/S or G_2/M phase depending on the cell type (Lara-Tejero and Galan, 2000, 2001; Whitehouse et al., 1998). This prevents the eukaryotic cell from entering mitosis, leading to cell death. CDT consists of three genes encoding the proteins CdtA, CdtB and CdtC (Lara-Tejero and Galan, 2001). All three proteins were found to be similar in amino acid content to the *E. coli* ortholog. Comparison of CDT ORFs from different *Campylobacter* strains and species shows an amino acid similarity range between 40 and 70%. This indicates that conservation exists between strains and species; however, differences in nucleotide composition do exist (Asakura et al., 2007).

The exact function of CDT is still not fully understood. However, what is known is that CdtB is the active subunit sharing homology with mammalian DNase I (Lara-Tejero and Galan, 2000), which functions by breaking down DNA. For CdtB to function, it must enter the nuclear compartment. It is believed that CdtA and CdtC allow this process to occur (Lara-Tejero and Galan, 2001; Lee et al., 2003). Studies have shown that CdtA and CdtC bind to HeLa cells; however, CdtB does not (Lee et al., 2003). The three CDT proteins are membrane associated and have been shown to be involved in stimulating pro-inflammatory cytokine IL-8 production (Hickey et al., 2000). Additionally, studies have shown CDT to elicit an inflamma-tory response in vivo using an NF-κB-deficient murine infection model (Fox et al., 2004). It is interesting to note that *C. jejuni* being a human pathogenic organism that often establishes long-term, asymptomatic associations with many hosts such as chickens has retained such a potentially potent toxin, although neutralizing anti-bodies to CDT are not produced on colonization with chickens, suggesting that the immune response of different hosts to *Campylobacter* differs (AbuOun et al., 2005).

3.3.5 Secretion Mechanism and Proteins

Protein secretion is extremely important for bacteria and is required for basic cel-lular functions as well as pathogenicity. There are at least five major secretion pathways in Gram-negative bacteria, of which type II and V are sec dependent with the others being sec independent. Type I secretion systems or ATP-binding cassette (ABC) exporters require three accessory proteins and are typified by the secretion of haemolysin HlyA of *E. coli* (Holland et al., 1990). In *C. jejuni* NCTC 11168 *cj0606–cj0609* are predicted to encode a type I secretion system (MacKichan et al., 2004), although a substrate has not been proposed. *C. fetus* is proposed to secrete surface layer proteins through a type I secretion system (Thompson et al., 1998).

Type II secretion is complex and requires the sec pathway for secretion across the inner membrane; an example is the secretion of PulA by *Klebsiella oxytoca* (Pugsley et al., 1991). *Campylobacter* genome sequencing projects have revealed genes with homology to those required for type II secretion systems (Fouts et al., 2005; Parkhill et al., 2000), and these genes have been shown in *C. jejuni* strain 81-176 to be required for natural transformation (Wiesner et al., 2003).

The type III secretion system has a needle-like structure for transporting proteins from the bacterial cytoplasm to the eukaryotic cell cytoplasm, an example of which is the secretion of *Yersinia* outer proteins (YOPs) by *Yersinia enterocolitica* (Wilharm et al., 2004). To date no components of a type III secretion system have been found in *Campylobacter*. However, *C. jejuni* has been proposed to use the flagellin export apparatus for secreting proteins (Konkel et al., 2004; Young et al., 1999). The flagellin export system is structurally and evolutionarily related to the type III secretion system (Gophna et al., 2003). Currently, 18 proteins that are not thought to contribute to flagellar assembly have been shown to be secreted through the flagellar export apparatus (Larson et al., 2008). These include the CiaB protein which is secreted post-adhesion and involved in invasion (Rivera-Amill and Konkel, 1999; Young et al., 1999); FlaC, also implicated in invasion (Song et al., 2004); and FspA which has been implicated in apoptosis (Poly et al., 2007). Thus, the flagellar export apparatus is an important secretion mechanism in *C. jejuni* and is required for host-cell invasion as well as motility (Guerry, 2007; Young et al., 2007).

Type IV secretion systems can secrete proteins, e.g. pertussis toxin of *Bordetella pertussis* (Weiss et al., 1993), or complexes of protein and single-stranded DNA, e.g. T-DNA from *Agrobacterium tumefaciens* (Christie, 1997). It is thought that type IV secretion systems evolved from bacterial conjugation systems. To date no chromosomally encoded type IV secretion systems have been found in *Campylobacter*, but they have been found on plasmids (see Section 3.3.6). When *comB3*, a component of the type IV secretion system, was mutated in *C. jejuni* 81-176 plasmid pVir a reduction in adherence, invasion and natural transformation was seen (Bacon et al., 2000), suggesting that this secretion system may be multifunctional. In contrast the type IV secretion system of *C. jejuni* 81-176 plasmid pTet is likely to be involved in conjugation (Batchelor et al., 2004).

Type V secretion systems or autotransporters, e.g. VacA of *H. pylori* (Reyrat et al., 1999), are known to have various virulence functions including adhesions, toxins and proteases (Henderson and Nataro, 2001). Autotransporters all possess an N-terminal sequence for secretion across the inner membrane via the sec pathway, a secreted mature protein (passenger domain) and a C-terminal domain (transport domain) which forms a pore through which the passenger portion of the protein is secreted across the outer membrane. The passenger domain may either remain attached to the outer membrane or be released into the extracellular environment (Henderson et al., 1998). The genome sequence of *C. jejuni* NCTC 11168 revealed putative secreted proteases (Cj1365), as well as CapA (Cj0628), which has been implicated in adhesion (see Section 3.3.3), and CapB (Cj1677) (Ashgar et al., 2007). Interestingly, components of a two-partner secretion system (TPS) have been found in *C. coli* RM2228, *C. lari* RM2100 and *C. jejuni* NCTC 11168 (Fouts et al., 2005), although whether these represent functional secretion systems remains to be seen. TPSs are like type V secretion systems but with the passenger and transporter domains being encoded by two separate proteins (Brown et al., 2004).

3.3.6 Plasmids

It is thought that between 19 and 53% of *C. jejuni* strains carry plasmids (Bacon et al., 2000). Plasmids ranging in size from 2 to 162 kb have been identified from *C. jejuni* and *C. coli* strains with some carrying multiple plasmids, although no single plasmid was common to all (Tenover et al., 1985). Despite the prevalence of plasmids within *C. jejuni* the first two published genome sequences, from strain NCTC 11168 and RM1221, did not contain plasmids. Strain 81-176 contains two plasmids, pVir (37 kb) (Bacon et al., 2002) and pTet (45 kb) (Batchelor et al., 2004), both of which encode type IV secretion systems, with pTet also containing the Tet(O) tetracycline resistance determinant. Recent sequencing initiatives have identified plasmids from different *Campylobacter* species (Fouts et al., 2005): *C. coli* RM2228 has an 178 kb plasmid pCC178, *C. lari* RM2100 a 46 kb plasmid pCL46 and *C. upsaliensis* RM3195 a 3.1 kb plasmid pCU3 and a 110 kb plasmid pCU110. Both plasmids pCU110 and pCL46 show coding strand bias; this was previously seen in pVir which encodes most CDSs on one strand with the exception of a single coding region. Interestingly, the larger plasmids also encode type IV secretion systems, but whether these systems are involved in conjugation, competence or virulence determinant secretion remains to be established.

Many plasmids have been associated with antibiotic resistance, and indeed the *C. coli* RM2228 plasmid pCC178 contains antibiotic resistance genes flanked by putative mobile genetic elements. The plasmid confers resistance to tetracycline and kanamycin, in association with a predicted transposable element similar to IS605 from *H. pylori*. Transposable element IS607 has also been associated with antibiotic resistance (Gibreel et al., 2004a); the *apha-3* gene, conferring resistance to kanamycin, was identified downstream of the apparent insertion sequence IS607 on large plasmids isolated from *C. jejuni*. The *apha-3* gene has also been identified on the 48 kb *C. coli* plasmid pIP1433, although previously this resistance determinant has only been found in Gram-positive cocci. Kanamycin resistance genes *apha-3* and *apha-7* have both been found on plasmids from *C. jejuni* (Alfredson and Korolik, 2007).

In addition to kanamycin resistance, tetracycline resistance has increased dramatically over the past 20–30 years with 55% of clinical *C. jejuni* isolates in

North America resistant to tetracycline and 95% of isolates in Thailand (Gibreel et al., 2004b). Tetracycline had been used prophylactically and therapeutically as a feed additive for poultry (Aarestrup, 1999; Avrain et al., 2003). Most tetracycline resistance is thought to be plasmid mediated, although a chromosomally mediated tetracycline resistance determinant has been reported for *C. coli* (Ng et al., 1987). Indeed, in tetracycline-resistant isolates from Canada 67% contained plasmids (Gibreel et al., 2004b). Tetracycline-resistant plasmids appear to be highly conserved between species with plasmids pCC31 from *C. coli* and pTet from *C. jejuni* sharing 94.3% amino acid identity (Batchelor et al., 2004).

Plasmids are not always associated with antibiotic resistance. The plasmid pVir was originally isolated from strain 81-176, but very similar plasmids have been found in other strains: 17% of clinical isolates from Canada contained pVir (Tracz et al., 2005). pVir has been implicated in virulence, with five genes from pVir found to affect in vitro invasion of epithelial cells (Bacon et al., 2002). The presence of pVir has been linked to bloody diarrhoea in some studies (Tracz et al., 2005), but this link has been refuted by others (Louwen et al., 2006).

3.4 Comparative Genomics

Following the publication of the NCTC 11168 genome sequence, a variety of approaches have been used to assess the genetic diversity of the *C. jejuni* genus. First, several more genomes have been sequenced or are in the process of being sequenced (Table 3.1); second, targeted approaches to sequence known variable regions have been undertaken (e.g. capsule, LOS) as well as conserved housekeeping genes (multilocus sequence typing, MLST) to assess population dynamics; and third comparative genome analysis of multiple strains using microarrays and subtractive hybridization to provide additional gene sequences absent from sequenced strains.

3.4.1 Comparative Phylogenomics

Comparative phylogenomics (whole genome comparisons of microbes using DNA microarrays combined with Bayesian phylogenies) provides a useful alternative to the classical typing methods. The advantage of comparative phylogenomics is the potential to identify genes/genetic loci specific to strain source or disease severity (Dorrell et al., 2005; Eisen and Wu, 2002).

In the original comparative phylogenomics study 111 *C. jejuni* strains from a variety of sources which included clinical human and different animal sources were analysed (Champion et al., 2005). The Bayesian phylogeny revealed two distinct clades (Fig. 3.6). Due to the distribution of strains, clades were termed "livestock" and "non-livestock". Surprisingly this study showed that the majority (39/70, 55.7%) of *C. jejuni* human isolates were found in the non-livestock clade, suggesting that most *C. jejuni* infections may be from non-livestock (and possibly non-agricultural) sources. Thus, the analysis provided useful insight into a potential

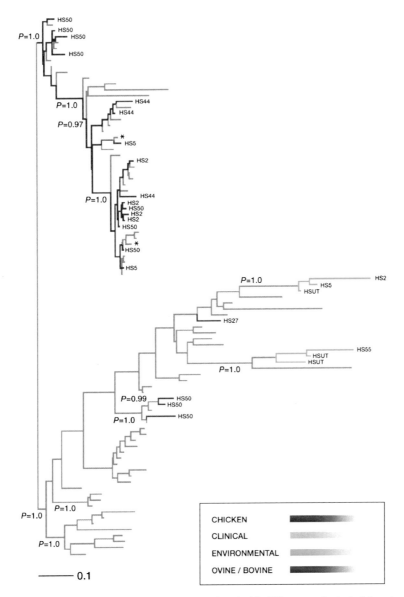

Fig. 3.6 Phylogenetic relationship of strains associated with different ecological niches. Image obtained from Champion et al. (2005). P = 1.0 represents 100% of all phylogenies showing a given topology. Penner serotypes, where available, are designated at the ends of branches; HSUT= untypeable

unidentified reservoir of *C. jejuni* infection that may have implications in disease control strategies. The comparative phylogenomics approach also illustrated a new robust methodological prototype that has been applied to other microbes (Dorrell et al., 2005; Howard et al., 2006; Stabler et al., 2006).

A different example using the comparative genomics approach was that demonstrated by Parker et al. (2006). Here, the authors identified four large genomic elements that were present in *C. jejuni* RM1221, but absent in *C. jejuni* NCTC 11168. A large-scale genomic comparative analysis of 67 *C. jejuni* and 12 *C. coli* strains was carried out. Strains were isolated from different geographical locations and also from a mixture of clinical and veterinary sources. The comparative genomics approach confirmed that 35 of the 67 *C. jejuni* strains have the genomic elements similar to those in strain RM1221 and also identified 18 other intraspecies hypervariable regions, such as the capsule and LOS biosynthesis regions. Finally, the authors also showed that in *C. jejuni* RM1221 genomic element 1 (CJIE1), a *Campylobacter* Mu-like phage was located differentially in other strains of *C. jejuni*, suggesting that random integration may be prevalent (Parker et al., 2006).

3.4.2 Multilocus Sequence Typing

Multilocus sequence typing (MLST) relies on the multiple sequencing of a small number of conserved housekeeping genes, which provides an evolutionary framework for population genetic studies. An MLST scheme for *Campylobacter* species has been developed, and *C. jejuni* has been shown to form weakly clonal populations based on sequence types (STs) (Dingle et al., 2001) (Fig. 3.7).

This typing scheme has been extended to include *C. coli* using the same loci but *C. coli*-specific primers (Dingle et al., 2005). Briefly, seven housekeeping genes

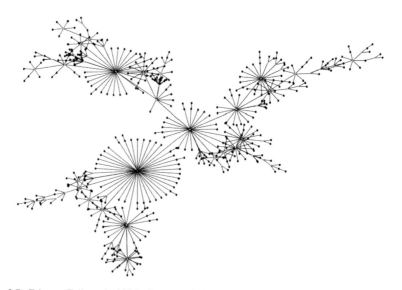

Fig. 3.7 E-burst (Feil et al., 2004) diagram of *Campylobacter jejuni* ST types generated online at http://eburst.mlst.net using a dataset containing only one example of each ST in the MLST databank. The *blue dot* represents the founder ST, *yellow dots* subgroup founders and *black dots* all other STs

were selected for analysis. For each housekeeping gene, the different sequences present are assigned as distinct alleles, and for each isolate, the alleles at each of the seven loci define the allelic profile or sequence type (ST). One of the advantages of using MLST is that allelic profiles can be obtained from clinical material by PCR amplification of the loci directly from cerebrospinal fluid or blood, where traditional culture methods may prove difficult. The comparison of *C. jejuni* and *C. coli* shows that they share about 86.5% nucleotide identity within the seven housekeeping genes, and there is evidence of a low rate of genetic exchange at these loci between the two species. The benefit of having a joint typing scheme for these two closely related pathogens is that it allows direct comparisons of epidemiology and population structures.

MLST has been used in long-term surveillance of *C. coli* in Denmark and association with different food sources (Litrup et al., 2007). By comparison of human isolates with isolates from different food sources it was determined that pigs are not a significant source of human campylobacteriosis in Denmark. Interestingly, the Danish *C. coli* isolates included 68% novel STs, showing a great diversity compared with studies from other countries. A recent study in the USA typed 488 *C. coli* strains isolated from food animals at different locations over a period of 6 years and found an association between ST and host (Miller et al., 2006). Swine and poultry isolates were found to be very diverse in contrast to cattle isolates which appeared to be largely clonal, with 83% of the isolates sharing the same ST. Interestingly, several turkey *C. coli* isolates contained the *aspA*103 allele which are more closely related to *C. jejuni aspA* alleles, suggesting lateral transfer of genes between the two species within poultry hosts. Indeed, it has recently been suggested that *C. jejuni* and *C. coli* are converging and will eventually de-speciate, due to an increase in genetic exchange, possibly mediated by human activity producing an overlap in previously distinct niches for these two species (Sheppard et al., 2008). In a recent study 11% of *C. coli* STs showed *C. jejuni* ancestry. *C. coli* haplotypes could be assigned to one of three clades, with *C. coli* Clade 1 showing the most genetic exchange with *C. jejuni*. This is likely to represent a recent change as the exchanged alleles are identical rather than being hybrids. Although the increase in recombination is bidirectional, it is more apparent that *C. coli* isolates have acquired *C. jejuni* alleles as opposed to the converse; this asymmetry may be the result of *C. coli* expanding into an environment where *C. jejuni* is more frequently present. Both species are isolated from chickens and cattle, providing an overlap in niches for recombination, possibly due to the advent of agriculture. In addition, it has been proposed that most transmission of *C. jejuni* is within species and that lineages may acquire a host signature and potentially adapt to the host (McCarthy et al., 2007).

3.4.3 Subtractive Hybridization

Subtractive hybridization has been used as a method for identifying genes expressed in one cell type but not in another by hybridizing cDNA to RNA. The method has since been adapted to identify DNA differences between bacterial strains and was evaluated as a method of comparing genomes using the two sequenced strains of *H. pylori* where it was shown to identify 95% of CDSs unique to one strain compared to the other (Agron et al., 2002). Ahmed et al. (2002) used subtractive hybridization as a technique to identify fragments in strain 81116 that were not present in strain NCTC 11168. Strain 81116 has been proposed to show greater colonization potential of chickens than strain NCTC 11168. In strain 81116, 24 fragments that were unique to this strain (less than 75% identity at the nucleotide level to NCTC 11168) were identified. One fragment was unique to 81116, and one was present in another nine strains, determined by DNA hybridization. Gene fragments identified included those with similarity to restriction–modification enzymes, arsenic resistance genes and cytochrome C oxidase III genes (Ahmed et al., 2002).

More recently Hepworth et al. (2007) used subtractive hybridization to assess genetic differences between MLST ST types from strains isolated from different hosts (Hepworth et al., 2007). A number of genes were identified from known plasticity regions, which include LOS, capsular biosynthesis and restriction–modification enzyme loci (Pearson et al., 2003; Taboada et al., 2004). Flagellar hook proteins (FlgE), flagellins, flagellar glycosylation, LOS biosynthesis and capsule genes were all identified in the screens. Variation was found to occur both within and between clonal complexes. However, the only clear class of genes related to ST type was associated with alternative terminal electron acceptors, and no class of genes was found to be related to host.

3.5 Post-genomic Era

In addition to sequence data other post-genomic techniques are being used to gain a holistic view of *C. jejuni* by adding layers of information, including transcript, proteome and metabolic data.

3.5.1 Transcriptomics

Although extensive studies have been carried out to assess the gene diversity between different strains and species of *C. jejuni* it should be noted that there are also differences in expression of these genes. For example, the genome-sequenced versus original clinical isolate of NCTC 11168 appears to differ in virulence-associated phenotypes (Gaynor et al., 2004), although no genetic difference was determined.

Often differences in expression are attributable to small genetic changes and also due to environmental conditions showing that *C. jejuni* has an aptitude for fine-tuning transcription to aid survival under different conditions. A study by Carrillo et al. (2004) showed that there were differences in colonization and motility in a passaged versus a non-passaged NCTC 11168 strain. These differences were later partly attributed to the flagellar regulatory system, and it was demonstrated that there is coordinate regulation of motility and virulence by *flhA* in *C. jejuni* (Carrillo et al., 2004). This highlights that small genetic differences and differences in regulation may have a marked impact on the virulence of certain strains. This is supported by a study by Malik-Kale et al. (2007) who demonstrated that a non-synonymous nucleotide change in *flgR*, part of a two-component system necessary for flagellar expression, was responsible for the differences in motility and virulence of two genetically matched *C. jejuni* isolates (Malik-Kale et al., 2007).

The study of transcript differences in gene expression under different environmental conditions especially with regard to colonization versus disease can be highly informative. Stintzi et al. (2005) investigated the changes in transcription associated with an intestinal lifestyle using a rabbit ileal loop model of infection (Stintzi et al., 2005). The resultant in vivo transcriptome of *C. jejuni* reflected the oxygen-limited, nutrient-poor and hyperosmotic conditions the bacterium encounters. Notably, the in vivo expression profile of *C. jejuni* suggests that the cell envelope is extensively remodelled; membrane proteins, peptidoglycan and glycosylation-associated genes are all differentially expressed. This is a common theme in *C. jejuni* expression studies. For example, changes in planktonic versus sessile growth were shown to partially overlap with profiles from in vivo expression studies. Sessile bacteria, thought to more closely reflect in vivo growth, undergo a shift in expression away from metabolic, motility and protein synthesis towards iron uptake, oxidative stress defence and membrane transport (Sampathkumar et al., 2006). Possible changes in membrane components were also suggested as the LOS-associated genes *waaC*, *waaD* and *lpxK* were all upregulated in sessile versus planktonic growth (Sampathkumar et al., 2006). Membrane constituents have also been shown to vary by transcriptional analysis as well as by Fourier transform infrared (FT-IR) spectroscopy under stress conditions (Moen et al., 2005), and transcription profiles of a temperature shift from 37 to 42°C show potential differences in surface structures (Bras et al., 1999; Stintzi, 2003). *C. jejuni* has also been shown to alter expression of electron transport genes and central metabolic pathways to adapt to colonization of the chick caecum (Woodall et al., 2005).

How *C. jejuni* regulates gene expression in response to these changes in environmental conditions is not fully understood; *C. jejuni* lacks several traditional stress response genes, but must survive hostile environments both in vivo and during transmission. A two-component regulatory system RacR–RacS has been shown to be involved in a temperature-dependent signalling pathway. Mutant analysis of the *racR* response regulator showed changes in the protein expression profile on temperature shift as well as a defect in chicken colonization (Bras et al., 1999). In addition, microarray-based screening of *C. jejuni* showed that upon infection of human epithelial cells, a stringent response is mounted. This stringent response is regulated by *spoT* and is required for stationary phase survival, growth and survival under increased oxygen tensions and rifampicin resistance. The stringent response was also necessary for adherence, invasion and intracellular survival in human epithelial cell culture models. *C. jejuni* was shown to contain a single *relA/spoT* homologue analogous to the Gram-positive paradigm (Gaynor et al., 2005).

3.5.2 Proteomics

Recently, the most comprehensive protein–protein interaction map for any organism was published for *C. jejuni* (Parrish et al., 2007). A yeast two-hybrid scheme was used to produce a static interaction map which includes interactions for 80% of the predicted proteins in strain NCTC 11168. Interaction data can be useful in defining biological pathways and may help to predict functions for previously uncharacterized proteins based on the functions of interacting proteins (Parrish et al., 2007). The map connects 663 poorly characterized proteins into sub-networks that may provide testable clues as to their function. The power of the system for mapping biological pathways was demonstrated using the chemotaxis pathway as an example. However, the interaction map is unlikely to provide data on protein interactions that require modifications such as phosphorylation as these will not occur in the yeast model.

A study by Cordwell et al. (2008) has identified membrane-associated proteins in a recent gastrointestinal isolate (JHH1) using sodium carbonate precipitation, ultracentrifugation, followed by 2-DE and MALDI-TOF MS as well as 2-DLC of trypsin digests coupled to MS/MS (Cordwell et al., 2008). In total 453 proteins were identified: 187 confirmed as membrane associated. This gastrointestinal isolate was then compared to a type strain (ATCC 700297) identifying 28 proteins present only, or at greater than twofold abundance, in JHH1. Of these, 22 were predicted to encode membrane-associated proteins. Conversely 19 proteins were unique to ATCC 700297 or found in higher abundance, 10 of which were predicted to be membrane associated. Chemotaxis proteins were found to vary, but interestingly only in the N-terminal region, suggesting a conserved C-terminal signal transduction domain but the propensity to respond to different environmental cues via a variable extracellular receptor domain. Proteins involved in flagellar motility also varied, and when the immunoreactivity of proteins was tested against convalescent human serum, flagellin proved to be the strongest antigen.

3.5.3 Metabolomics

The metabolome has been defined as the measurement of the change in relative concentrations of metabolites (Oliver et al., 1998). Metabolites are low molecular weight molecules such as amino acids, peptides or carbohydrates that are intermediates and products of gene expression or metabolism. The measurement of metabolites (metabolic profile) can give a snapshot of the physiology of the cell. This metabolomic data merged with transcriptomic and proteomic data can give a more global and complete image of living organisms.

The ultimate experiment would be to quantify all the metabolites in a cellular system (Goodacre et al., 2004). However, this is currently impossible due to the chemical complexity and heterogeneity of metabolites, the dynamic measuring range required, the throughput of measurements and the bias in extraction protocols (Goodacre et al., 2004). Analytical tools used in the separation of analytes include gas chromatography, high-performance liquid chromatography and capillary electrophoresis. Separation techniques are followed with measurement techniques such as mass spectrometry (MS) and nuclear magnetic resonance (NMR) (Dettmer and Hammock, 2004; Dettmer et al., 2007; Oldiges et al., 2007).

Metabolomics is an area of research that is gaining interest from diverse disciplines including functional genomics, integrative and systems biology, drug discovery and therapy monitoring (Fiehn, 2002; Goodacre et al., 2004). Examples of metabolomics being applied can be illustrated with metabolic profiling experiments, e.g. analysis of urine or blood samples to detect changes caused by a toxic effect on the patient body (Beckonert et al., 2007; Ebbels et al., 2007; Keun et al., 2002; Nicholls et al., 2003). Analysis of such alterations can allow patient diagnosis to be performed. Elucidation of toxicity is of particular importance to the pharmaceutical industry as identification of such compounds can lead to clinical trials being avoided.

C. jejuni has recently undergone initial subcellular metabolomics study (McNally et al., 2006a, 2007; Soo et al., 2004). The metabolomics approach by Soo et al. (2004) used novel scanning techniques in the identification of sugar nucleotides from wild-type and isogenic mutants allowing the determination of specific sugar nucleotides involved in the biosynthesis of pseudaminic acid, a sialic acid-like sugar previously observed on the flagellin of some pathogenic bacteria (Soo et al., 2004). McNally et al. (2006b) used a metabolomics approach to define the function of flagellin glycosylation genes in *C. jejuni* 81-176. This method has allowed a wealth of functional characterization data to be obtained for genes involved in flagellin glycosylation. The results demonstrate how a targeted approach within the bacterial metabolome can assign gene functions to genes and identify metabolic intermediates and novel biosynthetic pathways (McNally et al., 2006a).

3.6 Future Perspectives

Campylobacter genetics has flourished from being a data-poor to being a data-rich subject area in less than a decade. In parallel comprehensive transcriptome, metabolome, proteome and protein–protein interaction maps have been undertaken, making *C. jejuni* a model organism to study in the post-genome era. Future studies using this information will be directed at understanding how *C. jejuni* is a commensal in its main sources of infection such as poultry, but by contrast is highly infectious in humans. The future challenge will be to translate this wealth of new information into appropriate strategies to reduce *C. jejuni* from the food chain and subsequently reduce the burden of *C. jejuni*-associated disease.

References

Aarestrup FM (1999) Association between the consumption of antimicrobial agents in animal husbandry and the occurrence of resistant bacteria among food animals. Int J Antimicrob Agents 12(4):279–285

AbuOun M, Manning G, Cawthraw SA, Ridley A, Ahmed IH, Wassenaar TM, Newell DG (2005) Cytolethal Distending Toxin (CDT)-negative *Campylobacter jejuni* strains and anti-CDT neutralizing antibodies are induced during human infection but not during colonization in chickens. Infect Immun 73(5):3053–3062

Agron PG, Macht M, Radnedge L, Skowronski EW, Miller W, Andersen GL (2002) Use of subtractive hybridization for comprehensive surveys of prokaryotic genome differences. FEMS Microbiol Lett 211(2):175–182

Ahmed IH, Manning G, Wassenaar TM, Cawthraw S, Newell DG (2002) Identification of genetic differences between two *Campylobacter jejuni* strains with different colonization potentials. Microbiology 148:1203–1212

Alfredson DA, Korolik V (2007) Antibiotic resistance and resistance mechanisms in *Campylobacter jejuni* and *Campylobacter coli*. FEMS Microbiol Lett 277(2):123–132

Alm RA, Guerry P, Trust TJ (1993) The *Campylobacter* sigma 54 *flaB* flagellin promoter is subject to environmental regulation. J Bacterio 175(14):4448–4455

Altekruse SF, Stern NJ, Fields PI, Swerdlow DL (1999) *Campylobacter jejuni* – an emerging foodborne pathogen. Emerg Infect Dis 5(1):28–35

Asakura M, Samosornsuk W, Taguchi M, Kobayashi K, Misawa N, Kusumoto M, Nishimura K, Matsuhisa A, Yamasaki S (2007) Comparative analysis of cytolethal distending toxin (cdt) genes among *Campylobacter jejuni*, *C. coli* and *C. fetus* strains. Microb Pathog 42(5–6):174–183

Ashgar SSA, Oldfield NJ, Wooldridge KG, Jones MA, Irving GJ, Turner DPJ, Ala'Aldeen DAA (2007) CapA, an autotransporter protein of *Campylobacter jejuni*, mediates association with human epithelial cells and colonization of the chicken gut. J Bacteriol 189(5):1856–1865

Avrain L, Humbert F, L'Hospitalier R, Sanders P, Vernozy-Rozand C, Kempf I (2003) Antimicrobial resistance in *Campylobacter* from broilers: association with production type and antimicrobial use. Vet Microbiol 96(3):267–276

Bachtiar BM, Coloe PJ, Fry BN (2007) Knockout mutagenesis of the *kpsE* gene of *Campylobacter jejuni* 81116 and its involvement in bacterium-host interactions. FEMS Immunol Med Microbiol 49(1):149–154

Bacon DJ, Alm RA, Burr DH, Hu L, Kopecko DJ, Ewing CP, Trust TJ, Guerry P (2000) Involvement of a plasmid in virulence of *Campylobacter jejuni* 81–176. Infect Immun 68(8):4384–4390

Bacon DJ, Alm RA, Hu L, Hickey TE, Ewing CP, Batchelor RA, Trust TJ, Guerry P (2002) DNA sequence and mutational analyses of the pVir plasmid of *Campylobacter jejuni* 81–176. Infect Immun 70(11):6242–6250

Bacon DJ, Szymanski CM, Burr DH, Silver RP, Alm RA, Guerry P (2001) A phase-variable capsule is involved in virulence of *Campylobacter jejuni* 81–176. Mol Microbiol 40(3):769–777

Batchelor RA, Pearson BM, Friis LM, Guerry P, Wells JM (2004) Nucleotide sequences and comparison of two large conjugative plasmids from different *Campylobacter* species. Microbiology 150(10):3507–3517

Beckonert O, Keun HC, Ebbels TMD, Bundy J, Holmes E, Lindon JC, Nicholson JK (2007) Metabolic profiling, metabolomic and metabonomic procedures for NMR spectroscopy of urine, plasma, serum and tissue extracts. Nat Protoc 2(11):2692–2703

Bernatchez S, Szymanski CM, Ishiyama N, Li J, Jarrell HC, Lau PC, Berghuis AM, Young NM, Wakarchuk WW (2005) A single bifunctional UDP-GlcNAc/Glc 4-Epimerase supports the synthesis of three cell surface glycoconjugates in *Campylobacter jejuni*. J Biol Chem 280(6):4792–4802

Black RE, Levine MM, Clements ML, Hughes TP, Blaser MJ (1988) Experimental *Campylobacter jejuni* infection in humans. J Infect Dis 157(3):472–479

Blaser MJ, Engberg J (2008) Clinical aspects of Campylobacter jejuni and campylobacter coli infections. In: Nachamkin I, Szymanski C, Blaser MJ (eds.) Campylobacter, 3rd edn. ASM Press, Washington, DC, pp 99–121

Bolotin A, Quinquis B, Sorokin A, Ehrlich SD (2005) Clustered regularly interspaced short palindrome repeats (CRISPRs) have spacers of extrachromosomal origin. Microbiology 151(8):2551–2561

Bras AM, Chatterjee S, Wren BW, Newell DG, Ketley JM (1999) A Novel *Campylobacter jejuni* two-component regulatory system important for temperature-dependent growth and colonization. J Bacteriol 181(10):3298–3302

Brown NF, Logue CA, Boddey JA, Scott R, Hirst RG, Beacham IR (2004) Identification of a novel two-partner secretion system from *Burkholderia pseudomallei*. Mol Genet Genomics 272(2):204–215

Carrillo CD, Taboada E, Nash JHE, Lanthier P, Kelly J, Lau PC, Verhulp R, Mykytczuk O, Sy J, Findlay WA, Amoako K, Gomis S, Willson P, Austin JW, Potter A, Babiuk L, Allan B, Szymanski CM (2004) Genome-wide Expression Analyses of *Campylobacter jejuni* NCTC11168 reveals coordinate regulation of motility and virulence by *flhA*. J Biol Chem 279(19):20327–20338

CDC (2006) Surveillance for foodborne-disease outbreaks – United Stated, 1998–2002. Morb Mortal Wkly Rep Surveill Summ 55(SS-10):1–48

CDC (2008) Preliminary foodnet data on the incidence of infection with pathogens transmitted commonly through food – 10 states, 2007. Morbid Mortal Weekly Rep 57(14):366–370

Champion OL, Gaunt MW, Gundogdu O, Elmi A, Witney AA, Hinds J, Dorrell N, Wren BW (2005) Comparative phylogenomics of the food-borne pathogen *Campylobacter jejuni* reveals genetic markers predictive of infection source. Proc Natl Acad Sci USA 102(44): 16043–16048

Chen Y-H, Poly F, Pakulski Z, Guerry P, Monteiro MA (2008) The chemical structure and genetic locus of *Campylobacter jejuni* CG8486 (serotype HS:4) capsular polysaccharide: the identification of 6-deoxy-d-ido-heptopyranose. Carbohydr Res 343(6):1034–1040

Chiu CPC, Watts AG, Lairson LL, Gilbert M, Lim D, Wakarchuk WW, Withers SG, Strynadka NCJ (2004) Structural analysis of the sialyltransferase CstII from *Campylobacter jejuni* in complex with a substrate analog. Nat Struct Mol Biol 11(2):163–170

Christie PJ (1997) *Agrobacterium tumefaciens* T-complex transport apparatus: a paradigm for a new family of multifunctional transporters in eubacteria. J Bacteriol 179(10):3085–3094

Coker AO, Isokpehi RD, Thomas BN, Amisu KO, Obi CL (2002) Human campylobacteriosis in developing countries. Emerg Infect Dis 8(3):237–243

Cordwell SJ, Len ACL, Touma RG, Scott NE, Falconer L, Jones D, Connolly A, Crossett B, Djordjevic SP (2008) Identification of membrane-associated proteins from *Campylobacter jejuni* strains using complementary proteomics technologies. Proteomics 8(1):122–139

Coward C, Grant AJ, Swift C, Philp J, Towler R, Heydarian M, Frost JA, Maskell DJ (2006) Phase-Variable surface structures are required for infection of *Campylobacter jejuni* by bacteriophages. Appl Environ Microbiol 72(7):4638–4647

Crushell E, Harty S, Sharif F, Bourke B (2004) Enteric campylobacter: purging its secrets? Pediatr Res 55(1):3–12

Cui S, Ge B, Zheng J, Meng J (2005) Prevalence and antimicrobial resistance of *Campylobacter* spp. and *Salmonella* serovars in organic chickens from Maryland retail stores. Appl Environ Microbiol 71(7):4108–4111

Debruyne L, Gevers D, VanDamme P (2008) Taxonomy of the family campylobacteraceae. In: Nachamkin I, Szymanski C, Blaser MJ (eds) Campylobacter, 3rd edn. ASM Press, Washington, DC, pp 3–25

Dettmer K, Aronov PA, Hammock BD (2007) Mass spectrometry-based metabolomics. Mass Spectrom Rev 26(1):51–78

Dettmer K, Hammock BD (2004) Metabolomics – a new exciting field within the "omics" sciences. Environ Health Perspect 112(7):A

Dingle KE, Colles FM, Falush D, Maiden MCJ (2005) Sequence typing and comparison of population biology of *Campylobacter coli* and *Campylobacter jejuni*. J Clin Microbiol 43(1):340–347

Dingle KE, Colles FM, Wareing DRA, Ure R, Fox AJ, Bolton FE, Bootsma HJ, Willems RJL, Urwin R, Maiden MCJ (2001) Multilocus sequence typing system for *Campylobacter jejuni*. J Clin Microbiol 39(1):14–23

Doig P, Kinsella N, Guerry P, Trust TJ (1996) Characterization of a post-translational modification of *Campylobacter* flagellin: identification of a sero-specific glycosyl moiety. Mol Microbiol 19(2):379–387

Dorrell N, Hinchliffe SJ, Wren BW (2005) Comparative phylogenomics of pathogenic bacteria by microarray analysis. Curr Opin Microbiol 8(5):620–626

Ebbels TMD, Keun HC, Beckonert OP, Bollard ME, Lindon JC, Holmes E, Nicholson JK (2007) Prediction and classification of drug toxicity using probabilistic modeling of temporal metabolic data: the consortium on metabonomic toxicology screening approach. J Proteome Res 6(11):4407–4422

Eisen JA, Wu M (2002) Phylogenetic analysis and gene functional predictions: phylogenomics in action. Theor Popul Biol 61(4):481–487

Fagerquist CK, Yee E, Miller WG (2007) Composite sequence proteomic analysis of protein biomarkers of *Campylobacter coli*, *C. lari* and *C. concisus* for bacterial identification. Analyst 132(10):1010–1023

Feil EJ, Li BC, Aanensen DM, Hanage WP, Spratt BG (2004) eBURST: Inferring patterns of evolutionary descent among clusters of related bacterial genotypes from multilocus sequence typing data. J Bacteriol 186(5):1518–1530

Feldman MF, Wacker M, Hernandez M, Hitchen PG, Marolda CL, Kowarik M, Morris HR, Dell A, Valvano MA, Aebi M (2005) Engineering N-linked protein glycosylation with diverse O antigen lipopolysaccharide structures in *Escherichia coli*. Proc Natl Acad Sci USA 102(8):3016–3021

Fiehn O (2002) Metabolomics – the link between genotypes and phenotypes. Plant Mol Biol 48(1):155–171

Fitzgerald C, Whichard J, Nachamkin I (2008) Diagnosis and antimicrobial susceptibility of Campylobacter species. In: Nachamkin I, Szymanski C, Blaser MJ (eds) Campylobacter, 3rd edn. ASM Press, Washington, DC, pp 227–243

Fouts DE, Mongodin EF, Mandrell RE, Miller WG, Rasko DA, Ravel J, Brinkac LM, DeBoy RT, Parker CT, Daugherty SC, Dodson RJ, Durkin AS, Madupu R, Sullivan SA, Shetty JU, Ayodeji MA, Shvartsbeyn A, Schatz MC, Badger JH, Fraser CM, Nelson KE (2005) Major structural differences and novel potential virulence mechanisms from the genomes of multiple *Campylobacter* species. PloS Biol 3(1):72–85

Fox JG, Rogers AB, Whary MT, Ge Z, Taylor NS, Xu S, Horwitz BH, Erdman SE (2004) Gastroenteritis in NF-{kappa}B-Deficient mice is produced with wild-type *Camplyobacter jejuni* but not with *C. jejuni* lacking cytolethal distending toxin despite persistent colonization with both strains. Infect Immun 72(2):1116–1125

Fry BN, Feng S, Chen Y-Y, Newell DG, Coloe PJ, Korolik V (2000) The *galE* gene of *Campylobacter jejuni* is involved in lipopolysaccharide synthesis and virulence. Infect Immun 68(5):2594–2601

Fry BN, Korolik V, ten Brinke JA, Pennings MT, Zalm R, Teunis BJ, Coloe PJ, van der Zeijst BA (1998) The lipopolysaccharide biosynthesis locus of *Campylobacter jejuni* 81116. Microbiology 144(8):2049–2061

Garza AG, Harris-Haller LW, Stoebner RA, Manson MD (1995) Motility protein interactions in the bacterial flagellar motor. Proc Natl Acad Sci USA 92(6):1970–1974

Gaynor EC, Cawthraw S, Manning G, MacKichan JK, Falkow S, Newell DG (2004) The genome-sequenced variant of *Campylobacter jejuni* NCTC 11168 and the original clonal clinical isolate differ markedly in colonization, gene expression, and virulence-associated phenotypes. J Bacteriol 186(2):503–517

Gaynor EC, Wells DH, MacKichan JK, Falkow S (2005) The *Campylobacter jejuni* stringent response controls specific stress survival and virulence-associated phenotypes. Mol Microbiol 56(1):8–27

Gibreel A, Skold O, Taylor DE (2004a) Characterization of plasmid-mediated *aphA-3* kanamycin resistance in *Campylobacter jejuni*. Microbial Drug Resist-Mech Epidemiol Dis 10(2):98–105

Gibreel A, Tracz DM, Nonaka L, Ngo TM, Connell SR, Taylor DE (2004b) Incidence of antibiotic resistance in *Campylobacter jejuni* isolated in Alberta, Canada, from 1999 to 2002, with special reference to *tet(O)*-mediated tetracycline resistance. Antimicrob Agents Chemother 48(9):3442–3450

Gibson DL, White AP, Snyder SD, Martin S, Heiss C, Azadi P, Surette M, Kay WW (2006) Salmonella produces an O-antigen capsule regulated by AgfD and important for environmental persistence. J Bacteriol 188(22):7722–7730

Gilbert M, Godschalk PCR, Parker CT, Endtz HP, Wakarchuk WW (2004) Genetic basis for the variation in the lipopolysaccharide outer core of campylobacter jejuni and possible association of glycosyltransferase genes with post-infectious neuropathies. In: Ketley JM, Konkel ME (eds) Campylobacter: new perspectives in molecular and cellular biology. Horizon Bioscience, Norfolk, pp 219–248

Gilbert M, Karwaski M-F, Bernatchez S, Young NM, Taboada E, Michniewicz J, Cunningham A-M, Wakarchuk WW (2002) The genetic bases for the variation in the Lipo-oligosaccharide of the mucosal pathogen, *Campylobacter jejuni*. Biosynthesis of sialylated ganglioside mimics in the core oligosaccharide. J Biol Chem 277(1):327–337

Gilbert M, Mandrell RE, Parker CT, Li J, Vinogradov E (2007) Structural analysis of the capsular polysaccharide from *Campylobacter jejuni* RM1221. Chembiochem 8(6):625–631

Godde JS, Bickerton A (2006) The repetitive DNA elements called CRISPRs and their associated genes: evidence of horizontal transfer among prokaryotes. J Mol Evolution 62(6):718–729

Godschalk PCR, Kuijf ML, Li J, St. Michael F, Ang CW, Jacobs BC, Karwaski M-F, Brochu D, Moterassed A, Endtz HP, van Belkum A, Gilbert M (2007) Structural Characterization of *Campylobacter jejuni* lipooligosaccharide outer cores associated with Guillain-Barre and Miller Fisher syndromes. Infect Immun 75(3):1245–1254

Gogol E, Cummings C, Burns R, Relman D (2007) Phase variation and microevolution at homopolymeric tracts in *Bordetella pertussis*. BMC Genomics 8(1):122

Goodacre R, Vaidyanathan S, Dunn WB, Harrigan GG, Kell DB (2004) Metabolomics by numbers: acquiring and understanding global metabolite data. Trends Biotechnol 22(5):245–252

Goon S, Kelly JF, Logan SM, Ewing CP, Guerry P (2003) Pseudaminic acid, the major modification on *Campylobacter* flagellin, is synthesized via the Cj1293 gene. Mol Microbiol 50(2):659–671

Gophna U, Ron EZ, Graur D (2003) Bacterial type III secretion systems are ancient and evolved by multiple horizontal-transfer events. Gene 312:151–163

Guerry P (2007) *Campylobacter* flagella: not just for motility. Trends Microbiol 15(10):456–461

Guerry P, Alm RA, Power ME, Logan SM, Trust TJ (1991) Role of two flagellin genes in *Campylobacter* motility. J Bacteriol 173(15):4757–4764

Guerry P, Ewing CP, Hickey TE, Prendergast MM, Moran AP (2000) Sialylation of lipooligosaccharide cores affects immunogenicity and serum resistance of *Campylobacter jejuni*. Infect Immun 68(12):6656–6662

Guerry P, Ewing CP, Schoenhofen IC, Logan SM (2007) Protein glycosylation in *Campylobacter jejuni*: partial suppression of *pglF* by mutation of *pseC*. J Bacteriol 189(18):6731–6733

Guerry P, Szymanski CM, Prendergast MM, Hickey TE, Ewing CP, Pattarini DL, Moran AP (2002) Phase variation of *Campylobacter jejuni* 81-176 lipooligosaccharide affects ganglioside mimicry and invasiveness in vitro. Infect Immun 70(2):787–793

Gundogdu O, Bentley S, Holden M, Parkhill J, Dorrell N, Wren B (2007) Re-annotation and re-analysis of the *Campylobacter jejuni* NCTC11168 genome sequence. BMC Genomics 8(1):162

Hallet B (2001) Playing Dr Jekyll and Mr Hyde: combined mechanisms of phase variation in bacteria. Curr Opin Microbiol 4(5):570–581

Harrington CS, Thomson-Carter FM, Carter PE (1997) Evidence for recombination in the flagellin locus of *Campylobacter jejuni*: implications for the flagellin gene typing scheme. J Clin Microbiol 35(9):2386–2392

Henderson IR, Nataro JP (2001) Virulence functions of autotransporter proteins. Infect Immun 69(3):1231–1243

Henderson IR, Navarro-Garcia F, Nataro JP (1998) The great escape: structure and function of the autotransporter proteins. Trends Microbiol 6(9):370–378

Hepworth P, Leatherbarrow H, Hart CA, Winstanley C (2007) Use of suppression subtractive hybridisation to extend our knowledge of genome diversity in *Campylobacter jejuni*. BMC Genomics 8(1):110

Hickey TE, McVeigh AL, Scott DA, Michielutti RE, Bixby A, Carroll SA, Bourgeois AL, Guerry P (2000) *Campylobacter jejuni* cytolethal distending toxin mediates release of interleukin-8 from intestinal epithelial cells. Infect Immun 68(12):6535–6541

Hofreuter D, Tsai J, Watson RO, Novik V, Altman B, Benitez M, Clark C, Perbost C, Jarvie T, Du L, Galan JE (2006) Unique features of a highly pathogenic *Campylobacter jejuni* strain. Infect Immun 74(8):4694–4707

Holland IB, Kenny B, Blight M (1990) Hemolysin secretion from *Escherichia coli*. Biochimie 72(2–3):131–141

Howard SL, Gaunt MW, Hinds J, Witney AA, Stabler R, Wren BW (2006) Application of comparative phylogenomics to study the evolution of *Yersinia enterocolitica* and to identify genetic differences relating to pathogenicity. J Bacteriol 188(10):3645–3653

Hu L, Kopecko DJ (1999) *Campylobacter jejuni* 81-176 associates with microtubules and dynein during invasion of human intestinal cells. Infect Immun 67(8):4171–4182

Jansen R, van Embden JDA, Gaastra W, Schouls LM (2002) Identification of a novel family of sequence repeats among prokaryotes. OMICS 6(1):23–33

Jin S, Joe A, Lynett J, Hani EK, Sherman P, Chan VL (2001) JlpA, a novel surface-exposed lipoprotein specific to *Campylobacter jejuni*, mediates adherence to host epithelial cells. Mol Microbiol 39(5):1225–1236

Jin S, Song YC, Emili A, Sherman PM, Chan VL (2003) JlpA of Campylobacter jejuni interacts with surface-exposed heat shock protein 90alpha and triggers signalling pathways leading to the activation of NF-kappaB and p38 MAP kinase in epithelial cells. Cell Microbiol 5(3):165–174

Johnson WM, Lior H (1988) A new heat-labile cytolethal distending toxin (CLDT) produced by *Campylobacter* spp. Microb Pathog 4(2):115–126

Kanipes MI, Holder LC, Corcoran AT, Moran AP, Guerry P (2004) A deep-rough mutant of *Campylobacter jejuni* 81-176 is noninvasive for intestinal epithelial cells. Infect Immun 72(4):2452–2455

Kanipes MI, Papp-Szabo E, Guerry P, Monteiro MA (2006) Mutation of *waaC*, Encoding Heptosyltransferase I in *Campylobacter jejuni* 81-176, affects the structure of both lipooligosaccharide and capsular carbohydrate. J Bacteriol 188(9):3273–3279

Kanipes MI, Tan X, Akelaitis A, Li J, Rockabrand D, Guerry P, Monteiro MA (2008) Genetic analysis of lipo-oligosaccharide core biosynthesis in *Campylobacter jejuni* 81-176. J Bacteriol 190(5):1568–1574

Karlyshev AV, Everest P, Linton D, Cawthraw S, Newell DG, Wren BW (2004) The *Campylobacter jejuni* general glycosylation system is important for attachment to human epithelial cells and in the colonization of chicks. Microbiology 150(6):1957–1964

Karlyshev AV, Champion OL, Churcher C, Brisson J-R, Jarrell HC, Gilbert M, Brochu D, St Michael F, Li J, Wakarchuk WW, Goodhead I, Sanders M, Stevens K, White B, Parkhill J, Wren BW, Szymanski CM (2005a) Analysis of *Campylobacter jejuni* capsular loci reveals multiple mechanisms for the generation of structural diversity and the ability to form complex heptoses. Mol Microbiol 55(1):90–103

Karlyshev AV, Ketley JM, Wren BW (2005b) The *Campylobacter jejuni* glycome. FEMS Microbiol Rev 29(2):377–390

Karlyshev AV, Linton D, Gregson NA, Lastovica AJ, Wren BW (2000) Genetic and biochemical evidence of a *Campylobacter jejuni* capsular polysaccharide that accounts for penner serotype specificity. Mol Microbiol 35(3):529–541

Karlyshev AV, Linton D, Gregson NA, Wren BW (2002) A novel paralogous gene family involved in phase-variable flagella-mediated motility in *Campylobacter jejuni*. Microbiology 148(2):473–480

Karlyshev AV, McCrossan MV, Wren BW (2001) Demonstration of polysaccharide capsule in *Campylobacter jejuni* using electron microscopy. Infect Immun 69(9):5921–5924

Karlyshev AV, Wren BW (2001) Detection and initial characterization of novel capsular polysaccharide among diverse *Campylobacter jejuni* strains using alcian blue dye. J Clin Microbiol 39(1):279–284

Kervella M, Pages JM, Pei Z, Grollier G, Blaser MJ, Fauchere JL (1993) Isolation and characterization of two *Campylobacter* glycine-extracted proteins that bind to HeLa cell membranes. Infect Immun 61(8):3440–3448

Keun HC, Beckonert O, Griffin JL, Richter C, Moskau D, Lindon JC, Nicholson JK (2002) Cryogenic probe C-13 NMR spectroscopy of urine for metabonomic studies. Anal Chem 74(17):4588–4593

Kilcoyne M, Moran AP, Shashkov AS, Senchenkova SyN, Ferris JA, Corcoran AT, Savage AV (2006) Molecular origin of two polysaccharides of *Campylobacter jejuni* 81116. FEMS Microbiol Lett 263(2):214–222

Konkel ME, Garvis SG, Tipton SL, Anderson DE Jr, Cieplak W Jr (1997) Identification and molecular cloning of a gene encoding a fibronectin-binding protein (CadF) from *Campylobacter jejuni*. Mol Microbiol 24(5):953–963

Konkel ME, Kim BJ, Rivera-Amill V, Garvis SG (1999a) Bacterial secreted proteins are required for the internalization of *Campylobacter jejuni* into cultured mammalian cells. Mol Microbiol 32(4):691–701

Konkel ME, Kim BJ, Rivera-Amill V, Garvis SG (1999b) Identification of proteins required for the internalization of campylobacter jejuni into cultured mammalian cells. In: Paul PS, Francis DH (eds) Mechanisms in the pathogenesis of enteric diseases 2, vol 473. Kluwer Academic Publishers Group, New York, NY, pp 215–224

Konkel ME, Klena JD, Rivera-Amill V, Monteville MR, Biswas D, Raphael B, Mickelson J (2004) Secretion of virulence proteins from *Campylobacter jejuni* is dependent on a functional flagellar export apparatus. J Bacteriol 186(11):3296–3303

Lara-Tejero M, Galan JE (2000) A bacterial toxin that controls cell cycle progression as a deoxyribonuclease i-like protein. Science 290(5490):354–357

Lara-Tejero M, Galan JE (2001) CdtA, CdtB, and CdtC form a tripartite complex that is required for cytolethal distending toxin activity. Infect Immun 69(7):4358–4365

Larsen JC, Szymanski C, Guerry P (2004) N-Linked protein glycosylation is required for full competence in *Campylobacter jejuni* 81-176. J Bacteriol 186(19):6508–6514

Larson CL, Christensen JE, Pacheco SA, Minnich SA, Konkel ME (2008) Campylobacter jejuni secretes proteins via the flagellar type III secretion system that contribute to host cell invasion and gastroenteritis. In: Nachamkin I, Szymanski C, Blaser MJ (eds) Campylobacter, 2nd edn. ASM Press, Washington, DC, pp 315–332

Lastovica AJ, Skirrow MB (2000) Clinical significance of campylobacter and related species other than campylobacter jejuni and C. coli. In: Nachamkin I, Blaser MJ (eds) Campylobacter, 2nd edn. ASM Press, Washington, DC, pp 89–120

Leach SA (1997) Growth, survival and pathogenicity of enteric campylobacters. Rev Med Microbiol 8(3):113–124

Lee RB, Hassane DC, Cottle DL, Pickett CL (2003) Interactions of *Campylobacter jejuni* cytolethal distending toxin subunits CdtA and CdtC with HeLa cells. Infect Immun 71(9):4883–4890

Leon-Kempis MdR, Guccione E, Mulholland F, Williamson MP, Kelly DJ (2006) The *Campylobacter jejuni* PEB1a adhesin is an aspartate/glutamate-binding protein of an ABC transporter essential for microaerobic growth on dicarboxylic amino acids. Mol Microbiol 60(5):1262–1275

Linton D, Dorrell N, Hitchen PG, Amber S, Karlyshev AV, Morris HR, Dell A, Valvano MA, Aebi M, Wren BW (2005) Functional analysis of the *Campylobacter jejuni* N-linked protein glycosylation pathway. Mol Microbiol 55(6):1695–1703

Linton D, Gilbert M, Hitchen PG, Dell A, Morris HR, Wakarchuk WW, Gregson NA, Wren BW (2000) Phase variation of a beta-1,3 galactosyltransferase involved in generation of the ganglioside GM1-like lipo-oligosaccharide of *Campylobacter jejuni*. Mol Microbiol 37(3): 501–514

Litrup E, Torpdahl M, Nielsen EM (2007) Multilocus sequence typing performed on *Campylobacter coli* isolates from humans, broilers, pigs and cattle originating in Denmark. J Appl Microbiol 103(1):210–218

Louwen RPL, Van Belkum A, Wagenaar JA, Doorduyn Y, Achterberg R, Endtz HP (2006) Lack of association between the presence of the pVir plasmid and bloody diarrhea in *Campylobacter jejuni* enteritis. J Clin Microbiol 44(5):1867–1868

MacKichan JK, Gaynor EC, Chang C, Cawthraw S, Newell DG, Miller JF, Falkow S (2004) The *Campylobacter jejuni dccRS* two-component system is required for optimal in vivo colonization but is dispensable for in vitro growth. Mol Microbiol 54(5):1269–1286

Malik-Kale P, Raphael BH, Parker CT, Joens LA, Klena JD, Quinones B, Keech AM, Konkel ME (2007) Characterization of genetically matched isolates of *Campylobacter jejuni* reveals that mutations in genes involved in flagellar biosynthesis alter the organism's virulence potential. Appl Environ Microbiol 73(10):3123–3136

McCarthy ND, Colles FM, Dingle KE, Bagnall MC, Manning G, Maiden MCJ, Falush D (2007) Host-associated genetic import in *Campylobacter jejuni*. Emerg Infect Dis 13(2): 267–272

McNally DJ, Aubry AJ, Hui JPM, Khieu NH, Whitfield D, Ewing CP, Guerry P, Brisson J-R, Logan SM, Soo EC (2007) Targeted metabolomics analysis of *Campylobacter coli* VC167 reveals legionaminic acid derivatives as novel flagellar glycans. J Biol Chem 282(19): 14463–14475

McNally DJ, Hui JPM, Aubry AJ, Mui KKK, Guerry P, Brisson JR, Logan SM, Soo EC (2006a) Functional characterization of the flagellar glycosylation locus in *Campylobacter jejuni* 81-176 using a focused metabolomics approach. J Biol Chem 281(27):18489–18498

McNally DJ, Jarrell HC, Khieu NH, Li J, Vinogradov E, Whitfield DM, Szymanski CM, Brisson JR (2006b) The HS: 19 serostrain of *Campylobacter jejuni* has a hyaluronic acid-type capsular polysaccharide with a nonstoichiometric sorbose branch and O-methyl phosphoramidate group. FEBS J 273(17):3975–3989

McNally DJ, Jarrell HC, Li J, Khieu NH, Vinogradov E, Szymanski CM, Brisson J-R (2005) The HS:1 serostrain of *Campylobacter jejuni* has a complex teichoic acid-like capsular polysaccharide with nonstoichiometric fructofuranose branches and O-methyl phosphoramidate groups. FEBS J 272(17):4407–4422

McNally DJ, Lamoureux MP, Karlyshev AV, Fiori LM, Li J, Thacker G, Coleman RA, Khieu NH, Wren BW, Brisson J-R, Jarrell HC, Szymanski CM (2007) Commonality and biosynthesis of the O-methyl phosphoramidate capsule modification in *Campylobacter jejuni*. J Biol Chem 282(39):28566–28576

Mead PS, Slutsker L, Dietz V, McCaig LF, Bresee JS, Shapiro C, Griffin PM, Tauxe RV (1999) Food-related illness and death in the United States. Emerg Infect Dis 5(5):607–625

Miller WG, Bates AH, Horn ST, Brandl MT, Wachtel MR, Mandrell RE (2000) Detection on surfaces and in Caco-2 cells of *Campylobacter jejuni* cells transformed with New *gfp*, *yfp*, and *cfp* marker plasmids. Appl Environ Microbiol 66(12):5426–5436

Miller WG, Englen MD, Kathariou S, Wesley IV, Wang G, Pittenger-Alley L, Siletz RM, Muraoka W, Fedorka-Cray PJ, Mandrell RE (2006) Identification of host-associated alleles by multilocus sequence typing of *Campylobacter coli* strains from food animals. Microbiology 152(1): 245–255

Moen B, Oust A, Langsrud O, Dorrell N, Marsden GL, Hinds J, Kohler A, Wren BW, Rudi K (2005) Explorative multifactor approach for investigating global survival mechanisms of *Campylobacter jejuni* under environmental conditions. Appl Environ Microbiol 71(4): 2086–2094

Monteville MR, Yoon JE, Konkel ME (2003) Maximal adherence and invasion of INT 407 cells by *Campylobacter jejuni* requires the CadF outer-membrane protein and microfilament reorganization. Microbiology 149(1):153–165

Moser I, Schroeder W, Salnikow J (1997) *Campylobacter jejuni* major outer membrane protein and a 59-kDa protein are involved in binding to fibronectin and INT 407 cell membranes. FEMS Microbiol Lett 157(2):233–238

Nachamkin I, Allos BM, Ho T (1998) *Campylobacter* species and Guillain-Barré syndrome. Clin Microbiol Rev 11(3):555–567

Ng LK, Stiles ME, Taylor DE (1987) DNA probes for identification of tetracycline resistance genes in *Campylobacter* species isolated from swine and cattle. Antimicrob Agents Chemother 31(11):1669–1674

Nicholls AW, Mortishire-Smith RJ, Nicholson JK (2003) NMR spectroscopic-based metabonomic studies of urinary metabolite variation in acclimatizing germ-free rats. Chem Res Toxicol 16(11):1395–1404

Oldiges M, Lutz S, Pflug S, Schroer K, Stein N, Wiendahl C (2007) Metabolomics: current state and evolving methodologies and tools. Appl Microbiol Biotechnol 76(3):495–511

Oliver SG, Winson MK, Kell DB, Baganz F (1998) Systematic functional analysis of the yeast genome. Trends Biotechnol 16(9):373–378

Parker CT, Horn ST, Gilbert M, Miller WG, Woodward DL, Mandrell RE (2005) Comparison of *Campylobacter jejuni* lipooligosaccharide biosynthesis loci from a variety of sources. J Clin Microbiol 43(6):2771–2781

Parker CT, Quinones B, Miller WG, Horn ST, Mandrell RE (2006) Comparative genomic analysis of *Campylobacter jejuni* strains reveals diversity due to genomic elements similar to those present in *C. jejuni* strain RM1221. J Clin Microbiol 44(11):4125–4135

Parkhill J, Wren BW, Mungall K, Ketley JM, Churcher C, Basham D, Chillingworth T, Davies RM, Feltwell T, Holroyd S, Jagels K, Karlyshev AV, Moule S, Pallen MJ, Penn CW, Quail MA, Rajandream MA, Rutherford KM, van Vliet AHM, Whitehead S, Barrell BG (2000) The genome sequence of the food-borne pathogen *Campylobacter jejuni* reveals hypervariable sequences. Nature 403(6770):665–668

Parrish J, Yu J, Liu G, Hines J, Chan J, Mangiola B, Zhang H, Pacifico S, Fotouhi F, DiRita V, Ideker T, Andrews P, Finley R (2007) A proteome-wide protein interaction map for *Campylobacter jejuni*. Genome Biol 8(7):R130

Pearson BM, Gaskin DJH, Segers RPAM, Wells JM, Nuijten PJM, van Vliet AHM (2007) The Complete genome sequence of *Campylobacter jejuni* strain 81116 (NCTC11828). J Bacteriol 189(22):8402–8403

Pearson BM, Pin C, Wright J, I'Anson K, Humphrey T, Wells JM (2003) Comparative genome analysis of *Campylobacter jejuni* using whole genome DNA microarrays. FEBS Lett 554(1–2):224–230

Pei Z, Burucoa C, Grignon B, Baqar S, Huang X-Z, Kopecko DJ, Bourgeois AL, Fauchere J-L, Blaser MJ (1998) Mutation in the *peb1A* Locus of *Campylobacter jejuni* reduces interactions with epithelial cells and intestinal colonization of mice. Infect Immun 66(3):938–943

Pickett CL, Pesci EC, Cottle DL, Russell G, Erdem AN, Zeytin H (1996) Prevalence of cytolethal distending toxin production in *Campylobacter jejuni* and relatedness of *Campylobacter* sp. *cdtB* gene. Infect Immun 64(6):2070–2078

Poly F, Ewing C, Goon S, Hickey TE, Rockabrand D, Majam G, Lee L, Phan J, Savarino NJ, Guerry P (2007) Heterogeneity of a *Campylobacter jejuni* protein that is secreted through the flagellar filament. Infect Immun 75(8):3859–3867

Poly F, Read T, Tribble DR, Baqar S, Lorenzo M, Guerry P (2007) Genome sequence of a clinical isolate of *Campylobacter jejuni* from Thailand. Infect Immun 75(7):3425–3433

Pugsley AP, Kornacker MG, Poquet I (1991) The general protein-export pathway is directly required for extracellular pullulanase secretion in *Escherichia coli* K12. Mol Microbiol 5(2):343–352

Purdy D, Buswell CM, Hodgson AE, McAlpine K, Henderson I, Leach SA (2000) Characterisation of cytolethal distending toxin (CDT) mutants of *Campylobacter jejuni*. J Med Microbiol 49(5):473–479

Reyrat J-M, Pelicic V, Papini E, Montecucco C, Rappuoli R, Telford JL (1999) Towards deciphering the *Helicobacter pylori* cytotoxin. Mol Microbiol 34(2):197–204

Rivera-Amill V, Konkel ME (1999) Secretion of campylobacter jejuni Cia proteins is contact dependent. In: Paul PS, Francis DH (eds) Mechanisms in the pathogenesis of enteric diseases 2, vol 473. Kluwer Academic Publishers Group, New York, NY, pp 225–229

Robinson DA (1981) Infective dose of *Campylobacter jejuni* in milk. Br Med J 282(6276):1584

Salaun L, Snyder LAS, Saunders NJ (2003) Adaptation by phase variation in pathogenic bacteria. Adv Appl Microbiol 52:263–301

Sampathkumar B, Napper S, Carrillo CD, Willson P, Taboada E, Nash JHE, Potter AA, Babiuk LA, Allan BJ (2006) Transcriptional and translational expression patterns associated with immobilized growth of *Campylobacter jejuni*. Microbiology 152(2):567–577

Schirm M, Soo EC, Aubry AJ, Austin J, Thibault P, Logan SM (2003) Structural, genetic and functional characterization of the flagellin glycosylation process in *Helicobacter pylori*. Mol Microbiol 48(6):1579–1592

Schouls LM, Reulen S, Duim B, Wagenaar JA, Willems RJL, Dingle KE, Colles FM, Van Embden JDA (2003) Comparative genotyping of *Campylobacter jejuni* by amplified fragment length polymorphism, multilocus sequence typing, and short repeat sequencing: strain diversity, host range, and recombination. J Clin Microbiol 41(1):15–26

Serino L, Virji M (2002) Genetic and functional analysis of the phosphorylcholine moiety of commensal *Neisseria* lipopolysaccharide. Mol Microbiol 43(2):437–448

Sheppard SK, McCarthy ND, Falush D, Maiden MCJ (2008) Convergence of *Campylobacter* species: implications for bacterial evolution. Science 320(5873):237–239

Skirrow MB (1977) *Campylobacter* enteritis – new disease. Br Med J 2(6078):9–11

Solomon EB, Hoover DG (1999) *Campylobacter jejuni*: a bacterial paradox. J Food Saf 19(2): 121–136

Song YC, Jin S, Louie H, Ng D, Lau R, Zhang Y, Weerasekera R, Rashid SA, Ward LA, Der SD, Chan VL (2004) FlaC, a protein of *Campylobacter jejuni* TGH9011 (ATCC43431) secreted through the flagellar apparatus, binds epithelial cells and influences cell invasion. Mol Microbiol 53(2):541–553

Soo EC, Aubry AJ, Logan SM, Guerry P, Kelly JF, Young NM, Thibault P (2004) Selective detection and identification of sugar nucleotides by CE-electrospray-MS and its application to bacterial metabolomics. Anal Chem 76(3):619–626

St Michael F, Szymanski CM, Li J, Chan KH, Khieu NH, Larocque S, Wakarchuk WW, Brisson J-R, Monteiro MA (2002) The structures of the lipooligosaccharide and capsule polysaccharide of *Campylobacter jejuni* genome sequenced strain NCTC 11168. Eur J Biochem 269(21): 5119–5136

Stabler RA, Gerding DN, Songer JG, Drudy D, Brazier JS, Trinh HT, Witney AA, Hinds J, Wren BW (2006) Comparative phylogenomics of *Clostridium difficile* reveals clade specificity and microevolution of hypervirulent strains. J Bacteriol 188(20):7297–7305

Stintzi A (2003) Gene expression profile of *Campylobacter jejuni* in response to growth temperature variation. J Bacteriol 185(6):2009–2016

Stintzi A, Marlow D, Palyada K, Naikare H, Panciera R, Whitworth L, Clarke C (2005) Use of genome-wide expression profiling and mutagenesis to study the intestinal lifestyle of *Campylobacter jejuni*. Infect Immun 73(3):1797–1810

Szymanski CM, Logan SM, Linton D, Wren BW (2003a) *Campylobacter* – a tale of two protein glycosylation systems. Trends Microbiol 11(5):233–238

Szymanski CM, St Michael F, Jarrell HC, Li J, Gilbert M, Larocque S, Vinogradov E, Brisson J-R (2003b) Detection of conserved N-Linked glycans and phase-variable lipooligosaccharides and capsules from *Campylobacter* cells by mass spectrometry and high resolution magic angle spinning NMR spectroscopy. J Biol Chem 278(27):24509–24520

Szymanski CM, Wren BW (2005) Protein glycosylation in bacterial mucosal pathogens. Nat Rev Micro 3(3):225–237

Taboada EN, Acedillo RR, Carrillo CD, Findlay WA, Medeiros DT, Mykytczuk OL, Roberts MJ, Valencia CA, Farber JM, Nash JHE (2004) Large-Scale comparative genomics meta-analysis of *Campylobacter jejuni* isolates reveals low level of genome plasticity. J Clin Microbiol 42(10):4566–4576

Tam CC (2001) *Campylobacter* reporting at its peak year of 1998: don't count your chickens yet. Commun Dis Public Health 4(3):194–199

Tenover FC, Williams S, Gordon KP, Nolan C, Plorde JJ (1985) Survey of plasmids and resistance factors in *Campylobacter jejuni* and *Campylobacter coli*. Antimicrob Agents Chemother 27(1):37–41

Thibault P, Logan SM, Kelly JF, Brisson J-R, Ewing CP, Trust TJ, Guerry P (2001) Identification of the carbohydrate moieties and glycosylation motifs in *Campylobacter jejuni* flagellin. J Biol Chem 276(37):34862–34870

Thompson SA, Shedd OL, Ray KC, Beins MH, Jorgensen JP, Blaser MJ (1998) *Campylobacter fetus* surface layer proteins are transported by a type I secretion system. J Bacteriol 180(24):6450–6458

Tracz DM, Keelan M, Ahmed-Bentley J, Gibreel A, Kowalewska-Grochowska K, Taylor DE (2005) pVir and bloody diarrhea in *Campylobacter jejuni* enteritis. Emerg Infect Dis 11(6): 838–843

Wacker M, Linton D, Hitchen PG, Nita-Lazar M, Haslam SM, North SJ, Panico M, Morris HR, Dell A, Wren BW, Aebi M (2002) N-linked glycosylation in *Campylobacter jejuni* and its functional transfer into *E. coli*. Science 298(5599):1790–1793

Weiss AA, Johnson FD, Burns DL (1993) Molecular characterization of an operon required for pertussis toxin secretion. Proc Natl Acad Sci USA 90(7):2970–2974

Whitehouse CA, Balbo PB, Pesci EC, Cottle DL, Mirabito PM, Pickett CL (1998) *Campylobacter jejuni* cytolethal distending toxin causes a G2-Phase cell cycle block. Infect Immun 66(5):1934–1940

Wiesner RS, Hendrixson DR, DiRita VJ (2003) Natural transformation of *Campylobacter jejuni* requires components of a type II secretion system. J Bacteriol 185(18):5408–5418

Wilharm G, Lehmann V, Neumayer W, Trcek J, Heesemann J (2004) *Yersinia enterocolitica* type III secretion: evidence for the ability to transport proteins that are folded prior to secretion. BMC Microbiol 4(1):27

Woodall CA, Jones MA, Barrow PA, Hinds J, Marsden GL, Kelly DJ, Dorrell N, Wren BW, Maskell DJ (2005) *Campylobacter jejuni* gene expression in the chick cecum: evidence for adaptation to a low-oxygen environment. Infect Immun 73(8):5278–5285

van der Woude MW, Baumler AJ (2004) Phase and antigenic variation in bacteria. Clin Microbiol Rev 17(3):581–611

Young NM, Brisson J-R, Kelly J, Watson DC, Tessier L, Lanthier PH, Jarrell HC, Cadotte N, St Michael F, Aberg E, Szymanski CM (2002) Structure of the N-linked glycan present on multiple glycoproteins in the gram-negative bacterium, *Campylobacter jejuni*. J Biol Chem 277(45):42530–42539

Young KT, Davis LM, DiRita VJ (2007) *Campylobacter jejuni*: molecular biology and pathogenesis. Nat Rev Micro 5(9):665–679

Young GM, Schmiel DH, Miller VL (1999) A new pathway for the secretion of virulence factors by bacteria: the flagellar export apparatus functions as a protein-secretion system. Proc Natl Acad Sci USA 96(11):6456–6461

Zhao C, Ge B, De Villena J, Sudler R, Yeh E, Zhao S, White DG, Wagner D, Meng J (2001) Prevalence of *Campylobacter* spp., *Escherichia coli*, and *Salmonella* serovars in retail chicken, turkey, pork, and beef from the Greater Washington, D.C., area. Appl Environ Microbiol 67(12):5431–5436

Ziprin RL, Young CR, Stanker LH, Hume ME, Konkel ME (1999) The absence of cecal colonization of chicks by a mutant of *Campylobacter jejuni* not expressing bacterial fibronectin-binding protein. Avian Dis 43(3):586–589

Chapter 4
Genomics of *Clostridium*

Mark Joseph Jacobson and Eric A. Johnson

4.1 Introduction

The clostridia have a rich history and contemporary importance in industrial, environmental, and medical microbiology. Due to their ability to form endospores, clostridia are ubiquitous in nature and are found in many environments, especially in soils and the intestinal tract of animals including humans. Many clostridia cause devastating diseases of humans and animals, such as botulism, tetanus, and gas gangrene, through the production of protein toxins. The clostridia produce more protein toxins that are lethal for humans and animals than any other bacterial genus (Johnson, 2005; Van Heyningen, 1950). Other species are important in the formation of solvents and organic acids by anaerobic fermentations or as a source of unique enzymes for biocatalysis (Bradshaw and Johnson, 2010; Hatheway and Johnson, 1998). The discovery of the role of clostridial protein toxins in causing diseases contributed to the concept of vaccination with detoxified forms of toxins to produce immunity in animals and humans. The importance of clostridia in industrial fermentations was demonstrated before World War I when an industrial process was developed for butanol as a precursor of butadiene (for synthetic rubber) and of acetone used in weapons. Currently, clostridia are being investigated as sources of novel enzymes for biotransformations and as organisms for production of solvents and organic acids by industrial fermentation, as well as for their capabilities to cause diseases in humans and animals.

The genus *Clostridium* comprises a vast array of organisms, and this chapter provides an overview of the genomics of medically and industrially important species. The chapter also describes the limited studies of the transcriptome and proteome of selected *Clostridium* species. Additionally, future directions are suggested for advancing the field "ohmics" in clostridia to better understand the genus.

E.A. Johnson (✉)
Department of Bacteriology, University of Wisconsin, Madison, WI 53706, USA
e-mail: eajohnso@wisc.edu

M. Wiedmann, W. Zhang (eds.), *Genomics of Foodborne Bacterial Pathogens*,
Food Microbiology and Food Safety, DOI 10.1007/978-1-4419-7686-4_4,
© Springer Science+Business Media, LLC 2011

4.1.1 Genus Definition

The genus *Clostridium* consists of rod-shaped, gram-positive, spore-forming, fermentative Firmicutes (Anonymous, 2009; Hatheway and Johnson, 1998). Most species form endospores that swell the cells giving the characteristic spindle shape [Clos.tri'di.um. Gr. N. *closter* as a spindle (Anonymous, 2009)]. They are anaerobic but vary in oxygen sensitivity, being aerotolerant to strict anaerobes. Some species may grow but not sporulate in the presence of oxygen. Most species are motile generally by peritrichous flagella.

Members of the genus *Clostridium* are extremely heterogeneous in physiological properties. Most species are chemoorganotrophic, while some are chemoautrophic or chemolithotrophic (Anonymous, 2009). Species may be saccharolytic, proteolytic, both, or neither. The genus *Clostridium* contains species that can grow at temperatures ranging from below 4°C to above 50°C, while most prefer the range of 25–37°C. Certain species can grow below pH 4 or above pH 8, but most prefer pH from 6 to 7.5. Some clostridia can fix molecular nitrogen and convert it to organic nitrogen, which is important in agriculture.

Certain *Clostridium* groups have been classified on the basis of physiological properties such as *Clostridium botulinum* (Cato et al., 1986). *Clostridium* does not carry out dissimilatory sulfate reduction. The genus is also characterized by a low G+C DNA content (mol%) of 22–53%, while most pathogenic strains have low G+C contents of 22–32%. The type species is *Clostridium butyricum*, which was probably first identified by Pasteur during butyric fermentation in the mid-1800s (Willis, 1969).

According to Bergey's Manual, 9th edition, there are 168 validly published species of the genus *Clostridium*. DSMZ has listed 220 species. The majority of the species fall in cluster I (*Clostridium* sensu stricto) of the clostridia as defined rDNA sequences in the classic paper of Collins et al. (1994). Due to the heterogeneity, the strains currently assigned to the genus *Clostridium* may belong to other or new genera. The polyphyletic structure of the genus *Clostridium* has been inadequately described, and an in-depth taxonomic description based on phylogenetic properties is needed (Collins et al., 1994; Rainey and Stackebrandt, 1993; Stackebrandt et al., 1999).

4.1.2 The Importance of Sporulation

Clostridia are found in a vast array of environments from soil to the intestinal tract of animals by virtue of their ability to produce resistant endospores. The production of endospores is induced by stress (Scheeff et al., 2009), whereby the cell senses the environment and responds by expressing a set of genes necessary for sporogenesis. The production of the spore results in dormancy and promotes the ability to survive different external stresses including extreme heat, cold, and chemicals that would otherwise kill the vegetative cells. While in this spore state, the bacteria are also resistant to aerobic conditions and can remain viable for years even in atmospheric conditions. Many food laws including low-acid canned food laws and high- and low-acid food designation are designed to control toxigenic clostridial spores, particularly *C. botulinum* (Johnson, 2007). Interestingly, since endospore formation is dependent on more than 50 genes, the mutation of even a single gene can result in nonsporulating bacteria. It has been suggested that food-borne and medical pathogens such as *Listeria monocytogenes* may have been derived during loss of sporulation ability (Onyenwoke et al., 2004).

4.1.3 Genetic Elements

The genetics of *Clostridium* has recently been reviewed (Bradshaw and Johnson, 2010). Many species of clostridia have extrachromosomal elements. Most of the plasmids are considered cryptic, but some have been shown to carry antibiotic resistance genes, bacteriocins, virulence factors, or genes for solvent production (Fischetti, 2000; Rood, 2004; Rood and Cole, 1991).

Certain clostridia possess various classes of transposable elements (Bradshaw and Johnson, 2010). This has been most clearly demonstrated in *Clostridium perfringens* and *Clostridium difficile*, which each have several transposable elements. Their presence along with IS elements has been noted in the genomic sequencing projects for *C. difficile* and is theorized to produce a mosaic genome for the species by altering the positions of several genes through recombinatory events. These elements do not appear to be prevalent in many species, although extensive surveys have not been performed. Many clostridia contain bacteriophages, which can encode for toxins and enzymes.

4.1.4 Genomic Features and Tools

GenBank (National Centers for Biotechnology Information) currently lists 91 genome sequencing projects of strains from 40 different *Clostridium* spp. (Bradshaw and Johnson, 2010). A number of these sequences are complete, but most are still either draft sequences requiring additional efforts such as annotation to be complete or in process. Sequencing efforts on *Clostridium* species of interest identified in this chapter are presented in Table 4.1. Considerable variation exists among genome size even within the same species. *C. perfringens* is the best characterized *Clostridium* spp. and contains a circular chromosome of about 3,600 kbp, with essential genes, such as those encoding tRNA and rRNA operon having an arrangement similar to that found in *Bacillus subtilis* (Barbe et al., 2009). The solvent-producing species *Clostridium acetobutylicum* has a genome size of about 6,500 kbp, and several genes for the solvent fermentation lie on a large plasmid (Nolling et al., 2001). These examples emphasize the diversity of genome features in *Clostridium* including the diversity and the importance of plasmids in encoding essential and beneficial traits.

Table 4.1 Clostridial genome sequences for species identified in this chapter[a]

Species and strain	Form	Length (nt)	ORFs	RNAs	GC content (%)	Accession number	Status
Clostridium acetobutylicum ATCC 824	Chromosome	3,940,880	4,067	107	30	AE001437	Complete
	Plasmid pSOL1	192,000	178	None	30	AE001438	Complete
Clostridium botulinum							
Type A1 ATCC 3502	Chromosome	3,886,916	3,776	114	28	AM412317	Complete
	Plasmid pBOT3502	16,344	19	None	26	AM412318	Complete
Type A1 ATCC 19397	Chromosome	3,863,450	3,692	98	28	CP000726	Complete
Type A1 Hall	Chromosome	3,760,560	3,569	106	28	CP000727	Complete
Type A2 Kyoto	Chromosome	4,155,278	4,035	105	28	CP001581	Complete
Type A3 Loch Maree	Chromosome	3,992,906	3,776	108	28	CP000962	Complete
	Plasmid pCLK	266,785	329	None	25	CP000963	Complete
Type B Eklund 17B	Chromosome	3,800,327	3,586	116	27	CP001056	Complete
	Plasmid pCLL	47,642	54	None	24	CP001057	Complete
Type B1 Okra B	Chromosome	3,958,233	3,780	109	28	CP000939	Complete
	Plasmid pCLD	148,780	195	None	25	CP000940	Complete
Type Ba4 657	Chromosome	3,977,794	3,940	115	28	CP001083	Complete
	Plasmid pCLJ	270,022	310	1	25	CP001081	Complete
	Plasmid pCLJ2	9,953	6	None	24	CP001082	Complete
Type Bf	Chromosome	4,217,754	4,217	127	28	ABDP00000000	Draft assembly
Type C Eklund	Chromosome	2,961,186	2,954	126	28	ABDQ00000000	Draft assembly
Type D 1873	Chromosome	2,379,404	2,330	119	27	ACSJ00000000	Draft assembly
	Plasmid pCLG1	107,690	128	None	26	CP001659	Complete
	Plasmid pCLG2	54,152	48	None	25	CP001660	Complete
Type E1 Beluga	Chromosome	3,999,201	3,830	116	27	ACSC00000000	Draft assembly
Type E3 Alaska E43	Chromosome	3,659,644	3,381	117	27	CP001078	Complete
Type F Langeland	Chromosome	3,995,387	3,832	148	28	CP000728	Complete
	Plasmid pCLI	17,531	24	None	26	CP000729	Complete
Type G	Chromosome						In progress
	Plasmid						

Table 4.1 (continued)

Species and strain	Form	Length (nt)	ORFs	RNAs	GC content (%)	Accession number	Status
Type B(A) NCTC 2916	Chromosome	4,031,357	3,877	129	28	ABDO00000000	Draft assembly
Type E Iwanii							In progress
Type C2 203U28	Plasmid pC2C203U28	106,981	122	None	26	AP010934	Complete
Clostridium difficile							
Strain 630	Chromosome	4,290,252	3,971	185	29	AM180355	Complete
	Plasmid pCD630	7,881	11	None	27	AM180356	Complete
Strain ATCC 43255	Chromosome	4,204,780	3,959	54	28	ABKJ00000000	Draft assembly
Strain CD196							In progress
Strain CIP 107932	Chromosome	4,032,580	3,686	43	28	ABKK00000000	Draft assembly
Strain NAP07							In progress
Strain NAP08							In progress
Strain QCD-23m63	Chromosome	3,396,085	3,611	29	28	ABKL00000000	Draft assembly
Strain QCD-32g58	Chromosome	4,108,089	4,071	NA	28	AAML00000000	Draft assembly
Strain QCD-37x79	Chromosome	4,329,888	4,042	40	28	ABHG00000000	Draft assembly
Strain QCD-63q42	Chromosome	4,440,437	4,243	55	28	ABHD00000000	Draft assembly
Strain QCD-66c26	Chromosome	4,126,050	3,769	46	28	ABFD00000000	Draft assembly
Strain QCD-76w55	Chromosome	4,392,595	4,094	38	28	ABHE00000000	Draft assembly
Strain QCD-97b34	Chromosome	4,059,010	3,748	37	28	ABHF00000000	Draft assembly
Strain R20291							In progress
Strain gs							In progress
Strain CD (R8375)	Plasmid pCD6	6,830	5	None	24	AY350745	Complete
Clostridium novyi NT	Chromosome	2,547,720	2,427	112	28	CP000382	Complete
Clostridium perfringens							
Strain ATCC 13124	Chromosome	3,256,683	3,017	118	28	CP000246	Complete
Strain SM101	Chromosome	2,897,393	2,701	124	28	CP000312	Complete
	Plasmid 1	12,397	10	None	26	CP000313	Complete
	Plasmid 2	12,206	11	None	25	CP000314	Complete

Table 4.1 (continued)

Species and strain	Form	Length (nt)	ORFs	RNAs	GC content (%)	Accession number	Status
Strain 13	Chromosome	3,031,430	2,786	126	28	BA000016	Complete
	Plasmid pCP13	54,310	63	None	25	AP003515	Complete
B strain ATCC 3626	Chromosome	3,896,305	3,710	120	28	ABDV00000000	Draft assembly
C strain JGS1495	Chromosome	3,661,329	3,447	114	28	ABDU00000000	Draft assembly
CPE strain F4969	Chromosome	3,510,272	3,290	139	28	ABDX00000000	Draft assembly
	Plasmid	70,480	62	None	26	AB236336	Complete
D strain JGS1721	Chromosome	4,045,016	3,904	125	28	ABOO00000000	Draft assembly
E strain JGS1987	Chromosome	4,127,102	4,034	138	28	ABDW00000000	Draft assembly
Strain NCTC 8239	Chromosome	3,324,319	3,068	132	28	ABDY00000000	Draft assembly
Strain CW2	Plasmid pCW3	47,263	51	None	27	DQ366035	Complete
Strain F5603	Plasmid pBCNF5603	36,695	36	None	25	AB189671	Complete
Strain NCTC 8533	Plasmid pCP8533ext	64,753	63	None	25	AB444205	Complete
Clostridium tetani E88	Chromosome	2,799,251	2,373	72	28	AE15927	Complete
	Plasmid pE88	74,082	61	None	24	AF528097	Complete
Clostridium thermocellum							
Strain ATCC 27405	Chromosome	3,843,301	3,305	71	38	CP000568	Complete
Strain DSM 2360	Chromosome	3,454,608	3,146	55	39	ACVX00000000	Draft assembly
Strain DSM 4150	Chromosome	3,321,980	3,021	42	38	ABVG00000000	Draft assembly

[a]Source: Bradshaw and Johnson (2010).

Beyond efforts to sequence the genome for representatives of the genus, there has also been work done to sequence various plasmids that differ in size from 2.4 to 270 kb (Bradshaw and Johnson, 2010). Recently, conjugative plasmids have been demonstrated in *C. botulinum* (Marshall et al., 2010).

Using completed genome sequences, microarrays have been developed for analyzing genotypic diversity, gene expression, and gene function in *Clostridium*. These arrays have primarily been used for both genome and transcriptome analyses. An excellent example is the detailed analysis of the transcriptome of the life cycle of *C. acetobutylicum* (Jones et al., 2008; Papoutsakis, 2008). These studies described major cellular regulatory systems; sigma and sporulation factors, including activity assays for major sporulation factors based on canonical sets of genes from their regulons; assessment of expression intensities; and identification of putative histidine kinases that may phosphorylate SpoOA under different conditions (Jones et al., 2008). The broad range of subject matter of these studies demonstrates the power of transcriptome arrays as it allows for in-depth characterization of the species and its genes by providing high-resolution analyses on a global scale in a time-efficient manner.

Specialized arrays have largely been used for studies of specific properties of *Clostridium* spp., such as for the genotyping of *C. perfringens* toxins (Al-Khaldi et al., 2004) and for screening gram-positive bacteria for multiple antibiotic resistance genes at one time (Perreten et al., 2005). The completion of genome sequences from clostridia, including multiple genomes from the same species, will enable detailed studies of the evolution of the genus *Clostridium* and will facilitate functional studies of genes at the genome level either through computer modeling or through functional array analysis.

4.1.5 The Proteome

The proteome of limited *Clostridium* species has been analyzed using the standard procedure of 2D electrophoresis gels. Other approaches are sometimes used including mass spectrometry and computerized protein modeling. However, these methods are only now becoming current for studying the proteome. Mass spectrometry analysis requires detailed information on individual proteins to be effective, and this was unavailable for clostridia before the genomes were sequenced.

To better understand how all of these approaches can be used to study the genus *Clostridium*, individual species and the current status of the field for analyzing their genetic framework on genomic, transcriptomic, and proteomic levels are assessed. The pathogenic species *C. botulinum*, *C. difficile*, *C. perfringens*, and *Clostridium tetani* and the beneficial species *C. acetobutylicum*, *Clostridium novyi*, and *Clostridium thermocellum* are described due to their pathogenic and industrial importance and availability of genomic and physiological information.

4.1.6 Clostridium botulinum

The identification of *C. botulinum* as a species is based on its unique property to produce botulinum neurotoxin (BoNT), which is the most poisonous protein toxin known. There exist seven serotypes of *C. botulinum* (A–G) based on the immunologic properties of its BoNT produced. This is a heterogeneous and non-phyletic-based species since beginning in the late 1800s any bacterium shown to possess BoNT was originally termed *C. botulinum*, regardless of its genetic and physiological properties. This trend continued until strains of *C. butyricum* and *Clostridium baratii* were also shown to possess and express the toxin gene (Aureli et al., 1986; Suen et al., 1988). As these species were already well established and studied, they could not be renamed to reflect their toxin status and were considered to be outliers. Furthermore, *C. botulinum* type G was shown to be taxonomically identical to *Clostridium argentinense* and is described as the latter species in the current literature (Hatheway and Johnson, 1998).

There are four overall groups, Groups I–IV, which consist of Group I: proteolytic strains of serotypes A, B, and F; Group II: nonproteolytic strains of serotypes B, F, and all type E strains; Group III: animal pathogens with the BoNT/C and/D genes located on pseudolysogenic bacteriophages; and Group IV: type G, *C. argentinense*. Groups I–IV have diverse physiological properties including temperature and pH ranges of growth, spore resistance properties, and other aspects that have been thoroughly reviewed (Franciosa et al., 2003; Hauschild, 1993; Johnson, 2007). They also differ in their pathogenicity including host range and type of botulism caused (Goodnough and Johnson, 1996; Smith and Sugiyama, 1988). Recently, subtypes of BoNTs within serotypes have been identified (Arndt et al., 2006; Hill et al., 2007), which have different structural and functional properties. BoNT is extraordinarily dangerous to humans and animals because it blocks cholinergic transmission in the peripheral nervous system. As such, transmission at ganglionic synapses, postganglionic parasympathetic synapses, and neuromuscular junctions is impaired. The action of the toxin has been most closely studied at the neuromuscular junction, and data suggest that neurogenic release of acetylcholine is blocked (Simpson, 1981).

4.1.6.1 Use of Genomics in the Study of *Clostridium botulinum*

For *C. botulinum*, the study of the genomics of the bacteria is in its infancy and currently burgeoning. For nearly a century, the primary focus of studies on *C. botulinum* was on the neurotoxin and its associated cluster proteins, with little regard for the host bacteria. While some studies have been performed to evaluate genetic approaches to identify the diversity of the species, such as sequencing and alignment of 16S rRNA sequences (Hutson et al., 1993a, b), this was limited to evaluating relationships and did not address the global significance of the species in a manner of understanding the biological genome and pangenome.

This approach changed radically with the sequencing of the *C. botulinum* genome of strain ATCC 3502 (type A) (Sebaihia et al., 2007). The strain arose from the work of Professor Ivan C. Hall from the University of Colorado School of Medicine and Hospitals. Throughout the 1920s and 1930s, Professor Hall had been instrumental in the identification and characterization of many strains of *C. botulinum*, primarily proteolytic strains that were known to affect humans. During one of his studies involving a case of canned peas with *C. botulinum* growth, a strain was isolated and deposited with the American Type Culture Collection and given the name ATCC 3502. The strain produces BoNT/A1 and is a proteolytic strain. It is one of many strains designated "Hall A" that differ by strain analyses such as PFGE (Johnson et al., 2005). The genome possessed a chromosome of 3,886,916 bp and a plasmid that is 16,344 bp with 3,650 and 19 predicted genes from coding sequences (CDSs), respectively (Sebaihia et al., 2007).

The plasmid encoded for a bacteriocin similar to that found in *C. botulinum* type B (Dineen et al., 2000). Study of the genome properties suggested the absence of recently acquired DNA which differs from other sequenced genomes of certain other *Clostridium* spp. such as *C. difficile*, which appears to be highly fluid with its high content of transposable elements (Sebaihia et al., 2006). Additionally, there was a minimal amount of shared genes in *C. botulinum* ATCC 3502 with other clostridial genomes with only 568 of the CDSs identified being shared with other genomic sequences, with 1,571 CDSs out of 3,650 being unique to *C. botulinum* (Sebaihia et al., 2007). This study established a foundation for ensuing genomic studies of the highly diverse *C. botulinum* group.

Using the information attained in the genome study, a microarray was devised that could be applied to the study of genomics of proteolytic strains. It was shown that there was a core genome among the proteolytic strains tested, although it was limited to 19% of the total genome (Carter et al., 2009). This demonstrates the significant amount of diversity observed within the species, even within just Group I. Lindstrom et al. (2009) evaluated genome diversity in a limited number (32) of proteolytic type B strains by comparative genomic hybridization (Lindstrom et al., 2009). The strains separated into two distinct clusters, and these could be compared to the reference strain *C. botulinum* ATCC 3502 to identify novel genes present in each cluster. The clusters were shown to be homogenous among their own strains and to share 88% homology with ATCC 3502 (Lindstrom et al., 2009). The two clusters differed from each other by 145 CDSs of the total 6,350. The most obvious of these differences lied in genes related to resistance to the toxic compounds arsenic and cadmium. Additionally, variation was noted between the two clusters in genes involved in transport and binding, adaptive mechanisms, fatty acid biosynthesis, cell membrane formation, and the presence of bacteriophages and transposon-related elements (Lindstrom et al., 2009). Each set of strains was shown to produce BoNT/B2.

A further study on the transcriptome as affected by physiological conditions was performed (Artin et al., 2010). This study focused on effects of carbon dioxide on growth and toxin production of *C. botulinum* proteolytic strains (Artin et al., 2010). While this study was conceptually similar to a previous study by the same laboratory on the effect of carbon dioxide on neurotoxin expression in BoNT/E-producing strains (Artin et al., 2008), it differed by including the analysis of the whole-cell transcriptome using DNA microarrays produced from the genomic sequence of ATCC 3502 (Artin et al., 2010). Using these arrays in conjunction with real-time quantitative PCR, the researchers showed that there were 13 genes with expression profiles closely matching the expression profiles of the neurotoxin under these conditions (Artin et al., 2010). The study also highlighted the dearth of knowledge of the relation of the transcriptome to the physiology of clostridia. In fact, some of the CDSs identified are still considered putative genes and have not been tested for function. Additional work must therefore be performed to create a comprehensive transcriptome for the clostridia.

Compared to the transcriptome, the proteome of *C. botulinum* is an unknown abyss as BoNT has been the primary focus of the field to the detriment of the study of the other proteins produced. Without further studies, *C. botulinum* will continue to be undefined on the protein level and will be limited in its understanding of other clostridial species.

4.1.7 Clostridium difficile

C. difficile is a highly virulent pathogen, which has increased in significance during recent years due to the emergence of hypervirulent strains (designated NAP1/BI/027) that have caused numerous fatalities. *C. difficile* was given its name because of the difficulty in isolating and culturing the bacteria given its strict anaerobe nature and fastidious nutrient requirements. Despite these stringent growth requirements, it is able to produce endospores that can persist even in anaerobic environment until conditions change to favor growth. The survival of spores in hospital and assisted care facilities has contributed to its rise in prominence as a human pathogen and *C. difficile* infection (CDI) has steadily increased. The usual infection process involves colonization of the human colon upon compromise of the immune system, usually as a result of an unrelated bacterial infection and its treatment with antibiotics. Infections also occur in other circumstances including patients receiving antacid/proton pump inhibitors and nonsteroidal anti-inflammatory drugs (Dawson et al., 2009; Stabler et al., 2009). The patient's antimicrobial or anti-inflammatory drugs disrupt the bacterial composition and balance in the gut microbiome (Emerson et al., 2008). Since there are fewer competitors, *C. difficile* opportunistically germinates from the spore state and colonizes the colon with ensuing production of its large toxins, TcdA and TcdB, as well as a smaller ADP-ribosylating toxin. TcdA and TcdB perforate the intestinal wall, leading to the symptoms of CDI. Therefore, the known components contributing to virulence and CDI are the ability to colonize, adherence, and hyperproduction of toxins. While it is still possible to treat the infection using antibiotics, the available choices are becoming limited. This is because the genomic profiles of the epidemic strains, NAP1/BI/027, have a high degree of resistance to many antibiotics except for vancomycin and metronidazole, which limits the options for treatment (Dawson et al., 2009; Warny et al., 2005). Furthermore, infections are well recognized to relapse as some *C. difficile* survive treatment and recolonize the gut to renew the infection process.

4.1.7.1 Use of Genomics for Study of *Clostridium difficile*

There has been significant progress in applying genomic, transcriptomic, and proteomic tools to understand the diversity and pathogenesis of *C. difficile*. This has included obtaining a genome sequence, utilizing the genome for evaluation of the expression of certain genes, creating microarrays for analysis of expression patterns and phylogeny, and analyzing the network of proteins required for the bacteria to adhere to the host using proteomic tools.

The genetic analysis of *C. difficile* stems from the elucidation of the genome sequence of strain 630 (Sebaihia et al., 2006), which is a virulent and multidrug-resistant strain, but not an epidemic strain. The genome is 4,290,252 bp in length for the chromosome, and it contains a 7,881-bp plasmid. The chromosome contained 3,776 predicted coding sequences (Sebaihia et al., 2006). A particularly interesting aspect of the genome was the high content of mobile genetic elements as 11% of the total genome consisted of these elements, mainly conjugative transposons (Sebaihia et al., 2006). This high level of mobile elements creates a mosaic genome that was postulated to arise from many years of development and genetic recombination or due to changes that promoted the introduction of additional antibiotic resistance genes.

Given the high degree of diversity within the species and the shifting nature of the *C. difficile* genome, further studies of the bacteria and its phylogeny were needed. The completed genome sequence was used to develop DNA microarrays, which were used to study 75 diverse isolates including hypervirulent, toxin-variable, and animal strains (Stabler et al., 2006). This analysis identified four statistically distinct clusters. Additionally, the study identified that the core genome among all of the strains was minimal, as only 19.7% of the genes were shared (Stabler et al., 2006).

This study was followed by another study applying comparative genomic hybridization to the study of *C. difficile*. Using the established 630 genome, the core genome for the species had only 16% of the genes shared among the 73 strains tested (Janvilisri et al., 2009). These strains included the hypervirulent strain QCD-32g58 which is a representative of NAP1/BI/027.

An additional study was performed to assess the evolutionary characteristics of the species as a whole and the hypervirulent strains as a group (He et al., 2010). It was concluded that many of the genomic properties of the species resulted from horizontal gene transfer (HGT) and large-scale recombination to produce the core genes shared among all of the members during the past 1.1–85 million years (He et al., 2010). The hypervirulent strain, of which members appear to be highly homologous, has arisen from multiple lineages implying that its hypervirulence evolved independently, since it had a limited number of SNP differences among 25 isolates of this group and 90.4% of these SNPs were in only two of the isolates tested (He et al., 2010).

The genome sequences for some NAP1/BI/027 strains (both epidemic and nonepidemic) were completely sequenced to allow for comparison to the reference strain to identify CDSs specific to the hypervirulent strain (Stabler et al., 2009). While the two genomes are highly similar, 234 genes were identified to be NAP1/BI/027 specific. Additionally, epidemic forms of the group had five unique genetic regions that were not found in the other groups. These regions include a novel phage island, a two-component regulatory system, and transcriptional regulators (Stabler et al., 2009).

Using the established genome of strain 630, it was also possible to analyze the transcriptome, specifically the important step of colonization including factors that may affect this process including environmental stimuli and antibiotic exposure (Emerson et al., 2008). These studies identified that hypervirulent *C. difficile* is capable of responding to environmental stimuli such as aerobic exposure by expression of reactive oxygen species (ROS) defense proteins to promote survival and growth in a mildly aerobic environment. This included upregulation of 7.6% of the annotated genes involved in electron transport (Emerson et al., 2008). These findings suggest that different cascades are used to respond to various stimuli and may be instrumental in pathogenesis. These genes included enzymes involved in cell wall biosynthesis and cell surface modification and structural proteins, which were upregulated in the presence of amoxicillin, ampicillin, clindamycin, and metronidazole. Additionally, antibiotics affected expression of components of the protein secretion apparatus. Flagellar proteins were also differently regulated in the presence of antibiotics, but only when grown in the presence of amoxicillin and clindamycin, while metronidazole did not elicit a similar response (Emerson et al., 2008).

Because of their role in the development of CDI, spores are an important area of study, especially for analysis of the proteome compared to the normal growth state (Lawley et al., 2009). Spore proteins from *C. difficile* were isolated using two solublization steps, followed by SDS-PAGE and LC-MS-MS analyses. The analysis identified 336 proteins that comprise the *C. difficile* spore proteome for the reference 630 strain (Lawley et al., 2009). When compared against the genome sequence, it was found that the proteins were dispersed throughout the genome with only one set clustering near the origin of replication.

Other features of the proteome were that 25% of all the detected polypeptides had translation functions. Additionally, 15 of the 16 chaperone proteins in the genome were found to be present in the spore proteome, implying that proper protein folding is important for spore function. Also, stress response proteins were found, including those giving resistance against ruberythrins, tellurium, and ROS (Lawley et al., 2009).

Only a small number of the total proteins in the proteome, 88 out of 336, were conserved among all of the sequenced clostridial genomes (Lawley et al., 2009). Of the other proteins, 29 (9% of the proteome) are unique to *C. difficile*. Additional studies will be required to identify their functional importance.

Another area of importance in the proteome is the proteinaceous paracrystalline layer, or the S-layer of the cell, which enables cells to adhere to surfaces and colonize an environment. The S-layer primarily of *C. difficile* consists of a high-molecular-weight S-layer protein and its low-molecular-weight protein partner (Fagan et al., 2009; Qazi et al., 2009; Wright et al., 2005). The S-layer proteome also contains an additional 47 different proteins, as shown by two-dimensional electrophoresis and mass spectrometry (Wright et al., 2005). The high-molecular-weight S-layer and its low-molecular-weight protein partner form a noncovalent complex and have been shown to have specific locations by X-ray crystallography. This analysis identified regions of conservation in the low-molecular-weight protein that were likely necessary for structural function, as other regions exhibited high degrees of variation, and could be used to avoid immune neutralization (Fagan et al., 2009).

C. difficile is one of the most well-studied *Clostridium* species given its ability to produce potent toxins and its importance in human disease. As such, it is currently the gold standard for comparative genomic analyses of other *Clostridium* spp. Notwithstanding, future work is needed to elucidate mechanisms of antibiotic resistance in order to develop new treatments. To properly address this issue, it will be necessary to better understand the transcriptome and proteome of *C. difficile* at the different stages of infection.

4.1.8 Clostridium perfringens

C. perfringens is a common inhabitant of the gastrointestinal tract of humans and animals and is also widely found free living in the environment. Smith (1975) concluded based on review of environmental analyses that it is the most prevalent *Clostridium* spp. on earth (Smith, 1975; Songer, 2010). *C. perfringens* is the cause of several human and animal diseases (Cooper and Songer, 2009; Rood, 1998; Songer, 2010). Among these are clostridial myonecrosis (gas gangrene) and food poisoning and a rare form of enteritis. The latter disease symptoms include vomiting, diarrhea, severe abdominal pain, and the presence of blood in the stool. The toxin associated with this form of the disease is β-toxin which can be inactivated by trypsin from the gut (Rood, 1998). The disease occurs mainly in people who change from low-protein diets to high-protein diets in which their trypsin levels are low due to the initial diets. As a result, the toxin is left in its active form and is able to cause the disease. *C. perfringens* also causes a large array of animal diseases through production of a multitude of protein toxins (Cooper and Songer, 2009; Smith, 1975).

4.1.8.1 Use of Genomics in the Study of *Clostridium perfringens*

C. perfringens is the best characterized *Clostridium* spp. from a genetic perspective. It was the first species within the genus *Clostridium* to be analyzed by genetic manipulation and functional genomic methods. However, its study has become specialized for certain groups, due to the high degree of diversity in the species and that only certain strains are amenable to genetic analyses.

C. perfringens was the first gram-positive bacterium whose genome was genetically mapped (Canard and Cole, 1989). This study showed that the genome consisted of a circular 3.6-Mb chromosome using pulsed-field gel electrophoresis (PFGE), but was unable to evaluate the genome with anything but the most basic of resolution. The study did show that several genes encoding extracellular toxins and enzymes were clustered within a 250-kb region near *oriC* (Canard and Cole, 1989).

The genome was fully sequenced in 2002 using a representative strain of the group causing gas gangrene (Shimizu et al., 2002b). The sequence results showed a 3,031,430-bp circular chromosome comprised of 2,660 ORFs. This sequencing project also showed that many enzymes for amino acid biosynthesis, the enzymes of the tricarboxylic acid cycle, and enzymes common in the respiratory chain of aerobes were absent. This distinguishes *C. perfringens* from other clostridia whose genome sequences are available, such as *C. acetobutylicum* which has an incomplete set of TCA cycle enzymes (Shimizu et al., 2002b). Additionally, 61 genes related to sporulation and germination were identified based on previous studies of the *B. subtilis* genome. Eighty additional genes are found in *B. subtilis* that perform a role in these processes that were not present in *C. perfringens*. These include many genes for spore coat proteins, some germination-related proteins, along with the genes for the major phosphorelay system of *B. subtilis* (Anonymous, 1993). The absence of these genes is common among other clostridia including *C. botulinum*, *C. difficile*, *C. acetobutylicum*, and *C. thermocellum* (Shimizu et al., 2002b). The mechanism by which clostridia sense environmental signals and translate them to complete sporulation is unknown at this time.

This initial study of *C. perfringens* genomics was succeeded by the genomic sequencing of the type strains ATCC 13124 and SM101, an enterotoxin-producing food poisoning strain (Myers et al., 2006). Comparison of the three genomes identified more than 300 unique "genomic islands" (Myers et al., 2006). The genes in these islands were shown to correlate to differences in virulence and phenotypic characteristics. Furthermore, SM101 had 69 IS elements, which is significantly higher than strains 13 and ATCC 13124, which had 7 and 3, respectively (Myers et al., 2006). The overall significance of these elements has yet to be determined, but they probably contribute to horizontal gene transfer and recombination. Until additional analyses are performed, it is not known if additional food poisoning strains contain similar numbers of IS elements.

Transcriptional and proteome analysis of *C. perfringens* has focused on the two-component regulatory system VirR/VirS. The system consists of a membrane sensor which is a histidine kinase and a cytoplasmic response regulator. Shimizu et al. (2002a) studied the two-component VirR/VirS system two-dimensional gel electrophoresis (Shimizu et al., 2002a). Fifteen proteins were identified in the wild-type strain to a strain deficient in VirR (Shimizu et al., 2002a).

Although genomic studies are relatively advanced in select strains of *C. perfringens*, no comprehensive studies of the transcriptome or the proteome of the species have been performed. A primary focus in the study of *C. perfringens* has been genotyping and strain differentiation, which is rational considering the diversity of the strains and toxins that are produced and the need for rapid diagnosis of the variety of diseases.

4.1.9 Clostridium tetani

Similar to other clostridia, *C. tetani* lives primarily in soil, dust, and the intestinal tracts of animals (Bruggemann et al., 2003). *C. tetani* is well known as the cause of tetanus in animals and humans through action of tetanus neurotoxin (TeNT). Tetanus is one of the longest known diseases in human history, having been reported for over 24 centuries (Bruggemann et al., 2003). The disease is identified by rigid or spastic paralysis, which is essentially the opposite of the flaccid paralysis characteristic of botulism, although both diseases are caused by related neurotoxins. Despite the efforts made to immunize people against the disease, it is still prevalent throughout the world. Although contained in industrialized countries, it is still considered endemic in many developing countries, with an estimated 248,000 deaths in 1997 by neonatal tetanus infections (Alam et al., 2008; Bruggemann et al., 2003).

4.1.9.1 Use of Genomics for Study of *Clostridium tetani*

The "Harvard" (E88) strain of *C. tetani* used for TeNT toxin production was sequenced (Brüggemean et al., 2003). The strain is nonsporulating and produces high quantities of TeNT, which is analogous to certain nonsporulating strains of *C. botulinum*. It was demonstrated to have a 2,799,250-bp chromosome encoding 2,372 ORFs and a 74,082-bp plasmid containing 61 ORFs including that encoding for TeTx (Bruggemann et al., 2003). The genome was identified as having few transposable elements with only 16 estimated transposase genes being present, most of these inactive due to mutations. Additionally, it was shown that the genome has a significant number of similarities compared to the genome of *C. perfringens*. This includes not only the limited number of transposable elements but also the high degree of strand bias, whereby 82% of the predicted ORFs were in the same direction as DNA replication, analogous to *C. perfringens* in which 83% of its ORFs have the same orientation (Bruggemann et al., 2003). The two genomes were also similar in possessing homologous forms of tetanolysin O, hemolysin, and fibronectin-binding proteins. The two genomes also differed substantially as *C. tetani* have a more extensive array of surface layer proteins than *C. perfringens* (Bruggemann et al., 2003).

When compared to both *C. perfringens* and *C. acetobutylicum* genomes, 1,506 genes were identified to be present in the three genomes, suggesting a common clostridial genome backbone (Bruggemann and Gottschalk, 2004). The analysis indicated that 516 distinct genes were present in *C. tetani* compared to the other two species, and 199 common genes were present in *C. tetani* and *C. perfringens* but not in *C. acetobutylicum*.

Little work has been performed for analysis of the transcriptome of *C. tetani*. Certain indirect work has been performed that provides some insight into the proteome of *C. tetani* (Alam et al., 2008). Antibodies produced by intramuscular injection in mice against heat-killed *C. perfringens* also reacted with several proteins of *C. tetani* (Alam et al., 2008). The antibodies against *C. perfringens* conferred some protection to mice when challenged intramuscularly with *C. tetani*, but did not protect against intravenous TeTx intoxication.

In many ways, the genomic study performed on *C. tetani* mirrors that performed on *C. botulinum*. Each species has a neurotoxin that has defined the field for many years, while there is increased interest in understanding the role of the bacterial hosts in pathogenesis and physiological properties. As such, many of the same issues discussed for *C. botulinum* are applicable to *C. tetani*. The first is that a comprehensive analysis of the transcriptome is needed to create more efficient countermeasures against infections, particularly in neonatal tetanus cases as they are a cause of significant morbidity and mortality in developing countries. Additionally, proteome analysis of *C. tetani* needs further work to better understand its evolution, physiology, and their pathogenesis.

4.1.10 Clostridium acetobutylicum

C. acetobutylicum is a beneficial clostridial species that was used during the first half of 1900s to produce acetone, butanol, and ethanol (Nolling et al., 2001). The fermentative production of solvents was replaced by petrochemicals. However, public interest and government support have increased for the production of "natural" compounds with the "green" movement and needs for eventual supplementation of petroleum-based fuels. Specifically, ethanol and butanol have attracted much interest and government support as alternatives to petroleum-based materials for manufacturing of industrial chemicals and as a fuel supply. In addition to these proposed industrial utilities, *C. acetobutylicum* has served as a model organism for clostridia including pathogenic species as its genetic and genomic systems are relatively advanced compared to most species.

4.1.10.1 Use of Genomics for Study of *Clostridium acetobutylicum*

The genome for *C. acetobutylicum* ATCC 824 was sequenced in 2001 and found to have a 3,940,880-bp chromosome with a 192,000-bp megaplasmid (Nolling et al., 2001). The megaplasmid was shown to contain most of the genes encoding enzymes for production of solvents by *C. acetobutylicum*. The chromosome had 3,740 ORFs and 107 stable RNA genes covering 88% of the genome (Nolling et al., 2001). Additionally, the chromosome had 11 ribosomal operons that were clustered closely to *oriC*. Furthermore, the level of sequence similarity between *C. acetobutylicum* and *B. subtilis* was greater than *C. acetobutylicum* and any other bacterial genome sequenced at the time, including about 200 genes in operons encoding central cellular functions such as translation and transcription (Nolling et al., 2001).

This information was utilized by in silico analysis procedures to create a metabolic network to evaluate the solventogenic processes, and attempts were made to increase solvent production (Lee et al., 2008). The model approach was used since attempts to use classic strain development tools including chemical mutagenesis and the introduction of genes involved in solventogenesis into heterologous hosts such as *E. coli* were unsuccessful. The model comprised 502 reactions and 479 metabolites (Lee et al., 2008). Despite this extensive effort, the model was still limited in the enhancement of solvent production in *C. acetobutylicum*. One reason was that the annotation of *C. acetobutylicum* and the understanding of gene regulation and gene function were considerably less complete than other organisms, such as *B. subtilis*. A second model was also established (Senger and Papoutsakis, 2008a, b). The two networks differed as one was created through a semi-automated approach while the other was more manually curated (Lee et al., 2008). Further studies are needed to establish their validity for increase of solvent and acid production.

Initial analysis of the transcriptome in *C. acetobutylicum* utilized a microarray covering only 25% of the genome and was used to study various gene expression in various strains and growth conditions (Alsaker et al., 2004; Tomas et al., 2003a; b, 2004; Tummala et al., 2003). The sequencing of the genome enabled the development of a microarray that covered the entire genome (Alsaker et al., 2005). Given the high degree of A+T content for *C. acetobutylicum*, special procedures had to be developed to optimize the microarray signals. These procedures included minimization of the nonspecific binding of probes with spotted material and optimization signal intensity. The modified array showed with 95% confidence that 40–56% of the genome was expressed at any given time (Alsaker et al., 2005). Future applications for these arrays have yet to be published but should be valuable for establishing methods to improve expression of the solvent production pathways.

To further study the relationship between solvent production and sporulation, proteomic procedures have also been employed (Sullivan and Bennett, 2006). These were performed using two-dimensional gel electrophoresis and mass spectrometry using the wild-type ATCC 824 strain with and without a control plasmid, a Spo0A null strain, and a strain having a Spo0A overexpression plasmid. By this approach, 2,081 protein spots were analyzed, 23 selected for identification and 18 shown to be unique (Sullivan and Bennett, 2006). The 18 proteins consisted of proteins in the heat shock stress response, acid and solvent formation, and transcription and translation. When the *C. acetobutylicum* protein profile was compared between the acidogenic and the solventogenic phases, it was noted that stress proteins, certain solvent phase-inducible genes such as *adc*, and other genes were upregulated. The stress proteins were among the most abundant expressed and included GroEL, GroES, heat shock protein 18, and phage shock protein. The upregulation of Adc is reasonable as it is essential for solvent production and catalyzes the decarboxylation of acetoacetate to acetone (Sullivan and Bennett, 2006). It was also noted that overexpression of spo0A altered the expression of some proteins including the glycolytic proteins Pgi and Tpi, which decreased in concentration during growth, and the ribosomal protein RplL which increased in concentration (Sullivan and Bennett, 2006). It was concluded that overexpression of spo0A induces endospore and creates an intracellular environment that is conducive to solvent formation.

When biofuels become more feasible as an alternative to petrochemicals, the use of improved strains of *C. acetobutylicum* should increase. As such, in-depth analyses will be useful to enhance valuable traits such as alcohol resistance and increased expression of solventogenic genes. Strain development will need to be integrated with other facets of these processes, particularly engineering optimization, to boost contribution to global energy demand.

4.1.11 Clostridium novyi

C. novyi is the closest genetic relative to *C. botulinum* Group III. It differs from Group III as it does not possess a BoNT gene on a bacteriophage, but it does have pathogenic properties causing wound infections (Smith, 1975). Interestingly, progress has been made to eliminate the pathogenic properties so that the strain can be used for other purposes including the treatment of cancer (Bettegowda et al., 2006). It has been known that clostridial spores when administered within tissue show a tropism to tumors, where anaerobiosis is enhanced. Recent mouse studies have shown that upon intravenous injection, *C. novyi* spores grow as vegetative cells within tumors. Clostridial spores are also being conjugated to chemical anticancer compounds for delivery.

4.1.11.1 Use of Genomics in the Study of *Clostridium novyi*

The genome of *C. novyi* was sequenced and was shown to comprise a single circular chromosome of 2,547,720 bp without any extrachromosomal components (Bettegowda et al., 2006). Analysis of the genome predicted 2,325 CDSs, with 1,620 (70%) of these assigned putative functions; 566 of the remaining CDSs had similarity to other genomes, while 139 had no homology (Bettegowda et al., 2006). While many genes were similar to those from other *Clostridium* genomes, there were significant differences as *C. novyi* was enriched in genes involved in the cell envelope and transport and binding proteins. Additionally, only 27 copies of IS elements were found on the chromosome. These IS elements were similar to those found in other bacterial species, but not to those found in other clostridia (Bettegowda et al., 2006).

The sequences of the 2,325 CDSs were used to produce custom oligonucleotide arrays, which were utilized to assess the expression pattern of the cells emerging from the spore state. For many years, it had been assumed by some investigators that endospores of clostridia contained no RNA (Bettegowda et al., 2006). However, this study showed that they did possess RNA but that it was of different molecular size from RNA found in vegetative cells. Specifically, they showed that the rRNA present in spores was smaller than the normal 23S rRNA found in cells. Detailed evaluation showed that the 23S rRNA was 300 bp smaller because of a deletion in the 5′ region (Bettegowda et al., 2006). The mRNA profile was also different. In vegetative cells the most abundant transcripts are usually involved in protein synthesis, namely ribosomal proteins and translational factors. In spores, however, these transcripts were absent and genes with redox capabilities were most prevalent. When the nucleotide arrays were used to test gene expression during in vivo tumor growth, it was found that several extracellular proteins were expressed at high levels, most notably lipid-degrading proteins including phospholipase C and two lipases. The roles in pathogenesis of these differences in expression will need to be further investigated to more optimally use clostridia to combat tumor growth.

4.1.12 Clostridum thermocellum

C. thermocellum is a thermophilic strict anaerobe known for its potent cellulase activity. Among cellulolytic bacteria, it has the highest rate of cellulose utilization making it potentially useful for the production of biofuel from plant biomass (Bayer et al., 2008). The bacteria are able to hydrolyze crystalline cellulose using its large cell surface-bound protein complex called the cellulosome (Bayer et al., 2008).

The cellulosome consists of a central noncatalytic scaffolding protein (CipA) which binds to a type I dockerin domain. CipA then binds to the cell surface using a type II dockerin domain which is one of the three S-layer anchor proteins (Gold and Martin, 2007).

4.1.12.1 Use of Genomics for the Study of *Clostridium thermocellum*

The genome of *C. thermocellum* ATCC 27405 was sequenced in 2006 as a part of Department of Energy efforts at Oak Ridge National Laboratory. It was found to have a circular chromosome of 3,843,301 bp with a 38.9% G+C content. The genomic sequence was used to create a whole-genome oligonucleotide microarray that contained 96.7% of the 3,163 putative CDSs (Brown et al., 2007). The arrays were used to determine if cellular RNA transcripts isolated from early logarithmic growth during cellobiose fermentation could be detected (Brown et al., 2007).

Much effort has been applied to improve strains of *C. thermocellum* for increased cellulase activity and for alcohol resistance. Generally, ethanol concentrations above 1% (w/v) inhibit growth and fermentation of most bacteria. In comparison, certain yeast species, particularly *Saccharomyces cerevisiae*, have much higher alcohol tolerance and can withstand ethanol concentrations greater than 10% (Williams et al., 2007). An alcohol-resistant strain of *C. thermocellum* was adapted to tolerate up to 8% ethanol which is a considerable improvement for its use in biofuel production (Williams et al., 2007). It was postulated that this improvement was due to alteration in the plasma membrane to decrease disruption by ethanol (Williams et al., 2007).

A proteomic approach was used to determine gene expression in the resistant strain. Specifically, the plasma membrane was analyzed for its protein content using gel electrophoresis (Williams et al., 2007). The wild-type and the ethanol-adapted strains had 60% of their membrane proteins differentially expressed. In the ethanol-adapted strain, 73% of the proteins were downregulated (Williams et al., 2007). Using protein identification techniques, it was shown that several of the downregulated proteins were involved in carbohydrate transport and metabolism. The proteins upregulated in the ethanol-adapted strain were associated with chemotaxis and signal transduction (Williams et al., 2007). This interesting finding needs further study to understand the mechanism of ethanol tolerance.

Proteome analysis of the cellulosome has also been performed to assess its protein content (Gold and Martin, 2007). The method used included a combination of metabolic isotope-labeling approach and nano-liquid chromatography-electrospray ionization mass spectrometry peptide sequencing. Using this method, 41 proteins were detected including 36 type I dockerin-containing proteins (Gold and Martin, 2007). These proteins were then analyzed to determine if growth on different media produced resulted in different protein content in cellulosomes. This was based on previous studies which showed that cellulolytic activity was regulated by carbon source and growth rate. Differing levels of protein expression were observed (Gold and Martin, 2007). Specifically, growth on cellulose yielded an increased amount of the cell surface anchor protein OlpB, exoglucanases CelS and CelK, and the glycoside hydrolase family 9 endoglucanase CelJ. The decreased expression was observed for the glycoside hydrolase family 8 endoglucanase CelA, glyco-side hydrolase family 5 endoglucanases CelB, CelE, and CelG, and hemicellulases XynA, XynC, XynZ, and XghA (Gold and Martin, 2007).

A genome-scale model has been used for the study of global metabolism in *C. thermocellum* (Roberts et al., 2010). The iSR432 model consists of 577 reactions using 525 intracellular metabolites, 432 genes, and a proteomic-based represen-tation of the cellulosome. The model led to increased understanding of gene expression, particularly of 27 genes and identification of altered areas of metabolism (Roberts et al., 2010). The model appears to have broad applications in studying the global metabolism of *C. thermocellum* and could potentially be applied to other clostridial processes (Roberts et al., 2010).

4.2 Summary

Clostridium is a diverse genus that possesses species of extraordinary pathogenesis, as well as species that have considerable benefits for humans. Despite their medical and industrial importance, they have been understudied using advanced genomic and related "ohmic" approaches. There exists considerable potential to elucidate the mechanisms of pathogenesis and to develop industrial strains for beneficial processes using genomics-based methods.

References

Alam SI, Bansod S, Singh L (2008) Immunization against *Clostridium perfringens* cells elicits protection against *Clostridium tetani* in mouse model: identification of cross-reactive proteins using proteomic methodologies. BMC Microbiol 8:194

Al-Khaldi S, Myers K, Rasooly A, Chizhikov V (2004) Genotyping of *Clostridium perfringens* toxins using multiple oligonucleotide microarray hybridization. Mol Cell Probes 18:359–367

Alsaker KV, Paredes CJ, Papoutsakis ET (2005) Design, optimization and validation of genomic DNA microarrays for examining the *Clostridium acetobutylicum* transcriptome. Biotechnol Bioprocess Eng 10:432–443

Alsaker KV, Spitzer TR, Papoutsakis ET (2004) Transcriptional analysis of spo0A overexpression in *Clostridium acetobutylicum* and its effect on the cell's response to butanol stress. J Bacteriol 186:1959–1971

Anonymous (1993) In: Sonenshein AL, Hoch JA, Losick R (eds) *Bacillus subtilis* and other gram-positive bacteria biochemistry, physiology, and molecular genetics. American Society for Microbiology, Washington, DC

Anonymous (2009) Genus *Clostridium*. Bergey's Manual of Systematic Bacteriology: Volume 3: The *Firmicutes*, 2nd edn. Bergey's Manual Trust, pp 738–827

Arndt JW, Jacobson MJ, Abola EE, Forsyth CM, Tepp WH, Marks JD, Johnson EA, Stevens RC (2006) A structural perspective of the sequence variability within botulinum neurotoxin subtypes A1-A4. J Mol Biol 362:733–742

Artin I, Carter AT, Holst E, Lovenklev M, Mason DR, Peck MW, Radstrom P (2008) Effects of carbon dioxide on neurotoxin gene expression in nonproteolytic *Clostridium botulinum* type E. Appl Environ Microbiol 74:2391

Artin I, Mason DR, Pin C, Schelin J, Peck MW, Holst E, Radstrom P, Carter AT (2010) Effects of carbon dioxide on growth of proteolytic *Clostridium botulinum*, its ability to produce neurotoxin, and its transcriptome. Appl Environ Microbiol 76:1168

Aureli P, Fenicia L, Pasolini B, Gianfranceschi M, McCroskey LM, Hatheway CL (1986) Two cases of type E infant botulism caused by neurotoxigenic *Clostridium butyricum* in Italy. J Infect Dis 154:207–211

Barbe V, Cruveiller S, Kunst F, Lenoble P, Meurice G, Sekowska A, Vallenet D, Wang T, Moszer I, Medigue C (2009) From a consortium sequence to a unified sequence: the *Bacillus subtilis* 168 reference genome a decade later. Microbiology 155:1758

Bayer EA, Lamed R, White BA, Flint HJ (2008) From cellulosomes to cellulosomics. Chem Rec 8:364–377

Bettegowda C, Huang X, Lin J, Cheong I, Kohli M, Szabo SA, Zhang X, Diaz LA Jr, Velculescu VE et al (2006) The genome and transcriptomes of the anti-tumor agent *Clostridium novyi*-NT. Nat Biotechnol 24:1573–1580

Bradshaw M, Johnson EA (2010) Genetic manipulation of *Clostridium*. In: Baltz RH, Demain AL, Davies JE (eds) Manual of industrial microbiology and biotechnology, 3rd edn. ASM Press, Washington, DC, pp 238–261

Brown SD, Raman B, McKeown CK, Kale SP, He Z, Mielenz JR (2007) Construction and evaluation of a *Clostridium thermocellum* ATCC 27405 whole-genome oligonucleotide microarray. Appl Biochem Biotechnol 137–140:663–674

Bruggemann H, Baumer S, Fricke WF, Wiezer A, Liesegang H, Decker I, Herzberg C, Martinez-Arias R, Merkl R, Henne A, Gottschalk G (2003) The genome sequence of *Clostridium tetani*, the causative agent of tetanus disease. Proc Natl Acad Sci USA 100:1316–1321

Bruggemann H, Gottschalk G (2004) Insights in metabolism and toxin production from the complete genome sequence of *Clostridium tetani*. Anaerobe 10:53–68

Canard B, Cole S (1989) Genome organization of the anaerobic pathogen *Clostridium perfringens*. Proc Natl Acad Sci 86:6676

Carter AT, Paul CJ, Mason DR, Twine SM, Alston MJ, Logan SM, Austin JW, Peck MW (2009) Independent evolution of neurotoxin and flagellar genetic loci in proteolytic *Clostridium botulinum*. BMC Genomics 10:115

Cato E, George WL, Finegold S (1986) Genus Clostridium. Bergey's Manual of Systematic Bacteriology: Volume 2. Bergey's Manual Trust, pp 1141–1200

Collins M, Lawson P, Willems A, Cordoba J, Fernandez-Garayzabal J, Garcia P, Cai J, Hippe H, Farrow J (1994) The phylogeny of the genus *Clostridium*: proposal of five new genera and eleven new species combinations. Int J Syst Evol Microbiol 44:812

Cooper KK, Songer JG (2009) Necrotic enteritis in chickens: a paradigm of enteric infection by *Clostridium perfringens* type A. Anaerobe 15:55–60

Dawson LF, Valiente E, Wren BW (2009) *Clostridium difficile* – a continually evolving and problematic pathogen. Infect Genet Evol 9:1410–1417

Dineen SS, Bradshaw M, Johnson EA (2000) Cloning, nucleotide sequence, and expression of the gene encoding the bacteriocin boticin B from *Clostridium botulinum* strain 213B. Appl Environ Microbiol 66:5480–5483

Emerson JE, Stabler RA, Wren BW, Fairweather NF (2008) Microarray analysis of the transcriptional responses of *Clostridium difficile* to environmental and antibiotic stress. J Med Microbiol 57:757–764

Fagan RP, Albesa-Jove D, Qazi O, Svergun DI, Brown KA, Fairweather NF (2009) Structural insights into the molecular organization of the S-layer from *Clostridium difficile*. Mol Microbiol 71:1308–1322

Fischetti VA (2000) Surface proteins of gram-positive bacteria. In: Fischetti, Novick, Ferretti, Portnoy, Rood (eds) Gram-positive pathogens. American Society for Microbiology, pp 11–24

Franciosa G, Aureli P, Schechter R(2003) *Clostridium botulinum*. In: Miliotis MD, Bier JW (eds) International handbook of foodborne pathogens. Marcel Dekker, New York, NY, pp 62–89

Gold ND, Martin VJ (2007) Global view of the *Clostridium thermocellum* cellulosome revealed by quantitative proteomic analysis. J Bacteriol 189:6787–6795

Goodnough MC, Johnson EA (1996) Botulism. In: Anonymous (ed) Topley and Wilson's current topics in microbiology. Arnold Publishing, London

Hatheway CL, Johnson EA (1998) *Clostridium*: the spore-bearing anaerobes. In: Balows A, Duerden B (eds) Topley and Wilson's microbiology and microbial infections Vol. 2, Systematic Bacteriology, Arnold, London, pp 731–782

Hauschild AHW (1993) Epidemiology of human foodborne botulism *Clostridium botulinum*: ecology and control in foods. Marcel Dekker, New York, NY, pp 69–104

He M, Sebaihia M, Lawley TD, Stabler RA, Dawson LF, Martin MJ, Holt KE, Seth-Smith HM, Quail MA et al (2010) Evolutionary dynamics of *Clostridium difficile* over short and long time scales. Proc Natl Acad Sci USA 107:7527–7532

Hill KK, Smith TJ, Helma CH, Ticknor LO, Foley BT, Svensson RT, Brown JL, Johnson EA, Smith LA et al (2007) Genetic diversity among botulinum neurotoxin-producing clostridial strains. J Bacteriol 189:818–832

Hutson R, Thompson D, Collins M (1993a) Genetic interrelationships of saccharolytic *Clostridium botulinum* types B, E and F and related clostridia as revealed by small-subunit rRNA gene sequences. FEMS Microbiol Lett 108:103–110

Hutson R, Thompson D, Lawson P, Schocken-Itturino R, Böttger E, Collins M (1993b) Genetic interrelationships of proteolytic *Clostridium botulinum* types A, B, and F and other members of the *Clostridium botulinum* complex as revealed by small-subunit rRNA gene sequences. Antonie Van Leeuwenhoek 64:273–283

Janvilisri T, Scaria J, Thompson AD, Nicholson A, Limbago BM, Arroyo LG, Songer JG, Grohn YT, Chang YF (2009) Microarray identification of *Clostridium difficile* core components and divergent regions associated with host origin. J Bacteriol 191:3881–3891

Johnson EA (2005) *Clostridium botulinum* and *Clostridium tetani*. In: Borriello SP, Murray PR, Funke G (eds) Topley and Wilson's microbiology and microbial infections, 8th edn. Hodder Arnold, London, pp 1035–1088

Johnson EA (2007) *Clostridium botulinum*. In: Doyle MP, Beuchat LR (eds) Food microbiology: fundamentals and frontiers, 3rd edn. ASM Press, Washington, DC, pp 401–421

Johnson EA, Tepp WH, Bradshaw M, Gilbert RJ, Cook PE, McIntosh ED (2005) Characterization of *Clostridium botulinum* strains associated with an infant botulism case in the United Kingdom. J Clin Microbiol 43:2602–2607

Jones S, Paredes C, Tracy B, Cheng N, Sillers R, Senger R, Papoutsakis E (2008) The transcriptional program underlying the physiology of clostridial sporulation. Genome Biol 9:R114

Lawley TD, Croucher NJ, Yu L, Clare S, Sebaihia M, Goulding D, Pickard DJ, Parkhill J, Choudhary J, Dougan G (2009) Proteomic and genomic characterization of highly infectious *Clostridium difficile* 630 spores. J Bacteriol 191:5377–5386

Lee J, Yun H, Feist AM, Palsson BO, Lee SY (2008) Genome-scale reconstruction and in silico analysis of the *Clostridium acetobutylicum* ATCC 824 metabolic network. Appl Microbiol Biotechnol 80:849–862

Lindstrom M, Hinderink K, Somervuo P, Kiviniemi K, Nevas M, Chen Y, Auvinen P, Carter AT, Mason DR, Peck MW, Korkeala H (2009) Comparative genomic hybridization analysis of two predominant Nordic group I (proteolytic) *Clostridium botulinum* type B clusters. Appl Environ Microbiol 75:2643–2651

Marshall K, Bradshaw M, Johnson E, Bruggemann H (2010) Conjugative botulinum neurotoxin-encoding plasmids in *Clostridium botulinum*. PLoS One 5:e11087

Myers GS, Rasko DA, Cheung JK, Ravel J, Seshadri R, DeBoy RT, Ren Q, Varga J, Awad MM et al (2006) Skewed genomic variability in strains of the toxigenic bacterial pathogen, *Clostridium perfringens*. Genome Res 16:1031–1040

Nolling J, Breton G, Omelchenko MV, Makarova KS, Zeng Q, Gibson R, Lee HM, Dubois J, Qiu D et al (2001) Genome sequence and comparative analysis of the solvent-producing bacterium *Clostridium acetobutylicum*. J Bacteriol 183:4823–4838

Onyenwoke RU, Brill JA, Farahi K, Wiegel J (2004) Sporulation genes in members of the low G C gram-type-positive phylogenetic branch (*Firmicutes*). Arch Microbiol 182: 182–192

Papoutsakis ET (2008) Engineering solventogenic clostridia. Curr Opin Biotechnol 19:420–429

Perreten V, Vorlet-Fawer L, Slickers P, Ehricht R, Kuhnert P, Frey J (2005) Microarray-based detection of 90 antibiotic resistance genes of gram-positive bacteria. J Clin Microbiol 43: 2291–2302

Qazi O, Hitchen P, Tissot B, Panico M, Morris HR, Dell A, Fairweather N (2009) Mass spectrometric analysis of the S-layer proteins from *Clostridium difficile* demonstrates the absence of glycosylation. J Mass Spectrom 44:368–374

Rainey FA, Stackebrandt E (1993) 16S rDNA analysis reveals phylogenetic diversity among the polysaccharolytic clostridia FEMS Microbiol Lett 113:125–128

Roberts SB, Gowen CM, Brooks JP, Fong SS (2010) Genome-scale metabolic analysis of *Clostridium thermocellum* for bioethanol production. BMC Syst Biol 4:31

Rood JI (1998) Virulence genes of *Clostridium perfringens*. Annu Rev Microbiol 52:333–360

Rood JI (2004) Virulence plasmids of spore-forming bacteria. In: Funnell BE, Phillips GJ (eds) Plasmid biology. ASM Press, Washington, DC, pp 413–422

Rood J, Cole S (1991) Molecular genetics and pathogenesis of *Clostridium perfringens*. Microbiol Mol Biol Rev 55:621–648

Scheeff ED, Axelrod HL, Miller MD, Chiu HJ, Deacon AM, Wilson IA, Manning G (2009) Genomics, evolution, and crystal structure of a new family of bacterial spore kinases. Proteins: Structure, Function, and Bioinformatics 78(6):1470–1482

Sebaihia M, Peck MW, Minton NP, Thomson NR, Holden MT, Mitchell WJ, Carter AT, Bentley SD, Mason DR et al (2007) Genome sequence of a proteolytic (group I) *Clostridium botulinum* strain Hall A and comparative analysis of the clostridial genomes. Genome Res 17:1082–1092

Sebaihia M, Wren BW, Mullany P, Fairweather NF, Minton N, Stabler R, Thomson NR, Roberts AP, Cerdeno-Tarraga AM et al (2006) The multidrug-resistant human pathogen *Clostridium difficile* has a highly mobile, mosaic genome. Nat Genet 38:779–786

Senger RS, Papoutsakis ET (2008a) Genome-scale model for *Clostridium acetobutylicum*: part I. metabolic network resolution and analysis. Biotechnol Bioeng 101:1036–1052

Senger RS, Papoutsakis ET (2008b) Genome-scale model for *Clostridium acetobutylicum*: Part II. Development of specific proton flux states and numerically determined sub-systems. Biotechnol Bioeng 101:1053–1071

Shimizu T, Shima K, Yoshino K, Yonezawa K, Shimizu T, Hayashi H (2002a) Proteome and transcriptome analysis of the virulence genes regulated by the VirR/VirS system in *Clostridium perfringens*. J Bacteriol 184:2587–2594

Shimizu T, Ohtani K, Hirakawa H, Ohshima K, Yamashita A, Shiba T, Ogasawara N, Hattori M, Kuhara S, Hayashi H (2002b) Complete genome sequence of *Clostridium perfringens*, an anaerobic flesh-eater. Proc Natl Acad Sci USA 99:996–1001

Simpson L (1981) The origin, structure, and pharmacological activity of botulinum toxin. Pharmacol Rev 33:155–188

Smith LDS (1975) Clostridial wound infections. In: Smith LDS (ed) The pathogenic anaerobic bacteria, 2nd edn. Charles C. Thomas, Springfield IL, pp 321–324

Smith LDS, Sugiyama H (1988) Botulism. In: Barlows, Albert (eds) The organism, its toxins, the disease, 2nd edn. Charles C. Thomas, Springfield, IL

Songer JG (2010) Clostridia as agents of zoonotic disease. Vet Microbiol 140:399–404

Stabler RA, Gerding DN, Songer JG, Drudy D, Brazier JS, Trinh HT, Witney AA, Hinds J, Wren BW (2006) Comparative phylogenomics of *Clostridium difficile* reveals clade specificity and microevolution of hypervirulent strains. J Bacteriol 188:7297–7305

Stabler RA, He M, Dawson L, Martin M, Valiente E, Corton C, Lawley TD, Sebaihia M, Quail MA et al (2009) Comparative genome and phenotypic analysis of *Clostridium difficile* 027 strains provides insight into the evolution of a hypervirulent bacterium. Genome Biol 10:R102

Stackebrandt E, Kramer I, Swiderski J, Hippe H (1999) Phylogenetic basis for a taxonomic dissection of the genus *Clostridium*. FEMS Immunol Med Microbiol 24: 253–258

Suen J, Hatheway C, Steigerwalt A, Brenner D (1988) Genetic confirmation of identities of neurotoxigenic *Clostridium baratii* and *Clostridium butyricum* implicated as agents of infant botulism. J Clin Microbiol 26:2191

Sullivan L, Bennett GN (2006) Proteome analysis and comparison of *Clostridium acetobutylicum* ATCC 824 and SpoOA strain variants. J Ind Microbiol Biotechnol 33:298–308

Tomas CA, Welker NE, Papoutsakis ET (2003a) Overexpression of groESL in *Clostridium acetobutylicum* results in increased solvent production and tolerance, prolonged metabolism, and changes in the cell's transcriptional program. Appl Environ Microbiol 69:4951–4965

Tomas CA, Alsaker KV, Bonarius HP, Hendriksen WT, Yang H, Beamish JA, Paredes CJ, Papoutsakis ET (2003b) DNA array-based transcriptional analysis of asporogenous, nonsolventogenic *Clostridium acetobutylicum* strains SKO1 and M5. J Bacteriol 185:4539–4547

Tomas CA, Beamish J, Papoutsakis ET (2004) Transcriptional analysis of butanol stress and tolerance in *Clostridium acetobutylicum*. J Bacteriol 186:2006–2018

Tummala SB, Junne SG, Paredes CJ, Papoutsakis ET (2003) Transcriptional analysis of product-concentration driven changes in cellular programs of recombinant *Clostridium acetobutylicum* strains. Biotechnol Bioeng 84:842–854

Van Heyningen WE (1950) In: Anonymous (ed) Bacterial toxins. Blackwell Scientific Publications, Oxford

Warny M, Pepin J, Fang A, Killgore G, Thompson A, Brazier J, Frost E, McDonald LC (2005) Toxin production by an emerging strain of *Clostridium difficile* associated with outbreaks of severe disease in North America and Europe. The Lancet 366:1079–1084

Williams TI, Combs JC, Lynn BC, Strobel HJ (2007) Proteomic profile changes in membranes of ethanol-tolerant *Clostridium thermocellum*. Appl Microbiol Biotechnol 74:422–432

Willis AT (1969) In: Anonymous (ed) Clostridia of wound infection. Butterworths, London

Wright A, Wait R, Begum S, Crossett B, Nagy J, Brown K, Fairweather N (2005) Proteomic analysis of cell surface proteins from *Clostridium difficile*. Proteomics 5:2443–2452

Chapter 5
Genomics of *Escherichia* and *Shigella*

Nicole T. Perna

The laboratory workhorse *Escherichia coli* K-12 is among the most intensively studied living organisms on earth, and this single strain serves as the model system behind much of our understanding of prokaryotic molecular biology. Dense genome sequencing and recent insightful comparative analyses are making the species *E. coli*, as a whole, an emerging system for studying prokaryotic population genetics and the relationship between system-scale, or genome-scale, molecular evolution and complex traits like host range and pathogenic potential. Genomic perspective has revealed a coherent but dynamic species united by intraspecific gene flow via homologous lateral or horizontal transfer and differentiated by content flux mediated by acquisition of DNA segments from interspecies transfers. Acquisitions are offset by frequent losses that largely preserve net genome size across the species as a whole. Upon this stage of mutagenic processes, natural selection and population dynamics offer up strains playing roles as different as mutualists and deadly pathogens. Human perspective leads us to categorize *E. coli* by their impact on our health and disease-associated characteristics like patterns of adherence to human cells in culture. Merging perspectives provide understanding of the evolutionary dynamics behind human disease associated with *Escherichia* as well as the complexity of prokaryotic species. In this chapter, we briefly review the growing literature on *Escherichia* and *Shigella* genome sequencing and comparative genomics studies, with emphasis on providing context for understanding the genomes of the subset of these organisms associated with food-borne disease in humans.

5.1 Genome Sequences – Quantity and Quality

There are presently (as of 6 October 2010) at least 45 complete genome sequences (Table 5.1) from *Escherichia* and *Shigella* strains in the public domain. Another

N.T. Perna (✉)
Department of Genetics and the Genome Center of Wisconsin, University of Wisconsin, Madison, WI 53706, USA
e-mail: ntperna@wisc.edu

M. Wiedmann, W. Zhang (eds.), *Genomics of Foodborne Bacterial Pathogens*, Food Microbiology and Food Safety, DOI 10.1007/978-1-4419-7686-4_5, © Springer Science+Business Media, LLC 2011

Table 5.1 Complete genomes of *E. coli* and *Shigella* isolates

Name	Serotype	Replicon (GenBank accession number)	References	Isolation source	Associated disease	Pathovar
E. coli 042	O44:H18	Chromosome (FN554766); 1 plasmid (FN554767)	Chaudhuri et al. (2010)	Human	Diarrhea	EAEC
E. coli 11128	O111:H-	Chromosome (AP010960); 5 plasmids (AP010961–65)	Ogura et al. (2009)	Human	Diarrhea	EHEC
E. coli 11368	O26:H11	Chromosome (AP010953); 4 plasmids (AP010954–57)	Ogura et al. (2009)	Human	Diarrhea	EHEC
E. coli 12009	O103:H2	Chromosome (AP010958); 1 plasmid (AP010959)	Ogura et al. (2009)	Human	Diarrhea	EHEC
E. coli 536	O6:H31:K15	Chromosome (CP000247)	Brzuszkiewicz et al. (2006)	Human	Acute pyelonephritis	ExPEC
E. coli 55989	Unknown	Chromosome (CU928145); 1 plasmid (CU928159)	Touchon et al. (2009)	Human	Diarrhea	EAEC
E. coli APEC O1	O1:H7:K1	Chromosome (CP000468); 2 plasmids (DQ517526, DQ381420)	Johnson et al. (2006a, b, 2007)	Poultry	Colibacillosis	ExPEC
E. coli ATCC 8739	Unknown	Chromosome (CP000946)	None	Lab strain	Nonpathogen	
E. coli B REL606	Unknown	Chromosome (CP000819)	Jeong et al. (2009)	Lab strain	Nonpathogen (derived from B strain)	
E. coli BL21(DE3)	Unknown	Chromosome (AM946981)	None	Lab strain	Nonpathogen (derived from B strain)	
E. coli BL21(DE3)	Unknown	Chromosome (CP001665)	None	Lab strain	Nonpathogen (derived from B strain)	

Table 5.1 (continued)

Name	Serotype	Replicon (GenBank accession number)	References	Isolation source	Associated disease	Pathovar
E. coli BL21(DE3)	Unknown	Chromosome (CP001509)	Jeong et al. (2009)	Lab strain	Nonpathogen (derived from B strain)	
E. coli CB9615	O55:H7	Chromosome (CP001846); 1 plasmid (CP001847)	Zhou et al. (2010)	Human	Diarrhea (enteropathogenic)	EPEC
E. coli CFT073	O6:H1:K2	Chromosome (AE014075)	Welch et al. (2002)	Human	Acute pyelonephritis	ExPEC
E. coli E2348/69	O127:H6	Chromosome (FM180568); 2 plasmids (FM180569–70)	Iguchi et al. (2009)	Human	Diarrhea (enteropathogenic)	EPEC
E. coli E24377A	O139:H28	Chromosome (CP000800); 6 plasmids (CP000795–799; CP000801)	Rasko et al. (2008)	Human	Diarrhea (enterotoxigenic)	ETEC
E. coli EC4115	O157:H7	Chromosome (CP001164); 2 plasmids (CP001163, CP001165)	None	Unknown	Diarrhea	EHEC
E. coli ED1a	O81:H (unknown)	Chromosome (CU928162); 1 plasmid (CU928147)	Touchon et al. (2009)	Human	Commensal	
E. coli EDL933	O157:H7	Chromosome (AE005174); 1 plasmid (AF074613)	Burland et al. (1998), Perna et al. (2001)	Raw meat	Diarrhea	EHEC
E. coli HS	O9:H4	Chromosome (CP000802)	Rasko et al. (2008)	Human	Commensal	
E. coli 1A1	O8:H (unknown)	Chromosome (CU928160)	Touchon et al. (2009)	Human	Commensal	
E. coli 1A139	Unknown	Chromosome (CU928164)	Touchon et al. (2009)	Human	Pyelonephritis	ExPEC
E. coli IHE3034	O18:K1:H7	Chromosome (CP001969)	Moriel et al. (2010)	Human	Neonatal meningitis	ExPEC
E. coli K-12 BW2952	OR:H48:K-	Chromosome (CP001396)	Ferenci et al. (2009)	Lab strain	Nonpathogen (derived from K-12 strain)	ExPEC

Table 5.1 (continued)

Name	Serotype	Replicon (GenBank accession number)	References	Isolation source	Associated disease	Pathovar
E. coli K-12 DH1	OR:H48:K-	Chromosome (CP001637)	None	Lab strain	Nonpathogen (derived from K-12 strain)	
E. coli K-12 DH10B	OR:H48:K-	Chromosome (CP000948)	Durfee et al. (2008)	Lab strain	Nonpathogen (derived from K-12 strain)	
E. coli K-12 MG1655	OR:H48:K-	Chromosome (U00096)	Blattner et al. (1997)	Lab strain	Nonpathogen (derived from K-12 strain)	
E. coli K-12 W3110	OR:H48:K-	Chromosome (AP009048)	Hayashi et al. (2006)	Lab strain	Nonpathogen (derived from K-12 strain)	
E. coli S88	O45:K1:H7	Chromosome (CU928161); 1 plasmid (CU928146)	Touchon et al. (2009)	Human	Neonatal meningitis	ExPEC
E. coli Sakai	O157:H7	Chromosome (BA000007); 2 plasmids (AB011548-9)	Hayashi et al. (2001), Makino et al. (1998)	Human	Diarrhea	EHEC
E. coli SE11	O152:H28	Chromosome (AP009240); 6 plasmids (AP009241-46)	Oshima et al. (2008)	Human	Commensal	
E. coli SE15	O150:H5	Chromosome (AP009378); 1 plasmid (AP009379)	Toh et al. (2010)	Human	Commensal	
E. coli SMS-3-5	O19:H34	Chromosome (CP000970); 4 plasmids (CP000971-74)	Fricke et al. (2008)	Environmental isolate	Unknown	
E. coli TW14359	O157:H7	Chromosome (CP001368); 1 plasmid (CP001369)	Kulasekara et al. (2009)	Human	Diarrhea	EHEC

Table 5.1 (continued)

Name	Serotype	Replicon (GenBank accession number)	References	Isolation source	Associated disease	Pathovar
E. coli UMN026	O17:K52:H18	Chromosome (CU928163); 2 plasmids (CU928148–49)	Touchon et al. (2009)	Human	Cystitis	ExPEC
E. coli UTI89	O18:H7:K1	Chromosome (CP000243)	Chen et al. (2006)	Human	Cystitis	ExPEC
E. fergusonii ATCC 35469	Unknown	Chromosome (CU928158); 1 plasmid (CU928144)	Touchon et al. (2009)	Human	Unknown	
S. boydii 227	4	Chromosome (CP000036); 1 plasmid (CP000037)	Yang et al. (2005)	Human	Shigellosis	*Shigella*
S. boydii BS512	18	Chromosome (CP001063); 5 plasmids (CP001058–62)	None	Human	Shigellosis	*Shigella*
S. dysenteriae 197	1	Chromosome (CP000034); 2 plasmids (CP000035, CP000640)	Yang et al. (2005)	Human	Shigellosis	*Shigella*
S. flexneri 2002017	X	Chromosome (CP001383); 5 plasmids (CP001384–88)	Ye et al. (2010)	Human	Shigellosis	*Shigella*
S. flexneri 2457T	2a	Chromosome (AE014073)	Wei et al. (2003)	Human	Shigellosis	*Shigella*
S. flexneri 301	2a	Chromosome (AE005674); 1 plasmid (AF386526)	Jin et al. (2002)	Human	Shigellosis	*Shigella*
S. flexneri 8401	5b	Chromosome (CP000266)	Nie et al. (2006)	Human	Shigellosis	*Shigella*
S. sonnei Ss046	D	Chromosome (CP000038); 4 plasmids (CP000039, CP000641–43)	Yang et al. (2005)	Human	Shigellosis	*Shigella*

296 *Escherichia* or *Shigella* genome projects are registered as in progress at NCBI, and shotgun assembly sequence data, ranging from 2 to over 2,900 contigs, are already available for 72 strains. As with all genomes, even the most complete of these genomes should be regarded with a critical eye, since genome annotation is an error-prone process that at best reflects a snapshot of methods and data available at the time it was generated, and sequencing errors have been found in all cases where additional data are brought to bear on the subject. We somewhat belabor these potential sources of error here because *E. coli* genomes have served as a model in this regard as well, in studies aimed at evaluating sequence and annotation accuracy. Single-nucleotide and small insertion/deletion (indel) errors were found in the original reported sequences of *E. coli* K-12 strain MG1655, as well as both *E. coli* O157:H7 strains Sakai and EDL933 (Hayashi et al., 2006) (Leopold et al., 2009). These were among the earliest *E. coli* genome sequencing efforts. It is possible that subsequent efforts had higher accuracy, and new approaches that combine sequencing technologies may offer improvements; however, many of the existing and future *E. coli* genome sequences should be regarded as drafts, albeit some of exceedingly high quality. To be sure, quick and dirty (simple and cost-effective) genome sequencing is now a reality and on the rise as a popular and versatile research tool. High-throughput draft sequencing can be embraced in its own right as a means of sampling populations, validating laboratory experimentation (confirming mutants), profiling RNA transcription, and describing DNA-binding protein occupancy. More caution is required when making generalizations or comprehensive quantification (i.e., number of shared genes) from genome sequence comparisons. Missing or misassembled sequences and structural annotation biases of many sorts, including over-prediction and under-prediction of genes, as well as incorrectly and/or inconsistently identified gene boundaries, can have confounded many downstream comparative analyses.

The *E. coli* K-12 strain MG1655 has been resequenced with every major sequencing instrumentation advance (Blattner et al., 1997; Harris et al., 2009; Miller et al., 2008), and the annotation of the MG1655 genome has been heavily manually curated to reflect much, though by no means all, of the abundance of experimental data generated by decades of research bearing on both structural and functional annotations, both of which have been used as "gold standards" against which to measure accuracy of new predictive methods. Not surprisingly, the MG1655 genome has taught us some hard lessons as well. Proliferation of many independent and divergent annotation updates in different electronic resources can be confusing. A broad community initiative to gather this distributed effort into a single GenBank update in 2006 (Riley et al., 2006) and a federally funded initiative to provide a definitive resource for this organism reflecting community input provided some relief. Still, despite all this attention, genes remain unannotated or underannotated, including, but not limited to, those tagged as "hypothetical" even for this extremely well-studied genome. It is important to remember that few other regions of few other *E. coli* genomes have been examined in such depth, and all inferences drawn from genome analyses and comparisons are subject to these limitations. Errors in each individual genome are compounded in comparative analyses, where annotation differences translate to missed orthologs, leading to a tendency to overestimate the amount of lineage-specific content. The best and most recent large-scale comparative analyses of *E. coli* genomes have buffered against the effects of annotation differences between genomes using sequence similarity search strategies that include checking for unannotated but homologous genes or proteins in six-frame translations of the genome or rely more heavily on genome-scale alignment where the sequence is not portioned a priori into genic segments. The global analyses described below largely ignore the growing collection of less complete draft sequences. Where they shed additional light on pathogenicity of food-borne *E. coli* and *Shigella*, we include these data in discussions of individual pathovars.

5.2 Genome Size and Structure

All complete *E. coli* and *Shigella* genomes sequenced to date are composed of a single circular chromosome of length varying from strain to strain in the range from 4.5 to 5.5 million base pairs (Mbp), a size range anticipated from pre-genomic estimates based on physical mapping (Bergthorsson and Ochman, 1998). Many of these genomes also include one or more plasmids, and only some of which are reflected in the genome sequences. Whole-genome shotgun sequencing from total genomic DNA preparations should include resident plasmids, but they can be excluded from the genome sequence for a number of reasons, one of which is that some plasmids, like the *Shigella flexneri* 2457T virulence-associated plasmid, include a large number of repetitive elements that still pose a bioinformatics challenge for common genome assembly methods. All plasmids available as part of the genome sequences listed in Table 5.1 are noted.

Setting aside structural annotation differences discussed above, the number of genes predicted in each genome is largely a function of genome size, and gene density is high (>85% of the genome is protein coding), typical of prokaryotic genomes. Common ancestry of all *E. coli* and *Shigella* is evident in the obvious homology of large tracts of the chromosomal sequences where nucleotide sequence identity typically exceeds 98%. The content and evolutionary dynamics of gene flow in this conserved core are discussed in more detail below. Very few large-scale rearrangements disrupt collinearity of this core in nominal "*Escherichia*," with the exception of a handful of inversion events in individual lineages, largely anchored in the long repetitive sequences encoding rRNA subunits or prophages (Iguchi et al., 2006). In contrast, comparisons of the *Shigella* genomes to each other and even the most closely related *E. coli* genomes revealed numerous inversion events, many of which terminate in other shorter repetitive regions such as insertion sequence (IS) elements (Jin et al., 2002; Nie et al., 2006; Wei et al., 2003; Yang et al., 2005). Despite this *Shigella*-specific shuffling, the core chromosome remains detectable. Plasmids, as expected given their transient and transferable nature, are more variable with spotty homology to each other often attributable solely to plasmid maintenance functions.

5.3 Pan-Genome Variable and Core Fractions
of Genome Content

If not the first community to use the term "pan-genome," *E. coli* researchers have been leaders of the conceptual framework, as well as development and application of methods to partition the genome content of a species, and individual strains, into two categories. Variously named by different authors, these are the "core" and the "variable" genetic materials found in a genome, with the pan-genome encompassing the sum total of both fractions found in all members of the species. The concept arises from the realization that gene acquisition mediated by horizontal gene transfer from other species leads to lineage-specific content, while other regions of a chromosome are universally, or nearly universally, conserved among members of a species. It is clear that the evolutionary dynamics that governs the size and composition of the core and variable fractions of pan-genomes differs between species. In short, *E. coli* has what is now known as an "open pan-genome," where ongoing gene acquisition in different lineages is expected to perpetually increase the size and diversity of the set of genes found in *E. coli* strains. This is distinct from prokaryotic species which experience very little horizontal transfer because they are either mechanistically less capable or found in environments where donors are few and far between, such as inside eukaryotic cells. Recent comparisons of large sets of complete *E. coli* genomes to each other begin to define the rate of pan-genome expansion and conservation of the core. Here, after a brief review of early observations from *E. coli* genome comparisons, we focus on recent studies that maximize the number of genomes under comparison (Chaudhuri et al., 2010; Rasko et al., 2008; Touchon et al., 2009).

In the late 1990s, there were numerous reports of horizontal gene transfer, particularly of pathogenicity islands, in a wide variety of bacteria, including pathogenic *E. coli*. Around the same time, large segments of the *E. coli* K-12 MG1655 genome were being published as subsets of overlapping clones from large-insert libraries were completed and assembled (Blattner et al., 1993; Burland et al., 1995; Daniels et al., 1992; Plunkett et al., 1993; Sofia et al., 1994). Analyses of nucleotide composition and codon usage in these sequences pointed at the possibility that a large fraction of even the K-12 genome, possibly as much as 20%, was horizontally acquired from organisms with computational biases distinct from a typical *E. coli* gene (Lawrence and Ochman, 1997; Medigue et al., 1991). Comparison of that first K-12 genome to the few and distantly related genomes available at the time of publication of the complete genome confirmed that these anomalous segments were not, at the very least, universally conserved among prokaryotes (Blattner et al., 1997). Several years later, comparison of the *E. coli* O157:H7 genome to the existing MG1655 sequence provided a much finer scale resolution of hundreds of boundaries between shared and lineage-specific regions, and the discovery of close to 2 Mb of DNA sequences that differentiated between the two genomes, relative to a shared core or backbone of closer to 4 Mb (Hayashi et al., 2001; Perna et al., 2001). Though not labeled as such, the *E. coli* pan-genome had begun to grow. When a third *E. coli* genome (urinary tract pathogen CFT073) became available, new comparisons shrank the number of genes found in the core genome of all *E. coli* and the pan-genome expanded further with the addition of even more genes specific to this lineage (Welch et al., 2002). No doubt the exact numbers of shared and lineage-specific regions from these early reports are hampered by the annotation issues described above. Newer reports provide greater insight into the true size of the core and the expected rate of continued gene discovery as more *E. coli* genome sequences are added to the comparison.

Curves can be fitted to describe the pan-genome dynamics as a function of number of genomes under comparison using random sampling of available genomes to estimate error arising from order of the genomes sequenced (Tettelin et al., 2002). This mitigates the effect of sampling biases. For example, results would differ if we compared the estimated size of the core (or variable) fraction of the genome with the same number of very closely related strains, like different clonal isolates of *E. coli* O157:H7 rather than genomes from strains that reflect the full phylogenetic spectrum of the species. Similar analyses of the *E. coli* pan-genome are found in multiple recent papers, though the number and composition of strains vary (Chaudhuri et al., 2010; Rasko et al., 2008; Touchon et al., 2009). One study that include newly sequenced EAEC O42 predicts that the core genome size will approach 2,356 genes which is close to the number of genes conserved among all *E. coli* sequenced to date (Chaudhuri et al., 2010). The majority of these genes (2,173) are also conserved in two sequenced representatives of other *Escherichia* species (*Escherichia albertii* and *Escherichia fergusonii*). This estimate is slightly higher than previous core genome sizes, likely because it accommodates genes that are unannotated but present in some genomes that were previously overlooked as orthologs. These analyses reveal that the *E. coli* species core genome includes roughly half of the genes in an average individual strain. Reconstruction of the conserved core is not equivalent to reconstruction of the complete ancestral genome, which surely contained many additional genes that have been lost in one or more of the sampled extant strains. Touchon et al. describe the distribution of ortholog groups according to the number of strains in which they are conserved, demonstrating that most of the pan-genomes are found in either nearly all or, alternately, very few strains, with relatively few ortholog groups found in intermediate numbers of strains (Touchon et al., 2009). The nearly universal genes were almost certainly part of the ancestral genome. The composition and size of the core genome is not expected to change significantly as additional strains are sequenced.

In contrast to the stable core genome, all the same analyses suggest that the variable fraction, and hence the total pan-genome, will continue to increase in size as more genomes are sequenced. By most recent estimates, the pan-genome will grow by about 360 genes with each newly sequenced *E. coli* genome (Chaudhuri et al., 2010). Presently, enumerations include at least 17,868 non-orthologous genes in the *E. coli* species pan-genome (Touchon et al., 2009), a number that exceeds the number of genes found in many individual eukaryotic organisms. Even when IS elements and obvious prophage-like sequences are excluded, there are still over 10,000 distinct genes in the *E. coli* pan-genome.

5.4 Phylogeny

Early molecular analyses of relationships among *E. coli* strains used multi-locus enzyme electrophoresis to group, largely nonpathogenic, isolates into related clusters (Ochman and Selander, 1984a; b; Ochman et al., 1983; Whittam et al., 1983). Subsequent analyses using additional markers were largely congruent, and the groups defined in these studies, A, B1, B2, and D, and E, have provided a relatively stable phylogenetic framework for integration of genome-scale data and pathogenic isolates. The persistence of a phylogenetic signal in genome-scale analyses is notable because of the many obvious exceptions. Even prior to comparative genomics, individual taxa, individual genes, and sub-genic regions were described which bear clear signals of intraspecific recombination leading to extensive debate about the extent of clonality in this species (Bisercic et al., 1991; Dykhuizen and Green, 1991; Milkman and Bridges, 1993). Genome sequences provided an opportunity to examine this question at a much larger scale. Analyses of hundreds of genes from a handful of *E. coli* genomes revealed substantial incongruence (Daubin et al., 2003). That is, different genes supported different tree topologies. This suggested significant rates of horizontal gene transfer among strains of *E. coli*, mediated by homologous recombination, also referred to as allelic substitution and/or gene conversion. A later study of seven genomes suggested that the length and boundaries of substituted tracts were not well aligned with gene boundaries and indicated that at least 10% of the core genome was involved in this type of event (Mau et al., 2006). Recent analyses that make use of genome alignment rather than annotated genes and bring powerful evolutionary analysis tools to bear provide our best perspectives to date. Touchon et al. (2009) estimate that substituted regions are about 50 bp in length on average, though they caution against interpreting this figure as the length of incoming DNA fragments for exchange, given the possibility of recombination of subregions of donor fragments and/or subsequent transfers erasing the record of earlier longer events, particularly in areas of high recombination. They predict that a given base of an *E. coli* genome is 100 times more likely to experience a conversion event than to undergo a point mutation. Counter-intuitively, given this high rate of recombination, both simulation studies and the genome data itself demonstrate a detectable phylogenetic signal. Using a total evidence approach that applied ML phylogenetic analysis to concatenated alignment of over 1,800 core genes produced a robust phylogeny with strong bootstrap support for many key branches. Similarly, Chaudhuri et al. (2010) used their 2,173 conserved genes from the core of the pan-genome analysis in phylogenetic analyses that confirmed the monophyly of "phylogroups" A, B1, B2, and E. Group D is split into groups D1 and D2 by placement of the root by outgroups *E. albertii* and *E. fergusonii* (Touchon et al., 2009). A slightly different strain set was used to create the entirely compatible tree shown in Fig. 5.1. All of these genome-scale analyses support multiple origins of *Shigella* species, and all concur that similar pathovars of *E. coli* are found in multiple distinct lineages. Collectively, these analyses show that groups B2 and some D strains form the basal lineages, with groups A, E, and *Shigella* species emerging later.

Phylogeny based on SNPs in 2.9 Mb Core Genome

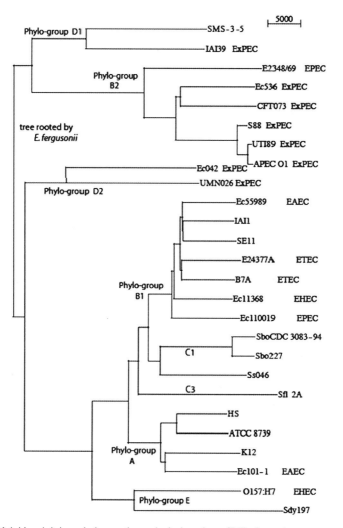

Fig. 5.1 Neighbor-joining phylogenetic analysis based on SNPs in regions conserved across all genomes under comparison using a Progressive Mauve alignment constructed with default parameters. Clades corresponding to previously described phylogroups are indicated along the relevant internal branches. All taxa are described in more detail in the text and Table 5.1. Pathovar designations, where appropriate, are shown after the isolate name

5.5 Enterohemorrhagic *E. coli* (EHEC)

The EHECs (or STECs) are deadly food- and waterborne pathogens that cause severe and often bloody diarrhea and hemolytic uremic syndrome (HUS). This pathovar is among the best represented with six complete genomes to date

(Table 5.1). Two serotype O157:H7 strains, EDL933 and Sakai, were early sequencing targets motivated by outbreaks in the 1990s in the United States, associated with contaminated hamburger, and in Japan, where contaminated radish sprouts sickened thousands (Hayashi et al., 2001; Perna et al., 2001). More recent high-profile outbreaks and large-scale food recalls have been deadly and costly and continue to fuel interest in O157:H7 genomics. A 2006 outbreak exhibiting a particularly high incidence of hemolytic uremic syndrome was linked to contaminated spinach, and a genome was sequenced of clinical isolate, TW14588, from that incident (Kulasekara et al., 2009). This group of strains is represented in Fig. 5.1 by the Sakai strain, and it clusters with *Shigella dysenteriae* Sd197 in phylogroup E. Extensive pre-genomics literature, much of which derives from the work of Thomas Whittam, supports the idea that serotype O157:H7 strains are a single recent clonal lineage of phylogroup E (Karch et al., 1993; Whittam et al., 1988; Whittam and Wilson, 1988; Whittam et al., 1993). Early stepwise reconstructions of the emergence of this lineage from a non-hemorrhagic lineage of EPEC O55:H7 included only small number of traits, like the serotype, a beta-glucuronidase deficiency, and importantly, the Shiga-like toxins themselves, for which the STECs are named (Feng et al., 1998; Whittam et al., 1993). A complete genome was recently sequenced from a representative of the O55:H7 lineage, strain CB9615, allowing a more thorough analysis of evolutionary events associated with the emergence of O157:H7 (Zhou et al., 2010). That analysis uses synonymous single-nucleotide polymorphisms (SNPs) carefully scrubbed clean of sites bearing the hallmarks of recombination rather than clonal mutation; to date the divergence of the O55:H7 and O157:H7 lineages to a mere 400 years ago is considerably more recent than previously thought. Through comparison to the common core of multiple O157:H7 genomes, using more distantly related *E. coli* and *Shigella* as outgroups, these authors enumerate gene content differences between the lineages, many of which lie in the numerous prophage and prophage remnants previously described from the O157:H7 genomes. Also notable is the absence in O157:H7 of a chromosomal type 2 secretion system (T2SS) presumed ancestral based on its widespread distribution in other *E. coli*. However, both the pO157 and the pO55 plasmids encode a distinct T2SS known to promote EHEC adherence and intestinal colonization (Ho et al., 2008). Most other virulence factors found on pO157 are not shared with pO55. SNPs, whether derived from recombination or clonal mutation number in the thousands and may either have direct effects on gene product functional divergence or impact regulation. Indeed, proteomic analysis revealed that at least 118 proteins are differentially expressed between CB9615 and O157:H7 (Zhou et al., 2010). Even for this relatively close comparison, it will be challenging to exhaustively characterize whether and how these genetic differences contribute to differences in pathogenicity. Genome-scale SNPs have also been employed to examine larger numbers of O157:H7 isolates, and in particular to seek correlates between population structure and severity of reported disease (Manning et al., 2008).

EHECs of other serotypes are a growing public health concern, and genome sequences are now available for three non-O157:H7 strains, 11128 (O111:H-), 11368 (O26:H11), and 12009 (O103:H2) (Ogura et al., 2009). Figure 5.1 includes strain 11368 to illustrate that these strains are not from phylogroup E, rather 11368 is a member of phylogroup B1, as are the other three of these non-O157 EHECs (Ogura et al., 2009), though it is not clear that these strains form a monophyletic cluster within B1. Of particular interest are genes shared by both the phylogroup B1 and E EHECs which are absent from strains more closely related to each group, as these represent possible convergent-independent acquisitions of determinants of shared elements of EHEC pathogenicity. Exact numbers of EHEC-specific genes vary depending on how many of the EHEC strains are required to encode predicted orthologs and how many and which non-EHECs are permitted to have orthologs, but 87 genes are reported to be common to all the phylogroup B1 and E EHECs and are not found in other strains except EPECs which share many aspects of pathogenicity. Many of the EHEC-/EPEC-specific factors are found among the prophages and include homologs of effector loci secreted via a type 3 secretion system (T3SS). The locus of enterocyte effacement (LEE) T3SS core region is conserved, albeit with some rearrangement in one non-O157 EHEC and differences in chromosomal insertion site consistent with previous reports for other EHEC and EPEC strains. Many EHECs contain strain-specific plasmids, but all sequenced EHECs include a related plasmid most generically called pEHECs, which encode a hemolysin and lipid A modification system, and other plasmid-borne genes linked to virulence in O157:H7 EHEC are found in various combinations in the non-O157 EHEC strains. As with the prophages and T3SS, sequence variation and structural differences in the homologous regions as well as contextual differences with respect to the surrounding chromosome or plasmid backbone point to complex evolutionary histories including non-clonal recombination events layered on top of independent acquisition of islands. Despite these complexities, these EHEC comparisons, coupled with the detailed dissection of the emergence of O157:H7, make EHECs the most thoroughly investigated pathovar genomes.

5.6 Enteropathogenic *E. coli* (EPEC)

EPECs are associated with severe watery diarrhea and are a major concern in developing parts of the world. Complete genomes are available for two EPEC strains, CB9615 and E2348/69 (Table 5.1). As detailed elsewhere, the CB9615 strain was targeted for sequencing because the EPEC lineage it represents is the closest known non-EHEC relative of the O157:H7 clone in phylogroup E (Zhou et al., 2010). The E2348/69 strain is a well-studied EPEC with copious associated direct experimental data on molecular mechanisms of virulence, as well as human volunteer and animal model studies. The genome sequence (Iguchi et al., 2009) provides a framework for this knowledge, and readers are advised to consult that paper for an enumeration of known virulence factors, which are beyond the scope of this chapter. E2348/69 is a member of phylogroup B2, one of the most basal distinct lineages, that includes human ExPEC as well as avian pathogens. Although E2348/69 shares the LEE T3SS, LEE-encoded effectors, and all of its non-LEE-encoded effectors with EHECs, this EPEC strain has fewer than half the number of effectors compared to O157:H7 Sakai. Notably, the Sakai effectors that are not associated with prophages are missing from E2348/69, and a second T3SS that is absent from clade B2 but found nearly universally among other *E. coli* strains suggests that the effectors missing from E2348/69 may be associated with the alternate T3SS. Why the LEE-secreted effectors should be phage encoded and the others not is unclear. The CB9615 genome was published after E2348/69 and includes a comparison of the two EPECs (Zhou et al., 2010). CB9615 has a larger secreted effector repertoire compared to E2348/69 that is more similar to its phylogenetically close neighbor O157:H7 than it is to the EPEC pathovar. This trend extends beyond the effectors, with EPEC CB9615 sharing more virulence-associated genes with its phylogroup than its pathovar.

5.7 Enterotoxigenic *E. coli* (ETEC)

The disease associated with ETECs is commonly referred to as traveler's diarrhea. The only complete genome available for an ETEC is from strain E24377A (Rasko et al., 2008); However, a good-quality draft sequence for a second ETEC, B7A, was also described in the same publication along with other genomes discussed elsewhere. These isolates were selected as verified human pathogens known to express the characteristic stable and labile toxins for which the ETECs are named. Both sequenced ETEC cluster in phylogroup B1. These ETEC genomes include an unusually large number of insertion sequences, making them a particular challenge to assemble. At 4.9 Mb, the E24377A chromosome is small compared to other strains, but the genome also includes a relatively large number of plasmids, six of which were reported alongside the chromosome sequence. Virulence factors distributed across all six of these plasmids in E24377A are found in a single plasmid (pCOO) in other strains (Froehlich et al., 2005), suggesting that the ETEC genome architecture is in flux. One interesting observation about the E24377A genome content is the presence of a propanediol utilization gene cluster unusual among *E. coli* strains but well characterized in *Salmonella typhimurium*. Rasko et al. (2008) analyzed the distribution of well-conserved, strain-specific, and pathovar-specific genes for each pathovar. The ETECs show among the largest number of strain-specific genes and the fewest group-specific genes, a pattern similar to EPEC and EAEC pathogroups. It is particularly notable for the ETECs given their relatively close phylogenetic relationship, which one might expect to elevate shared content. This suggests that the commonalities in pathogenicity are not strongly impacted by ETEC-specific genes.

5.8 Enteroaggregative *E. coli* (EAEC)

EAECs are a pathovar that causes significant diarrhea in the developed world. This group is named for characteristic adherence phenotype on cultured human epithelial cells (HEp-2). Two complete EAEC genomes are now available. *Escherichia coli* 042 was chosen as a representative EAgEC that has been demonstrated to elicit diarrhea in volunteers (Chaudhuri et al., 2010). Strain 55989 is one of the genomes sequenced by Touchon et al. (2009). These two EAECs cluster in different phylogroups, D2 and B1, respectively. A good-quality draft sequence is also available for atypical EAEC strain 101-1, which lacks a characteristic ETEC plasmid (Rasko et al., 2008), and clusters in group A.

EAEC 042 has a relatively large chromosome (5.24 Mb) and a single plasmid pAA (0.11 Mb) of the IncFIIA family. Seventeen genomic regions were described that differentiate EAEC 042 from other *E. coli* genomes, accounting for 24% of the chromosome. A conjugative transposon carries multiple antibiotic resistance genes and Biolog phenotype microarrays demonstrate that EAEC 042 shows diverse resistance phenotypes that distinguish it from *E. coli* K-12 MG1655. EAEC 042 also outperforms MG1655 at iron sequestration which may be linked to increased sensitivity to reactive oxygen species. Comparison of the three EAEC genomes and HS revealed only 210 EAEC-specific genes, only 13 of which were assessed likely to contribute to virulence (Touchon et al., 2009). Very few of these EAEC-specific genes are shared between 101–1 and 042, suggesting that convergent acquisition of content via horizontal transfer is not a major factor in EAEC pathogenicity (Rasko et al., 2008).

5.9 *Shigella* and Enteroinvasive *E. coli* (EIEC)

There are no sequenced genomes from EIEC isolates, but similarities in pathogenicity and phylogenetic analyses support the idea that EIECs may be close relatives of at least some *Shigella*. Both EIEC and all four species of *Shigella* cause bacterial dysentery, with the *Shigella* alone responsible for 160 million episodes annually around the world. The distribution of the four *Shigella* species exhibits geographic and economic biases. There are eight complete *Shigella* genomes, including two *Shigella boydii*, four *S. flexneri*, one *Shigella sonnei*, and one *S. dysenteriae* (Jin et al., 2002; Nie et al., 2006; Ogura et al., 2009; Wei et al., 2003; Yang et al., 2005; Ye et al., 2010). The lack of monophyly for this genus shown in Fig. 5.1 is well documented, with abundant support for multiple evolutionary trajectories leading to what we call *Shigella*. This sometimes contested idea was first championed by Pupo et al. (2000) and has been substantiated by genome sequence comparisons (Yang et al., 2005), multi-locus phylogenetic analyses of larger numbers of isolates (Yang et al., 2007), and comparative genome hybridization (Peng et al., 2006). All *Shigella* genome strains other than *S. dysenteriae* Sd197 cluster near phylogroup B1 with *S. flexneri* in phylogroup C3 and *S. boydii* and *S. sonnei* in phylogroup C1. *Shigella dysenteriae* 197 and other isolates of this species, cluster with the O157:H7 EHEC in phylogroup E.

Yang et al. compares five *Shigella* genomes, including representatives of all four species (Yang et al., 2007). All sequenced *Shigella* genomes include a virulence plasmid and the plasmids of different strains share many features including nearly identical replication origins, and most contain the Mxi-Spa T3SS. There is sufficient variation among other regions of the plasmids to question whether they are all derived from a single acquisition of the plasmid, and the important cell entry region is bracketed by insertion sequence (IS) elements, suggesting potential for transfer. This is not surprising given the overall high frequency of IS in all *Shigella*. IS elements account for 27–38% of the plasmids and 7–12% of the chromosomes of each genome. The IS elements are largely responsible for a relatively high number of pseudogenes (210–372) in all *Shigella* relative to other *E. coli* like K-12 strain MG1655. Many of the same genes have lost function in multiple *Shigella* such that the sum total number of affected orthologous gene sets is 313. In some cases, events that create a pseudogene are ancestral to two or more genomes, but there are also examples of the same gene becoming a pseudogene in two or more lineages through independent insertional mutation events. IS elements also account for many deletions that reduce the chromosome size in all four species and the larger number genome rearrangements that distinguish *Shigella* genomes, in particular Sd197 and Sb227, from other *E. coli*, which maintain collinearity for the most part. The contribution that any of these changes make to differences in the severity of pathogenesis and global distribution of infections remains largely unknown.

5.10 Extraintestinal *E. coli* (ExPEC)

Eight complete genome sequences are available for ExPEC. Five strains are urinary tract pathogens. Both CFT073 (Welch et al., 2002) and 536 (Brzuszkiewicz et al., 2006) are associated with acute pyelonephritis, while IAI39 (Touchon et al., 2009) is linked to less severe pyelonephritis. Strains UTI89 (Chen et al., 2006) and UMN026 (Touchon et al., 2009) cause cystitis. Two of the genome sequences IHE3034 (Moriel et al., 2010) and S88 (Touchon et al., 2009) are K1 strains that cause neonatal meningitis. One is an avian pathogen, APEC O1 (Johnson et al., 2007), included here because it clusters in phylogroup B1 with many of these ExPEC (Fig. 5.1). Exceptions are UMN026, which is a phylogroup D2 strain, and IAI39, which is in phylogroup D1. These are the most basal lineages of the *E. coli* tree, which may explain the elevated levels of strain-specific genes observed in some analyses (Rasko et al., 2008).

5.11 Laboratory Strains and Commensal *E. coli*

Genome sequences are available for multiple derivatives of two *E. coli* strains in widespread use in research and industry. Five genomes are derived from *E. coli* K-12, including BW2592, DH1, DH10B, MG1655, and W3110 (Blattner et al., 1997; Durfee et al., 2008; Ferenci et al., 2009; Hayashi et al., 2006; Jeong et al., 2009). Four are derivatives of a strain called B, including B REL606, and three are independent determinations (two unpublished) of the sequence from strain BL21(DE3) (Jeong et al., 2009). Both K-12 and B were isolated from humans without symptoms of *E. coli*-associated disease and can be considered nonpathogens. Similarly, strain ATCC8739 is a lab strain that will not be discussed here further because there is not yet an associated publication. Given long laboratory histories, with repeated mutagenic treatments and serial passages, these strains are sometimes distinguished from more recently isolated commensals, presently represented by five genomes. These are strains ED1a, HS, IAI1, SE11, and SE15 from phylogroups B2, A, B1, B1, and B2, respectively (Oshima et al., 2008; Rasko et al., 2008; Toh et al., 2010; Touchon et al., 2009). These commensal isolate genomes are good comparators for identifying potential virulence factors in other genomes, especially from the same phylogroup, but it is important to remember that commensal strains, like HS, are able to colonize human hosts and should be expected to encode genes related to this phenotype. Based on phylogeny and gene content, it has been argued that extraintestinal virulence of B2 pathogens may be a by-product of commensalism (Le Gall et al., 2007).

5.12 Other *Escherichia*

A complete genome is available for one environmental isolate of *E. coli*, strain SMS-3-5, from industrially polluted water (Fricke et al., 2008). This strain shows unusual antibiotic resistance, determinants of which were identified on both the chromosome and one of several plasmids. SMS-3-5 contains homologs of virulence factors from several other groups of *E. coli*, particularly APEC and ExPEC. It is a member of the D1 lineage, one of the most basal groups of *E. coli*. More extensive genome sampling of environmental isolates has the potential to illuminate pathogen reservoirs. The only complete genome for other *Escherichia* species comes from the type strain of *Escherichia fergusonii*, ATCC 35469 (Touchon et al., 2009). This genome is especially useful as an outgroup for phylogenetic analyses of *E. coli* and *Shigella*, and analyses of ancestral states (with the exception of genome organization due to many derived rearrangements), as it is much more closely related than *Salmonella*. A mature draft sequence is also available for *E. albertii*, but there is no publication to date.

5.13 Summary

Escherichia coli and *Shigella* are among the most densely sampled groups of prokaryotic genomes. Comparative analyses within this group and with a closely related species, *E. fergusonii*, have revealed the evolutionary dynamics behind the divergence of genome content, phylogenetic lineages, and disease potential. Similar pathovars have emerged from the larger group multiple times, through largely independent convergent events. Despite the wealth of data and some excellent evolutionary genetic studies, precise linkage between individual evolutionary events and virulence-associated phenotypes remains elusive in all but a few cases. Better understanding may lie in the nearly 300 additional genome projects for this group already in progress.

References

Bergthorsson U, Ochman H (1998) Distribution of chromosome length variation in natural isolates of *Escherichia coli*. Mol Biol Evol 15:6–16

Bisercic M, Feutrier JY, Reeves PR (1991) Nucleotide sequences of the gnd genes from nine natural isolates of *Escherichia coli*: evidence of intragenic recombination as a contributing factor in the evolution of the polymorphic gnd locus. J Bacteriol 173:3894–3900

Blattner FR, Burland V, Plunkett G 3rd, Sofia HJ, Daniels DL (1993) Analysis of the *Escherichia coli* genome. IV. DNA sequence of the region from 89.2 to 92.8 minutes. Nucleic Acids Res 21:5408–5417

Blattner FR, Plunkett G 3rd, Bloch CA, Perna NT, Burland V, Riley M, Collado-Vides J, Glasner JD, Rode CK, Mayhew GF et al (1997) The complete genome sequence of *Escherichia coli* K-12. Science 277:1453–1462

Brzuszkiewicz E, Bruggemann H, Liesegang H, Emmerth M, Olschlager T, Nagy G, Albermann K, Wagner C, Buchrieser C, Emody L et al (2006) How to become a uropathogen: comparative genomic analysis of extraintestinal pathogenic *Escherichia coli* strains. Proc Natl Acad Sci USA 103:12879–12884

Burland V, Plunkett G 3rd, Sofia HJ, Daniels DL, Blattner FR (1995) Analysis of the *Escherichia coli* genome VI: DNA sequence of the region from 92.8 through 100 minutes. Nucleic Acids Res 23:2105–2119

Burland V, Shao Y, Perna NT, Plunkett G, Sofia HJ, Blattner FR (1998) The complete DNA sequence and analysis of the large virulence plasmid of *Escherichia coli* O157:H7. Nucleic Acids Res 26:4196–4204

Chaudhuri RR, Sebaihia M, Hobman JL, Webber MA, Leyton DL, Goldberg MD, Cunningham AF, Scott-Tucker A, Ferguson PR, Thomas CM et al (2010) Complete genome sequence and comparative metabolic profiling of the prototypical enteroaggregative *Escherichia coli* strain 042. PLoS One 5:e8801

Chen SL, Hung CS, Xu J, Reigstad CS, Magrini V, Sabo A, Blasiar D, Bieri T, Meyer RR, Ozersky P et al (2006) Identification of genes subject to positive selection in uropathogenic strains of *Escherichia coli*: a comparative genomics approach. Proc Natl Acad Sci USA 103:5977–5982

Daniels DL, Plunkett G 3rd, Burland V, Blattner FR (1992) Analysis of the *Escherichia coli* genome: DNA sequence of the region from 84.5 to 86.5 minutes. Science 257:771–778

Daubin V, Moran NA, Ochman H (2003) Phylogenetics and the cohesion of bacterial genomes. Science 301:829–832

Durfee T, Nelson R, Baldwin S, Plunkett G 3rd, Burland V, Mau B, Petrosino JF, Qin X, Muzny DM, Ayele M et al (2008) The complete genome sequence of *Escherichia coli* DH10B: insights into the biology of a laboratory workhorse. J Bacteriol 190:2597–2606

Dykhuizen DE, Green L (1991) Recombination in *Escherichia coli* and the definition of biological species. J Bacteriol 173:7257–7268

Feng P, Lampel KA, Karch H, Whittam TS (1998) Genotypic and phenotypic changes in the emergence of *Escherichia coli* O157:H7. J Infect Dis 177:1750–1753

Ferenci T, Zhou Z, Betteridge T, Ren Y, Liu Y, Feng L, Reeves PR, Wang L (2009) Genomic sequencing reveals regulatory mutations and recombinational events in the widely used MC4100 lineage of *Escherichia coli* K-12. J Bacteriol 191:4025–4029

Fricke WF, Wright MS, Lindell AH, Harkins DM, Baker-Austin C, Ravel J, Stepanauskas R (2008) Insights into the environmental resistance gene pool from the genome sequence of the multidrug-resistant environmental isolate *Escherichia coli* SMS-3-5. J Bacteriol 190: 6779–6794

Froehlich B, Parkhill J, Sanders M, Quail MA, Scott JR (2005) The pCoo plasmid of enterotoxigenic *Escherichia coli* is a mosaic cointegrate. J Bacteriol 187:6509–6516

Harris DR, Pollock SV, Wood EA, Goiffon RJ, Klingele AJ, Cabot EL, Schackwitz W, Martin J, Eggington J, Durfee TJ et al (2009) Directed evolution of ionizing radiation resistance in *Escherichia coli*. J Bacteriol 191:5240–5252

Hayashi T, Makino K, Ohnishi M, Kurokawa K, Ishii K, Yokoyama K, Han CG, Ohtsubo E, Nakayama K, Murata T et al (2001) Complete genome sequence of enterohemorrhagic *Escherichia coli* O157:H7 and genomic comparison with a laboratory strain K-12. DNA Res 8:11–22

Hayashi K, Morooka N, Yamamoto Y, Fujita K, Isono K, Choi S, Ohtsubo E, Baba T, Wanner BL, Mori H et al (2006) Highly accurate genome sequences of *Escherichia coli* K-12 strains MG1655 and W3110. Mol Syst Biol 2:2006.0007

Ho TD, Davis BM, Ritchie JM, Waldor MK (2008) Type 2 secretion promotes enterohemorrhagic *Escherichia coli* adherence and intestinal colonization. Infect Immun 76:1858–1865

Iguchi A, Iyoda S, Terajima J, Watanabe H, Osawa R (2006) Spontaneous recombination between homologous prophage regions causes large-scale inversions within the *Escherichia coli* O157:H7 chromosome. Gene 372:199–207

Iguchi A, Thomson NR, Ogura Y, Saunders D, Ooka T, Henderson IR, Harris D, Asadulghani M, Kurokawa K, Dean P et al (2009) Complete genome sequence and comparative genome analysis of enteropathogenic *Escherichia coli* O127:H6 strain E2348/69. J Bacteriol 191: 347–354

Jeong H, Barbe V, Lee CH, Vallenet D, Yu DS, Choi SH, Couloux A, Lee SW, Yoon SH, Cattolico L et al (2009) Genome sequences of *Escherichia coli* B strains REL606 and BL21(DE3). J Mol Biol 394:644–652

Jin Q, Yuan Z, Xu J, Wang Y, Shen Y, Lu W, Wang J, Liu H, Yang J, Yang F et al (2002) Genome sequence of *Shigella flexneri* 2a: insights into pathogenicity through comparison with genomes of *Escherichia coli* K12 and O157. Nucleic Acids Res 30:4432–4441

Johnson TJ, Johnson SJ, Nolan LK (2006a) Complete DNA sequence of a ColBM plasmid from avian pathogenic *Escherichia coli* suggests that it evolved from closely related ColV virulence plasmids. J Bacteriol 188:5975–5983

Johnson TJ, Kariyawasam S, Wannemuehler Y, Mangiamele P, Johnson SJ, Doetkott C, Skyberg JA, Lynne AM, Johnson JR, Nolan LK (2007) The genome sequence of avian pathogenic *Escherichia coli* strain O1:K1:H7 shares strong similarities with human extraintestinal pathogenic *E. coli* genomes. J Bacteriol 189:3228–3236

Johnson TJ, Wannemuehler YM, Scaccianoce JA, Johnson SJ, Nolan LK (2006b) Complete DNA sequence, comparative genomics, and prevalence of an IncHI2 plasmid occurring among extraintestinal pathogenic *Escherichia coli* isolates. Antimicrob Agents Chemother 50:3929–3933

Karch H, Bohm H, Schmidt H, Gunzer F, Aleksic S, Heesemann J (1993) Clonal structure and pathogenicity of Shiga-like toxin-producing, sorbitol-fermenting *Escherichia coli* O157:H. J Clin Microbiol 31:1200–1205

Kulasekara BR, Jacobs M, Zhou Y, Wu Z, Sims E, Saenphimmachak C, Rohmer L, Ritchie JM, Radey M, McKevitt M et al (2009) Analysis of the genome of the *Escherichia coli* O157:H7 2006 spinach-associated outbreak isolate indicates candidate genes that may enhance virulence. Infect Immun 77:3713–3721

Lawrence JG, Ochman H (1997) Amelioration of bacterial genomes: rates of change and exchange. J Mol Evol 44:383–397

Le Gall T, Clermont O, Gouriou S, Picard B, Nassif X, Denamur E, Tenaillon O (2007) Extraintestinal virulence is a coincidental by-product of commensalism in B2 phylogenetic group *Escherichia coli* strains. Mol Biol Evol 24:2373–2384

Leopold SR, Magrini V, Holt NJ, Shaikh N, Mardis ER, Cagno J, Ogura Y, Iguchi A, Hayashi T, Mellmann A et al (2009) A precise reconstruction of the emergence and constrained radiations of *Escherichia coli* O157 portrayed by backbone concatenomic analysis. Proc Natl Acad Sci USA 106:8713–8718

Makino K, Ishii K, Yasunaga T, Hattori M, Yokoyama K, Yutsudo CH, Kubota Y, Yamaichi Y, Iida T, Yamamoto K et al (1998) Complete nucleotide sequences of 93-kb and 3.3-kb plasmids of an enterohemorrhagic *Escherichia coli* O157:H7 derived from Sakai outbreak. DNA Res 5:1–9

Manning SD, Motiwala AS, Springman AC, Qi W, Lacher DW, Ouellette LM, Mladonicky JM, Somsel P, Rudrik JT, Dietrich SE et al (2008) Variation in virulence among clades of *Escherichia coli* O157:H7 associated with disease outbreaks. Proc Natl Acad Sci USA 105:4868–4873

Mau B, Glasner JD, Darling AE, Perna NT (2006) Genome-wide detection and analysis of homologous recombination among sequenced strains of *Escherichia coli*. Genome Biol 7:R44

Medigue C, Rouxel T, Vigier P, Henaut A, Danchin A (1991) Evidence for horizontal gene transfer in *Escherichia coli* speciation. J Mol Biol 222:851–856

Milkman R, Bridges MM (1993) Molecular evolution of the *Escherichia coli* chromosome. IV. Sequence comparisons. Genetics 133:455–468

Miller JR, Delcher AL, Koren S, Venter E, Walenz BP, Brownley A, Johnson J, Li K, Mobarry C, Sutton G (2008) Aggressive assembly of pyrosequencing reads with mates. Bioinformatics 24:2818–2824

Moriel DG, Bertoldi I, Spagnuolo A, Marchi S, Rosini R, Nesta B, Pastorello I, Corea VA, Torricelli G, Cartocci E et al (2010) Identification of protective and broadly conserved vaccine antigens from the genome of extraintestinal pathogenic *Escherichia coli*. Proc Natl Acad Sci USA 107:9072–9077

Nie H, Yang F, Zhang X, Yang J, Chen L, Wang J, Xiong Z, Peng J, Sun L, Dong J et al (2006) Complete genome sequence of *Shigella flexneri* 5b and comparison with *Shigella flexneri* 2a. BMC Genomics 7:173

Ochman H, Selander RK (1984a) Evidence for clonal population structure in *Escherichia coli*. Proc Natl Acad Sci USA 81:198–201

Ochman H, Selander RK (1984b) Standard reference strains of *Escherichia coli* from natural populations. J Bacteriol 157:690–693

Ochman H, Whittam TS, Caugant DA, Selander RK (1983) Enzyme polymorphism and genetic population structure in *Escherichia coli* and *Shigella*. J Gen Microbiol 129:2715–2726

Ogura Y, Ooka T, Iguchi A, Toh H, Asadulghani M, Oshima K, Kodama T, Abe H, Nakayama K, Kurokawa K et al (2009) Comparative genomics reveal the mechanism of the parallel evolution of O157 and non-O157 enterohemorrhagic *Escherichia coli*. Proc Natl Acad Sci USA 106:17939–17944

Oshima K, Toh H, Ogura Y, Sasamoto H, Morita H, Park SH, Ooka T, Iyoda S, Taylor TD, Hayashi T et al (2008) Complete genome sequence and comparative analysis of the wild-type commensal *Escherichia coli* strain SE11 isolated from a healthy adult. DNA Res 15: 375–386

Peng J, Zhang X, Yang J, Wang J, Yang E, Bin W, Wei C, Sun M, Jin Q (2006) The use of comparative genomic hybridization to characterize genome dynamics and diversity among the serotypes of *Shigella*. BMC Genomics 7:218

Perna NT, Plunkett G 3rd, Burland V, Mau B, Glasner JD, Rose DJ, Mayhew GF, Evans PS, Gregor J, Kirkpatrick HA et al (2001) Genome sequence of enterohaemorrhagic *Escherichia coli* O157:H7. Nature 409:529–533

Plunkett G 3rd, Burland V, Daniels DL, Blattner FR (1993) Analysis of the *Escherichia coli* genome. III. DNA sequence of the region from 87.2 to 89.2 minutes. Nucleic Acids Res 21:3391–3398

Pupo GM, Lan R, Reeves PR (2000) Multiple independent origins of Shigella clones of *Escherichia coli* and convergent evolution of many of their characteristics. Proc Natl Acad Sci USA 97:10567–10572

Rasko DA, Rosovitz MJ, Myers GS, Mongodin EF, Fricke WF, Gajer P, Crabtree J, Sebaihia M, Thomson NR, Chaudhuri R et al (2008) The pangenome structure of *Escherichia coli*: comparative genomic analysis of *E. coli* commensal and pathogenic isolates. J Bacteriol 190:6881–6893

Riley M, Abe T, Arnaud MB, Berlyn MK, Blattner FR, Chaudhuri RR, Glasner JD, Horiuchi T, Keseler IM, Kosuge T et al (2006) *Escherichia coli* K-12: a cooperatively developed annotation snapshot – 2005. Nucleic Acids Res 34:1–9

Sofia HJ, Burland V, Daniels DL, Plunkett G 3rd, Blattner FR (1994) Analysis of the *Escherichia coli* genome. V. DNA sequence of the region from 76.0 to 81.5 minutes. Nucleic Acids Res 22:2576–2586

Tettelin H, Masignani V, Cieslewicz MJ, Donati C, Medini D, Ward NL, Angiuoli SV, Crabtree J, Jones AL, Durkin AS et al (2005) Genome analysis of multiple pathogenic isolates of *Streptococcus agalactiae*: implications for the microbial "pan-genome". Proc Natl Acad Sci USA 102:13950–13955

Toh H, Oshima K, Toyoda A, Ogura Y, Ooka T, Sasamoto H, Park SH, Iyoda S, Kurokawa K, Morita H et al (2010) Complete genome sequence of the wild-type commensal *Escherichia coli* strain SE15, belonging to phylogenetic group B2. J Bacteriol 192:1165–1166

Touchon M, Hoede C, Tenaillon O, Barbe V, Baeriswyl S, Bidet P, Bingen E, Bonacorsi S, Bouchier C, Bouvet O et al (2009) Organised genome dynamics in the *Escherichia coli* species results in highly diverse adaptive paths. PLoS Genet 5:e1000344

Wei J, Goldberg MB, Burland V, Venkatesan MM, Deng W, Fournier G, Mayhew GF, Plunkett G 3rd, Rose DJ, Darling A et al (2003) Complete genome sequence and comparative genomics of *Shigella flexneri* serotype 2a strain 2457T. Infect Immun 71:2775–2786

Welch RA, Burland V, Plunkett G 3rd, Redford P, Roesch P, Rasko D, Buckles EL, Liou SR, Boutin A, Hackett J et al (2002) Extensive mosaic structure revealed by the complete genome sequence of uropathogenic *Escherichia coli*. Proc Natl Acad Sci USA 99:17020–17024

Whittam TS, Ochman H, Selander RK (1983) Geographic components of linkage disequilibrium in natural populations of *Escherichia coli*. Mol Biol Evol 1:67–83

Whittam TS, Wachsmuth IK, Wilson RA (1988) Genetic evidence of clonal descent of *Escherichia coli* O157:H7 associated with hemorrhagic colitis and hemolytic uremic syndrome. J Infect Dis 157:1124–1133

Whittam TS, Wilson RA (1988) Genetic relationships among pathogenic *Escherichia coli* of serogroup O157. Infect Immun 56:2467–2473

Whittam TS, Wolfe ML, Wachsmuth IK, Orskov F, Orskov I, Wilson RA (1993) Clonal relationships among *Escherichia coli* strains that cause hemorrhagic colitis and infantile diarrhea. Infect Immun 61:1619–1629

Yang J, Nie H, Chen L, Zhang X, Yang F, Xu X, Zhu Y, Yu J, Jin Q (2007) Revisiting the molecular evolutionary history of *Shigella* spp. J Mol Evol 64:71–79

Yang F, Yang J, Zhang X, Chen L, Jiang Y, Yan Y, Tang X, Wang J, Xiong Z, Dong J et al (2005) Genome dynamics and diversity of *Shigella* species, the etiologic agents of bacillary dysentery. Nucleic Acids Res 33:6445–6458

Ye C, Lan R, Xia S, Zhang J, Sun Q, Zhang S, Jing H, Wang L, Li Z, Zhou Z et al (2010) Emergence of a new multidrug-resistant serotype X variant in an epidemic clone of *Shigella flexneri*. J Clin Microbiol 48:419–426

Zhou Z, Li X, Liu B, Beutin L, Xu J, Ren Y, Feng L, Lan R, Reeves PR, Wang L (2010) Derivation of *Escherichia coli* O157:H7 from its O55:H7 precursor. PLoS One 5:e8700

Chapter 6
Listeria Genomics

Didier Cabanes, Sandra Sousa, and Pascale Cossart

6.1 Introduction

6.1.1 Listeriology

Listeria monocytogenes is a gram-positive bacterium discovered by Murray et al. (1926) following an epidemic affecting rabbits and guinea pigs in animal care houses in Cambridge (England) (Fig. 6.1). Human cases were reported in 1929 and listeriosis was long considered as a zoonosis. The first human listeriosis outbreak directly linked to the consumption of *Listeria*-contaminated foodstuffs was only reported by Schlech et al. (1983). *L. monocytogenes* is now recognized as a foodborne pathogen. Since this first large outbreak, *L. monocytogenes* is also recognized as a public health problem, and there has been a tremendous interest in elucidating the epidemiology of this organism in order to protect the consumer against listeriosis. The bacterium possesses properties of a foodborne pathogen: it is resistant to acid and high salt concentrations; it grows at low temperature,

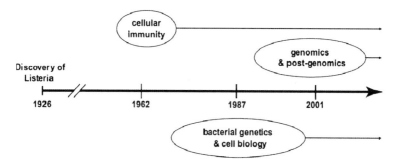

Fig. 6.1 Key dates of the listeriology (modified from Cossart, 2007)

D. Cabanes (✉)
Group of Molecular Microbiology, Institute for Molecular and Cell Biology, Porto, Portugal
e-mail: didier@ibmc.up.pt

M. Wiedmann, W. Zhang (eds.), *Genomics of Foodborne Bacterial Pathogens*,
Food Microbiology and Food Safety, DOI 10.1007/978-1-4419-7686-4_6,
© Springer Science+Business Media, LLC 2011

down to freezing point, growing in refrigerated foods. Additionally, *L. monocyto-genes* readily produces biofilm in food production plants. Because of its ubiquitous nature, *L. monocytogenes* commonly contaminates raw products and, through cross-contamination, other food items. Thus, all human beings are routinely exposed to *L. monocytogenes*. Despite this, listeriosis is a relatively rare disease in humans.

However, mainly in immunocompromised individuals, newborn babies, elderly and pregnant women, *L. monocytogenes* infection may lead to a severe disease with clinical features that vary from septicemias, meningitis, meningo-encephalitis to abortions. In healthy individuals, a severe gastroenteritis can also occur after ingestion of highly contaminated food (Swaminathan and Gerner-Smidt, 2007).

6.1.2 Listeria *Infectious Process*

The knowledge of the *L. monocytogenes* infectious process mostly derives from epidemiological data in humans and animals and from infection studies in various animal models, in particular the mouse model (Cabanes et al., 2008; Disson et al., 2008; Lecuit, 2007). Bacteria after crossing the intestinal barrier reach, via the lymph and the blood, the liver in which they replicate in hepatocytes and also the spleen. Then bacteria via hematogenous dissemination can reach the brain and the placenta (Fig. 6.2). The disease is thus due to the original property of *L. monocytogenes* to be able to cross three host barriers: the intestinal barrier, the blood–brain barrier, and the maternofetal barrier. It is also due to the capacity of *Listeria* to resist intracellular killing when phagocytosed by macrophages and to invade many types of cells that are normally non-phagocytic.

During infection, *Listeria* is thus mostly intracellular. How *Listeria* enters into non-phagocytic cells and spreads from cell to cell has been investigated in great detail (Cabanes et al., 2008; Cossart et al., 2003; Cossart and Sansonetti, 2004; Cossart and Toledo-Arana, 2008; Pizarro-Cerda and Cossart, 2006; Seveau et al.,

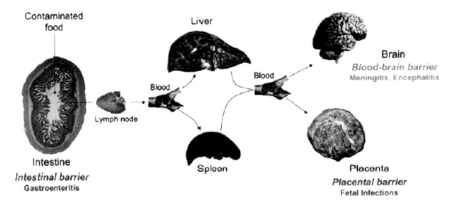

Fig. 6.2 Successive steps of human listeriosis (Lecuit et al. 2007)

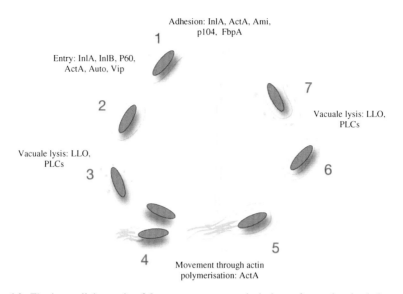

Fig. 6.3 The intracellular cycle of *L. monocytogenes* and virulence factors involved. *L. monocytogenes* is able to adhere to cells and to induce its entry into target cells (*1*). The bacteria are initially trapped within a phagocytic vacuole (*2*), but through the activity of phospholipases, the bacteria-containing compartment is lysed (*3*). Once in the cytoplasm, free bacteria are able to multiply and also to polymerize cellular actin (*4*). *L. monocytogenes* moves and encounters the plasma membrane of the infected cell: the bacteria then push this membrane creating an invagination in the membrane of a neighboring cell, invading it (*5*). The bacteria are then found in a double membrane-bound compartment (*6*): these membranes are lysed again (*7*), and *L. monocytogenes* is ready to start a new infection cycle

2007) (Fig. 6.3). Internalization results from the tight apposition of the plasma membrane around the incoming microbe. After entry into cells, bacteria are entrapped into a vacuole from which they escape after about 30 min. Once inside the cytosol they replicate and start to recruit and polymerize cellular actin, generating a network of branched filaments. This polymerization process at one pole of the bacteria produces energy to propel bacteria that move into the cytoplasm at a speed of about 10 mm per minute. When reaching the plasma membrane, bacteria induce the formation of protrusions which may invade neighboring cells and generate a two-membrane vacuole from which bacteria escape, allowing a new cycle of replication to take place in a second infected cell. It is believed that this phenomenon of direct cell-to-cell spread allows dissemination of the bacteria in various infected tissues while being protected from host defenses. The major bacterial genes and factors critical for the successive steps of the cell infectious process were identified in most cases by genetic means (Fig. 6.3) (Dussurget et al., 2004). Adherence involves a number of surface proteins, including InlA, ActA, Ami, p104, and FbpA (Alvarez-Dominguez et al., 1997; Dramsi et al., 2004; Mengaud et al., 1996; Milohanic et al., 2001; Pandiripally et al., 1999; Suarez et al., 2001). Internalin A (InlA) and InlB were the first listerial factors identified as mediating bacterial invasion into different target cell types. Their cellular receptors have also been identified, and the molecular signaling cascades triggered during *Listeria* entry into host cells are being characterized in detail (Bonazzi et al., 2009; Seveau et al., 2007; Vazquez-Boland et al., 1992). More recent findings suggest that other molecules are also necessary for internalization (P60, ActA, Auto, Vip) (Cabanes et al., 2004, 2005; Pilgrim et al., 2003; Suarez et al., 2001), revealing a complex dialogue between *Listeria* and eukaryotic cells during the early phases of the infection cycle. After internalization, *Listeria* resides in a vacuole that is lysed by listeriolysin O (LLO) in concert with two phospholipases C (PLCs) (Schnupf and Portnoy, 2007; Vazquez-Boland et al., 1992). The actin-based motility is mediated by the surface protein ActA (Domann et al., 1992; Kocks et al., 1992). Most of the *Listeria* virulence genes involved in these infectious steps are regulated by the transcriptional activator PrfA (Scortti et al., 2007).

6.2 The *Listeria* Genome

6.2.1 Listeria *Genome Sequences*

L. monocytogenes is one of the eight species of the genus *Listeria*, a genus belonging to the Firmicutes division, characterized by a low GC content and closely related to *Bacillus subtilis* and staphylococci. Only *L. monocytogenes* and *L. ivanovii* are pathogenic, *L. ivanovii* being strictly an animal pathogen. Currently, the complete genome sequences of *L. monocytogenes* EGDe (serotype 1/2a) (Glaser et al., 2001), *L. monocytogenes* F2365 (serotype 4b) (Nelson et al., 2004), *Listeria innocua* CLIP 11262 (serotype 6a) (Glaser et al., 2001), *Listeria welshimeri* SLCC 5334 (serotype 6b) (Hain et al., 2006), and *Listeria seeligeri* (Steinweg et al., 2010) are published. Additionally, the incomplete genomes of *L. monocytogenes* F6854 (serotype 1/2a) and *L. monocytogenes* H7858 (serotype4b) are also published (Nelson et al., 2004). The genome sequences of *Listeria grayi*, *L. ivanovii* and of the very recently isolated *Listeria rocourtiae* (Leclercq et al., 2009) and *Listeria marthii* (Graves et al., 2010), as well as an impressive number of *L. monocytogenes* of diverse serotypes and isolated from different sources are currently being sequenced and are partially available (http://www.genomesonline.org).

6.2.2 *General Features of the* Listeria *Genome*

All the listerial genomes sequenced to date are circular chromosomes with sizes that vary between 2.7 and 3.0 Mb in length. Some *Listeria* strains also contain plasmids. Listeriae are members of the low G+C group of bacteria and have an average G+C content of 38%. The genome sequence encodes approximately 2,900 putative protein-coding genes, of which ≈65% genes have an assigned function. The analysis of the expression of the entire *L. monocytogenes* EGDe genome in several in vitro and in vivo conditions using tiling arrays revealed that *Listeria* expresses more than 98% of its ORFs in any condition (Toledo-Arana et al., 2009). Taking

Fig. 6.4 Genomic map showing the localization of sRNAs of *L. monocytogenes. Yellow arrows* indicate sRNAs absent in *L. innocua* (Toledo-Arana et al., 2009)

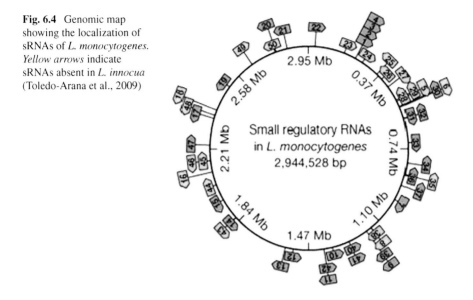

into account all conditions tested, this approach allowed also the identification of 517 polycistronic operons encoding 1,719 genes that represent 60% of the annotated genome. Half of the operons are bicistronic and more than 80% contain less than five cistrons (Toledo-Arana et al., 2009). In addition, 50 small RNAs (Fig. 6.4), at least two of which are involved in virulence in mice, several antisense RNAs, long overlapping 5′- and 3′-untranslated region, and riboswitches were discovered (Toledo-Arana et al., 2009). Analysis of *Listeria* genomes allows common and particular features of Listeriae to be determined, as well as differences between pathogenic and a nonpathogenic *Listeria* strain. Although both strain and serotype-specific genes were identified, all *Listeria* genomes revealed a highly conserved synteny in gene organization and content. An unexpected synteny with the genomes of the nonpathogenic bacterium *B. subtilis*, as well as the pathogenic bacterium *Staphylococcus aureus* was also observed. Interestingly, the whole-genome sequence of *L. welshimeri* revealed common steps in genome reduction with *L. innocua* as compared to *L. monocytogenes*, suggesting similar genome evolutionary paths from an ancestor (Fig. 6.5) (Hain et al., 2006, 2007). The smaller size of this genome is the result of deletions in genes involved in virulence and of "fitness" genes required for intracellular survival, transcription factors, LPXTG- and LRR-containing proteins as well as genes involved in carbohydrate transport and metabolism.

L. monocytogenes EGDe genome analysis reveals an exceptionally large number of genes encoding putative transport systems, transcriptional regulators, and surface and secreted proteins, consistent with the ability of *Listeria* to colonize a broad range of ecosystems (Glaser et al., 2001). It encodes 331 transport proteins, including 39 putative phosphotransferase sugar-uptake systems, i.e., roughly twice as many as *Escherichia coli*. The *L. monocytogenes* coding capacity dedicated to

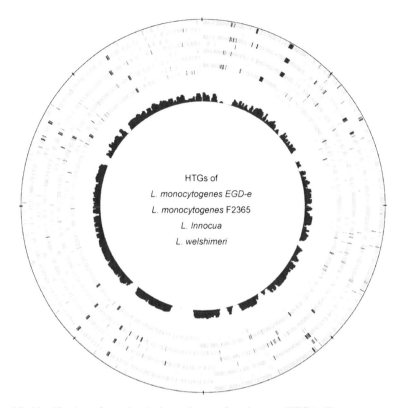

Fig. 6.5 Identification of putative horizontally transferred genes (HTGs). The *outermost circle* represents the scale in kilobytes starting with the origin of replication at position 0. The following *double circles* show in *gray* the distribution of coding sequences (CDS) of *L. monocytogenes* (*second circle*), *L. monocytogenes* F2365 (*third circle*), *L. innocua* (*fourth circle*), and *L. welshimeri* (*fifth circle*). rRNA operons are *colored* in *blue*, a putative prophage region in *black*, and HTGs in *red*. The *innermost circle* displays the conservation on the level of nucleotides among the above-mentioned genomes (Hain et al., 2007)

transcriptional regulators (7.3%) provides a means for coordinate gene expression in response to the diverse environmental conditions outside as well as inside the host. One hundred and thirty-three surface proteins and 86 secreted proteins are encoded by the genome of *L. monocytogenes*. Among the 133 surface proteins, 41 belong to the LPXTG protein family (Bierne and Cossart, 2007; Cabanes et al., 2002). Thirty of the 133 genes encoding surface proteins and 23 of the 86 genes encoding secreted proteins are absent from the genome of *L. innocua* (i.e., 22.6 and 26.7%, respectively), whereas 270 of the 2,853 genes encoded by the whole *L. monocytogenes* genome are absent from *L. innocua* (i.e., 10.5%). The high proportion of surface or secreted protein-encoding genes in the *L. monocytogenes* genome and their over-representation compared to *L. innocua* probably reflects the capacity of the bacterium to interact with a large variety of surfaces in the environment and to colonize many cell types within the host.

6.3 Comparative Genomics

The goal of comparative genomics is to identify genetic differences across entire genomes, to correlate those differences to biological function, and to gain insight into selective evolutionary pressures and patterns of gene transfer or loss, particularly within the context of virulence in pathogenic species.

6.3.1 Comparative Genomics of Listeria *Species: Biodiversity*

A general feature of the *Listeria* genomes is the very strong conservation in genome organization with no inversions or shifts of large genome segments (Buchrieser, 2007). A nearly perfect conservation of the order as well as the relative orientation of these orthologous genes was identified, indicating a high stability and a close phylogenetic relationship of listerial genomes (Buchrieser et al., 2003). This conserved genome organization may be related to the low occurrence of IS elements, suggesting that IS transposition or IS-mediated deletions are not key evolutionary mechanisms in *Listeria*. Phylogenomic analysis based on whole-genome sequences reveals the same phylogenetic relationship as determined by 16S-rRNA sequencing of *L. monocytogenes, L. innocua, L. welshimeri, L. seeligeri, L. ivanovii*, and *L. grayi* which branches the genus into three main groups: the first group consists of the closely related species *L. monocytogenes, L. innocua*, and *L. welshimeri* in which *L. welshimeri* reveals the deepest branching of this group. *L. seeligeri* and *L. ivanovii* exhibit the second group and *L. grayi* seems to be very distant from both groups (Fig. 6.6). In addition, 16S-rRNA gene sequence analysis recently confirmed the close phylogenetic relatedness of *L. marthii* to *L. monocytogenes* and *L. innocua* and more distant relatedness to *L. welshimeri, L. seeligeri, L. ivanovii* and *L. grayi* (Graves et al., 2010).

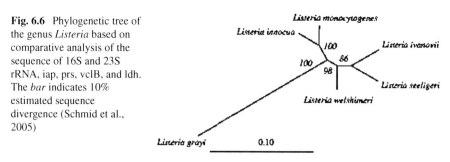

Fig. 6.6 Phylogenetic tree of the genus *Listeria* based on comparative analysis of the sequence of 16S and 23S rRNA, iap, prs, vclB, and ldh. The *bar* indicates 10% estimated sequence divergence (Schmid et al., 2005)

Among the 13 serovars of *L. monocytogenes*, the serovars 1/2a, 1/2b, and 4b are responsible for 95% of listeriosis reported in humans (Swaminathan and Gerner-Smidt, 2007). Different studies have defined three distinct phylogenetic lineages of *L. monocytogenes*, with lineage I strains containing isolates that have caused epidemics in humans (including serovars 1/2b and 4b), lineage II strains containing isolates responsible for sporadic human disease (including serovars 1/2a and

1/2c), and lineage III strains comprising mostly animal pathogens (Jeffers et al., 2001). Recent phylogenetic analysis has revealed that within lineage I, serovar 4b evolved from serovar 1/2b (Ragon et al., 2008). The genome of *L. monocytogenes* CLIP80459, a serovar 4b epidemic isolate, has been partially sequenced and compared with the sequences of *L. monocytogenes* EGDe (serovar 1/2a) and *L. innocua* (Doumith et al., 2004). The most striking finding is the high genetic diversity within *L. monocytogenes* strains. Approximately 8% of the CLIP80459 genes are absent from the EGDe genome whereas 10.5% of the EGDe genes are absent from the *L. innocua* genome. Thus, the genetic diversity among *L. monocytogenes* isolates is close to that between *Listeria* species.

The genome content of 113 strains representing all serovars of *L. monocytogenes* and all species of the genus *Listeria* has been studied using DNA macroarray (Doumith et al., 2004). The genomic variability among these strains is mainly due to differences in genes encoding surface proteins and genes involved in sugar metabolism, some of which could be fitness factors necessary for adaptation to various environments, others could be virulence factors necessary for the infectious cycle within the host.

More recent in silico analysis of draft and completed sequences from *Listeria* species and various typed strains of *L. monocytogenes* has revealed a number of regions of difference (RDs) between *Listeria* species and strains (Milillo et al., 2009). The study identified 15 RDs present in *L. monocytogenes* but absent from other *Listeria* species, 3 RDs that are present in lineage I strains and absent from lineage II strains of *L. monocytogenes*, and 4 RDs present in lineage II strains and absent from lineage I strains.

6.3.2 DNA-Based Microarrays for Listeria Detection and Identification

The availability of several published genome sequences for *Listeria* allowed the generation of several DNA microarrays for the detection, identification, discrimination, and subtyping of *Listeria* strains.

6.3.2.1 Detection and Identification

The accurate detection of foodborne pathogens such as *L. monocytogenes* is an important step forward to ensure food safety and proper control of industrial processes in food technology. Several multipathogen DNA-based microarrays were developed in order to detect and identify pathogens for environmental and biodefense applications (Call et al., 2003b; Lin et al., 2005; Sergeev et al., 2004). To detect the common intestinal pathogenic bacteria quickly and accurately, a rapid experimental procedure was set up based upon the gene chip technology (Jin et al., 2005). Target genes were amplified and hybridized by oligonucleotide microarrays. One hundred and seventy strains of bacteria in pure culture belonging to 11 genera, including *Listeria,* were successfully discriminated. *Listeria* is also part of the 15 species identifiable using oligonucleotide arrays designed for the identification of bacteria responsible for bacteremia (Hou et al., 2008). A microarray technique using multiple virulence genes was also developed for the detection and identification of enteropathogenic bacteria at the species and subspecies levels, and the target bacteria included *L. monocytogenes* (You et al., 2008). This technique appears as an attractive diagnostic tool for rapidly and simultaneously identifying multiple enteropathogenic pathogens in clinical practice, especially in patients with infectious diarrhea.

6.3.2.2 Discrimination and Subtyping

A rapid microarray-based assay was developed for the reliable detection and discrimination of six species of the *Listeria* genus: *L. monocytogenes, L. ivanovii, L. innocua, L. welshimeri, L. seeligeri,* and *L. grayi* (Volokhov et al., 2002). Results of the microarray analysis of 53 reference and clinical isolates of *Listeria* spp. demonstrated that this method allowed unambiguous identification of all six *Listeria* species based on sequence differences in the *iap* gene. DNA microarrays provide in addition a new format to resolve genetic differences among *Listeria* isolates and identify specific genes associated with the infecting pathogen. Several mixed genome microarrays were constructed to allow discrimination among *L. monocytogenes* serovars (Borucki et al., 2003, 2004). A suspension microarray approach was also developed allowing direct and high-throughput subtyping of *L. monocytogenes* from genomic DNA (Borucki et al., 2005). *Listeria* DNA microarray-based strategies were also developed to allow phylogenetic lineages between listerial species (Call et al., 2003a; Zhang et al., 2003). Collectively, these results indicate that many genes have diverged between lineages of *L. monocytogenes*.

6.3.3 Comparative Genomics and Virulence Genes Identification

A main goal of comparative genomics is the identification of new virulence factors. The identification of virulence genes through comparative genomics relies primarily on the identification in a given species of a gene or other genetic signature or characteristics consistent with a role in pathogenicity. Functional analysis is then necessary to decipher the role of potential virulence genes and to determine their function in the establishment and maintenance of infection. Analysis of genomes from closely related *Listeria* species allowed the identification of several new virulence factors.

6.3.3.1 Sortases

The genome of *L. monocytogenes* (strain EGDe) contains 41 genes that encode LPXTG proteins (Cabanes et al., 2002). LPXTG proteins are anchored to the peptidoglycan by transpeptidases named sortases (Schneewind et al., 1992). Two genes encoding putative sortases were identified by genome sequence analysis (Bierne et al., 2002; Cabanes et al., 2005; Pucciarelli et al., 2005; Sabet et al., 2005). *srtA* encodes a sortase responsible for the anchoring of InlA, InlJ, Vip (Bierne et al., 2002; Cabanes et al., 2005; Pucciarelli et al., 2005; Sabet et al., 2005), and at least 12 other proteins to the peptidoglycan (Bierne et al., 2002; Cabanes et al., 2005;

Pucciarelli et al., 2005; Sabet et al., 2005). *L. monocytogenes* mutants lacking *srtA* are defective in internalizing into human cells, as well as in colonizing the liver and spleen of mice infected orally. The *Listeria* genome encodes another sortase, SrtB, which has a limited number of substrates containing a C-terminal NXXTN sorting motif. SrtB does not seem to contribute to virulence in mice after intravenous inoculation (Bierne et al., 2004).

6.3.3.2 *InlJ*

Data generated with a listerial DNA macroarray (Doumith et al., 2004) identified *inlJ* as a LPXTG protein-encoding gene present in the genomes of *L. monocytogenes* strains and absent from all other *Listeria* species. An *inlJ* deletion mutant is significantly attenuated in virulence after intravenous infection of mice or oral inoculation of transgenic mice expressing human E-cadherin in the intestine (Sabet et al., 2005). *InlJ* encodes a leucine-rich repeat (LRR) protein that is structurally related to the listerial invasion factor InlA. The crystal structure of the internalin domain of InlJ was resolved (Bublitz et al., 2008). InlJ was recently proposed to act as a *L. monocytogenes* sortase-anchored adhesin specifically expressed during infection in vivo (Sabet et al., 2008).

6.3.3.3 Vip

vip was identified by comparative genomics as another LPXTG protein-encoding gene absent from nonpathogenic *Listeria* species (Cabanes et al., 2005). *vip* is positively regulated by PrfA, the transcriptional activator of the major *Listeria* virulence factors. Vip is anchored to the *Listeria* cell wall by sortase A and is required for entry into some mammalian cells. The cellular receptor for Vip was identified as the endoplasmic reticulum (ER) resident chaperone Gp96. Comparative infection studies using oral and intravenous inoculation of nontransgenic and transgenic mice expressing human E-cadherin demonstrated a role for Vip in *Listeria* virulence. Vip thus appeared as a new virulence factor exploiting Gp96 as a receptor for cell invasion and/or signaling events that may interfere with the host immune response in the course of the infection.

6.3.3.4 Auto

Comparative genomics led also to the identification of *aut*, a gene absent from the genome of *L. innocua*, which encodes Auto, a GW surface protein with an autolytic activity (Cabanes et al., 2004). Auto appears to be necessary but not sufficient for entry. Inactivation of *aut* indicated that Auto probably plays a role at both early and late stages of the infectious process. It is the first autolysin absent from *L. innocua* that plays a key role in pathogenicity, possibly by controlling the bacterial surface architecture and/or composition or by modifying the peptidoglycan hydrolysis in turn modulating the host innate immune response.

6.3.3.5 Bsh

Comparison of *L. monocytogenes* and *L. innocua* genomes revealed the existence of a gene absent in *L. innocua* that encodes a bile salt hydrolase (BSH), an enzyme that deconjugates conjugated bile salts (Dussurget et al., 2002). The *Listeria bsh* is controlled by sigma B and positively regulated by PrfA. BSH activity is increased at low oxygen tension, a condition that *Listeria* encounters after ingestion. Reduced oxygen tension in the host could be a signal sensed by bacteria to express the *bsh* gene and other virulence genes when and where necessary. BSH was shown to be a novel type of virulence factor involved in both the intestinal and hepatic phases of listeriosis. The *bsh* gene and BSH activity are absent from *L. innocua* but are present in all human pathogenic *Listeria* species, establishing a link between *Listeria* resistance to bile salts and its capacity to colonize and infect humans (Dussurget et al., 2002).

6.4 Functional Genomics

DNA microarrays are powerful tools that allow easy genomic comparisons between different strains allowing genomic comparison of non-sequenced strains with a sequenced reference strain. Additionally, they are used for transcriptional profiling, monitoring gene expression of every gene in the genome, by determining levels of transcripts. The use of microarrays has enhanced the understanding of *Listeria* evolution and pathogenesis by comparing genome content of pathogenic and non-pathogenic strains. They have also been instrumental in understanding virulence gene expression of pathogens during infection.

6.4.1 Comparative Transcriptome Analysis of Listeria Strains

Among the 13 *L. monocytogenes* serovars described, human listeriosis is mostly associated with strains of serovars 4b, 1/2b, and 1/2a. As described before, within the species *L. monocytogenes*, three phylogenetic lineages are described. Serovar 1/2a belongs to phylogenetic lineage I, while serovars 4b and 1/2b group in phylogenetic lineage II. To explore the role of gene expression in the adaptation of *L. monocytogenes* strains of these two major lineages to different environments, as well as in virulence, a whole-genome expression profiling of six *L. monocytogenes* isolates of serovars 4b, 1/2b, and 1/2a of distinct origins was performed using a *Listeria* multigenome DNA array (Severino et al., 2007). The expression profiles of two strains having distinct 50% lethal doses, as assessed in the mouse model, were further analyzed. Analysis of the differentially expressed genes identified differences in protein-, nucleic acid-, carbon metabolism-, and virulence-related gene expression. Comparison of the expression profiles of the core genomes of all strains revealed differences between the two lineages with respect to cell wall synthesis, the stress-related σB regulon, and virulence-related genes. These findings suggest different patterns of interaction with host cells and the environment, key factors for host colonization and survival in the environment.

6.4.2 Transcriptome Analysis in Listeria Mutants Deficient in Regulators

Genome-wide transcriptome studies have been used to catalog and analyze genes belonging to several *Listeria* regulons (PrfA, σB, σ54, VirR, MogR, CtsR, HrcA, RelA-CodY, and DegU) and to examine overlapping regulatory interactions.

6.4.2.1 PrfA

PrfA is the major regulator of *Listeria* virulence gene expression. The expression profiles of the wild-type EGDe and a *prfA*-deleted mutant were the first analyzed by transcriptomics (Milohanic et al., 2003). Three groups of genes that are regulated differently were identified. Group I comprises 11 genes both regulated and preceded by a putative PrfA box. Group II comprises eight negatively regulated genes. Group III comprises 53 genes, of which only 2 are preceded by a putative PrfA box. This study revealed that PrfA can act as an activator or a repressor and suggested that PrfA may directly or indirectly activate different sets of genes in association with different sigma factors. The relative contributions of σB and PrfA to transcript levels of genes previously identified as differentially regulated by PrfA were recently approached by quantitative RT-PCR and subgenomic microarray analyses (McGann et al., 2008; Ollinger et al., 2008). These analyses suggest that the regulatory interactions between PrfA and σB contribute to the predominant role of PrfA as a direct regulator of virulence genes, while σB regulates a wider range of virulence and stress response genes.

6.4.2.2 SigB

The general stress response, controlled by the alternative sigma factor (σB), has an important role for bacterial survival both in the environment and during infection. Whole-genome-based transcriptional profiling was used to identify σB-dependent genes at different growth phases (Hain et al., 2008). A total of 105 genes appear σB-positively regulated and 111 genes appear to be under negative control of σB. Genes comprising the σB regulon encode solute transporters, novel cell wall proteins, universal stress proteins, transcriptional regulators and include those involved in osmoregulation, carbon metabolism, ribosomefunction and envelope function, as well as virulence and niche-specific survival genes such as those involved in bile resistance and exclusion.

The composition and functions of the *L. monocytogenes* and *L. innocua* σB regulons were previously hypothesized to differ due to their virulence differences. Transcript levels in stationary-phase cells and in cells exposed to salt stress were characterized by microarray analyses for both species (Raengpradub et al., 2008).

Both *L. monocytogenes* and *L. innocua* Δ*sigB*-null mutants have increased motility compared to the respective isogenic parent strains, suggesting that σB affects motility and chemotaxis. Although *L. monocytogenes* and *L. innocua* differ in σB-dependent acid stress resistance and have species-specific σB-dependent genes, the *L. monocytogenes* and *L. innocua* σB regulons show considerable conservation, with a common set of at least 49 genes that are σB dependent in both species.

6.4.2.3 CtsR

While σB positively regulates the transcription of class II stress response genes, HrcA and CtsR negatively regulate class I and III stress response genes, respectively. To identify interactions between σB and CtsR, *L. monocytogenes* Δ*ctsR* and Δ*ctsR* Δ*sigB* strains were analyzed by microarray transcriptomics (Hu et al., 2007b). This study identified 42 CtsR-repressed genes, 22 genes with lower transcript levels in the Δ*ctsR* strain, and at least 40 genes coregulated by both CtsR and σB, including genes encoding proteins with confirmed or plausible roles in virulence and stress response. Statistical analyses also confirmed interactions between the *ctsR* and the *sigB* null mutations in both heat resistance and invasion phenotypes.

6.4.2.4 HrcA

To define the HrcA regulon and identify interactions between HrcA, CtsR, and σB, *L. monocytogenes* Δ*hrcA*, Δ*ctsR* Δ*hrcA*, and Δ*hrcA* Δ*sigB* strains, along with previously described Δ*sigB*, Δ*ctsR*, and Δ*ctsR* Δ*sigB* strains, were characterized using phenotypic assays and whole-genome transcriptome analysis of the Δ*hrcA* strain (Hu et al., 2007a). The *hrcA* and *sigB* deletions had significant effects on heat resistance. While the *hrcA* deletion had no significant effect on acid resistance or invasion efficiency in Caco-2 cells, a significant effect of interactions between the *hrcA* deletion and the *ctsR* deletion on invasiveness was observed. Microarray-based transcriptome analyses identified 25 HrcA-repressed genes, and 36 genes that appear to be indirectly upregulated by HrcA. A number of genes were found to be coregulated by either HrcA and CtsR, HrcA, and σB or all three regulators. These data suggest that σB, CtsR, and HrcA form a regulatory network that contributes to the transcription of a number of *L. monocytogenes* genes.

6.4.2.5 Sigma 54

The role of the alternative σ54 factor, encoded by the *rpoN* gene, was investigated in *L. monocytogenes* by comparing the global gene expression of the wild-type and an *rpoN* mutant (Arous et al., 2004). Seventy-seven genes, whose expression was modulated in the *rpoN* mutant as compared to the wild-type strain, were identified. Most of the modifications were related to carbohydrate metabolism and in particular to pyruvate metabolism. However, only the *mptACD* operon was shown to be directly controlled by σ54. σ54 seems to be mainly involved in the control of carbohydrate metabolism in *L. monocytogenes* via direct regulation of PTS activity, alteration of the pyruvate pool, and modulation of carbon catabolite regulation.

6.4.2.6 *VirR*

virR is a gene encoding a putative response regulator of a two-component system identified by signature-tagged mutagenesis (STM) (Mandin et al., 2005). Deletion of *virR* severely decreased virulence in mice as well as invasion in cultured cell lines. Using a transcriptomic approach, 12 genes regulated by VirR were identified, including the *dlt* operon, previously reported to be important for *L. monocytogenes* virulence. However, a strain lacking *dltA* was not as impaired in virulence as the Δ*virR* strain, suggesting a role in virulence for other members of the *vir* regulon. This was verified by the involvement of another VirR-regulated gene, *mprF*, in *Listeria* resistant to cationic antimicrobial peptides and virulence (Thedieck et al., 2006). VirR thus appears to control virulence by a global regulation of surface component modifications. These modifications may affect interactions with host cells, including components of the innate immune system.

6.4.2.7 DegU

An isogenic mutant of *L. monocytogenes* with a deletion of the response regulator gene *degU* showed a lack of motility due to the absence of flagella. Microarray analyses were used to identify the listerial genes that depend on DegU for expression (Williams et al., 2005). The two *L. monocytogenes* operons encoding flagella-specific genes and the monocistronically transcribed *flaA* gene were shown to be positively regulated by DegU at 24°C, but are not expressed at 37°C.

6.4.2.8 MogR

In *L. monocytogenes* MogR tightly represses expression of flagellin (FlaA) during extracellular growth at 37°C and during intracellular infection. MogR is also required for full virulence in a murine model of infection, the severe virulence defect of MogR-negative bacteria being proposed to be due to overexpression of FlaA. Microarray analyses revealed that MogR represses transcription of all known flagellar motility genes by binding directly to recognition sites positioned within promoter regions such that RNA polymerase binding is occluded (Shen and Higgins, 2006). MogR repression of transcription is antagonized in a temperature-dependent manner by the DegU response regulator and DegU further regulates FlaA levels through a post-transcriptional mechanism. Using tiling arrays two different promoters were identified for MogR (Toledo-Arana et al., 2009). P1 generates a long 5′-UTR that overlaps three genes required for the synthesis of the flagellum, and the P1 transcript is not affected by the temperature, it is overexpressed in stationary phase and is SigB dependent. P2 is constitutively expressed (Toledo-Arana et al., 2009).

6.4.2.9 CodY

L. monocytogenes encodes a functional member of the CodY family of global regulatory proteins that is responsive to both GTP and branched chain amino acids. By transcript analyses the CodY regulon was shown to comprise genes involved in amino acid metabolism, nitrogen assimilation as well as genes involved in sugar uptake and incorporation, indicating a role for CodY in *L. monocytogenes* in both carbon and nitrogen assimilation (Bennett et al., 2007). A *relA* mutant reduced the expression of the CodY regulon in early stationary phase, and introduction of a *codY* mutation into a *relA* mutant restored virulence. This indicates that the activity of a CodY-type protein influences pathogenesis and provides new information on the physiological adaptation of *L. monocytogenes* to post-exponential phase growth and virulence.

6.4.3 Transcriptomics to Study Listeria Adaptation to Stress Conditions

Whole-genome transcriptional profiling of *Listeria* also permitted new insights into adaptive responses activated by the bacteria when growing in different environments, in particular under stress conditions.

6.4.3.1 Cold Shock

Whole-genome microarray experiments were performed to define the *L. monocytogenes* cold growth regulon and to identify genes differentially expressed during growth at 4 or 37°C (Chan et al., 2007). Microarray analysis revealed that a large number of *L. monocytogenes* genes are differentially expressed at 4 and 37°C, with more genes showing higher transcript levels than lower transcript levels at 4°C. *L. monocytogenes* genes with higher transcript levels at 4°C include a number of genes and operons with previously reported or plausible roles in cold adaptation. *L. monocytogenes* genes with lower transcript levels at 4°C include a number of virulence and virulence-associated genes as well as some heat-shock genes.

6.4.3.2 Heat Shock

To investigate the heat-shock response of *L. monocytogenes*, whole-genome expression profiles of cells that were grown at 37°C and exposed to 48°C were examined using DNA microarrays (van der Veen et al., 2007). After 3 min, 25% of the genes were differentially expressed, while after 40 min only 2% of genes showed differential expression, indicative of the transient nature of the heat-shock response. Several heat-shock-induced genes are part of the SOS response in *L. monocytogenes*. Furthermore, numerous differentially expressed genes that have roles in the cell division machinery or cell wall synthesis were downregulated. This expression pattern is in line with the observation that heat shock results in cell elongation and prevention of cell division.

6.4.3.3 Alkaline Stress

In order to gain a more comprehensive perspective on the physiology and regulation of the alkali-tolerance response (AlTR) of *Listeria*, the differential gene expression of cells adapted at pH 9.5 and unadapted cells (pH7.0) was evaluated using DNA microarray (Giotis et al., 2008). *L. monocytogenes* was shown to exhibit a significant AlTR following exposure to mild alkali (pH 9.5), which is capable of protecting cells from subsequent lethal alkali stress (pH 12.0). Alkali pH provides therefore *L. monocytogenes* with nonspecific multiple stress resistance that may be vital for survival in the human gastrointestinal tract as well as within food processing systems where alkali conditions prevail. The AlTR in *L. monocytogenes* appears thus to function to minimize excess alkalization and energy expenditures while mobilizing available carbon sources.

6.4.3.4 Hydrostatic Pressure

High hydrostatic pressure processing (HPP) is currently being used as a treatment for certain foods to control the presence of foodborne pathogens, such as *L. monocytogenes*. Genomic microarray analysis was performed to determine the effects of HPP on *L. monocytogenes* in order to understand how it responds to mechanical stress injury (Bowman et al., 2008). HPP induced increased expression of genes associated with DNA repair mechanisms, transcription and translation protein complexes, the septal ring, the general protein translocase system, flagella assemblage and chemotaxis, and lipid and peptidoglycan biosynthetic pathways. On the other hand, HPP appears to suppress a wide range of energy production and conversion, carbohydrate metabolism, and virulence-associated genes accompanied by strong suppression of the SigB and PrfA regulons. HPP also affected genes controlled by the pleotrophic regulator CodY. HPP-induced cellular damage appears to lead to increased expression of genes linked to sections of the cell previously shown in bacteria to be damaged or altered during HPP exposure and suppression of gene expression associated with cellular growth processes and virulence.

6.4.3.5 Glycerol

L. monocytogenes is able to efficiently utilize glycerol as a carbon source. Comparative transcriptome analyses of *L. monocytogenes* showed high-level transcriptional upregulation of the genes known to be involved in glycerol uptake and metabolism (*glpFK* and *glpD*) in the presence of glycerol (compared to that in the presence of glucose and/or cellobiose) (Joseph et al., 2008). Transcriptional downregulation in the presence of glycerol was observed for several genes that are positively regulated by glucose, including genes involved in glycolysis, N-metabolism, and the biosynthesis of branched-chain amino acids. The highest level of transcriptional upregulation was observed for all PrfA-dependent genes during early and late logarithmic growth in glycerol.

6.4.3.6 Anti-listerial Action of Complex Microbial Consortia

Some complex red-smear cheese microbial multispecies consortia exclude *L. monocytogenes* through yet unknown interactions. In order to characterize this inhibitory action, microarray experiments have been performed (Hain et al., 2007). Contact with anti-listerial microbial consortia causes extensive transcriptome changes.

It was found that nearly 400 genes are up- or downregulated upon contact with the anti-listerial consortium. Among the strongly induced genes are genes involved in energy supply and uptake systems. Under these conditions, *Listeria* cells are severely stressed, which is demonstrated by the induction of general stress proteins. The cells react by inducing genes involved in cell wall synthesis and by changing the lipid composition. A further large group of highly induced genes includes those involved in DNA repair and maintenance. Interestingly, genes involved in bacteriocin production are induced which could be evidence for a counterattack mounted by *L. monocytogenes*. In addition, several detoxification systems are induced.

6.4.3.7 Milk

To study how *L. monocytogenes* survives and grows in UHT skim milk, microarray technology was used to monitor the gene expression profiles of strain F2365 in UHT skim milk (Liu and Ream, 2008). Compared to *L. monocytogenes* grown in brain heart infusion (BHI) broth for 24 h at 4°C, 26 genes were upregulated in UHT skim milk, whereas 14 genes were downregulated. The upregulated genes included genes encoding for transport and binding proteins, transcriptional regulators, proteins in amino acid biosynthesis and energy metabolism, protein synthesis, cell division, and hypothetical proteins. The downregulated genes included genes that encode for transport and binding proteins, protein synthesis, cellular processes, cell envelope, energy metabolism, a transcriptional regulator, and an unknown protein. This represents the first study of global transcriptional gene expression profiling of *L. monocytogenes* in a liquid food.

6.4.4 Transcriptomics and Listeria *Adaptation to Host Cell Cytosol*

A successful transition of *L. monocytogenes* from the extracellular to the intracellular environment requires a precise adaptation response to conditions encountered in the host milieu. DNA microarray was used to investigate the transcriptional profile of intracellular *L. monocytogenes* following epithelial cell or macrophage infection.

6.4.4.1 Transcriptomics in Epithelial Cells

Following epithelial cell infection, approximately 19% of the *Listeria* genes were differentially expressed relative to their level of transcription when grown in BHI medium, including genes encoding transporter proteins essential for the uptake of carbon and nitrogen sources, factors involved in anabolic pathways, stress proteins, transcriptional regulators, and proteins of unknown function (Chatterjee et al., 2006). This study of the *Listeria* genome expression in epithelial cells indicates that *L. monocytogenes* can use alternative carbon sources like phosphorylated glucose and glycerol and nitrogen sources like ethanolamine during replication in epithelial cells and that the pentose phosphate cycle, but not glycolysis, is the predominant pathway of sugar metabolism in the host environment. Additionally, the synthesis of arginine, isoleucine, leucine, and valine, as well as a species-specific phosphoenolpyruvate-dependent phosphotransferase system, seems to play a major role in the intracellular growth of *L. monocytogenes*.

6.4.4.2 Transcriptomics in Macrophages

Murine macrophage cell line P388D1 was infected with *L. monocytogenes* EGDe and the gene expression profile of *L. monocytogenes* inside the vacuolar and cytosolic environments of the host cell was examined by using whole-genome microarray and mutant analyses (Joseph et al., 2006). Seventeen percent of the total genome was mobilized to enable adaptation for intracellular growth. Intracellularly expressed genes showed responses typical of glucose limitation within bacteria, with a decrease in the amount of mRNA-encoding enzymes in the central metabolism and a temporal induction of genes involved in alternative carbon source utilization pathways and their regulation. Adaptive intracellular gene expression involved genes that are associated with virulence, the general stress response, cell division, and changes in cell wall structure and included many genes with unknown functions. A total of 41 genes were species specific, being absent from the genome of the nonpathogenic *L. innocua* CLIP 11262 strain. Twenty-five genes were strain specific, i.e., absent from the genome of the previously sequenced *L. monocytogenes* F2365 serotype 4b strain, suggesting heterogeneity in the gene pool required for intracellular survival of *L. monocytogenes* in host cells.

6.4.5 *In Vivo Transcriptomics:* Listeria *Host Adaptation and Pathogenesis*

Virulence being a trait that only manifests in a susceptible host, in order to gain a better understanding of the nature of host–pathogen interactions, the *L. monocytogenes* genome expression was studied during mouse infection (Camejo et al., 2009). In the spleen of infected mice, <20% of the *Listeria* genome is differentially expressed, essentially through gene activation, as compared to exponential growth in BHI. During infection, *Listeria* is in an active multiplication phase, as revealed by the high expression of genes involved in replication, cell division, and multiplication. In vivo bacterial growth requires increased expression of genes involved in adaptation of the bacterial metabolism and stress responses, in particular to oxidative stress. *Listeria* interaction with its host induces cell wall metabolism and surface expression of virulence factors (Fig. 6.7). During infection, *L. monocytogenes* also activates subversion mechanisms of host defenses, including resistance to cationic peptides, peptidoglycan modifications, and release of muramyl peptides. The in vivo differential expression of the *Listeria* genome is coordinated by a complex regulatory network, with a central role for the PrfA–SigB interplay. In particular, *L. monocytogenes* upregulates in vivo the two major virulence regulators, PrfA and VirR, and their downstream effectors. Mutagenesis of in vivo-induced genes allowed the identification of novel *L. monocytogenes* virulence factors (Camejo et al., 2009).

Fig. 6.7 In vivo expression
of virulence genes.
Expression during mouse
spleen infection of the 31
known virulence genes
differentially regulated
in vivo. A peak of expression
was observed for the majority
of these virulence genes 48 h
p.i. All measurements are
relative to BHI. Genes were
selected for this analysis
when their expression
deviated from BHI by at least
a factor of 2.0 in at least one
time point (Camejo et al.,
2009)

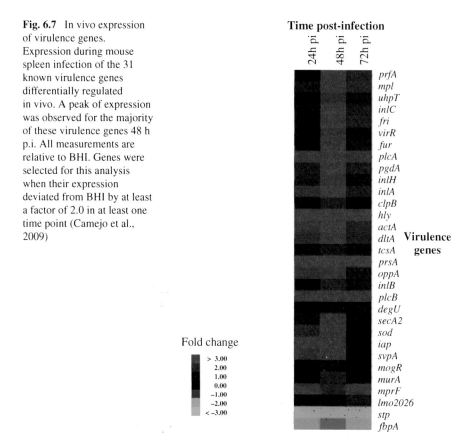

To understand how *L. monocytogenes* switches from a saprophyte to a pathogen, the transcription of its entire genome was investigated using tiling microarrays – made of 25-base oligonucleotides with a 9-nucleotide overlap – covering both strands of *L. monocytogenes* chromosome uniformly (Toledo-Arana et al., 2009). Transcripts of EGDe (as well as deletion mutants for *prfA* and *sigB*) grown to exponential phase at 37°C in rich medium were compared to *Listeria* RNAs extracted from bacteria present in the intestinal lumen of mice 24 h after oral inoculation or from bacteria incubated in blood from healthy human donors. Comparison of gene expression levels in the intestine and in BHI revealed an extensive transcriptional reshaping, with 437 upregulated and 769 downregulated genes. Many were related to bacterial stress responses and hypoxic lifestyle. Interestingly, 232 genes were under the control of SigB. Some of these genes encode the invasion proteins, InlA and InlB, and surface proteins, for example, InlH, Lmo0610, and Lmo2085, which could have a role in adhesion to and/or crossing of the intestinal barrier. Analysis also revealed σB-dependent upregulation of the bile salt hydrolase bsh gene required for survival in vivo (Dussurget et al., 2002). These results establish the crucial role

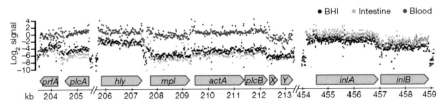

Fig. 6.8 Tiling maps of the virulence locus and the *inlAB* locus (Toledo-Arana et al., 2009)

of SigB for *Listeria* intestinal adaptation. Strikingly, PrfA did not have a significant role in the control of most genes differentially expressed in the intestine. In contrast to what was observed in the intestinal lumen, the virulence gene locus was highly overexpressed in blood (Fig. 6.8), in agreement with the function of these genes. A total of 1,261 genes were differentially expressed in blood: 545 genes showed higher transcript levels whereas 716 showed lower transcript levels. Transcription of the virulence locus and other virulence genes (Uhtp, InlA, InlB, InlC) was PrfA dependent and SigB independent (Toledo-Arana et al., 2009). In both intestine and blood, PrfA- and SigB-independent differential gene expression occurs predicting other regulatory pathways in the coordination of *Listeria* infection.

6.5 *Listeria* Proteomics

Proteomics is an important tool to complete whole-genome expression studies and help to understand the microorganism physiology.

6.5.1 Comparative Proteomics

A partially annotated proteome reference map of *L. monocytogenes* was developed in 2003 for exponentially growing cells under standardized conditions (Ramnath et al., 2003). The map contained 33 identified proteins representing the four main protein functional classes and was used to compare the total proteins of different *Listeria* strains, constituting a starting point for analyzing the proteomes of different *Listeria* isolates. To investigate the diversity of *L. monocytogenes*, the intracellular and extracellular proteins of 12 strains were also analyzed (Dumas et al., 2008). The clustering of the strains in two groups based on proteomic patterns is in agreement with the *L. monocytogenes* phylogenetic lineages. A total of 26 and 21 protein spots were shown to be significantly overexpressed and underexpressed, respectively, in the six strains of serovar 1/2a (lineage II) compared to strains of serovar 1/2b or 4b. To better understand the differences that exist between different genetic lineages/serovars of *L. monocytogenes*, the global protein expression of the serotype 1/2a strain EGD and the serotype 4b strain F2365 was more recently analyzed (Donaldson et al., 2009). A total of 1,754 proteins from EGD and 1,427 proteins from F2365 were identified, of which 1,077 were common to both. Analysis of proteins that had significantly altered expression between strains revealed potential biological differences between these two *L. monocytogenes* strains. In particular, the strains differed in expression of proteins involved in cell wall physiology and flagellar biosynthesis, as well as DNA repair proteins and stress response proteins.

6.5.1.1 Proteomics of Surface Proteins

The surface subproteome of *L. monocytogenes* that includes many proteins already known to be involved in virulence and interaction with host cells has been characterized and 55 proteins were identified (Schaumburg et al., 2004). About 16% of these proteins are of unknown function and three proteins have no orthologue in the nonpathogenic *L. innocua*. Remarkably, a relatively high number of proteins with a function in the cytoplasmic compartment were identified in this surface proteome, like enolase. Enolase was found to be able to bind human plasminogen, indicating a possible role of this protein as receptor for human plasminogen on the bacterial cell surface. The *Listeria* cell wall proteome was also characterized using a nonelectrophoretic approach (Calvo et al., 2005). A total of 30 protein species were unequivocally identified in cell wall extracts of the genome strains *L. monocytogenes* EGDe (19 proteins) and *L. innocua* CLIP11262 (11 proteins). Among them, 20 proteins bearing an LPXTG motif recognized for covalent anchoring to the peptidoglycan were identified. Other proteins detected included peptidoglycan lytic enzymes, a penicillin-binding protein, and proteins bearing an NXZTN motif recently proposed to direct protein anchoring to the peptidoglycan.

L. monocytogenes lipoproteins were also systematically analyzed by proteomics using a mutant defective in lipoprotein diacylglyceryl transferase (Lgt), an enzyme involved in lipoprotein processing (Baumgartner et al., 2007). This study revealed new findings concerning the roles of Lgt- and lipoprotein-specific signal peptidase II (Lsp) during lipoprotein processing, identified 26 of the 68 lipoproteins predicted in the initial annotation of the *L. monocytogenes* EGDe genome, and evidenced that a few of these lipoproteins are regulated by PrfA.

6.5.1.2 Proteomics of Secreted Proteins

Extracellular proteins of bacterial pathogens play a crucial role in the infection of the host. A first comprehensive validation of the secretory subproteome of *Listeria* was performed in 2005 (Trost et al., 2005). Proteomic analysis has identified 105 proteins in the culture supernatant of *L. monocytogenes*. Among these are all the currently known virulence factors with a signal peptide. The comparison between the *L. monocytogenes* and the nonpathogenic species *L. innocua* was performed to reveal proteins probably involved in pathogenicity and/or the adaptation to their respective lifestyles. In addition to the eight known virulence factors, all of which have no orthologous genes in *L. innocua*, eight additional proteins have been identified that exhibit the typical key feature defining the known listerial virulence factors. Further significant differences between the two species are evident in the group of cell wall and secretory proteins that warrant further study. In order to take into account the biodiversity of *L. monocytogenes* species, a new exoproteomic analysis was recently carried out on 12 representative strains (Dumas et al., 2009). A total of 151 spots were identified and corresponded to 60 non-orthologous proteins. Most of them were actually predicted as secreted via the Sec translocon. While the core exoproteome included most proteins related to bacterial virulence, cell wall biogenesis, as well as proteins secreted by unknown pathways, a slight variation in the protein members of these categories was observed and constituted the variant exoproteome.

6.5.2 Functional Proteomics

6.5.2.1 Growth Phase Proteomics

Proteomic approach was used to investigate the response of *L. monocytogenes* Scott A cultures to the transition from exponential to stationary growth phase (Weeks and Lupfer, 2004). Global analysis of the proteome revealed that the expression levels of more than 50% of all proteins observed changed significantly during this transition phase. The proteome of *L. monocytogenes* EGDe cells in the exponential or stationary phase of growth was also analyzed and allowed to characterize 161 variations in protein spot intensity, of which 38 were identified, including ribosomal proteins and proteins involved in cellular metabolism or in stress adaptation (Folio et al., 2004).

6.5.2.2 Proteomics of Biofilms

Biofilm formation may be important in the colonization of the food processing environment by *Listeria* and represents a serious problem for food safety. The proteome of a *L. monocytogenes* isolated from a food plant was first investigated to study the differential protein pattern expressed by biofilms and planktonic bacteria (Tremoulet et al., 2002). The proteomes of *L. monocytogenes* expressed in suspension and biofilm state were also analyzed in the presence and absence of a carbon source, revealing induced proteins that contribute to the mechanisms of starvation survival in both conditions (Helloin et al., 2003). To gain insight into the physiological state of cells in adherent multicellular layers formed by *L. monocytogenes* 568, biofilm- and planktonic-grown cells from the same cultures were compared (Hefford et al., 2005). Nineteen proteins were found to be more expressed in biofilm-grown cells and were involved in stress response, envelope and protein synthesis, biosynthesis, energy generation, and regulatory functions. In order to identify proteins that might be associated with biofilm production and chlorine tolerance, the protein expression of cultural variants of *L. monocytogenes* Scott A was compared (Folsom and Frank, 2007). Hypochlorous acid-tolerant variants produced greater amounts of biofilm, with an upregulation of a sugar-binding protein and downregulation of the 50S ribosomal protein.

6.5.2.3 Proteomic Adaptation to Stress Conditions

Listeria have developed many sophisticated strategies to withstand hostile environmental conditions, and as a result the organisms become more resistant to further stress. In order to better understand these mechanisms, the proteome of *Listeria* was analyzed in response to different stresses.

The *L. monocytogenes* proteome was first studied in response to acidic conditions even before the publication of the first *Listeria* genome sequence (Phan-Thanh and Mahouin, 1999). In the absence of genome sequence data, this study was focused on the comparison of the response to different levels of acid stress, revealing a correlation between the stress level and the number of proteins induced. Protein variations in *L. monocytogenes* were then analyzed in response to salt (Duche et al., 2002), alkali (Giotis et al., 2008), and heat-shock (Agoston et al., 2009) stresses. After an osmotic shock, the synthesis rate of 59 proteins was altered by salinity including Ctc and GbuA (Duche et al., 2002). The proteomic analysis of the alkali-tolerance response (AlTR) of *L. monocytogenes* revealed modification in proteins that are associated with virulence, the general stress response, cell division, and changes in cell wall structure (Giotis et al., 2008). Comparison of the proteome of *L. monocytogenes* under heat-shock conditions showed 18 proteins differentially expressed at 60°C. DnaN, a previously identified stress protein, TcsA, a lipoprotein (CD4 T-cell-stimulating antigen), and Gap (glyceraldehyde-3-phosphate-dehydrogenase) were selectively expressed under heat-shock conditions (Agoston et al., 2009).

Proteomics was also used to analyze the response of *Listeria* to specific culture conditions. Changes in proteins found were studied in the presence of salts of organic acids generally recognized as safe substances, revealing modification in the expression level of an oxidoreductase, a lipoprotein, a DNA-binding protein, an alpha-amylase, and two SecA proteins (Mbandi et al., 2007). Protein expression by *L. monocytogenes* grown on ready-to-eat (RTE) meat matrix was also studied, identifying proteins more expressed under these conditions and known to be involved in virulence and stress adaptation such as ClpB, ClpC, ClpP, and surface antigen (Mujahid et al., 2008). The effect of liquid smoke on the proteomic pattern of *L. monocytogenes* was investigated. The proteome of cytosoluble proteins was slightly modified after incubation with liquid smoke but no protein already characterized in response to other known stresses was induced, except the protease ClpP (Guilbaud et al., 2008).

6.5.2.4 Proteomics of *Listeria* Mutants

In *L. monocytogenes* the alternative sigma factor sigmaB plays important roles in both virulence and stress tolerance. Proteomics was used to define components of the SigB regulon. Thirty-eight proteins whose expression was sigmaB dependent were identified; 17 of these proteins were found to require the presence of *sigB* for full expression, while 21 were expressed at a higher level in the *sigB* mutant (Abram et al., 2008b). The proteomic data reaffirmed a role for sigmaB in the general stress response and highlighted a probable role for sigmaB in metabolism, especially in the utilization of alternative carbon sources. The impact of a *sigB* deletion on the proteome of *Listeria* grown in a defined medium both in the presence and in the absence of osmotic stress was also investigated (Abram et al.,

2008a, b). Eleven proteins were found to be differentially expressed in the *sigB* mutant. These proteins had putative functions related to carbon utilization or are known to be stress inducible. Deletion of genes encoding some of these proteins revealed that they play a role in osmotolerance and acid stress tolerance to various degrees.

Sortases are enzymes that anchor surface proteins to the cell wall of gram-positive bacteria by cleaving a sorting motif located in the C-terminus of the protein substrate. A non-gel proteomic approach was used to identify surface proteins recognized by the two sortases of *L. monocytogenes*, SrtA and SrtB (Pucciarelli et al., 2005). A total of 13 LPXTG-containing proteins were identified exclusively in strains having a functional SrtA. In contrast, two surface proteins were identified when only SrtB was active. The analysis of the peptides identified in these proteins suggests that SrtB may recognize two different sorting motifs.

IspC is a cell wall-anchored peptidoglycan hydrolase (autolysin), capable of degrading the cell wall peptidoglycan of the bacterium. To determine if this enzyme has any biological functions and/or plays a role in virulence, an *ispC* deletion mutant was generated (Wang and Lin, 2008). Analysis of the mutant phenotype supports a role for IspC as an adhesin in virulence. Proteomic analysis showed a reduced surface expression of some known or putative virulence factors (e.g., ActA, InlC2, and FlaA) due to IspC deficiency. This study demonstrates that IspC is not important for cell division or separation but is essential for full virulence in vivo.

6.6 Conclusions

Genome sequencing has revolutionized how we approach the problem of understanding *Listeria* pathogenesis. Comparative genomics allowed important progresses in our comprehension of *Listeria* biodiversity, but also provided powerful tools for *Listeria* identification and discrimination. Additionally, comparative genomics has especially increased the pace of identification and characterization of *Listeria* virulence factors. Genomic microarray-based methods allowed the study of all the *Listeria* gene expression under various physiological or genetic states of interest. In particular, postgenomic studies have moved us closer to understand which genes are essential for *Listeria* growth and survival in different environments, including inside the host. Tiling microarrays coupled to high-speed sequencing open now new perspectives for comprehensive epidemiologic, evolution, adaptation, expression, and regulation studies. In particular, this approach revealed the importance of non-coding sequences. Indeed, RNAs (riboswitches, 5′-untranslated regions of mRNAs, and small non-coding RNAs) emerge as crucial regulators, enabling the pathogen to adapt its metabolic needs during infection and to express its virulence genes when required. The analysis of regions outside open reading frames should provide major advances not only in the understanding of the *Listeria* pathogenesis but also in the development of novel strategies to prevent foodborne diseases.

References

Abram F, Starr E, Karatzas KA, Matlawska-Wasowska K, Boyd A, Wiedmann M, Boor KJ, Connally D, O'Byrne CP (2008a) Identification of components of the sigma B regulon in *Listeria monocytogenes* that contribute to acid and salt tolerance. Appl Environ Microbiol 74:6848–6858

Abram F, Su WL, Wiedmann M, Boor KJ, Coote P, Botting C, Karatzas KA, O'Byrne CP (2008b) Proteomic analyses of a *Listeria monocytogenes* mutant lacking sigmaB identify new components of the sigmaB regulon and highlight a role for sigmaB in the utilization of glycerol. Appl Environ Microbiol 74:594–604

Agoston R, Soni K, Jesudhasan PR, Russell WK, Mohacsi-Farkas C, Pillai SD (2009) Differential expression of proteins in *Listeria monocytogenes* under thermotolerance-inducing, heat shock, and prolonged heat shock conditions. Foodborne Pathog Dis 6:1133–1140

Alvarez-Dominguez C, Vazquez-Boland JA, Carrasco-Marin E, Lopez-Mato P, Leyva-Cobian F (1997) Host cell heparan sulfate proteoglycans mediate attachment and entry of *Listeria monocytogenes*, and the *Listerial* surface protein ActA is involved in heparan sulfate receptor recognition. Infect Immun 65:78–88

Arous S, Buchrieser C, Folio P, Glaser P, Namane A, Hebraud M, Hechard Y (2004) Global analysis of gene expression in an rpoN mutant of *Listeria monocytogenes*. Microbiology 150:1581–1590

Baumgartner M, Karst U, Gerstel B, Loessner M, Wehland J, Jansch L (2007) Inactivation of Lgt allows systematic characterization of lipoproteins from *Listeria monocytogenes*. J Bacteriol 189:313–324

Bennett HJ, Pearce DM, Glenn S, Taylor CM, Kuhn M, Sonenshein AL, Andrew PW, Roberts IS (2007) Characterization of relA and codY mutants of *Listeria monocytogenes*: identification of the CodY regulon and its role in virulence. Mol Microbiol 63:1453–1467

Bierne H, Cossart P (2007) *Listeria monocytogenes* surface proteins: from genome predictions to function. Microbiol Mol Biol Rev 71:377–397

Bierne H, Garandeau C, Pucciarelli MG, Sabet C, Newton S, Garcia-del Portillo F, Cossart P, Charbit A (2004) Sortase B, a new class of sortase in *Listeria monocytogenes*. J Bacteriol 186:1972–1982

Bierne H, Mazmanian SK, Trost M, Pucciarelli MG, Liu G, Dehoux P, Jansch L, Garcia-del Portillo F, Schneewind O, Cossart P, Consortium TELG (2002) Inactivation of the *srtA* gene in *Listeria monocytogenes* inhibits anchoring of surface proteins and affects virulence. Mol Microbiol 43:464–489

Bonazzi M, Lecuit M, Cossart P (2009) *Listeria monocytogenes* internalin and E-cadherin: from structure to pathogenesis. Cell Microbiol 1:a003087

Borucki MK, Kim SH, Call DR, Smole SC, Pagotto F (2004) Selective discrimination of *Listeria monocytogenes* epidemic strains by a mixed-genome DNA microarray compared to discrimination by pulsed-field gel electrophoresis, ribotyping, and multilocus sequence typing. J Clin Microbiol 42:5270–5276

Borucki MK, Krug MJ, Muraoka WT, Call DR (2003) Discrimination among *Listeria monocytogenes* isolates using a mixed genome DNA microarray. Vet Microbiol 92:351–362

Borucki MK, Reynolds J, Call DR, Ward TJ, Page B, Kadushin J (2005) Suspension microarray with dendrimer signal amplification allows direct and high-throughput subtyping of *Listeria monocytogenes* from genomic DNA. J Clin Microbiol 43:3255–3259

Bowman JP, Bittencourt CR, Ross T (2008) Differential gene expression of *Listeria monocytogenes* during high hydrostatic pressure processing. Microbiology 154:462–475

Bublitz M, Holland C, Sabet C, Reichelt J, Cossart P, Heinz DW, Bierne H, Schubert WD (2008) Crystal structure and standardized geometric analysis of InlJ, a *Listerial* virulence factor and leucine-rich repeat protein with a novel cysteine ladder. J Mol Biol 378:87–96

Buchrieser C (2007) Biodiversity of the species *Listeria monocytogenes* and the genus *Listeria*. Microbes Infect 9:1147–1155

Buchrieser C, Rusniok C, Kunst F, Cossart P, Glaser P (2003) Comparison of the genome sequences of *Listeria monocytogenes* and *Listeria* innocua: clues for evolution and pathogenicity. FEMS Immunol Med Microbiol 35:207–213

Cabanes D, Dehoux P, Dussurget O, Frangeul L, Cossart P (2002) Surface proteins and the pathogenic potential of *Listeria monocytogenes*. Trends Microbiol 10:238–245

Cabanes D, Dussurget O, Dehoux P, Cossart P (2004) Auto, a surface associated autolysin of *Listeria monocytogenes* required for entry into eukaryotic cells and virulence. Mol Microbiol 51:1601–1614

Cabanes D, Lecuit M, Cossart P (2008) Animal models of *Listeria* infection. Curr Protoc Microbiol (Chapter 9:Unit9B 1), 9B.1.1–9B.1.17

Cabanes D, Sousa S, Cebria A, Lecuit M, Garcia-del Portillo F, Cossart P (2005) Gp96 is a receptor for a novel *Listeria monocytogenes* virulence factor, Vip, a surface protein. Embo J 24:2827–2838

Call DR, Borucki MK, Besser TE (2003a) Mixed-genome microarrays reveal multiple serotype and lineage-specific differences among strains of *Listeria monocytogenes*. J Clin Microbiol 41:632–639

Call DR, Borucki MK, Loge FJ (2003b) Detection of bacterial pathogens in environmental samples using DNA microarrays. J Microbiol Methods 53:235–243

Calvo E, Pucciarelli MG, Bierne H, Cossart P, Albar JP, Garcia-Del Portillo F (2005) Analysis of the *Listeria* cell wall proteome by two-dimensional nanoliquid chromatography coupled to mass spectrometry. Proteomics 5:433–443

Camejo A, Buchrieser C, Couvé E, Carvalho F, Reis O, Ferreira P, Sousa S, Cossart P, Cabanes D (2009) In vivo transcriptional profiling of *Listeria monocytogenes* and mutagenesis identify new virulence factors involved in infection. PloS Pathogens 5:e1000449. doi:1000410.1001371/journal.ppat.1000449

Chan YC, Raengpradub S, Boor KJ, Wiedmann M (2007) Microarray-based characterization of the *Listeria monocytogenes* cold regulon in log- and stationary-phase cells. Appl Environ Microbiol 73:6484–6498

Chatterjee SS, Hossain H, Otten S, Kuenne C, Kuchmina K, Machata S, Domann E, Chakraborty T, Hain T (2006) Intracellular gene expression profile of *Listeria monocytogenes*. Infect Immun 74:1323–1338

Cossart P (2007) Listeriology (1926–2007): the rise of a model pathogen. Microbes Infect 9(10):1143–1146

Cossart P, Pizarro-Cerda J, Lecuit M (2003) Invasion of mammalian cells by *Listeria monocytogenes*: functional mimicry to subvert cellular functions. Trends Cell Biol 13:23–31

Cossart P, Sansonetti PJ (2004) Bacterial invasion: the paradigms of enteroinvasive pathogens. Science 304:242–248

Cossart P, Toledo-Arana A (2008) *Listeria monocytogenes*, a unique model in infection biology: an overview. Microbes Infect 10:1041–1050

Disson O, Grayo S, Huillet E, Nikitas G, Langa-Vives F, Dussurget O, Ragon M, Le Monnier A, Babinet C, Cossart P, Lecuit M (2008) Conjugated action of two species-specific invasion proteins for fetoplacental listeriosis. Nature 455:1114–1118

Domann E, Wehland J, Rohde M, Pistor S, Hartl M, Goebel W, Leimeister-Wachter M, Wuenscher M, Chakraborty T (1992) A novel bacterial virulence gene in *Listeria monocytogenes* required for host cell microfilament interaction with homology to the proline-rich region of vinculin. Embo J 11:1981–1990

Donaldson JR, Nanduri B, Burgess SC, Lawrence ML (2009) Comparative proteomic analysis of *Listeria monocytogenes* strains F2365 and EGD. Appl Environ Microbiol 75:366–373

Doumith M, Cazalet C, Simoes N, Frangeul L, Jacquet C, Kunst F, Martin P, Cossart P, Glaser P, Buchrieser C (2004) New aspects regarding evolution and virulence of *Listeria monocytogenes* revealed by comparative genomics and DNA arrays. Infect Immun 72:1072–1083

Dramsi S, Bourdichon F, Cabanes D, Lecuit M, Fsihi H, Cossart P (2004) FbpA, a novel multifunctional *Listeria monocytogenes* virulence factor. Mol Microbiol 53:639–649

Duche O, Tremoulet F, Namane A, Labadie J (2002) A proteomic analysis of the salt stress response of *Listeria monocytogenes*. FEMS Microbiol Lett 215:183–188

Dumas E, Desvaux M, Chambon C, Hebraud M (2009) Insight into the core and variant exoproteomes of *Listeria monocytogenes* species by comparative subproteomic analysis. Proteomics 9:3136–3155

Dumas E, Meunier B, Berdague JL, Chambon C, Desvaux M, Hebraud M (2008) Comparative analysis of extracellular and intracellular proteomes of *Listeria monocytogenes* strains reveals a correlation between protein expression and serovar. Appl Environ Microbiol 74:7399–7409

Dussurget O, Cabanes D, Dehoux P, Lecuit M, Buchrieser C, Glaser P, Cossart P (2002) *Listeria monocytogenes* bile salt hydrolase is a PrfA-regulated virulence factor involved in the intestinal and hepatic phases of listeriosis. Mol Microbiol 45:1095–1106

Dussurget O, Pizarro-Cerda J, Cossart P (2004) Molecular determinants of *Listeria Monocytogenes* virulence. Annu Rev Microbiol 58:587–610

Folio P, Chavant P, Chafsey I, Belkorchia A, Chambon C, Hebraud M (2004) Two-dimensional electrophoresis database of *Listeria monocytogenes* EGDe proteome and proteomic analysis of mid-log and stationary growth phase cells. Proteomics 4:3187–3201

Folsom JP, Frank JF (2007) Proteomic analysis of a hypochlorous acid-tolerant *Listeria monocytogenes* cultural variant exhibiting enhanced biofilm production. J Food Prot 70:1129–1136

Giotis ES, Muthaiyan A, Blair IS, Wilkinson BJ, McDowell DA (2008) Genomic and proteomic analysis of the alkali-tolerance response (AlTR) in *Listeria monocytogenes* 10403S. BMC Microbiol 8:102

Glaser P, Frangeul L, Buchrieser C, Rusniok C, Amend A, Baquero F, Berche P, Bloecker H, Brandt P, Chakraborty T, Charbit A, Chetouani F, Couve E, de Daruvar A, Dehoux P, Domann E, Dominguez-Bernal G, Duchaud E, Durant L, Dussurget O, Entian KD, Fsihi H, Portillo FG, Garrido P, Gautier L, Goebel W, Gomez-Lopez N, Hain T, Hauf J, Jackson D, Jones LM, Kaerst U, Kreft J, Kuhn M, Kunst F, Kurapkat G, Madueno E, Maitournam A, Vicente JM, Ng E, Nedjari H, Nordsiek G, Novella S, de Pablos B, Perez-Diaz JC, Purcell R, Remmel B, Rose M, Schlueter T, Simoes N, Tierrez A, Vazquez-Boland JA, Voss H, Wehland J, Cossart P (2001) Comparative genomics of *Listeria* species. Science 294:849–852

Graves LM, Helsel LO, Steigerwalt AG, Morey RE, Daneshvar MI, Roof SE, Orsi RH, Fortes ED, Milillo SR, den Bakker HC, Wiedmann M, Swaminathan B and Sauders BD (2010) *Listeria marthii* sp. Nov., isolated from the natural environment, finger lakes national forest. Int J Syst Evol Microbiol 60(Pt 6):1280–1288

Guilbaud M, Chafsey I, Pilet MF, Leroi F, Prevost H, Hebraud M, Dousset X (2008) Response of *Listeria monocytogenes* to liquid smoke. J Appl Microbiol 104:1744–1753

Hain T, Chatterjee SS, Ghai R, Kuenne CT, Billion A, Steinweg C, Domann E, Karst U, Jansch L, Wehland J, Eisenreich W, Bacher A, Joseph B, Schar J, Kreft J, Klumpp J, Loessner MJ, Dorscht J, Neuhaus K, Fuchs TM, Scherer S, Doumith M, Jacquet C, Martin P, Cossart P, Rusniock C, Glaser P, Buchrieser C, Goebel W, Chakraborty T (2007) Pathogenomics of *Listeria* spp. Int J Med Microbiol 297:541–557

Hain T, Hossain H, Chatterjee SS, Machata S, Volk U, Wagner S, Brors B, Haas S, Kuenne CT, Billion A, Otten S, Pane-Farre J, Engelmann S, Chakraborty T (2008) Temporal transcriptomic analysis of the *Listeria monocytogenes* EGD-e sigmaB regulon. BMC Microbiol 8:20

Hain T, Steinweg C, Kuenne CT, Billion A, Ghai R, Chatterjee SS, Domann E, Karst U, Goesmann A, Bekel T, Bartels D, Kaiser O, Meyer F, Puhler A, Weisshaar B, Wehland J, Liang C, Dandekar T, Lampidis R, Kreft J, Goebel W, Chakraborty T (2006) Whole-genome sequence of *Listeria welshimeri* reveals common steps in genome reduction with *Listeria* innocua as compared to *Listeria monocytogenes*. J Bacteriol 188:7405–7415

Hefford MA, D'Aoust S, Cyr TD, Austin JW, Sanders G, Kheradpir E, Kalmokoff ML (2005) Proteomic and microscopic analysis of biofilms formed by *Listeria monocytogenes* 568. Can J Microbiol 51:197–208

Helloin E, Jansch L, Phan-Thanh L (2003) Carbon starvation survival of *Listeria monocytogenes* in planktonic state and in biofilm: a proteomic study. Proteomics 3:2052–2064

Hou XL, Jiang HL, Cao QY, Zhao LY, Chang BJ, Chen Z (2008) Using oligonucleotide suspension arrays for laboratory identification of bacteria responsible for bacteremia. J Zhejiang Univ Sci B 9:291–298

Hu Y, Oliver HF, Raengpradub S, Palmer ME, Orsi RH, Wiedmann M, Boor KJ (2007a) Transcriptomic and phenotypic analyses suggest a network between the transcriptional regulators HrcA and sigmaB in *Listeria monocytogenes*. Appl Environ Microbiol 73:7981–7991

Hu Y, Raengpradub S, Schwab U, Loss C, Orsi RH, Wiedmann M, Boor KJ (2007b) Phenotypic and transcriptomic analyses demonstrate interactions between the transcriptional regulators CtsR and Sigma B in *Listeria monocytogenes*. Appl Environ Microbiol 73:7967–7980

Jeffers GT, Bruce JL, McDonough PL, Scarlett J, Boor KJ, Wiedmann M (2001) Comparative genetic characterization of *Listeria monocytogenes* isolates from human and animal listeriosis cases. Microbiology 147:1095–1104

Jin LQ, Li JW, Wang SQ, Chao FH, Wang XW, Yuan ZQ (2005) Detection and identification of intestinal pathogenic bacteria by hybridization to oligonucleotide microarrays. World J Gastroenterol 11:7615–7619

Joseph B, Mertins S, Stoll R, Schar J, Umesha KR, Luo Q, Muller-Altrock S, Goebel W (2008) Glycerol metabolism and PrfA activity in *Listeria monocytogenes*. J Bacteriol 190:5412–5430

Joseph B, Przybilla K, Stuhler C, Schauer K, Slaghuis J, Fuchs TM, Goebel W (2006) Identification of *Listeria monocytogenes* genes contributing to intracellular replication by expression profiling and mutant screening. J Bacteriol 188:556–568

Kocks C, Gouin E, Tabouret M, Berche P, Ohayon H, Cossart P (1992) *Listeria monocytogenes*-induced actin assembly requires the *actA* gene product, a surface protein. Cell 68:521–531

Leclercq A, Clermont D, Bizet C, Grimont PA, Le Fleche-Mateos A, Roche SM, Buchrieser C, Cadet-Daniel V, Le Monnier A, Lecuit M and Allerberger F (2009) *Listeria rocourtiae* sp. Nov. Int J Syst Evol Microbiol 60(Pt 9):2210–2214

Lecuit M (2007) Human listeriosis and animal models. Microbes Infect 9:1216–1225

Lin MC, Huang AH, Tsen HY, Wong HC, Chang TC (2005) Use of oligonucleotide array for identification of six foodborne pathogens and Pseudomonas aeruginosa grown on selective media. J Food Prot 68:2278–2286

Liu Y, Ream A (2008) Gene expression profiling of *Listeria monocytogenes* strain F2365 in UHT skim milk. Appl Environ Microbiol 74:6859–6866

Mandin P, Fsihi H, Dussurget O, Vergassola M, Milohanic E, Toledo-Arana A, Lasa I, Johansson J, Cossart P (2005) VirR, a response regulator critical for *Listeria monocytogenes* virulence. Mol Microbiol 57:1367–1380

Mbandi E, Phinney BS, Whitten D, Shelef LA (2007) Protein variations in *Listeria monocytogenes* exposed to sodium lactate, sodium diacetate, and their combination. J Food Prot 70:58–64

McGann P, Raengpradub S, Ivanek R, Wiedmann M, Boor KJ (2008) Differential regulation of *Listeria monocytogenes* internalin and internalin-like genes by sigmaB and PrfA as revealed by subgenomic microarray analyses. Foodborne Pathog Dis 5:417–435

Mengaud J, Ohayon H, Gounon P, Mège RM, Cossart P (1996) E-cadherin is the receptor for internalin, a surface protein required for entry of *Listeria monocytogenes* into epithelial cells. Cell 84:923–932

Milillo SR, Badamo JM, Wiedmann M (2009) Contributions to selected phenotypic characteristics of large species- and lineage-specific genomic regions in *Listeria monocytogenes*. Food Microbiol 26:212–223

Milohanic E, Glaser P, Coppee JY, Frangeul L, Vega Y, Vazquez-Boland JA, Kunst F, Cossart P, Buchrieser C (2003) Transcriptome analysis of *Listeria monocytogenes* identifies three groups of genes differently regulated by PrfA. Mol Microbiol 47:1613–1625

Milohanic E, Jonquieres R, Cossart P, Berche P, Gaillard JL (2001) The autolysin ami contributes to the adhesion of *Listeria monocytogenes* to eukaryotic cells via its cell wall anchor. Mol Microbiol 39:1212–1224

Mujahid S, Pechan T, Wang C (2008) Protein expression by *Listeria monocytogenes* grown on a RTE-meat matrix. Int J Food Microbiol 128:203–211

Murray EGD, Webb RA, Swann MBR (1926) A disease of rabbits characterized by a large mononuclear leucocytosis, caused by a hitherto undescribed bacillus bacterium monocytogenes (n.sp.). J Pathol Bacteriol 29:407–439

Nelson KE, Fouts DE, Mongodin EF, Ravel J, DeBoy RT, Kolonay JF, Rasko DA, Angiuoli SV, Gill SR, Paulsen IT, Peterson J, White O, Nelson WC, Nierman W, Beanan MJ, Brinkac LM, Daugherty SC, Dodson RJ, Durkin AS, Madupu R, Haft DH, Selengut J, Van Aken S, Khouri H, Fedorova N, Forberger H, Tran B, Kathariou S, Wonderling LD, Uhlich GA, Bayles DO, Luchansky JB, Fraser CM (2004) Whole genome comparisons of serotype 4b and 1/2a strains of the food-borne pathogen *Listeria monocytogenes* reveal new insights into the core genome components of this species. Nucleic Acids Res 32:2386–2395

Ollinger J, Wiedmann M, Boor KJ (2008) SigmaB- and PrfA-dependent transcription of genes previously classified as putative constituents of the *Listeria monocytogenes* PrfA regulon. Foodborne Pathog Dis 5:281–293

Pandiripally VK, Westbrook DG, Sunki GR, Bhunia AK (1999) Surface protein p104 is involved in adhesion of *Listeria monocytogenes* to human intestinal cell line, Caco-2. J Med Microbiol 48:117–124

Phan-Thanh L, Mahouin F (1999) A proteomic approach to study the acid response in *Listeria monocytogenes*. Electrophoresis 20:2214–2224

Pilgrim S, Kolb-Maurer A, Gentschev I, Goebel W, Kuhn M (2003) Deletion of the gene encoding p60 in *Listeria monocytogenes* leads to abnormal cell division and loss of actin-based motility. Infect Immun 71:3473–3484

Pizarro-Cerda J, Cossart P (2006) Bacterial adhesion and entry into host cells. Cell 124:715–727

Pucciarelli MG, Calvo E, Sabet C, Bierne H, Cossart P, Garcia-Del Portillo F (2005) Identification of substrates of the *Listeria monocytogenes* sortases A and B by a non-gel proteomic analysis. Proteomics 5:4808–4817

Raengpradub S, Wiedmann M, Boor KJ (2008) Comparative analysis of the sigma B-dependent stress responses in *Listeria monocytogenes* and *Listeria* innocua strains exposed to selected stress conditions. Appl Environ Microbiol 74:158–171

Ragon M, Wirth T, Hollandt F, Lavenir R, Lecuit M, Le Monnier A, Brisse S (2008) A new perspective on *Listeria monocytogenes* evolution. PLoS Pathog 4:e1000146

Ramnath M, Rechinger KB, Jansch L, Hastings JW, Knochel S, Gravesen A (2003) Development of a *Listeria monocytogenes* EGDe partial proteome reference map and comparison with the protein profiles of food isolates. Appl Environ Microbiol 69:3368–3376

Sabet C, Lecuit M, Cabanes D, Cossart P, Bierne H (2005) The LPXTG protein InlJ, a new internalin involved in *Listeria monocytogenes* virulence. Infect Immun 73:6912–6922

Sabet C, Toledo-Arana A, Personnic N, Lecuit M, Dubrac S, Poupel O, Gouin E, Nahori MA, Cossart P, Bierne H (2008) The *Listeria monocytogenes* virulence factor InlJ is specifically expressed in vivo and behaves as an adhesin. Infect Immun 76:1368–1378

Schaumburg J, Diekmann O, Hagendorff P, Bergmann S, Rohde M, Hammerschmidt S, Jansch L, Wehland J, Karst U (2004) The cell wall subproteome of *Listeria monocytogenes*. Proteomics 4:2991–3006

Schlech WF 3rd, Lavigne PM, Bortolussi RA, Allen AC, Haldane EV, Wort AJ, Hightower AW, Johnson SE, King SH, Nicholls ES, Broome CV (1983) Epidemic listeriosis – evidence for transmission by food. N Engl J Med 308:203–206

Schmid MW, Ng EY, Lampidis R, Emmerth M, Walcher M, Kreft J, Goebel W, Wagner M, Schleifer KH (2005) Evolutionary history of the genus Listeria and its virulence genes. Syst Appl Microbiol 28(1):1–18

Schneewind O, Model P, Fischetti VA (1992) Sorting of protein A to the staphylococcal cell wall. Cell 70:267–281

Schnupf P, Portnoy DA (2007) Listeriolysin O: a phagosome-specific lysin. Microbes Infect 9:1176–1187

Scortti M, Monzo HJ, Lacharme-Lora L, Lewis DA, Vazquez-Boland JA (2007) The PrfA virulence regulon. Microbes Infect 9:1196–1207

Sergeev N, Distler M, Courtney S, Al-Khaldi SF, Volokhov D, Chizhikov V, Rasooly A (2004) Multipathogen oligonucleotide microarray for environmental and biodefense applications. Biosens Bioelectron 20:684–698

Seveau S, Pizarro-Cerda J, Cossart P (2007) Molecular mechanisms exploited by *Listeria monocytogenes* during host cell invasion. Microbes Infect 9:1167–1175

Severino P, Dussurget O, Vencio RZ, Dumas E, Garrido P, Padilla G, Piveteau P, Lemaitre JP, Kunst F, Glaser P, Buchrieser C (2007) Comparative transcriptome analysis of *Listeria monocytogenes* strains of the two major lineages reveals differences in virulence, cell wall, and stress response. Appl Environ Microbiol 73:6078–6088

Shen A, Higgins DE (2006) The MogR transcriptional repressor regulates nonhierarchal expression of flagellar motility genes and virulence in *Listeria monocytogenes*. PLoS Pathog 2:e30

Steinweg C, Kuenne CT, Billion A, Mraheil MA, Domann E, Ghai R, Barbuddhe SB, Kärst U, Goesmann A, Pühler A, Weisshaar B, Wehland J, Lampidis R, Kreft J, Goebel W, Chakraborty T, Hain T (2010) Complete genome sequence of *Listeria seeligeri*, a nonpathogenic member of the genus *Listeria*. J Bacteriol 192(5):1473–1474

Suarez M, Gonzalez-Zorn B, Vega Y, Chico-Calero I, Vazquez-Boland JA (2001) A role for ActA in epithelial cell invasion by *Listeria monocytogenes*. Cell Microbiol 3:853–864

Swaminathan B, Gerner-Smidt P (2007) The epidemiology of human listeriosis. Microbes Infect 9:1236–1243

Thedieck K, Hain T, Mohamed W, Tindall BJ, Nimtz M, Chakraborty T, Wehland J, Jansch L (2006) The MprF protein is required for lysinylation of phospholipids in *Listeria* membranes and confers resistance to cationic antimicrobial peptides (CAMPs) on *Listeria monocytogenes*. Mol Microbiol 62:1325–1339

Toledo-Arana A, Dussurget O, Nikitas G, Sesto N, Guet-Revillet H, Balestrino D, Loh E, Gripenland J, Tiensuu T, Vaitkevicius K, Barthelemy M, Vergassola M, Nahori MA, Soubigou G, Regnault B, Coppee JY, Lecuit M, Johansson J, Cossart P (2009) The *Listeria* transcriptional landscape from saprophytism to virulence. Nature 459:950–956

Tremoulet F, Duche O, Namane A, Martinie B, Labadie JC (2002) Comparison of protein patterns of *Listeria monocytogenes* grown in biofilm or in planktonic mode by proteomic analysis. FEMS Microbiol Lett 210:25–31

Trost M, Wehmhoner D, Karst U, Dieterich G, Wehland J, Jansch L (2005) Comparative proteome analysis of secretory proteins from pathogenic and nonpathogenic *Listeria* species. Proteomics 5:1544–1557

Vazquez-Boland JA, Kocks C, Dramsi S, Ohayon H, Geoffroy C, Mengaud J, Cossart P (1992) Nucleotide sequence of the lecithinase operon of *Listeria monocytogenes* and possible role of lecithinase in cell-to-cell spread. Infect Immun 60:219–230

van der Veen S, Hain T, Wouters JA, Hossain H, de Vos WM, Abee T, Chakraborty T, Wells-Bennik MH (2007) The heat-shock response of *Listeria monocytogenes* comprises genes involved in heat shock, cell division, cell wall synthesis, and the SOS response. Microbiology 153:3593–3607

Volokhov D, Rasooly A, Chumakov K, Chizhikov V (2002) Identification of *Listeria* species by microarray-based assay. J Clin Microbiol 40:4720–4728

Wang L, Lin M (2008) A novel cell wall-anchored peptidoglycan hydrolase (autolysin), IspC, essential for *Listeria monocytogenes* virulence: genetic and proteomic analysis. Microbiology 154:1900–1913

Weeks M, Lupfer MB (2004) Complicating race: the relationship between prejudice, race, and social class categorizations. Pers Soc Psychol Bull 30:972–984

Williams T, Joseph B, Beier D, Goebel W, Kuhn M (2005) Response regulator DegU of *Listeria monocytogenes* regulates the expression of flagella-specific genes. FEMS Microbiol Lett 252:287–298

You Y, Fu C, Zeng X, Fang D, Yan X, Sun B, Xiao D, Zhang J (2008) A novel DNA microarray for rapid diagnosis of enteropathogenic bacteria in stool specimens of patients with diarrhea. J Microbiol Methods 75:566–571

Zhang C, Zhang M, Ju J, Nietfeldt J, Wise J, Terry PM, Olson M, Kachman SD, Wiedmann M, Samadpour M, Benson AK (2003) Genome diversification in phylogenetic lineages I and II of *Listeria monocytogenes*: identification of segments unique to lineage II populations. J Bacteriol 185:5573–5584

Chapter 7
Genomics of *Salmonella* Species

Rocio Canals, Michael McClelland, Carlos A. Santiviago, and Helene Andrews-Polymenis

7.1 Introduction

Progress in the study of *Salmonella* survival, colonization, and virulence has increased rapidly with the advent of complete genome sequencing and higher capacity assays for transcriptomic and proteomic analysis. Although many of these techniques have yet to be used to directly assay *Salmonella* growth on foods, these assays are currently in use to determine *Salmonella* factors necessary for growth in animal models including livestock animals and in in vitro conditions that mimic many different environments. As sequencing of the *Salmonella* genome and microarray analysis have revolutionized genomics and transcriptomics of salmonellae over the last decade, so are new high-throughput sequencing technologies currently accelerating the pace of our studies and allowing us to approach complex problems that were not previously experimentally tractable. Studies on such complex problems as, for example, the genetic adaptation of *Salmonella* serotypes to particular hosts and large-scale analysis of the *Salmonella* proteome under a given set of conditions will allow us insight into the biology of this diverse group of organisms on a scale that was not previously possible. In this chapter, we review the major genomic, transcriptomic, and proteomic studies of salmonellae and we outline what we have learned through these studies about the factors necessary for *Salmonella* survival and growth in many diverse environments including livestock and foods.

7.2 General Aspects of *Salmonella* and Salmonellosis

The genus *Salmonella* is a heterogeneous group of Gram-negative bacteria, differentiable by biochemical and serological properties. The Kauffmann–White serotyping scheme distinguishes more than 2,500 *Salmonella* serotypes (commonly referred to

H. Andrews-Polymenis (✉)
Department of Microbial and Molecular Pathogenesis, College of Medicine, Texas A&M University System Health Science Center, College Station, TX 77843-11114, USA
e-mail: handrews@medicine.tamhsc.edu

M. Wiedmann, W. Zhang (eds.), *Genomics of Foodborne Bacterial Pathogens*,
Food Microbiology and Food Safety, DOI 10.1007/978-1-4419-7686-4_7,
© Springer Science+Business Media, LLC 2011

as "serovars") based on antigenic variations in the lipopolysaccharide (O-antigen) and flagella (H-antigen).

Salmonella is currently classified into two species: *Salmonella bongori* (formerly classified as the subspecies V) and *Salmonella enterica*. *S. enterica* is further subdivided into six subspecies that are designated either by names (subspecies *enterica, salamae, arizonae, diarizonae, houtenae,* and *indica*) or by roman numerals (subspecies I, II, IIIa, IIIb, IV, and VI, respectively). It is worth mentioning that a third *Salmonella* species (i.e., *Salmonella subterranea*) was recently described (Shelobolina et al., 2004); however, MLST analyses have revealed that the bacteria described is quite far from *Salmonella* and may represent a novel species of *Enterobacter* or *Escherichia* (Sylvain Brisse, personal communication).

Serovars within *S. enterica* subspecies *enterica* (ssp. I) are responsible for ~99% of all salmonellosis in warm-blooded animals. All other subspecies of *S. enterica* and *S. bongori* are primarily commensal organisms in cold-blooded hosts, such as reptiles and snakes, but can cause sporadic human infections. *S. enterica* serovars show considerable variability in severity and characteristics of the diseases they cause, ranging from asymptomatic infections and mild diarrhea to severe systemic disease resulting in death of the host. In addition, *S. enterica* serovars present extremely different host ranges. Broad-host range serovars (host *"generalists"*), like Typhimurium and Enteritidis, cause disease in a wide variety of animal hosts, including humans. Other serovars are host *"specialists"* and can only infect a very narrow spectrum of hosts. The most prominent examples of *Salmonella* serovars presenting a narrow-host range are Typhi and Gallinarum, which cause systemic disease only in humans and fowl, respectively.

Salmonellae cause ~1.4 million cases of food-borne diarrheal disease annually in the United States, associated with approximately 500 deaths (Mead et al., 1999; Voetsch et al., 2004). Worldwide, these organisms are responsible for hundreds of millions of cases of salmonellosis and hundreds of thousands of deaths. Typhoid fever is a serious systemic illness caused by typhoidal serotypes of *Salmonella* (i.e., Typhi, Paratyphi A, and Paratyphi B) and is essentially eradicated in the United States, although foreign travelers returning from endemic areas in Asia, Africa, or South America import a small number of cases (<1,000/year) (Mead et al., 1999). Worldwide, typhoid fever is estimated to affect more than 21 million people annually, and these infections are associated with more than 200,000 deaths (Crump et al., 2004). Typhi infection is characterized by sustained high fever, weakness, abdominal pain and may also be accompanied by a rash and diarrhea (only in one-third of patients) (Hornick et al., 1970a, b). Typhoid patients develop a low-level bacteremia and high numbers of organism are found in the intestinal lymphoid tissue as well as in systemic sites including the spleen, liver, mesenteric lymph nodes.

The most common disease syndrome caused by non-typhoidal *Salmonella* serotypes in humans in the United States is enterocolitis and results from infection by 1 of over 2,000 serotypes. The four most common implicated serotypes, Typhimurium, Enteritidis, Newport, and Heidelberg, account for 52% of confirmed cases in the United States (CDC, 2005). Serotype Typhimurium, a non-typhoidal serotype, has been the most frequently isolated serotype from human sources since 1997 in the United States and was responsible for 19.3% of cases in 2005 (CDC,

2005). Multiple antibiotic resistance is becoming highly prevalent in Typhimurium, with 45% of isolates resistant to one or more drugs and 26% of isolates resistant to five drugs in a 2003 national survey (CDC, 2005). In humans, enterocolitis as a result of Typhimurium infection results in diarrhea, abdominal cramps, and fever of 4–7 days duration (CDC, 2008) that are usually self-limiting in humans in the developed world.

Systemic illness and bacteremia can also result from infection with non-typhoidal salmonellae (NTS). Children under the age of 5 are most frequently infected with diarrheal NTS (20% of isolations in the United States in 2005) and along with the elderly and immunocompromised are at high risk for the development of fatal systemic salmonellosis after NTS infection. Although NTS are the most frequent cause of death associated with food-borne infectious agents in the United States, the total number of such cases in this country is relatively low. However, around the world where large portions of the population are immunocompromised (especially in sub-Saharan Africa where HIV infection rates have risen dramatically) infection with non-typhoidal salmonellae causes systemic disease with high frequency and with high rates of fatality (Alausa et al., 1977; Gordon et al., 2001, 2002; Kankwatira et al., 2004). Furthermore, non-typhoidal salmonellae are a major cause of high morbidity and death in children under the age of 5 years old, especially those from desperately poor circumstances (Kariuki et al., 2006). Children with non-typhoidal *Salmonella* bacteremia present with fever, cough, diarrhea, hepatosplenomegaly, elevated hepatic enzymes such as LDH, and tachypnea (Cheesbrough et al., 1997; Bar-Meir et al., 2005; Brent et al., 2006), but frequently present without gastrointestinal signs including diarrhea (Kariuki et al., 2006). Nearly 60% of children infected with NTS present with bacteremia, and serotype Typhimurium is responsible for 59% of cases examined (Kariuki et al., 2006).

The molecular mechanisms of systemic infection and gastroenteritis caused by salmonellae have been heavily studied and have concentrated largely on the type III secretion system (TTSS) encoded by *Salmonella* pathogenicity islands 1 and 2 (SPI-1, SPI-2) and their associated effectors as well as a few other "pathogenicity islands" containing groups of virulence-associated genes. The high prevalence of non-typhoidal *Salmonella* in the food supply worldwide, leading to the high number of cases of non-typhoidal salmonellosis, the increasing potential for deadly systemic infection, and the increasing prevalence of multi-drug resistant *Salmonellae* (CDC, 2005) drive the need to develop a better understanding of the ability of this organism to survive and grow in systemic sites and the need to develop novel treatments to specifically target this organism during deadly acute systemic infections.

7.3 Virulence Factors

Characterization of the molecular mechanisms that support the interactions of salmonellae with their animal hosts has advanced over the past decade, mainly through the study of *S. enterica* serovar Typhimurium (*S.* Typhimurium) in tissue culture and animal models of infection. Thus far, over 75 genes have been implicated in *Salmonella* virulence, including genes located in at least 6 pathogenicity islands

(SPI-1 to SPI-5 and the *spv* locus in the *Salmonella* virulence plasmid) (Galán and Curtiss, 1989; Galán et al., 1992; Gulig et al., 1992; Gulig and Doyle, 1993; Hensel et al., 1995, 1997). Recently an additional three potential pathogenicity islands have been identified, including SPI-14, SPI-15, and SPI-16 (Vernikos and Parkhill, 2006; Bogomolnaya et al., 2008). In addition to these pathogenicity islands, genes encoding lipopolysaccharide-related functions (biosynthesis and modification) are critical for the virulence of this organism both in the intestine and during systemic infection. We will briefly review the roles of the type III secretion system (TTSS), encoded by SPI-1 and SPI-2, that are critical for infection and are heavily studied.

SPI-1 and SPI-2 encode two independent TTSS that inject effector proteins directly into host cells during infection and are critical during various stages of infection (Galán and Curtiss, 1989; Hensel et al., 1998). TTSS-1 was originally identified as important during the intestinal phase of disease for the invasion of host cells (Galán and Curtiss, 1989), and these are now known to include M-cells, intestinal epithelial cells, and macrophages (Alpuche-Aranda et al., 1994; Jones et al., 1994). During the intestinal phase of infection, the TTSS-1 is involved in the invasion of cells lining the small intestine using several effectors (including SipA, SopE, SopE2 and SopB) to induce actin rearrangement that results in membrane ruffling and macropinocytosis (Hardt et al., 1998; Patel and Galan, 2006, 2008). In addition, the TTSS-1 action elicits a strong intestinal inflammatory response consisting primarily of neutrophils, CXC chemokine production, and eliciting fluid accumulation in the lumen of the intestine (Zhang et al., 2003a,b). The effector proteins SipA, SopA, SopB, SopD, and SopE2 have been shown to be critical for the ability of non-typhoidal *Salmonella* to elicit these responses in ligated ileal loops in the calf (Zhang et al., 2003a,b). Although it was originally thought that the TTSS-1 was only important during the intestinal phase of disease, several studies examining low-level systemic persistence in *Salmonella*-resistant mice recently have suggested that it is also important during systemic colonization (Monack et al., 2004; Lawley et al., 2006). In humans, both typhoidal and non-typhoidal *Salmonella* can breach the intestinal epithelial barrier and cause systemic disease. Some circumstances, including an impaired IL-17 axis during HIV infection (Raffatellu et al., 2008), readily allow non-typhoidal *Salmonella* including Typhimurium to breach the intestinal epithelial barrier and cause acute systemic disease. After crossing this barrier, *S.* Typhimurium enters macrophages by SPI-1-mediated and SPI-1-independent macropinocytosis and disseminates to systemic sites of the reticuloendothelial system inside this cell type (Worley et al., 2006). Several TTSS-1 effectors including SipA, SopB, SopD, and SopE2 have recently been recognized to contribute to intracellular stages of the infection in macrophages (Hernandez et al., 2004; Jiang et al., 2004; Drecktrah et al., 2005; Brawn et al., 2007).

Once inside the macrophage, *Salmonella* is located inside a spacious vacuole that shrinks over time and is called the "*Salmonella*-containing vacuole" (SCV). *S.* Typhimurium replicates inside the SCV and its ability to grow inside macrophages is critical for virulence during systemic infection in animal models (Fields et al., 1986; Richter-Dahlfors et al., 1997). During *S.* Typhimurium infection of *Salmonella*-susceptible mice, the TTSS-2 encoded by SPI-2 is critical for the ability of *Salmonella* to cause systemic infection, and it is now known to be critical for establishment of the SCV and intracellular replication (Hensel et al., 1995). This secretion system was identified by signature-tagged mutagenesis during acute systemic infection using *Salmonella*-susceptible murine models, and over 20 effectors injected into the host cell by this system have been identified to date that are critical for establishing an intracellular niche for replication of the organism (Haraga et al., 2008). The biogenesis of this compartment and its interaction with the endocytic pathway have been extensively studied, it fuses with lysosomes, acidifies, and interacts with both the early and late endocytic pathways (Carrol et al., 1979; Alpuche Aranda et al., 1992; Rech et al., 1996; Steele-Mortimer et al., 1999; Drecktrah et al., 2007). In addition to growth in macrophages, recent work indicates that *Salmonella* resides inside neutrophils, secretes SPI-2 effectors there, and replicates in this niche during murine infection (Geddes et al., 2007). Although it was previously thought that the TTSS-2 was only important during systemic infection, recent work has shown that it is expressed in the intestine as well (Brown et al., 2005), perhaps in preparation for intracellular survival and growth inside host cells.

7.4 Forward Genetics in Salmonellae

In the last 30 years, forward genetics has been the single most highly used technique to study microbial life, including bacterial pathogens such as *Salmonella* (Fields et al., 1986; Gulig and Curtiss, 1987; Kukral et al., 1987; Gulig and Curtiss, 1988; Galán and Curtiss, 1989; Miller et al., 1989a, b; Bäumler et al., 1994; Hensel et al., 1995; Lodge et al., 1995; Bowe et al., 1998; Turner et al., 1998; Tsolis et al., 1999; Bispham et al., 2001; Lichtensteiger and Vimr, 2003; Morgan et al., 2004; Chan et al., 2005; Ku et al., 2005; Bearson et al., 2006; Lawley et al., 2006; Carnell et al., 2007). The vast majority of forward genetic studies in salmonellae have used libraries of random transposon insertions that are labor intensive to screen, limiting their use in such complex and expensive models as livestock. In spite of this, the most significant virulence factors in *Salmonella*, including SPI-1 and SPI-2, were identified by forward genetic analyses (Galán and Curtiss, 1989; Hensel et al., 1995). Because many forward genetic studies were performed in the pre-genomic era, surveys of the *Salmonella* genome were generally not saturating. With our current knowledge of complete genome sequences for salmonellae, new techniques for analysis of highly complex transposon libraries, rapid generation of deletion mutants, use of microarray analysis, and the advent of high-throughput sequencing offer a unique opportunity for functional genomics of *Salmonella* in many different environments including foods and livestock models. Identification of genes necessary for growth under a given set of conditions is the first step in the development of novel interventions and therapeutics to prevent *Salmonella* growth. Indeed, mutants in several of the genes described as important for growth during infection in livestock have been tested as potential vaccine candidates.

7.4.1 Signature-Tagged Mutagenesis in Salmonella

The introduction of signature-tagged mutagenesis (STM), a methodology that combined the advantages of transposon mutagenesis and negative selection, was a breakthrough in the analysis of banks of mutants (Hensel et al., 1995). STM provided an approach that could be used during infection of a host animal, as individual mutants were tracked based on the addition of unique sequence tags to the transposon. This approach has been widely used to identify candidate *Salmonella* genes necessary in many hosts including livestock. Initially used to identify *S.* Typhimurium genes involved in systemic infection in mice using small pools of uniquely tagged mutants (Hensel et al., 1995), these experiments led to the discovery of *Salmonella* pathogenicity island 2 (SPI-2) while studying mutants in ~25% of the genome. Mutants in genes necessary for LPS (*rfb*) and purine (*pur*) biosynthesis, genes in the *spv* operon, and the EnvZ/OmpR two-component system were identified as selected against during systemic infection in mice (Hensel et al., 1995). SPI-2 was subsequently mapped and determined to encode a type III secretion system (TTSS-2) required for *Salmonella* growth during systemic infection (Shea et al., 1996).

A few studies have used STM to screen collections of random transposon mutants in *S.* Typhimurium for virulence defects in different animal hosts, including livestock animals that are vectors for *Salmonella* transmission to humans (Tsolis et al., 1999; Morgan et al., 2004; Carnell et al., 2007). Genes required for *Salmonella* growth in the mouse and in the cow were identified using STM by screening only 260 mutants. Three genes necessary during murine infection and one gene necessary for bovine infection were identified (Tsolis et al., 1999). One of the genes required only in mice, *slrP* (for *Salmonella* leucine-rich repeat protein), encodes an effector protein secreted by both the SPI-1 and SPI-2 TTSSs (Miao and Miller, 2000; Geddes et al., 2005). The functions of the other two mouse-specific genes identified remain elusive (Tsolis et al., 1999). STM screens have also been used in more extensive studies to identify genes necessary for colonization of livestock.

Screening of 1,045 STM mutants led to the identification of *S.* Typhimurium candidate mutants that are unable to colonize the intestine of either calves or chickens (Morgan et al., 2004). The nature of the candidate genes identified in this study suggested that genes encoded on SPI-1, SPI-2, and SPI-4 are important for colonization in calves but not in chickens, suggesting a fundamental difference in the mechanism of infection in birds versus mammals. In addition to this observation, cell surface polysaccharides, cell envelope proteins, and many "housekeeping" genes and genes of unknown function were found to be important for survival in and colonization of the intestine (Morgan et al., 2004).

Candidate genes required by *S.* Typhimurium to colonize porcine intestines have also been identified using STM (Carnell et al., 2007). At least 95 candidate genes were identified as important for porcine intestinal colonization, most encoding known or putative secreted or surface-exposed molecules. Many attenuating mutations in SPI-1 and SPI-2 were identified, confirming that these regions play an important role in intestinal colonization in swine. Genes encoded in other pathogenicity islands, including the *saf* fimbrial operon in SPI-6 encoding *Salmonella* atypical fimbriae, were also implicated in intestinal colonization in swine (Carnell et al., 2007). In additional work in this study, the efficacy of secreted protein vaccines from wild-type and a *prgH* mutant of *S.* Typhimurium was evaluated following intramuscular vaccination of pigs. In both cases, a significant reduction in bacterial shedding was observed during the acute phase of infection indicating that protection was not reliant on SPI-1-secreted proteins (Carnell et al., 2007).

STM has been also used to identify mutants presenting colonization defects in different strains of mice, as an alternative to testing different animal hosts. An STM library of 960 *S.* Typhimurium mutants was used to infect wild-type and immunodeficient (NOS2$^{-/-}$, gp$^{91phox-/-}$, and gp$^{91phox-/-}$ NOS2$^{-/-}$) C57BL/6 mice to identify genes required for the bacterium to resist the detrimental effects of host phagocyte oxidase during infection (Hisert et al., 2005). A gene, named *cdgR* (*STM1344*), cyclic diguanylate (c-diGMP) regulator, that encodes a predicted c-diGMP phosphodiesterase was identified in these studies (Hisert et al., 2005). The phenotype observed for the *cdgR* mutant indicates that bacterial c-diGMP regulates host–pathogen interactions involving antioxidant defense and cytotoxicity (Hisert et al., 2005).

A number of additional studies have also identified genes required for host-specific *Salmonella* serovars Dublin, Choleraesuis, and Gallinarum to colonize their respective livestock hosts (calves, pigs, and chickens, respectively). In the first of these studies, a bank of 5,280 signature-tagged mutants was constructed in *S.* Dublin, and pools of 96 mutants were used to infect mice and calves via the intravenous route (Bispham et al., 2001). In this study, mutants in *sseD*, encoding a protein secreted by the SPI-2-encoded TTSS-2, were subsequently shown to be attenuated in calves and mice following infection either by the intravenous route or by the oral route. Notably, both the *sseD* mutant and a mutant in a second SPI-2-encoded gene, *ssaT*, induced weaker secretory and inflammatory responses in bovine ligated ileal loops than the parental wild-type strain, indicating that *S.* Dublin may require SPI-2 for the induction of both systemic and enteric salmonellosis in calves (Bispham et al., 2001).

Two studies have used STM to determine the genes necessary for colonization of swine by the swine-specific serotype Choleraesuis (Lichtensteiger and Vimr, 2003; Ku et al., 2005). Analysis of only 45 STM mutants in Choleraesuis during infection in pigs indicated that a mutant in *hilA*, a transcriptional regulator that controls the expression of many genes necessary for virulence (Bajaj et al., 1996), failed to colonize by either oral or intraperitoneal route (Lichtensteiger and Vimr, 2003). *hilA* mutants were attenuated in pigs after oral but not intraperitoneal inoculation in competitive infections, indicating that *hilA* plays a role in enteric but not systemic infections in swine (Lichtensteiger and Vimr, 2003). In a second study, a library of 960 STM mutants of *S.* Choleraesuis was screened for attenuation in pigs (Ku et al., 2005). Thirty-three candidate mutants with attenuated growth in swine were identified, and growth attenuation during infection was confirmed for 20 of these mutants (Ku et al., 2005). Growth-attenuated mutants identified in this study had mutations in genes encoding proteins with diverse functions including the integrity of the outer membrane, type III secretion, transport of macromolecules, LPS biosynthesis. Five of the mutants identified in this screen, *ssaV*, *dgoT*, *ssaJ*, *spiA*, and an insertion in *SC2632* within prophage Gifsy-1, were evaluated as attenuated live vaccines in pigs and two of them (*ssaV* and *SC2632*) performed better than the commercially available live vaccine (Ku et al., 2005).

Finally, genes of poultry-specific serotype Gallinarum necessary for growth in chicks were studied using a PCR-based STM assay (Shah et al., 2005). Mutants in 20 presumptive attenuated genes were identified after screening ~25% of the Gallinarum genome and 18 of these mutants were subsequently confirmed using in vitro and in vivo competition assays (Shah et al., 2005). Mutants in SPI-1, SPI-2, SPI-10, and *envZ* were selected against in the chicken during infection. Furthermore, two novel pathogenicity islands were described in this study: SPI-13 and SPI-14, including genes homologous to *STM3117–STM3134* and *STM0854–STM0859* in *S.* Typhimurium, respectively (Shah et al., 2005). Phenotypes for mutants in genes within SPI-13 and SPI-14 have been also reported during systemic infection of *S.* Typhimurium in mice, and genes encoded on SPI-13 in Typhimurium are known to be exquisitely upregulated in macrophages (Chan et al., 2005; Shi et al., 2006; Haneda et al., 2009; Santiviago et al., 2009).

The genetic requirements for *Salmonella* growth and survival have also been examined for less complex conditions that are relevant to food safety using STM. Such attempts have screened for mutants that are negatively selected in vitro under conditions that mimic a single niche. One such study led to the identification of genes important for the survival of *S.* Typhimurium in the swine gastric environment using ex vivo swine stomach contents (SSC) (Bearson et al., 2006). In this study, 1,600 STM mutants were analyzed and 19 were identified as defective in the SSC assay. Remarkably, 13 of these mutants were at least 10 times more sensitive in the SSC assay than the wild-type organism, but only 3 mutants displayed the same level of sensitivity following an acidic challenge at pH 3.0. Further examination determined that, although the lethal effects of the SSC are pH dependent, low pH is not the only killing mechanism in the stomach environment. This study unveiled a role for several *S.* Typhimurium genes in gastric survival, indicating that this environment is defined by more than low pH (Bearson et al., 2006).

7.4.2 Genomic DNA Microarrays

Complete genome sequences have allowed the development of open reading frame (ORF) microarrays (McClelland et al., 2001; Parkhill et al., 2001), complete tiling arrays for *Salmonellae* (Navarre et al., 2006), and other customized oligonucleotide arrays (Santiviago et al., 2009). Complete genomic hybridization, transcriptional studies, forward genetics, and mapping of transposon insertion locations are all possible on a larger scale through the use of microarray analysis. We discuss the use of microarrays to survey genetic content, for forward genetics, and for mapping transposon insertions in the following section (use of arrays for transcriptional profiling is in Section 7.5).

7.4.2.1 Microarray Analysis for Genetic Content

Microarrays have been used to define the genes present in clinical and epidemic *Salmonella* isolates implicated in food-borne outbreaks that have not yet been sequenced, to determine the complete genetic content of these strains (Andrews-Polymenis et al., 2004; Porwollik et al., 2005; Reen et al., 2005; Kang et al., 2006). This work has shown that most serovars consist of strains that are highly similar to each other and differ in gene content from other serovars, with some exceptions. Particular serovars contain isolates that differ substantially from each other in genetic content, while some serovars contain isolates with nearly identical gene content (Porwollik et al., 2004). DNA-based classification of strains, or the classification of isolates by "genovars," will allow us to further refine the host range and disease symptoms associated with particular genome variants.

7.4.2.2 Microarray-Based Screenings of Transposon Mutants

Although signature-tagged mutagenesis presented an improvement over individual mutant screening in animals, this approach remains labor intensive and has only allowed screening of a fraction of the *Salmonella* genome in animal models. In order to increase the throughput in the analysis of mutants, new methods have been developed. For example, the internally tagged transposons used in STM have been replaced by the use of transposons bearing outwardly facing promoters that can be used to generate transcripts unique to each mutant by virtue of their unique genomic location. Thus, changes in the distribution of transposon mutants under selection can be monitored by comparative hybridizations of these transcripts on a microarray. This strategy is commonly called transposon-associated site hybridization (TRaSH) (Badarinarayana et al., 2001; Sassetti et al., 2001; Baldwin and Salama, 2007; Winterberg and Reznikoff, 2007).

In these techniques, a transposon library is generated using a transposon containing an outward-facing promoter at one or both ends (Sassetti et al., 2001; Chan et al., 2003; Lawley et al., 2006; Baldwin and Salama, 2007; Winterberg and Reznikoff, 2007; Chaudhuri et al., 2009). The library is then grown in vitro and saved as a reference (input) while a second sample is subjected to selection (output). Genomic DNA from both the input and the output libraries is obtained and used to specifically amplify junction fragments containing the T7 promoter. The resulting amplified product from input and output samples is used for T7 in vitro transcription and differentially labeled by incorporation of fluorescent dyes. Mutants under selection are determined by the comparison of the labeled transcripts in the input pool to the labeled transcripts in the output pool using a *Salmonella*-based genomic microarray (Chan et al., 2005; Lawley et al., 2006; Chaudhuri et al., 2009).

Salmonella mutants under negative selection have been identified during infection of murine macrophages in vitro (Chan et al., 2005), an important niche for *Salmonella* growth during systemic infection (Fields et al., 1986). Using a library containing ~50,000 mini-Tn*10* transposon insertions in the virulent *S.* Typhimurium isolate SL1344, mutants under selection after serial passage in RAW264.7 macrophages were identified using an ORF microarray (Chan et al., 2005). Genes that were selected against included 25 genes on the pathogenicity-associated region of SPI-2 (Chan et al., 2005). In addition, genes in the putative operon *STM0854–STM0859*, induced in RAW264.7 macrophages in vitro (Eriksson et al., 2003), were negatively selected in mice (Chan et al., 2005). In serotype Gallinarum these genes are located on a pathogenicity island termed SPI-14, and mutants in several of these genes (corresponding to *STM0855* and *STM0859*) are avirulent in chicks (Shah et al., 2005). Surprisingly, mutants in genes within SPI-1 were also identified as important for *Salmonella* survival and growth within macrophages (Chan et al., 2005). Candidate genes necessary during systemic infection of *Salmonella*-susceptible BALB/c mice have also been defined using this approach (Chan et al., 2005). The microarray-based comparative analysis of the input and output samples confirmed the observations made in the original STM screen (Hensel et al., 1995), implicating SPI-2 and genes necessary for LPS biosynthesis as important for systemic colonization in this murine systemic infection model (Chan et al., 2005).

A similar transposon-based strategy was used to identify *Salmonella* genes required for long-term systemic infection in *Salmonella*-resistant 129SvJ mice after intraperitoneal infection (Lawley et al., 2006). Analysis was performed on the mutant pools recovered from the spleen and liver at 7, 14, 21, and 28 days post-infection to identify mutants selected during infection (Lawley et al., 2006). Survival of *Salmonella* in the spleen and liver of infected animals did not appear to require unique sets of genes (Lawley et al., 2006). Second, progressive selection against certain mutants was observed depending on the duration of the infection, suggesting that different classes of genes are required at distinct stages of infection as the pathogen encounters a variety of environments inside the host. By 28 days post-infection, 118 candidate genes were identified that contribute to long-term systemic colonization of mice. Of these genes, 64 genes (53% of the observations made) encode hypothetical proteins or proteins with unknown function, suggesting that we have much to learn about the genetic requirements for long-term colonization by salmonellae. Mutants under negative selection in this system included several regions predicted to be horizontally acquired (SPI-1 to SPI-6, prophages, and the virulence plasmid). In addition, mutations in fimbrial genes, PhoP-regulated genes, and genes involved in LPS biosynthesis were also negatively selected in mice. A major finding of this study was that genes in SPI-1 and SPI-4 contribute to long-term systemic infection in mice, an observation that challenges the classic description of *Salmonella* pathogenesis. These novel observations were confirmed by competitive assays in which the relative fitness of an individual mutant was compared to that of the wild-type strain within a single mouse (Lawley et al., 2006).

In the studies cited earlier, the microarray hybridizations were conducted on a spotted array of PCR products amplified from *S.* Typhimurium strain SL1344, representing more than 4,100 open reading frames annotated in the genome of *S.*

Typhimurium strain LT2 (Chan et al., 2003). Using this platform the hybridization signals from different insertions in any given gene are reported only by one or two spots, making precise mapping of each insertion to a given open reading frame possible. Finer mapping of transposon insertions in such a library has recently been addressed by the development of a quantitative screen for insertion mutants called transposon-mediated differential hybridization (TMDH) (Chaudhuri et al., 2009). TMDH employs customized Tn5 and Mu transposons that include outward-facing T7 and SP6 promoters. Labeled transcripts generated from both promoters are hybridized to whole-genome tiling microarrays and data analyzed using a novel bioinformatics algorithm to accurately map the insertions present in the transposon library (Charles and Maskell, 2001). After intravenous inoculation into BALB/c mice, mutants were recovered from infected spleen and liver 48 h post-infection. Mutants attenuated for growth during infection are identified by comparison of hybridization data from the library grown in vitro with data obtained after passage of the library through mice. Using TMDH, 6,108 insertions were unequivocally mapped to 2,824 genes, from a total of 10,368 mutants analyzed. As expected, the most highly attenuated mutants identified in this study included mutations in genes on SPI-2, genes involved in biosynthesis of LPS genes necessary for biosynthesis of aromatic compounds and purines, and *spv* genes on the virulence plasmid. Interestingly, a series of mutants presenting a hypervirulent phenotype ("hypercompetitive mutants") was defined and included mutants in flagellar biosynthesis in the *flg* and *fli* operons. Defined deletion mutants of novel targets identified in the TMDH were generated and assayed as attenuated live vaccines. Thus, *atpA* (encoding a subunit of ATP synthase) and *trxA* (encoding a subunit of the thioredoxin system) mutants protected against challenge with the wild-type *S.* Typhimurium strain (Chaudhuri et al., 2009).

7.4.2.3 Microarray-Based Screening of Targeted Deletion Mutants

Transposons have been used for many forward genetic screens in *Salmonella*, but such methods have several limitations. First, screening of transposon mutants requires the use of very large numbers of mutants for complete coverage of the surveyed genome. "Bottlenecks", or circumstances where the population of mutants experiences random loss during selection, limit the usefulness of transposon libraries of high complexity for screening in some instances. Salmonellae encounter such bottlenecks, for example, when they pass from the gut across the epithelial barrier, a situation where only a small number of organisms can successfully cross and reach systemic sites. Second, random transposon insertions can have polar effects on neighboring gene expression. With the advent of techniques to target individual genes for deletion in enterobacteria, complete genome sequence information can be used to generate ordered libraries of targeted non-polar deletion mutants of individual genes using Lambda Red recombination (Datsenko and Wanner, 2000). Using this approach, we developed a library of targeted deletion mutants in *S.* Typhimurium ATCC 14028, as well as an array-based selection technique to assay this mutant collection.

Many genetic methods have been developed and used to generate defined mutants of interest, but the appearance of one in particular revolutionized the genetics studies in *Escherichia coli, Salmonella*, and other Gram-negative bacteria. Known as "recombineering" (recombination-mediated genetic engineering) (Yu et al., 2000; Ellis et al., 2001), this technology exploits homologous recombination mediated by the Red recombination system encoded by Lambda phage (for an overview and applications see Karlinsey (2007), and Sawitzke et al. (2007). In the most popular versions of this technique, referred to as the Lambda Red-swap method (Datsenko and Wanner, 2000), PCR primers 60–65 bases long are synthesized with 40–45 bases on the 5′-ends corresponding to the ends of the desired deletion and 3′-ends homologous to a plasmid template. PCR is used to amplify the desired sequences from the plasmid template, including an antibiotic resistance cassette and flanking sequences, generating a PCR product with ends homologous to the gene targeted for deletion. Genes are deleted from the strain of interest after recombination of this linear PCR product into the genome and the corresponding swap of the target gene out of the genome, catalyzed by the Lambda Red recombinase.

Using this technique (see Fig. 7.1), a partial library containing mutants in nearly ~1,000 genes specific to salmonellae and not present in closely related organisms has been generated in *S.* Typhimurium ATCC 14028 (Santiviago et al., 2009). This library has been used to identify genes necessary for systemic infection in BALB/c mice, using a novel and inexpensive oligonucleotide microarray in a new assay called array-based analysis of cistrons under selection (ABACUS) (Santiviago et al., 2009). The findings of this study underscore the results of previous high-throughput transposon-based studies in this model of infection and in *Salmonella*-resistant murine models of long-term infection (Chan et al., 2005; Lawley et al., 2006). However, using the ABACUS methodology mutants in a number of additional genes, including six sRNA genes not previously implicated during systemic infection, were determined to be candidate genes necessary for systemic infection in BALB/c mice. Thirteen of these mutants, encoding three sRNAs (*sroA, istR,* and *oxyS*), a rare tRNA-LEU (*leuX*), four putative membrane proteins (*STM1131, STM2215, STM2303,* and *STM2120*), four genes of unknown function (*STM3120, STM3121,* and *STM731, STM1760*), and TatC (*STM3975*), have been confirmed using competitive intraperitoneal infections in BALB/c mice. Defects in systemic colonization have been confirmed using non-polar unmarked deletions of eight of these mutants (*STM1131, STM2303, STM3120, STM3121, leuX, istR, sroA,* and *oxyS*). Unmarked mutants in *STM2303, STM3121* and mutants in four additional genes have been complemented *in trans*, restoring wild-type systemic colonization by these mutants (Santiviago et al., 2009; Andrews-Polymenis, unpublished results). The molecular role of these genes during systemic infection is currently under investigation. It is remarkable that this assay identified novel genes necessary for systemic infection, as systemic infection of BALB/c mice by *Salmonella* is a very highly studied system.

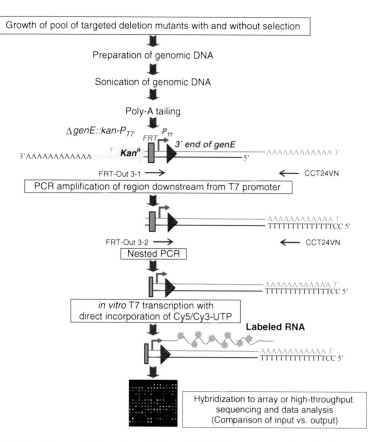

Fig. 7.1   Protocol for monitoring of pool of targeted deletion mutants under selection

7.4.3 Recent Advances in High-Throughput DNA Sequencing

Advances in sequencing technology have made sequencing cheaper and faster, and the development of sequencing platforms based on different technology has fostered the growth of innovative uses for this technology. The first wave of novel high-throughput sequencing platforms included a variety of technologies such as 454 pyrosequencing (Roche Applied Sciences), sequencing by synthesis with reversible terminators generally called Solexa sequencing (Illumina/Solexa), and massively parallel sequencing by hybridization and ligation implemented in the supported oligonucleotide ligation and detection (SOLiD) system (Applied Biosystems). All of these systems are generically referred to as the "second-generation" or "next-generation" sequencing systems. One of the main advantages of these platforms is the parallel determination of sequence data directly from amplified single DNA fragments, without the need for cloning. These new sequencing technologies offer dramatic increases in cost-effective sequence throughput at the expense of read length. Huge amounts of sequence data can rapidly be produced using these next-generation technologies, compensating for reduced read length.

High-throughput sequencing approaches are beginning to be used to understand many aspects of *Salmonella* biology, not only to complete many new entire genome sequences and do transcriptome analysis, but also to map insertions across collections of mutants, and as a detection method in forward genetic analysis to replace some array-based applications. We will outline the major new technologies (see Table 7.1 for summary) and then summarize how these technologies are being implemented to understand *Salmonella* biology.

Table 7.1 Available high-throughput sequencing (HTS) methodologies

Technique	Company	References
"Second generation" or "next generation"		
454 pyrosequencing	Roche Applied Sciences	http://www.454.com
Illumina sequencing	Illumina/Solexa	http://www.illumina.com
SOLiD (supported oligonucleotide ligation and detection)	Applied Biosystems	http://www3.appliedbiosystems.com
"Third generation" or "next–next generation"		
True single-molecule sequencing (tSMS)	Helicos Biosciences	http://www.helicosbio.com
SMRT sequencing (single-molecule real time)	Pacific Biosciences	http://www.pacificbiosciences.com
Hybridization-assisted nanopore sequencing (HANS)	NABsys	http://www.nabsys.com
Design polymer-assisted nanopore sequencing	Ling Vitae	http://www.lingvitae.com
Transmission electron microscopy-assisted DNA sequencing	ZS Genetics	http://zsgenetics.com

7.4.3.1 454 pyrosequencing

This *sequencing-by-synthesis* method was the first next-generation technology released in late 2005. Sequencing libraries are constructed by fragmenting the DNA and ligating specific adapters to the 3′- and 5′-end of the sheared fragments. Fragments are attached to agarose beads and a complex mixture of these beads is subjected to a process known as *emulsion PCR single-molecule amplification*. In this process, sonication is used to produce an emulsion of oil with water droplets containing a unique bead and PCR reagents. These droplets act like individual amplification reactors during PCR amplification to produce beads decorated with $\sim 10^6$–10^7 copies of the unique DNA template in each droplet. This amplification step is necessary to produce signals intense enough to be reliably detected during the *sequencing-by-synthesis* reaction step. After denaturation, each single bead is placed into a well in an optical picotiter plate facing a highly sensitive CCD camera. Specific primers and beads containing polymerase and other enzymes required for the downstream pyrosequencing reaction are added to each well. Solutions of dNTPs are added one at a time into the system so that synthesis of a complementary strand can occur. The incorporation of a following base by the polymerase during DNA synthesis produces pyrophosphate, which initiates a series of downstream reactions that ultimately results in light production by luciferase. Light emission is recorded for each well in the picotiter plate by the CCD camera after each round of dNTP added, allowing hundreds of thousands of pyrosequencing reactions to be carried out in parallel and massively increasing the throughput of the whole sequencing process. The latest 454 platform released by Roche Applied Sciences (*Genome Sequencer FLX Titanium Series*) generates more than 1 million individual reads of 400 bases (i.e., 400–600 Mb of sequence data) per 10-h instrument run (information on the 454 technology can be found at http://www.454.com).

7.4.3.2 Illumina/Solexa Sequencing

The Illumina platform, launched in 2006, is based on massively parallel sequencing-by-synthesis using proprietary reversible terminator nucleotides labeled with fluorescent dyes. DNA to be sequenced is fragmented and adapters are ligated to both ends of each fragment. After denaturation, the fragments are attached to an optically transparent planar surface densely coated with adapters and complementary adapters and subjected to *solid-phase bridge amplification*. In this process, each single-stranded fragment, attached by one end to the surface, bends over and creates a bridge structure by hybridizing with its free end to a complementary adapter on the surface of the support. A mixture containing PCR reagents is supplied to the system and the adapters act as the primers in a following PCR amplification. After several amplification cycles, an ultra-high density sequencing flow cell with hundreds of millions of random clusters, each containing $\sim 1{,}000$ copies of the same template, is produced. These clusters are also known as DNA "*polonies*" as they resemble bacterial colonies. The flow cell is then supplied with sequencing reagents, including specific primers, DNA polymerase, and all four nucleotides each labeled with different fluorescent dyes. The labeled nucleotides include a blocking group at the 3'-OH such that each incorporation occurs as a unique event. After each incorporation step, the terminator nucleotide is detected and identified via its fluorescent dye by a CCD camera, and the 3'-blocking group is chemically removed to prepare the DNA for a next round of incorporation. The latest platform released by Illumina/Solexa (*Genome Analyzer IIx*) generates ~ 300 million reads of 100 bases (i.e., ~ 30 Gb of sequence data) per run (~ 10 days) (information on the *Illumina/Solexa* technology can be found at http://www.illumina.com).

7.4.3.3 SOLiD Sequencing

This platform based on hybridization and ligation chemistry was developed by Applied Biosystems and launched in 2007. As in 454 methodology, the sequencing process starts with the construction of sequencing libraries by an *emulsion PCR single-molecule amplification* step on magnetic beads. The beads containing the amplification products are transferred to a glass surface where sequencing occurs by sequential rounds of hybridization and ligation of random oligonucleotides containing known dinucleotides at the 3'-end. Combinations of 16 dinucleotides labeled by four different fluorescent dyes are used. Using a four-color encoding scheme, each interrogated position is effectively probed two times so that the identity of the site is determined analyzing the color resulting from two successive ligation reactions. The latest platform released (*SOLiD 3 Plus System*) is advertised to generate more than 1 billion individual reads of 50 bases (i.e., 50–60 Gb of sequence data) per run in 12–14 days (further information on SOLiD technology can be found at http://www3.appliedbiosystems.com).

7.4.3.4 The Next Wave of Sequencing Methods

Most recently, a group of ultra-high-throughput sequencing methods collectively known as the "third-generation" or "next–next generation" sequencing systems has been developed, all of which exploit *single-molecule DNA sequencing* (SMS) technologies. Among others, third-generation methods includes technologies like *true single-molecule sequencing* (*HeliScope* by Helicos Biosciences), *SMRT sequencing* (by Pacific Biosciences), *hybridization-assisted nanopore sequencing* (*HANS* by NABsys), *design polymer-assisted nanopore sequencing* (by LingVitae), and *transmission electron microscopy-assisted DNA sequencing* (by ZS Genetics). Although some of these SMS technologies are in the proof-of-principle step of development, it is anticipated that they will be much faster and cheaper that any other method developed so far. For recent reviews on second- and third-generation high-throughput sequencing systems and their applications in functional genomics and microbial genetics see Gupta (2008), Mardis (2008), Morozova and Marra (2008), Ansorge (2009), and MacLean et al. (2009).

7.4.3.5 Impact of Novel High-Throughput DNA Sequencing Techniques in *Salmonella* Functional Genomics

High-throughput sequencing has been used to speed the pace of complete genome sequencing of the salmonellae. The genome sequences of isolates representing six serovars have recently been published (McClelland et al., 2001; Parkhill et al., 2001; Deng et al., 2003; McClelland et al., 2004; Chiu et al., 2005; Thomson et al., 2008), and additional genomes in several additional serovars are complete and available (for recent review see Andrews-Polymenis et al., 2009). 454 and Illumina/Solexa have been used for sequence comparison of Typhi genomes (Holt et al., 2008) and are now being used to the study of non-typhoidal salmonellae (Jarvik et al., 2010). A summary of completed and current *Salmonella* sequencing projects is provided in Table 7.2.

Table 7.2 Completed and current *Salmonella* sequencing projects

Species	Subspecies	Serovar/ serotype	Strain	Method	Sequencing center	GenBank accession number	Status	Citations
S. bongori	–		12,149	ABI	Sanger Institute		Complete	
S. enterica	*arizonae*	62:z4,z23:-	ATCC BAA-731, SARC5	ABI	Washington University	CP000880.1	Complete	
S. enterica	*enterica*	Agona	SL483	ABI	J. Craig Venter Institute	CP001138.1	Complete	
S. enterica	*enterica*	Choleraesuis	SC-B67	ABI	Chang Gung University	AE017220.1	Complete	Chiu et al. (2005)
S. enterica	*enterica*	Dublin	CT_02021853	ABI	J. Craig Venter Institute	CP001144.1	Complete	
S. enterica	*enterica*	Enteritidis	P125109	ABI	Sanger Institute	AM933172.1	Complete	Thomson et al. (2008)
S. enterica	*enterica*	Gallinarum	287/91	ABI	Sanger Institute	AM933173.1	Complete	Thomson et al. (2008)
S. enterica	*enterica*	Hadar	–	ABI	Sanger Institute		Complete	
S. enterica	*enterica*	Heidelberg	SL476	ABI	J. Craig Venter Institute	CP001120.1	Complete	
S. enterica	*enterica*	Infantis	–	ABI	Sanger Institute		Complete	
S. enterica	*enterica*	Newport	SL254	ABI	J. Craig Venter Institute	CP001113.1	Complete	
S. enterica	*enterica*	Paratyphi A	AKU_12601	ABI	Sanger Institute	FM200053.1	Complete	Holt et al. (2009)
S. enterica	*enterica*	Paratyphi A	ATCC 9150, SARB42	ABI	Washington University	CP000026.1	Complete	McClelland et al. (2004)
S. enterica	*enterica*	Paratyphi B	SPB7, SGSC4150	ABI	Washington University	CP000886.1	Complete	

Table 7.2 (continued)

Species	Subspecies	Serovar/serotype	Strain	Method	Sequencing center	GenBank accession number	Status	Citations
S. enterica	enterica	Paratyphi C	RKS4594, SARB49	ABI	Peking University	NC_012125.1	Complete	Liu et al. (2009)
S. enterica	enterica	Schwarzengrund	CVM19633	ABI	J. Craig Venter Institute	CP001127.1	Complete	
S. enterica	enterica	Typhi	E98-0664	454	Sanger Institute	NZ_CAAU00000000	Complete	Holt et al. (2008)
S. enterica	enterica	Typhi	150(98)S	Illumina GS	Sanger Institute		Complete	Holt et al. (2008)
S. enterica	enterica	Typhi	404ty	454/Illumina GS	Sanger Institute	NZ_CAAQ00000000	Complete	Holt et al. (2008)
S. enterica	enterica	Typhi	8(04)N	Illumina GS	Sanger Institute		Complete	Holt et al. (2008)
S. enterica	enterica	Typhi	AG3	454/Illumina GS	Sanger Institute	NZ_CAAY00000000	Complete	Holt et al. (2008)
S. enterica	enterica	Typhi	CT18	ABI	Sanger Institute	AL513382.1	Complete	Parkhill et al. (2001)
S. enterica	enterica	Typhi	CT18	Illumina GS	Sanger Institute		Complete	Holt et al. (2008)
S. enterica	enterica	Typhi	E00-7866	454	Sanger Institute	NZ_CAAR00000000	Complete	Holt et al. (2008)
S. enterica	enterica	Typhi	E01-6750	454	Sanger Institute	NZ_CAAS00000000	Complete	Holt et al. (2008)
S. enterica	enterica	Typhi	E02-1180	454	Sanger Institute	NZ_CAAT00000000	Complete	Holt et al. (2008)
S. enterica	enterica	Typhi	E02-2759	Illumina GS	Sanger Institute		Complete	Holt et al. (2008)
S. enterica	enterica	Typhi	E03-4983	Illumina GS	Sanger Institute		Complete	Holt et al. (2008)

Table 7.2 (continued)

Species	Subspecies	Serovar/ serotype	Strain	Method	Sequencing center	GenBank accession number	Status	Citations
S. enterica	enterica	Typhi	E03-9804	Illumina GS	Sanger Institute		Complete	Holt et al. (2008)
S. enterica	enterica	Typhi	E98-2068	454	Sanger Institute	NZ_CAAV00000000	Complete	Holt et al. (2008)
S. enterica	enterica	Typhi	E98-3139	454/Illumina GS	Sanger Institute	NZ_CAAZ00000000	Complete	Holt et al. (2008)
S. enterica	enterica	Typhi	ISP-03-07467	Illumina GS	Sanger Institute		Complete	Holt et al. (2008)
S. enterica	enterica	Typhi	ISP-04-06979	Illumina GS	Sanger Institute		Complete	Holt et al. (2008)
S. enterica	enterica	Typhi	J185SM	454	Sanger Institute	NZ_CAAW00000000	Complete	Holt et al. (2008)
S. enterica	enterica	Typhi	M223	454	Sanger Institute	NZ_CAAX00000000	Complete	Holt et al. (2008)
S. enterica	enterica	Typhi	Ty2	ABI	U. Wisconsin	AE014613.1	Complete	Deng et al. (2003)
S. enterica	enterica	Typhi	Ty2	Illumina GS	Sanger Institute		Complete	Holt et al. (2008)
S. enterica	enterica	Typhimurium	ATCC 14028s[a]	454/Illumina GS/SOLiD	U. Arizona	CP001363.1	Complete	Jarvik et al. (2010)
S. enterica	enterica	Typhimurium	CDC 60-6516[b]	454	U. Arizona		Draft sequence complete	Jarvik et al. (2010)
S. enterica	enterica	Typhimurium	ATCC 14028s-o	454	U. Arizona		Draft sequence complete	Jarvik et al. (2010)
S. enterica	enterica	Typhimurium	ATCC 14028r (rough)	454	U. Arizona		Draft sequence complete	Jarvik et al. (2010)

Table 7.2 (continued)

Species	Subspecies	Serovar/ serotype	Strain	Method	Sequencing center	GenBank accession number	Status	Citations
S. enterica	enterica	Typhimurium	ATCC 14028s[a]	454	Washington University		Draft sequence complete	Santiviago et al. (2009)
S. enterica	enterica	Typhimurium	D23580	ABI	Sanger Institute	FN424405.1	Complete	Kingsley et al. (2009)
S. enterica	enterica	Typhimurium	A130	454	Sanger Institute		Draft sequence complete	Kingsley et al. (2009)
S. enterica	enterica	Typhimurium	5579	454	Sanger Institute		Draft sequence complete	Kingsley et al. (2009)
S. enterica	enterica	Typhimurium	DT104	ABI	Sanger Institute		Complete	
S. enterica	enterica	Typhimurium	LT2	ABI	Washington University	AE006468.1	Complete	McClelland et al. (2001)
S. enterica	enterica	Typhimurium	SL1344	ABI	Sanger Institute		Complete	
S. enterica	arizonae	41:z4, z23:-	ATCC BAA-1577, CDC 05-0715	454	Washington University		In progress	
S. enterica	diarizonae	61:1, v:1.5	ATCC BAA-639, CDC 01-005	ABI	Washington University		In progress	
S. enterica	diarizonae	48:i:z	ATCCBAA-1579, CDC 05-0625	454	Washington University		In progress	
S. enterica	houtenae	50:z4,z23:-	ATCC BAA-1580, CDC 99-0125	454	Washington University		In progress	
S. enterica	houtenae	48:g,z51:-	ATCC BAA-1581, CDC 05-0642	454	Washington University		In progress	

Table 7.2 (continued)

Species	Subspecies	Serovar/serotype	Strain	Method	Sequencing center	GenBank accession number	Status	Citations
S. enterica	indica	11:b:e,n,x	ATCC BAA-1576, SARC14	ABI	Washington University		In progress	
S. enterica	indica	45:a:e,n,x	ATCC BAA-1578, SARC13	ABI	Washington University		In progress	
S. enterica	salamae	47:b:1,5	ATCC BAA-1583, CDC 05-0626	454	Washington University		In progress	
S. enterica	salamae	58:d:z6	ATCC BAA-1582, SARC3	454	Washington University		In progress	
S. enterica	enterica	4,[5],12:i:-	CVM23701	ABI	J. Craig Venter Institute	NZ_ABAO00000000	In progress	
S. enterica	enterica	Abortusovis	SSM0041	454	Washington University		In progress	
S. enterica	enterica	Bovismorbificans	01-05481 PT13	454	Washington University		In progress	
S. enterica	enterica	Braenderup	S-500	454	Washington University		In progress	
S. enterica	enterica	Brandenburg	KMR12	ABI	Korea		In progress	
S. enterica	enterica	Dublin	–	ABI	U. Illinois at Urbana-Champaign		In progress	
S. enterica	enterica	Enteritidis	48-0811	Illumina GS	Washington University		In progress	
S. enterica	enterica	Enteritidis	LK5	ABI	U. Illinois at Urbana-Champaign		In progress	
S. enterica	enterica	Enteritidis	SARB17	454	Washington University		In progress	

Table 7.2 (continued)

Species	Subspecies	Serovar/serotype	Strain	Method	Sequencing center	GenBank accession number	Status	Citations
S. enterica	enterica	Enteritidis	SARB19	Illumina GS	Washington University		In progress	
S. enterica	enterica	Hadar	RI_05P066	ABI	J. Craig Venter Institute	NZ_ABFG00000000	In progress	
S. enterica	enterica	Heidelberg	SL486	ABI	J. Craig Venter Institute	NZ_ABEL00000000	In progress	
S. enterica	enterica	Indiana	KMR53	ABI	Korea		In progress	
S. enterica	enterica	Infantis	SARB27	454	Washington University		In progress	
S. enterica	enterica	Javiana	GA_MM04042433	ABI	J. Craig Venter Institute	NZ_ABEH00000000	In progress	
S. enterica	enterica	Kentucky	CDC 191	ABI	J. Craig Venter Institute	NZ_ABEI00000000	In progress	
S. enterica	enterica	Kentucky	CVM29188	ABI	J. Craig Venter Institute	NZ_ABAK00000000	In progress	
S. enterica	enterica	Miami	ATCC BAA-1586, 02-3341	454	Washington University		In progress	
S. enterica	enterica	Montevideo	SARB30	454	Washington University		In progress	
S. enterica	enterica	Muenchen	SARB32	454	Washington University		In progress	
S. enterica	enterica	Muenchen	SARB34	454	Washington University		In progress	
S. enterica	enterica	Muenster	ATCC BAA-1575, 0065-00	ABI	Washington University		In progress	
S. enterica	enterica	Newport	CVM36720	ABI	University of Maryland, IGS		In progress	

Table 7.2 (continued)

Species	Subspecies	Serovar/ serotype	Strain	Method	Sequencing center	GenBank accession number	Status	Citations
S. enterica	*enterica*	Newport	SL317	ABI	J. Craig Venter Institute	NZ_ABEW00000000	In progress	
S. enterica	*enterica*	Panama	KMR64	ABI	Korea		In progress	
S. enterica	*enterica*	Paratyphi B	ATCC BAA-1585	ABI	Washington University		In progress	
S. enterica	*enterica*	Paratyphi B	SARB47	454	Washington University		In progress	
S. enterica	*enterica*	Paratyphi B, tartrate (+)	ATCC BAA-1584, S-1241	454	Washington University		In progress	
S. enterica	*enterica*	Paratyphi C	RKS4594, SARB49	Illumina GS	Washington University		In progress	
S. enterica	*enterica*	Poona	SGSC4934	454	Washington University		In progress	
S. enterica	*enterica*	Pullorum	–	ABI	U. Illinois at Urbana-Champaign		In progress	
S. enterica	*enterica*	Saintpaul	SARA23	ABI	J. Craig Venter Institute	NZ_ABAM00000000	In progress	
S. enterica	*enterica*	Saintpaul	SARA29	ABI	J. Craig Venter Institute	NZ_ABAN00000000	In progress	
S. enterica	*enterica*	Schwarzengrund	KMR78	ABI	Korea		In progress	
S. enterica	*enterica*	Schwarzengrund	SL480	ABI	J. Craig Venter Institute	NZ_ABEJ00000000	In progress	
S. enterica	*enterica*	Sendai	55-2461	454	Washington University		In progress	
S. enterica	*enterica*	Senftenberg	SARB59	454	Washington University		In progress	

Table 7.2 (continued)

Species	Subspecies	Serovar/serotype	Strain	Method	Sequencing center	GenBank accession number	Status	Citations
S. enterica	enterica	Stanley	SARB60	454	Washington University	NZ_ACBF00000000	In progress	
S. enterica	enterica	Tennessee	CDC07-0191	ABI	CDC		In progress	
S. enterica	enterica	Thompson	SARB62	454	Washington University		In progress	
S. enterica	enterica	Typhi	SGSC2661	Illumina GS	Washington University		In progress	
S. enterica	enterica	Typhi	Ty21a	ABI	Naval Medical Research Center		In progress	
S. enterica	enterica	Typhimurium	DT2	ABI	Sanger Institute		In progress	
S. enterica	enterica	Virchow	SL491	ABI	J. Craig Venter Institute	NZ_ABFH00000000	In progress	
S. enterica	enterica	Weltevreden	HI_N05-537	ABI	J. Craig Venter Institute	NZ_ABFF00000000	In progress	

[a] Contemporary isolate
[b] Ancestor of all ATCC 14028 strains

The high-throughput sequencing (HTS) technologies that have primarily been used in *Salmonella* sequencing are 454 pyrosequencing, Illumina/Solexa, and, most recently, SOLiD (Holt et al., 2008; Jarvik et al., 2010). The use of these tools provides a new window into genetic diversity within and between *Salmonella* serovars, and within and between epidemic and nonepidemic isolates, at a level that has not previously been possible. These sequences also provide a scaffold for functional genomic studies of *Salmonella* in particular environments, including livestock models and other sources of food-borne infection. Relevant examples of the impact produced by the application of these technologies in understanding *Salmonella* biology are discussed below. These studies include the generation of complete genome sequences for many strains in a single project, high-throughput sequencing of whole transcriptomes and analysis of complex mixtures of mutants.

HTS for Complete Genomic Sequence Comparisons

High-throughput sequencing is now being used for complete genomic sequence comparisons of both clinical isolates of typhoidal *Salmonella* isolates and highly studied laboratory strains and clinical isolates of non-typhoidal salmonellae. HTS can differentiate closely related typhoidal isolates by single nucleotide polymorphism (SNP) typing. *S. enterica* serovar Typhi (*S.* Typhi) is considered a monomorphic organism as the genomes of individual isolates are highly conserved and clonally related (Roumagnac et al., 2006). In support of this notion, DNA sequencing of 199 gene fragments from a collection of 105 isolates of *S.* Typhi detected only 82 single nucleotide polymorphisms (SNPs) (Roumagnac et al., 2006). Because of the low level of genetic variation in monomorphic organisms, further differentiation of Typhi isolates requires a whole-genome approach.

High-throughput sequencing using a combination of 454 pyrosequencing and Illumina/Solexa technologies was recently employed to study genome variation and evolution in *S.* Typhi (Holt et al., 2008). The complete genome sequence of 19 strains of *S.* Typhi (including the previously sequenced strains Ty2 and CT18) was obtained and analyzed in a single study (Holt et al., 2008). Strains were selected for sequencing because they represent major nodes in the phylogeny of this serovar (Roumagnac et al., 2006). The comparative analysis of the genome sequences allowed the determination of complete allele data for 1,787 SNPs, all of which followed the same phylogenetic tree previously defined. This analysis revealed little evidence of adaptative selection, antigenic variation, or recombination among *S.* Typhi isolates. Rather, evolution in Typhi seems to be characterized by continuing loss of gene function, consistent with a small effective population size (Holt et al., 2008). Antigenic variation in Typhi does not appear to be driven by immune selection, but adaptive selection for mutations that result in antibiotic resistance is very strong. The genetic isolation and drift patterns in Typhi underscore the importance of asymptomatic carriers of *S.* Typhi as a reservoir for this organism, underscoring the need for identification of carrier individuals and new methods to treat the carrier state (Holt et al., 2008).

HTS is also being employed to study genetic variation in non-typhoidal salmonellae. Despite its wide use in genetic analysis, *S.* Typhimurium strain LT2 is highly attenuated for virulence both in vitro and in vivo. The most commonly studied virulent strain of *S.* Typhimurium is ATCC 14028, isolated in 1960 from a pool of chicken organs (Patti Fields, personal communication). A nearly complete draft sequence of the ATCC 14028 genome was obtained using the 454 pyrosequencing technology in order to design primers for the generation of a comprehensive collection of defined deletion mutants in *S.* Typhimurium (Santiviago et al., 2009). When the draft sequence of ATCC 14028 was compared to the completed genome of the laboratory strain LT2, over 95% of the two genomes were orthologous presenting less than 1% of divergence. The ATCC 14028 and LT2 genomes differ only by a few hundred single base mutations, the absence of the two Fels phages in ATCC 14028, and other insertions and deletions encompassing less than 40 kb (Santiviago et al., 2009).

More recently, a combination of Roche 454-FLX pyrosequencing, Illumina/Solexa, and ABI SOLiD technologies has been used to obtain the complete genomic sequences of a contemporary isolate of strain ATCC 14028 (Jarvik et al., 2010). In addition, 454-FLX Titanium technology was used to obtain the genomic sequences of three strains that form the lineage leading to the contemporary isolate of ATCC 14028 (Jarvik et al., 2010). In agreement with previous observations (Santiviago et al., 2009), the comparison of genome sequences from strains ATCC 14028 and LT2 indicated that these genomes are 98% identical. The greatest difference between the two strains was the distribution of four prophages: ATCC 14028 harbors prophages Gifsy-3 and ST64B, but lacks Fels-1 and Fels-2, while LT2 harbors both Fels phages and lacks Gifsy-3 and ST64B (Jarvik et al., 2010). Numerous other insertion/deletion events differentiate both genomes, including a 5.1 kb deletion that removed at least four genes (*STM3256-STM3259*). Notably, several mutations were detected that might be responsible for differences in the virulence potential between strains ATCC 14028 and LT2. These mutations were mapped to genes *bigA*, *mviM*, *srfB*, *rpoS*, and *slyA*. On the other hand, the comparison of the sequence from the four ATCC 14028 derivatives revealed sequence changes that have accumulated during laboratory passage of these organisms (Jarvik et al., 2010).

Most non-typhoidal *Salmonella* (NTS) infections are associated with self-limiting gastroenteritis, but can also be associated with systemic disease in immuno-compromised hosts in the developed world. In contrast, an increasing number of invasive disease cases associated with NTS have been reported in the developing world, especially in sub-Saharan Africa (Alausa et al., 1977; Gordon et al., 2001, 2002; Kankwatira et al., 2004). Notably, a significant proportion of these cases in Kenya and Malawi appear to have been caused by an invasive *S.* Typhimurium isolate of a distinct phylogenetic lineage (Kingsley et al., 2009). High-throughput sequencing is being used to determine the genetic differences between these invasive NTS isolates and standard laboratory reference isolates, to gain insight into the mutations that are associated with host adaptation.

Initial multilocus sequence typing (MLST) analysis of these invasive *S.* Typhimurium strains from Malawi and Kenya revealed a dominant sequence type (ST313) that is rarely reported outside Africa (Kingsley et al., 2009). To gain insights into similarities and differences between ST313 isolates and previously sequenced *S.* Typhimurium isolates, the genomic sequence of a multidrug-resistant (MDR) representative of ST313 (*S.* Typhimurium isolate D23580) was determined using classic dye-terminator chemistry on ABI3730 automated sequencers (Kingsley et al., 2009). In addition, draft sequences were generated for two other ST313 representatives using the 454-FLX platform. The sequence revealed a distinct prophage repertoire and a complex genetic element encoding MDR genes in the virulence plasmid. Remarkably, the analysis of genome sequences suggests that ST313 *S.* Typhimurium may have undergone partial selective genome degradation by pseudogene formation and chromosomal deletions (Kingsley et al., 2009). This pattern of genome degradation resembles that observed for host-adapted *Salmonella* serovars that cause invasive disease such as *S.* Typhi, *S.* Paratyphi A, and *S.* Gallinarum (Parkhill et al., 2001; McClelland et al., 2004; Holt et al., 2008; Thomson et al., 2008; Holt et al., 2009). Genome analysis of other epidemic ST313 isolates from Malawi and Kenya provided evidence for microevolution and clonal replacement in the field, likely driven by the introduction of antibiotic treatment (Kingsley et al., 2009).

7.4.3.6 Transposon-Directed Insertion-Site Sequencing (TraDIS)

Highly accurate mapping of mutants in a transposon library has recently been accomplished using high-throughput sequencing technology combined with analysis of transposon mutants in *Salmonella* (Langridge et al., 2009). Deep sequencing has recently been used to identify insertion transposon insertion sites in a highly complex transposon library containing $\sim$1.1 million individual transposon mutants in a derivative of *S.* Typhi isolate Ty2 (Langridge et al., 2009). Total DNA from the library was used to specifically amplify the region adjacent to each insertion, and this material was subjected to deep sequencing using the Illumina technology (Langridge et al., 2009). This state-of-the-art approach, called *transposon-directed insertion-site sequencing* (TraDIS), was used to unambiguously map over 370,000 insertion sites (an average of more than 80 sites per gene) in the genome of *S.* Typhi. The data obtained was analyzed to comprehensively identify genes essential for in vitro growth under standard laboratory conditions. The 365 essential genes identified by TraDIS include those related with fundamental biological processes like cell division, DNA replication, transcription, and translation. Notably, 256 of these genes have been also reported as essentials in *E. coli* (Baba et al., 2006).

Using TraDIS, fewer insertions were detected in 274 genes after six rounds of in vitro passage under standard laboratory conditions than before selection. Thus, these genes were defined as "advantageous" for growth (Langridge et al., 2009). However, enrichment in sequence reads was detected for a small group of genes that were defined as "disadvantageous" for growth. This later group included at least 30 genes involved in the synthesis and assembly of flagella, as in the case of the "hypercompetitive mutants" identified by TMDH (Chaudhuri et al., 2009).

A comparison of mutants in the library in vitro grown in the presence or absence of bile allowed the comprehensive assay of every *S.* Typhi gene for its contribution to bile tolerance. Understanding the molecular basis for bile resistance is relevant in

S. Typhi as this trait allows this pathogen to cause long-term infections in the gall bladder resulting in chronic carriage (Crawford et al., 2010). Among the 169 genes identified in the screen as implicated in bile resistance were those related to LPS biosynthesis and modification (e.g., *waa, wba, pagP*) and many other previously implicated in bile tolerance (e.g., *acrA, acrB, tolC, rob, phoP, phoQ*). Novel observations included genes encoding membrane-associated proteins (*mrcA, mrcB, sanA*) and more than 30 genes encoding hypothetical proteins (Langridge et al., 2009).

Finally, using a similar HTS strategy we have developed a technique in which the ligation steps used in TraDIS above are replaced by T-tailing (Canals, unpublished results). We applied this method to identify Typhimurium genes that can be disrupted by a transposon and remain viable in Luria broth. We also have examined pools of transposon mutants after passage through mouse spleen after intraperitoneal injection, revealing mutants that are less fit in this model. Lower complexity libraries, such as the library of 5000 mutants in each individual gene in *Salmonella*, which we are constructing, need relatively little sequencing capacity. Thus, many experiments can be multiplexed into the same lane. For example, if an average of 50 reads are desired from each mutant then these 5000 mutants can be sampled in 250,000 reads. One lane of a sequencer yields 5 million reads, indicating that 20 experiments could be multiplexed in the same lane at a total sequencing cost of $2000, or $80 per sample. As the sequencers improve, the cost will come down further. Fitness assays of pools of mutants are likely to primarily use HTS rather than microarrays in the future.

7.4.3.7 HTS Transcriptome Analysis in *Salmonella*

Genome-wide high-throughput cDNA sequencing has been used to study transcription in eukaryotes (for example, see Nagalakshmi et al., 2008; Wilhelm et al., 2008). Recently, a similar approach was developed and optimized to study complete transcriptomes of prokaryotes using Illumina/Solexa technology for deep sequencing single-stranded cDNA produced from total RNA samples (Croucher et al., 2009). This improvement over protocols using double-stranded cDNA allows keeping the information on the direction of transcription in the samples. The quality and resolution of the sequence data produced permits precise mapping of known transcripts and identification of novel transcripts in a genome-wide scale, aiding in the validation of the current annotation available for the genomes surveyed. Also, data produced using this method are appropriate for its use in quantitative studies of global transcription in prokaryotes and eukaryotes. A proof-of-principle study describes the successful application of this technology in *S. bongori*, *Streptococcus pneumoniae*, and in the yeast *Schizosaccharomyces pombe* to reveal a clear, unbiased picture of their corresponding transcriptomes at a specific point during growth in vitro (Croucher et al., 2009).

This strand-specific cDNA sequencing methodology, referred to as "ssRNA-seq," was used in a study to analyze the transcriptome of *S.* Typhi (Perkins et al., 2009). In this global analysis, the transcriptional template strands for both coding and non-coding sequences in the genome of *S.* Typhi strain Ty2, including transcripts from annotated pseudogenes, genes within prophages and related elements, complete regions previously not annotated in the genome, and $3'$- or $5'$-untranslated region in many annotated genes, were identified. Furthermore, 40 novel non-coding RNAs were identified.

A combination of proteomic analyses and transcriptome data was employed to confirm observations made for many hypothetical genes and to refine the available genome annotation (Perkins et al., 2009). Furthermore, the ssRNA-seq methodology was employed to compare the transcriptomes of the Ty2 wild-type strain to that of an otherwise-isogenic *ompR* mutant (Perkins et al., 2009). Novel genomic regions under the control of this global transcriptional regulator, as well as the confirmation of known members of the OmpR regulon, such as the entire *viaB* locus (including operons *tviABCDE* and *vexABCDE*, involved in the biosynthesis and export of the Vi antigen), *ssrAB*, *ompC*, and *ompS*, were identified (Perkins et al., 2009). Although many observations were confirmed by RT-PCR, some discrepancies were observed when comparing sequencing data to data from microarray hybridization of the original RNA samples sequenced. Nevertheless, the combination of both technologies allowed the identification of the complete OmpR regulon in *S*. Typhi (Perkins et al., 2009).

7.4.4 Future Prospects

Forward genetic methods in *Salmonella* have been continuously refined so that we now have the ability to screen for fitness defects in a large range of environments. The continued development of targeted deletion libraries of single and double deletion mutants and high-throughput screening methods will be instrumental in beginning to understand the "interactome" of this important pathogen, as is beginning to occur in other Enterobacteriaceae (Typas et al., 2008). Finally, it seems clear that high-throughput sequencing will contribute to many methods used to study every step in *Salmonella* biology, from transcription to fitness, allowing the ordering of regulatory interactions and the production of regulatory hierarchies.

7.5 *Salmonella* Transcriptomic Studies

7.5.1 Technical Approaches for the Study of Gene Expression

Multiple approaches have been used to study gene expression in particular growth conditions and environments. In addition to genetic approaches described in the previous section, microarray technology has been extensively applied to investigate transcriptomic patterns using bacteria either grown in culture or isolated from host fluids, tissues, or cells. Other methods have emerged to identify, for instance, promoters that are induced under specific environments, such as in vivo expression technology (IVET) and differential fluorescence induction (DFI). Furthermore, new high-throughput pyrosequencing (HTPS) strategies have been developed and adapted to identify RNA bound to cellular proteins (summary of the techniques used to study gene expression in *Salmonellae* is provided in Table 7.3).

Table 7.3 Techniques used to study gene expression in salmonellae

Method	Serovar/strain	Environment	Identified genes	References
Microarray technology				
cDNAs and microarray analysis	S. enterica sv. Typhimurium 14028	LB broth, swarm agar (0.6% agar) and solid agar (1.5% agar)	See reference	Wang et al. (2004)
cDNAs and microarray analysis	S. enterica sv. Typhimurium 14028	LB broth, swarm agar	See reference	Wang et al. (2006)
cDNAs and microarray analysis	S. enterica sv. Typhimurium LT2		See reference	Frye et al. (2006)
cDNAs and microarray analysis	S. enterica sv. Typhimurium SL1344	Planktonic growth and biofilm growth on HEp-2 cells	See reference	Ledeboer et al. (2006)
SCOTS (selective capture of transcribed sequences)-cDNAs and microarray analysis	S. enterica sv. Typhi ISP1820	Cultured human macrophages (THP-1)	See reference	Faucher et al. (2006)
cDNAs and microarray analysis	S. enterica sv. Typhimurium ATCC 14028	Anaerobic conditions	See reference	Fink et al. (2007)
cDNAs and microarray analysis	S. enterica sv. Typhimurium SL1344	LB and addition of adrenaline	See reference	Karavolos et al. (2008)
IVET (in vivo expression technology)				
IVET, promoterless purA-lacZY (ΔpurA)	S. enterica sv. Typhimurium ATCC 14028	BALB/c mice (IP, spleen)	carAB operon, pheST himA operon, rfb operon	Mahan et al. (1993)
IVET, promoterless cat-lacZY	S. enterica sv. Typhimurium ATCC 14028	BALB/c mice (IP, spleen)	fadB	Mahan et al. (1995)

Table 7.3 (continued)

Method	Serovar/strain	Environment	Identified genes	References
IVET, promoterless *purA-lacZY* (Δ*purA*), and promoterless *cat-lacZY*	*S. enterica* sv. Typhimurium ATCC 14028	BALB/c mice (IG, small intestine, spleen, liver; IP, spleen)	*phoP*, *pmrB*, *cadC*, *iviXIII* (ChvD-like), *vacB/vacC*, *spvB*, *cfa*, *otsA*, *recD*, *hemA*, *entF*, *fhuA*, *cirA*, *mgtA/mgtB*, *iviX* (heavy metal transport), *ndk*, *ivi-VIA* (Tia/Hra1-like), *ivi-VIB* (PfEMP1-like)	Heithoff et al. (1997)
IVET, promoterless *purA-lacZY* (Δ*purA*)	*S. enterica* sv. Typhimurium ATCC 14028	BALB/c mice (oral, small intestine; IP, spleen)	*gipA*	Stanley et al. (2000)
RIVET (recombinase-based in vivo expression technology)				
RIVET, promoterless *cre* (*loxP-sacB-npt-loxP*)	*S. enterica* sv. Typhimurium ATCC 14028	Cultured epithelial cells (Hep-2)	Chromosomal DNA fragment including 4 ORFs	Altier and Suyemoto (1999)
RIVET, promoterless *tmpR* (*res1-tetAR-res1*)	*S. enterica* sv. Typhimurium ATCC 14028	BALB/c mice (PO or IP, distal ileum lumenal space, ileal Peyer's patches, mesenteric lymph nodes, spleen, liver)	*pagP*, *pmrH* (fusion constructions)	Merighi et al. (2005)

Table 7.3 (continued)

Method	Serovar/strain	Environment	Identified genes	References
RIVET, promoterless *cre* (*loxP-sacB-npt-loxP*)	S. enterica sv. Typhimurium ATCC 14028	Pigs (oral, tonsils, intestine)	*wecC/rffG*, *bcfA*, *fdx-hscA*, *yciR*, *hydH*, *hpaB/hpaR*, *metL*, *cysQ*, *sppA*, *cbiF/cbiG*, STM0611/0612/0613, *oadB*, *rlgA*, STM2689, STM1731, STM2755/2756/2757, *yiiG*, STM0325, STM4489, STM4320/*pheR*, *ybbP*, STM1634/1635, *ybgH*, *ygiK*, STM1368, *yciA/yciB*, *sitC*, *kduD*, *parA*,	Huang et al. (2007)
DFI (differential fluorescence induction)				
DFI, *gfp*	S. enterica sv. Typhimurium SL1344	Acidic environment (in vitro acid shock)	[a]FAI-1 (*aas*), FAI-2 (*dps*), FAI-3 (*marR*), FAI-4 (*pagA*), FAI-5 (*rma*). FAI-15/23 (*pbpA*), FAI-25 (*emrR*)	Valdivia and Falkow (1996)
DFI, *gfp*	S. enterica sv. Typhimurium	Cultured murine macrophages (RAW 264.7)	[b]*mig-1* (*aas*), *mig2* (*pagA/ugd*), *mig-3*, *mig-4* (*phoS*), *mig-5*, *mig-7* (*yjbA*), *mig-10* (*ssaH*), *mig-13* (*orf.f198*), *mig-14*, *mig-20*, *mig-23* (*himA*), *mig-26* (*exc*), *mig-29* (*hslU*), *mig-30*	Valdivia and Falkow (1997)

Table 7.3 (continued)

Method	Serovar/strain	Environment	Identified genes	References
DFI, *gfp*	*S. enterica* sv. Typhimurium SL1344	BALB/c mice (IV, spleen)	*pipB, sifA, aroQ*	Bumann (2002)
Others				
HTPS (high-throughput pyrosequencing) and coIP-on-Chip, Hfq-bound mRNAs	*S. enterica* sv. Typhimurium SL1344	Lennox (L) broth	New sRNAs, *hilD* and *flhDC* mRNAs	Sittka et al. (2008)
Computational screening and experimental characterization	*S. enterica* sv. Typhimurium SL1344	Exponential and stationary phase in LB and stress conditions (low oxygen, anaerobic conditions, starvation, extreme acid conditions, oxidative stress, iron limitation, magnesium limitation, osmotic shock, cold shock, heat shock)	19 novel island-encoded sRNAs (*isr* genes)	Padalon-Brauch et al. (2008)

[a]FAI, fluorescent acid-inducible promoters
[b]*mig*, macrophage-inducible genes

7.5.1.1 Microarray Technology

Gene expression in *Salmonella* has been widely studied by using microarray technology, but there are only a few examples of studies performed in food. Survival and growth of *Salmonella* in food often occurs under nutrient starvation and stress conditions of temperature, pH, osmolarity, and oxidation. Therefore, the data obtained from gene expression studies using these in vitro conditions may be relevant for food safety. Other interesting approaches have been applied to more complex environments, including the host cells and the infected hosts themselves. Gene expression analyses in livestock are of special interest to avoid contamination of the chain food.

The use of DNA microarrays for bacterial transcriptional studies during infection has some limitations, including the difficulty in obtaining enough bacterial RNA from infected cell culture experiments, the short half-life of bacterial mRNA, and the contamination of bacterial mRNA with ribosomal and host RNA (Hinton et al., 2004). Faucher et al. developed a selective capture of transcribed sequences (SCOTS) method and subsequent cDNA hybridization on microarrays to obtain high-quality *S.* Typhi transcript profiles from infected host cells (Faucher et al., 2006).

7.5.1.2 Promoter Fusions: IVET, RIVET, DFI

Technologies based on reporter gene fusions have been developed to identify promoters induced in the host environments. In vivo expression technology (IVET) uses a promoter-trap strategy to assay the transcription of a promoterless reporter gene under the control of a random library of bacterial DNA fragments. This method involves a screening of promoters that are active under in vivo conditions (Mahan et al., 1993; Rediers et al., 2005).

Mahan et al. developed an IVET approach for the identification of *S.* Typhimurium genes that are induced in animal tissues. An auxotrophic mutant in the *purA* gene was used for the selection (Mahan et al., 1993). In a later work, a construction of a new IVET vector permitted the use of antibiotic resistance genes as reporter genes. The *fadAB* operon involved in the metabolism of fatty acids was obtained and characterized after infection of a murine model (Mahan et al., 1995).

Recombinase-based in vivo expression technology (RIVET) is a more recent variation of the original IVET (Camilli and Mekalanos, 1995). RIVET was developed to overcome the inability of the two previous methods to identify weak and transient gene expression. This technology is based on the activity of a promoterless site-specific recombinase mediating the excision of a DNA fragment placed between two specific target sequences by recombination. A system based on the phage P1 recombinase *cre* was developed and applied to identify *S.* Typhimurium genes specifically expressed when bacteria invade cultured epithelial cells (Altier and Suyemoto, 1999).

Differential fluorescence induction (DFI) is another promoter-trap method that uses the green fluorescent protein (GFP) as a reporter. Fluorescence-activated cell sorting (FACS) is used to monitor promoter activity and permits a high-throughput screening of gene expression (Valdivia and Falkow, 1996; Rediers et al., 2005). This approach was applied to identify acid-inducible genes (Valdivia and Falkow, 1996) and genes of *S.* Typhimurium induced inside phagocytic cells (Valdivia and Falkow, 1997). Gene expression of *S.* Typhimurium has been analyzed in murine Peyer's patches using two-color flow cytometry (Bumann, 2002).

7.5.2 In Vitro Conditions in Transcriptomic Studies

The transcriptional response of *S.* Typhimurium to acidic conditions has been studied using DFI (Valdivia and Falkow, 1996). Eight acid-inducible promoters were isolated and characterized based on their homologies to previously known promoter regions. Four of the regions identified in this study are involved in the structure and maintenance of the bacterial cell surface, including the upstream region of the *E. coli aas* (2-acylglycerophosphoethanolamine acyltransferase) gene, the *pagA* locus, and the *pbpA* promoter. Two DNA inserts were homologous to the promoter regions of genes in *E. coli*, *dps* and *rna*, that are associated with the stress response. The final two promoters were homologous to regions upstream of the *E. coli mar* and *emr* loci, which are both involved in the expression of efflux pumps (Valdivia and Falkow, 1996). Intra-phagosomal survival of *S.* Typhimurium had been associated with the activation of genes differentially regulated at low pH (Rathman et al., 1996). The levels of gene expression from these promoters inside macrophages did not correlate with the in vitro acid shock induction (Valdivia and Falkow, 1996), suggesting the existence of additional signals, apart from the low pH in the phagosomal environment, that may regulate expression of genes important for intracellular survival.

Fnr is a global anaerobic regulator in *S.* Typhimurium ATCC 14028s and has been shown by transcriptional profiling to differentially regulate at least 311 genes (Fink et al., 2007). Transcriptional profiles of *fnr* mutant and wild-type salmonellae were compared by microarray analysis. *fnr* mutants were attenuated during infection in a murine model and they were also non-motile and lacked flagella, implicating Fnr in virulence and motility in serovar Typhimurium (Fink et al., 2007). *fnr* mutants are also highly attenuated in macrophages, indicating that Fnr may be functional inside the phagosome (Fink et al., 2007).

The microarray data generated in this transcriptional analysis identified components of the Fnr modulon which are involved in diverse processes including ethanolamine utilization (*eut* operon), type III secretion system-1 regulation (TTSS-1), motility, chemotaxis, and numerous genes involved in metabolic functions (Fink et al., 2007). These findings agree with previous work demonstrating that *Salmonella* grown under oxygen-limited growth conditions are more adherent and invasive than when they are grown in aerobic conditions (Bajaj et al., 1996; Klein et al., 2000).

The survival and growth on surfaces have been associated with surface colonization and host invasion (Harshey, 2003). Global gene expression profiles of *S.* Typhimurium 14028 in Luria–Bertani (LB) broth and on two different surfaces, swarming agar and solid agar, may increase our understanding of *Salmonella* growth

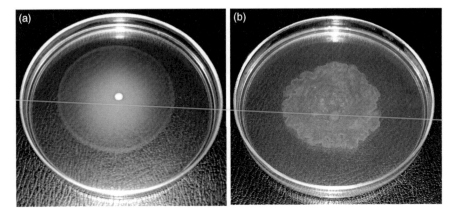

Fig. 7.2 *Salmonella enterica* sv. Typhimurium 14028 on swim agar (**a**) and on swarm agar (**b**) (7 h at 37°C)

and movement on solid surfaces (Wang et al., 2004) (see Fig. 7.2 for illustration of swimming and swarming motility). *Salmonellae* in the swarmer cell state are hyperflagellated and they move within a mixture of "slime" of polysaccharides, surfactants, and proteins that they produce to facilitate movement on a solid surface (Toguchi et al., 2000). Flagella and lipopolysaccharide are essential for swarming motility (Harshey and Matsuyama, 1994; Toguchi et al., 2000). Several reports have addressed the mechanisms that regulate swarming (Harshey, 2003), but the external factors that induce the differentiation to swarmer cells are poorly understood.

Comparative transcriptional studies show that one-third of the *Salmonella* functional genome was differentially regulated in swarming versus non-swarming conditions (Wang et al., 2004). Among the genes upregulated after growth on surface were genes involved in the synthesis of the flagellar filament (*fliC*, *fljB*, *fljA*, and *fliD*), lipopolysaccharide biosynthesis, iron uptake (*sitABCD* system on SPI-1) metabolism, and the type III secretion system (SPI-1 and SPI-2). *S.* Typhimurium swarmer cells show elevated resistance to multiple antibiotics and this phenotype is linked to the upregulation of the *pmrHFIJKLM* operon (Wang et al., 2004). This operon is involved in resistance to antimicrobial peptides, including polymixin, via modifications of the lipid A component of LPS (Kim et al., 2003).

New flagellar genes that affect surface motility were recently identified by microarray analyses in *S.* Typhimurium mutants (Wang et al., 2006). Among these newly discovered flagellar genes, two genes, *btuC* and *yhiH*, were involved in swimming motility. Twenty-eight additional genes, including *STM1301*, *STM1485*, *STM1657*, *STM3154*, *STM3156*, *STM3604*, and *ymdA* (Wang et al., 2006), were identified as new motility genes. The chemotaxis system promotes migration in swimming motility depending on chemical gradients and it enhances swarming motility by promoting optimal surface wetness (Mariconda et al., 2006). Two new *S. enterica* chemoreceptors were also identified, McpB and McpC, in addition to the known receptors (Tar, Tap, Tsr, Tcp, and Aer) (Wang et al., 2006).

Transcriptional analysis using microarrays has also been used to screen for novel genes coregulated with known flagellar genes in S. Typhimurium LT2 (Frye et al., 2006). Total RNA from different flagellar mutants was compared to total RNA from the isogenic wild-type strain, and the microarray results were verified using Mud-*lac* operon fusions. Based on this work, the *srfABC* operon is part of the flagellar class 2 operons, although it is not required for neither flagellar construction nor motility (Frye et al., 2006). The *srfABC* operon was originally identified as a putative horizontally acquired DNA under the control of SsrB (*srf*, SsrB-regulated *factor*) (Worley et al., 2000), a regulator encoded within SPI-2 (Waterman and Holden, 2003). However, recent work shows that the *srfABC* operon is expressed in SPI-1-inducing conditions but not in SPI-2 conditions and repressed by both RcsB and PhoP (Garcia-Calderon et al., 2007). Furthermore, expression of four new chemotaxis genes unique to *Salmonellae* clusters with the flagellar class 3 genes: *cheV*, *mcpA*, *mcpB*, and *mcpC*. Several of these genes, *cheV*, *mcpB*, and *mcpC*, as well as the *aer* gene, are dependent of sigma 28, required for the transcription of the flagellar class 3 genes. *fliB*, involved in methylation of the flagellar lysine residues, was determined to be a flagellar class 2 gene and it was also suggested to be required for virulence (Frye et al., 2006).

The ability of *Salmonella* to form biofilms may play a role in intestinal carriage in livestock such as poultry, swine, and cattle (Althouse et al., 2003; Morgan et al., 2004). Biofilm formation on non-biological surfaces may also be considered a source of food contamination, particularly in the food processing industry (Joseph et al., 2001). Extracellular polysaccharides, such as cellulose and colanic acid, and fimbriae are significant components of biofilms formed on HEp-2 cells and chicken intestinal epithelium (Ledeboer and Jones, 2005; Ledeboer et al., 2006). However, the composition of biofilms formed by *Salmonella* spp. varies depending on environmental conditions (Prouty and Gunn, 2003; Ledeboer and Jones, 2005; Ledeboer et al., 2006). Quorum sensing plays a critical role in this group behavior related to the dependence on cell density (Surette et al., 1999; Prouty et al., 2002). The gene expression profile of bacteria growing in biofilms on HEp-2 cells was compared to the gene expression profile of bacteria growing planktonically in liquid medium (Ledeboer et al., 2006). In this comparison, 2.2% of the S. Typhimurium genome was upregulated. Four types of fimbriae, Lpf, Pef, Tafi, and type 1 fimbriae, are important for *Salmonella* biofilm formation on eukaryotic cells (Boddicker et al., 2002; Ledeboer et al., 2006). The phenotypes of these fimbrial mutants were confirmed on chicken intestinal epithelia and on tissue culture-coated plastic (non-biological surface). In addition to genes involved in fimbrial biosynthesis, genes encoding proteins involved in conjugation (*traI*, *traR*, *trbE*, *traO*, *traK*, *traD*, *traQ*, *traJ*, *traA*, and *traM*), in antibiotic resistance (*marA* and *marB*), in intracellular survival (*ssaM*, *ssaI*, *pagD*, *msgA*, *pipB*, *spvR*, *ssaS*, *sifA*, *pipD*, and *ssaL*), and many genes with no known function were induced by biofilm growth (Ledeboer et al., 2006). Thus, fimbriae, and numerous other functions, play a critical role in S. Typhimurium colonization or biofilm formation on eukaryotic cell surfaces.

7.5.3 Regulators of Gene Expression

S. Typhimurium modifies components of the bacterial surface to counteract mechanisms of the host innate immune defenses (Miller et al., 1989b; Miller, 1991; Guina et al., 2000). The PhoP/PhoQ two-component system is a regulator of many genes including those involved in the resistance to cationic antimicrobial peptides (CAMP), such as the PmrA/PmrB two-component system (Gunn and Miller, 1996). PmrA/PmrB regulates the addition of 4-amino-4-deoxy-L-arabinose to the lipid A, and this modification is required for resistance to CAMPs (Zhou et al., 2001). Using a RIVET approach, gene expression of some PhoP- and PmrA-regulated genes was studied during infection of S. Typhimurium in BALB/c mice (Merighi et al., 2005). The PhoP/PhoQ and PmrA/PmrB regulons were expressed in bacteria located extracellularly in the intestine of BALB/c mice and not only intracellularly inside macrophages as previously thought (Alpuche Aranda et al., 1992). In vitro data revealed that low levels of magnesium downregulate PhoP-activated genes by activating PhoQ. In addition, high concentrations of iron (III) activate PmrB, and acidic pH activates both PhoP and PmrA regulons. However, the activation of these two regulons in the intestine was found not to be dependent on acidic pH or iron levels (Merighi et al., 2005). A DFI approach permitted the identification of eight promoters that required the PhoP/PhoQ two-component regulatory system for their induction in macrophages (Valdivia and Falkow, 1997). In that work, 14 promoters of S. Typhimurium were reported to be induced in phagocytic cells.

The new high-throughput pyrosequencing (HTPS) technology was recently applied to identify *Salmonella* RNAs that are targets of the Hfq regulator (Sittka et al., 2008). Hfq can associate with regulatory small non-coding RNAs (sRNA) and mRNAs to stabilize interactions and act as a post-transcriptional regulator (Zhang et al., 2003c). Transcriptomic mRNA profiles of S. Typhimurium wild type and a mutant in *hfq* were compared. Hfq regulates at least 18% of the *Salmonella* genome at the transcriptional level, including four pathogenicity islands (SPI-1, SPI-2, SPI-3, and SPI-5), the flagellar regulon, and two stress sigma factors, RpoS and RpoE (Sittka et al., 2008). To identify Hfq targets, the HTPS technology was used to sequence hundreds of thousands of cDNAs obtained from RNAs co-immunoprecipitated (coIP) with the chromosomally epitope-tagged Hfq protein (Sittka et al., 2008). This analysis revealed new sRNAs and increased the number of identified sRNA in S. Typhimurium to 64. Combining the transcriptomic and co-immunoprecipitation data, Hfq controls the expression of the master regulators of the SPI-1 and the flagellar regulon, HilD and $FlhD_2C_2$, respectively (Sittka et al., 2008).

7.5.4 Virulence Genes

Macrophages and neutrophils are able to produce and release adrenaline and noradrenaline when they interact with the bacterial lipopolysaccharide (Flierl et al., 2007). The effects of adrenaline on the S. Typhimurium SL1344 transcriptome were recently reported (Karavolos et al., 2008). Microarray data showed that 0.6% of the transcriptome was significantly regulated by adrenaline. Genes involved in uptake of siderophores and microcins (*fhuA* and *fhuC*, *exbBD*), manganese transport (*sitAB*), iron transport (*feoAB*), resistance to the early oxygen-dependent microbicidal mechanism of phagocytes (*sodA*), and oxidative stress (*oxyR*) were found to be upregulated. The expression of the *pmrHFIJKLM* operon, involved in modifications of the lipid A that confer resistance to the cationic antimicrobial peptides, was decreased upon exposure to adrenaline. Thus, the net effects of adrenaline exposure on *Salmonellae* were a reduction in antimicrobial peptide resistance and an activation of mechanisms for protection from oxidative stress (Karavolos et al., 2008).

A DFI two-color flow cytometry technique has been used to identify *Salmonella* SL1344 promoters induced during infection of BALB/c mice (Bumann, 2002). The promoters of the *sicA-sipBCDA-iacP* operon (P_{sicA}) in SPI-1, the *ssaHIJKLMV* operon in SPI-2 (P_{ssaH}), and the PhoP-regulated *pagC* (P_{pagC}) were studied during infection. In vitro, P_{sicA} is induced under low oxygen and high osmolarity conditions, P_{ssaH} under acidic conditions, and P_{pagC} with low magnesium levels. The SPI-1 secreted effectors were expressed during the initial stages of intestinal invasion, while genes from SPI-2 were found to be expressed in the infected host tissues. The *pagC* promoter was induced in most of the infected host cells. In addition, four *Salmonella* promoters were upregulated in the spleen of infected mice: the promoters of *pipB*, within the SPI-V; *sifA*, secreted by the type III secretion system encoded by SPI-2; *aroQ*, involved in the biosynthesis of aromatic amino acids; and a promoter that showed limited homology to the *S.* Typhimurium LT2 strain genome (Bumann, 2002).

S. Typhimurium 14028 promoters induced during infection of BALB/c mice and/or murine cultured macrophages were identified using IVET approaches (Heithoff et al., 1997). Environmental signals during host infection trigger the induction of a wide range of regulatory, metabolic, and virulence *Salmonella* genes that contribute to its pathogenicity. Several in vivo-induced genes (*ivi*) encode regulatory proteins, including *phoP*, *pmrB*, and *cadC*. The PhoPQ regulatory system is activated after intragastric and intraperitoneal infection in mice (Heithoff et al., 1997). Another group of genes induced during infection are involved in metabolic functions: *recBD* and *hemA*; *entF*, *fhuA*, and *cirA* (induced under iron-limiting conditions in vitro); *mgtA* and *mgtB* (induced under low magnesium levels in vitro). Some genes regulated by sigma S (sigma 38) encoded by *rpoS*, including *spvB*, and some adhesin- and invasin-like genes were also detected as *ivi* genes (Heithoff et al., 1997). The *gipA* gene, within the Gifsy-1 locus, was induced during the early stages of infection in the small intestine of mice in a more recent work (Stanley et al., 2000). *gipA* mutants were unable to survive in the Peyer's patches, and this finding is in agreement with the finding that these mutants are attenuated after oral but not after intraperitoneal infection (Stanley et al., 2000).

Many sRNA act as regulators by binding to proteins to regulate their activity or by binding to mRNAs, the latter interaction is mediated by the Sm-like RNA-binding protein, Hfq (Valentin-Hansen et al., 2004). Factors that permit *Salmonella* to adapt to the acidic environment of the stomach and to the intracellular environment of macrophages are considered to contribute to its virulence, particularly after ingestion of contaminated food. The expression of 19 novel island-encoded sRNAs (*isr* genes) of *S.* Typhimurium SL1344 was studied under these and other stress conditions (Padalon-Brauch et al., 2008). Most of these genes were found to be expressed when *Salmonella* resides inside the macrophage environment (Padalon-Brauch et al., 2008). For example, IsrJ, encoding a sRNA, was expressed under low oxygen and low magnesium levels and affected the translocation of effector proteins into non-phagocytic cells (Padalon-Brauch et al., 2008). Several additional sRNAs, including IsrA, IsrG, IsrH, can also modulate the expression of their flanking genes in *cis* and are also affected by those same genes, usually in response to environmental changes for *Salmonella* adaptation (Padalon-Brauch et al., 2008).

7.5.5 Livestock

One important aspect for human food safety is the asymptomatic carriage of salmonellae in animals that can be a source of food contamination. The mechanisms that permit *Salmonella* to survive and persist in clinically healthy animal hosts with persistent shedding of bacteria in feces are poorly understood. A combination of the RIVET technology and a signature-tagging approach was used to identify *S.* Typhimurium promoters induced during infection of swine (Huang et al., 2007). Thirty-one genes were identified after isolation of a genomic library of random DNA fragments from tonsils and/or intestine of pigs. Several of the induced genes are known to be involved in bacterial adhesion and colonization, including *bcfA*, *wecC*/*rffG*, *hscA*, and *yciR*, and in virulence, *metL*. Mutants in the *Salmonella* pathogenicity islands SPI-1 and SPI-2 had previously been reported to maintain their ability to colonize the lumen of chicken intestine, another important asymptomatic carrier of *Salmonella* (Morgan et al., 2004). Thus, the finding that no genes within the SPI-1 and SPI-2 were isolated from this more recent study may be in agreement with that previous result. Only elevated temperature and increased osmolarity were found to induce some of the new genes originally as expressed during infection. These two factors may act as environmental signals to induce gene expression in pigs (Huang et al., 2007).

7.5.6 Current Limitations and Future Prospects

Transcriptomics has revealed considerable amount of information about the genes expressed in a variety of environments. However, expression of an RNA does not necessarily mean that it, or a protein product of the transcript, is required in the circumstance under study. The real strength of RNA profiling is still to come and will arise from integrating RNA profiles into functional regulatory units, as well as integrating this information with fitness information discussed in Section 7.4.

7.6 *Salmonella* Proteomic Studies

Gene transcription and protein levels do not necessarily correlate due to post-transcriptional regulation. A combination of genomic and proteomic studies is essential to fully understand the mechanisms that modulate virulence and pathogenesis in host–pathogen interactions. In recent years, new approaches in proteomic analyses, including multiplexed gels, mass spectrometry, and protein arrays have been developed and applied to the study of host–pathogen interactions. *Salmonella* proteins associated with virulence, as well as proteins altered in host response to infection, may define novel therapeutic targets, new vaccine development approaches, and new strategies for interfering with *Salmonella* growth in foods (Zhang et al., 2005) (summary of proteomic studies in salmonellae in Table 7.4).

7.6.1 Technical Approaches for Proteomic Characterization

In the last few years, mass spectrometry (MS)-based proteomics has become a powerful tool to identify *Salmonella* proteins present under particular conditions (Aebersold and Mann, 2003; Coldham and Woodward, 2004; Adkins et al., 2006; Rodland et al., 2008). *S.* Typhimurium LT2 contains 4,489 coding sequences (CDS/ORFs) including 39 pseudogenes and a plasmid encoding 108 ORFs (McClelland et al., 2001). Approximately 233 *S.* Typhimurium cytosolic proteins were identified using two-dimensional gel electrophoresis combined with MS for a comparative analysis with the serovar Pullorum proteomic profile (Encheva et al., 2005). Two-dimensional chromatography was used to detect 816 proteins from *S.* Typhimurium grown in LB broth (Coldham and Woodward, 2004).

Recently, many improvements in available proteomics techniques have resulted in improved sensitivity and resolution, data analysis, and throughput. For example, the "bottom-up" approach, reversed-phase chromatographic separation of peptide mixtures, has been one of the most highly used MS-based methods for the study of proteomes. Proteins are isolated from bacteria grown under conditions of interest and then denatured and digested with a site-specific protease. The resulting peptides are separated by reversed-phase liquid chromatography in conjunction with MS (Wolters et al., 2001). A liquid chromatography (LC)-MS-based bottom-up proteomic approach was used to evaluate the *S.* Typhimurium proteome and a total of 2,343 proteins were identified (Adkins et al., 2006). Tandem mass spectrometry (MS/MS) has been used to characterize proteins by selecting and fragmenting the peptides of interest (Nunn et al., 2006). An accurate mass and time (AMT) tag approach was developed to identify *S.* Typhimurium proteins isolated from infected macrophages (Shi et al., 2006), improving the throughput and sensitivity of the LC-MS/MS bottom-up approaches (Rodland et al., 2008).

Multiple methodologies are usually necessary to fully understand the global proteomic complement of a system. The improvements in mass spectrometry-based methods for proteomic analysis should be accompanied by optimization of sample

Table 7.4 Proteomics methods used in salmonellae

Technique	Study	Serovar/strain	Environment	Identified proteins	References
Two-dimensional electrophoresis (2-DE) and peptide mass fingerprinting and nanoelectrospray mass spectrometry (MS)	Detection of protease activities and identification of the proteolytic cleavage sites, *phoQ*-constitutive mutant	*S. enterica* sv. Typhimurium SL1344	LB and a Mg^{2+}-depleted growth medium	See reference	Adams et al. (1999)
2-DE and peptide mass fingerprinting and nanoelectrospray MS	Identification of acid-regulated proteins	*S. enterica* sv. Typhimurium SL1344	Acidic growth conditions (logarithmic-phase acid tolerance response)	See reference	Adams et al. (2001)
2-DE	SPI-2 encoded proteins, recombinant expression in *E. coli* and overexpression of the SsrAB regulatory system	*S. enterica* sv. Typhimurium ATCC14028	Phosphate starvation, acidified medium, and M9 medium	2-DE map of SPI-2 proteins	Deiwick et al. (2002)
Two-dimensional high-performance liquid chromatography (HPLC)-MS	Characterization of the proteome of *S. enterica* sv. Typhimurium	*S. enterica* sv. Typhimurium SL1344	LB broth	816 proteins	Coldham and Woodward (2004)
2-DE and MS (peptide mass fingerprinting by MALDI-TOF analysis)	Effects of a *dsbA*-null mutant on periplasmic extracts	*S. enterica* sv. Typhi 5866	LB broth	65 proteins	Agudo et al. (2004)

Table 7.4 (continued)

Technique	Study	Serovar/strain	Environment	Identified proteins	References
Two-dimensional gel electrophoresis (2-DGE) and tandem MS (LC-MS/MS)	Comparison of proteomes (cytosolic proteins)	*S. enterica* sv. Typhimurium and Pullorum	Columbia blood agar	233 *S. enterica* sv. Typhimurium proteins	Encheva et al. (2005)
Liquid chromatography (LC)-MS (bottom-up approach)	*S. enterica* sv. Typhimurium global proteome analysis under laboratory conditions and phagosome-mimicking culture conditions	*S. enterica* sv. Typhimurium LT2 and ATCC14028	Logarithmic- and stationary-phase cultures in rich medium and an acidic and magnesium-depleted minimal medium (MgM)	2,121 LT2 proteins and 2,296 ATCC14028 proteins	Adkins et al. (2006)
Shotgun proteomics on LCQ-DUO and LTQ-FT	Comparison of a high-performance mass spectrometer (LTQ-FT) and a lower performance mass spectrometer (LCQ-DUO). Characterization of a PhoP constitutive strain	*S. enterica* sv. Typhimurium	LB broth	655 proteins	Nunn et al. (2006)
LC-MS analysis and an accurate mass and time (AMT) tag method	Detection of changes in protein abundance at different times after infection	*S. enterica* sv. Typhimurium 14028	Cultured murine macrophages (RAW 264.7)	315 *S. enterica* sv. Typhimurium proteins and 371 macrophage proteins	Shi et al. (2006)

Table 7.4 (continued)

Technique	Study	Serovar/strain	Environment	Identified proteins	References
Flow cytometry (gfp) and MS	In vivo enzyme expression	S. enterica sv. Typhimurium SL1344	BALB/c mice (IV, cecum in the enteritis model and spleen in the typhoid fever model)	370 proteins in the typhoid fever model and 835 proteins in the enteritis model	Becker et al. (2006)
Comparative peptidomics using LC-MS/MS and LC-Fourier transform ion cyclotron resonance (FTICR)-MS	Comparison of Salmonella cultured in a rich medium and in a phagosome-mimicking medium	S. enterica sv. Typhimurium 14028 and LT2	Rich medium and an acidic, low magnesium, and minimal nutrient medium	682 proteins	Manes et al. (2007)
2-DGE and tandem mass spectrometry (LC-MS/MS)	Comparison of protein expression patterns in closely related Salmonella serovars	S. enterica sv. Typhimurium, Pullorum, Enteritidis, Choleraesuis and Dublin	Columbia blood agar	See reference	Encheva et al. (2007)
Stable isotope labeling with amino acids in cell culture (SILAC) and quantitative MS	Identification of host targets of the SPI-1 effector SopB/SigD	S. enterica sv. Typhimurium SL1344	HEK293, HeLa, and Raw264.7 cultured cells	See reference	Rogers et al. (2008)
Two-dimensional HPLC-MS	Analysis of the proteomes of variants of S. enterica sv. Typhimurium obtained after prolonged exposure to farm disinfectants	S. enterica sv. Typhimurium SL1344	Prolonged exposure to farm disinfectants	See reference	Karatzas et al. (2008)

Table 7.4 (continued)

Technique	Study	Serovar/strain	Environment	Identified proteins	References
LC-MS	*S. enterica* sv. Typhi proteome analysis under laboratory conditions and phagosome-mimicking culture conditions	*S. enterica* sv. Typhi strain Ty2	Logarithmic and stationary growth conditions and an acidic and low magnesium medium (MgM)	2,066 proteins	Ansong et al. (2008)
HPLC-MS/MS	Optimization of sample preparation for MS analysis	*S. enterica* sv. Typhimurium	Cultured murine macrophages (RAW 264.7)	See reference	Mottaz-Brewer et al. (2008)
Histidine-biotin-histidine protein (HBH) tag, in vivo cross-linking with formaldehyde and LC-MS/MS	Protein-protein interactions: HimD, PhoP, PduB	*S. enterica* sv. Typhimurium 14028	MgM medium	See reference	Chowdhury et al. (2009)
LC-MS	Effects of Hfq and SmpB on the global proteome	*S. enterica* sv. Typhimurium 14028	Logarithmic- and stationary-phase cultures, and in two variations of an acidic, low magnesium minimal media (AMM)	1,621 proteins	Ansong et al. (2009)
Two-dimensional difference gel electrophoresis (2-D DIGE)	Global proteome under laboratory conditions and in vivo-mimicking culture conditions	*S. enterica* sv. Typhimurium SL1344	LB and low oxygen tension and high osmolarity conditions (gut environment)	250 differentially expressed proteins, but only 66 identified proteins among them	Sonck et al. (2009)

Table 7.4 (continued)

Technique	Study	Serovar/strain	Environment	Identified proteins	References
LC-MS/MS	Macrophage response to *Salmonella* infection	*S. enterica* sv. Typhimurium 14028	Cultured murine macrophages (RAW 264.7)	1,006 *S. enterica* sv. Typhimurium proteins and 115 macrophage proteins	Shi et al. (2009)
Reverse-phase protein array (RPA)	Host cell response to the effector proteins SopE/E2 and SigD (SopB)	*S. enterica* sv. Typhimurium SL1344	Cultured epithelial cells (HeLa)	See reference	Molero et al. (2009)
2-DGE and MS	Comparison of proteomes (cytosolic proteins)	*S. enterica* sv. Gallinarum and Enteritidis	LB agar	34 differentially expressed proteins	Osman et al. (2009)

preparation procedures, such as looking at sample location and post-translational modifications or maximizing protein recovery (Mottaz-Brewer et al., 2008). Agudo et al. (2004) developed a direct method to identify disulfide proteins from periplasmic extracts of *S.* Typhi by two-dimensional electrophoresis (2-DE) in order to study the effects of a *dsbA*-null mutant. The *dsbA* gene encodes a periplasmic disulfide-bond oxidoreductase and *dsbA* mutants lack flagellin, glucose-1-phosphatase, AI-2 autoinducer-producing protein LuxS, in addition to lacking DsbA (Agudo et al., 2004). A 2-DE map of *S.* Typhimurium SPI-2 proteins was obtained using recombinant expression in *E. coli*, as well as overexpression of the SsrAB regulatory system in *Salmonella* (Deiwick et al., 2002). This approach, in addition to sub-cellular fractionation and pulse labeling, permitted the analysis of the type III secretion system proteins encoded by SPI-2, proteins that are barely detectable by direct proteomic analyses. When sample amount is not limiting traditional approaches are adequate to determine the proteome under a given condition. For example, shotgun proteomics using an old, inexpensive, ion-trap mass spectrometer (LCQ-DUO) has been used to characterize PhoP constitutive strain of *S.* Typhimurium (Nunn et al., 2006). To increase the proteome coverage from the LCQ-DUO to the level of a high-performance mass spectrometer as LTQ-FT, protein pre-fractioning steps and gas-phase fractionation (GFP) are recommended (Nunn et al., 2006).

Mass spectrometry has also been used for the identification of transient interactions within living cells. A method to study protein–protein interactions in *S.* Typhimurium has recently been developed (Chowdhury et al., 2009). A histidine–biotin–histidine protein tag was used in in vivo cross-linking to stabilize interactions under physiological conditions and facilitate subsequent purification under denaturing conditions. The dynamic interactions of three virulence factors (HimD, PhoP, and PduB) were studied under in vitro conditions that mimic systemic infection in mice. This analysis revealed interactions between PhoP and Hfq and perhaps this interaction explains the mechanism of PhoP post-transcriptional regulation of SsrA (Chowdhury et al., 2009). Host targets of the SPI-1 effector SopB were determined by stable isotope labeling with amino acids in cell culture (SILAC) approach (Rogers et al., 2008).

In the last few years, these proteomic advances have led to the necessity to develop new analytical and bioinformatic tools and databases to manage the vast amount of information generated by large-scale proteomics experiments. One example is the SQL-LIMS database that was developed for integration and storage of proteomics data emerging from different kinds of analyses and applied to the proteome of *S.* Typhimurium, as well as the proteomes of *Mycobacterium tuberculosis* and *Helicobacter pylori*, and protein complexes as 20S proteasome (Schmidt et al., 2009). Post-processed data obtained from different approaches, such as two-dimensional gel electrophoresis, LC-electrospray ionization (ESI)-MS, and matrix-assisted laser desorption/ionization (MALDI)-MS, are stored in SQL-LIMS (Schmidt et al., 2009). Combining proteomics with genomic and transcriptomic results is one of the goals of systems biology in order to integrate the data to create models that could simulate the bacterial physiology.

7.6.2 Post-transcriptional Regulation

The comparison of transcriptional and translational profiles provides a handle to understand post-transcriptional regulation. In bacteria, only a few post-transcriptional regulators have been identified as RNA-binding protein factors including CsrA, Hfq, and SmpB (Ansong et al., 2009). A mutation in any of these *Salmonella* genes results in a loss of virulence in mice and the inability to survive in macrophages (Altier et al., 2000; Julio et al., 2000; Sittka et al., 2007; Ansong et al., 2009). Hfq is a major post-transcriptional regulator that modulates the translation of mRNAs by small non-coding RNAs (sRNAs) (Sittka et al., 2007). SmpB, another RNA-binding protein, is required for SsrA function (also known as tmRNA) by interaction with this sRNA (Karzai et al., 1999).

The effects of Hfq and SmpB on transcriptional and translational profiles of *S.* Typhimurium strain 14028 have been determined by microarray analysis and by LC-MS-based proteomics, respectively (Ansong et al., 2009). A comparison of protein abundance between *smpB* and *hfq* mutants and a wild-type strain has been determined under logarithmic- and stationary-phase growth conditions and in acidic minimal media (AMM), mimicking the intra-phagosomal environment (Ansong et al., 2009). 1,621 *S.* Typhimurium proteins were identified, that represented ~36% coverage of the predicted protein coding regions.

Although Hfq and SmpB showed modest effects on *Salmonella* transcript levels using microarray analysis, proteomic analysis revealed significant differences in protein abundances (Ansong et al., 2009). Protein levels of HtrA, a stress protein involved in *Salmonella* virulence (Lewis et al., 2009), were strongly increased in the *hfq* mutant grown to logarithmic phase in LB, and two additional proteins, OsmY and STM1513, were strongly down-regulated in an *hfq⁻* background during growth in LB to stationary phase and in in vitro conditions mimicking phagosomal growth (Ansong et al., 2009). In an *smpB* negative background, FliC was down-regulated in LB during logarithmic phase, while an upregulation of YciF occurred in in vitro conditions mimicking the phagosome (Ansong et al., 2009).

These findings showed that Hfq and SmpB directly or indirectly regulate expression of at least 20 and 4%, respectively, of the *Salmonella* proteome (Ansong et al., 2009). Hfq plays a significant role in central metabolism, stress response, expression of several ribosomal proteins as well as virulence factors, including proteins involved in the LPS biosynthesis, two-component systems (SsrA/B, PhoP/Q, OmpR/EnvZ), motility, and invasion (SPI-1 TTSS). Finally, Hfq strongly regulates the propanediol utilization (*pdu*) operon, required for replication within macrophages (Klumpp and Fuchs, 2007), by a post-transcriptional mechanism (Ansong et al., 2009).

SmpB may function as a general post-transcriptional regulator in *S.* Typhimurium (Ansong et al., 2009). SmpB plays a role in motility and histidine biosynthesis by affecting some chemotaxis proteins, Tsr and several flagellar proteins, and HisDGCBHAF (Ansong et al., 2009). Post-transcriptional and post-translational regulation may reflect a necessity to more rapidly adapt to changes in the environmental conditions.

7.6.3 **Salmonella** *Protein Expression Under Diverse Growth Conditions*

Limitations such as low abundance of some proteins, transient expression, and contamination by host proteins make the characterization of the *Salmonella* proteome under in vivo conditions very difficult. Therefore in vitro conditions thought to mimic the host environment are being used to estimate how the *Salmonella* proteome changes during infection. *S.* Typhimurium cultures grown to logarithmic phase in rich medium are known to express genes needed for invasion of host cells, such as components of SPI-1 (Adkins et al., 2006; Thompson et al., 2006). Stationary phase cultures of *S.* Typhimurium express genes involved in systemic infection and replication within macrophages, such as components of SPI-2 (Adkins et al., 2006; Thompson et al., 2006). Bacterial growth in acidic and magnesium-depleted minimal media has been used to mimic the conditions of the phagocytic vacuole within infected macrophages, as low pH and low Mg^{2+} are known to induce expression of proteins required for macrophage infection, such as the TTSS encoded by SPI-2 (Alpuche Aranda et al., 1992; Garcia Vescovi et al., 1996; Rappl et al., 2003). Comparison of *Salmonella* proteomic profiles grown in different culture conditions represents a useful approach for understanding the molecular mechanisms involved in pathogenesis.

The use of antibiotics in livestock has been associated with the emergence of antibiotic-resistant strains of *Salmonella* (Bauer-Garland et al., 2006). The comparison of protein expression between multiple antibiotic-resistant (MAR) strains and non-antibiotic-resistant isolates may lead to the detection of biomarkers for multiple antibiotic resistances. Analysis of the proteome of the *S.* Typhimurium SL1344 using two-dimensional high-performance liquid chromatography (HPLC) mass spectrometry identified 816 individual proteins including several proteins regulated by the *mar* locus (Coldham and Woodward, 2004). These *mar*-regulated proteins included key effectors such as the efflux pump components AcrA and TolC and the porin OmpF (Coldham and Woodward, 2004).

Exposure to farm disinfectants can select for antibiotic-resistant strains (Karatzas et al., 2008). Such antibiotic-resistant variants show low levels of outer membrane proteins (OmpC and OmpF), linked to reduced susceptibility to antimicrobial compounds, and membrane-bound F_1F_0 ATP synthase subunits, explaining the reduced invasiveness (Karatzas et al., 2008). Several proteins were overexpressed in all antibiotic-resistant variants, including TolB and ElaB, proteins related to the metabolism of purine and pyrimidine nucleotides (Udp, GuaA, and GcvP), AcrAB and TolC, and stress proteins (ClpX). Prolonged treatment with disinfectants might result in reduced susceptibility to several antibiotics and antimicrobials related to low levels of outer membrane proteins and overexpression of AcrAB-TolC (Karatzas et al., 2008).

Tryptic mass fingerprinting in conjunction with nanoelectrospray mass spectrometry was used to detect protease activities and to identify the proteolytic cleavage sites of one of them in *S.* Typhimurium strain SL1344 (Adams et al., 1999). This strategy uses the whole cell lysate to detect protease activity without the necessity of purifying the protease of interest. At least two proteases are induced under Mg^{2+}-depleted growth conditions and both of them were regulated by the two-component regulatory system PhoP–PhoQ (Adams et al., 1999). Furthermore, the proteolytic site on the substrates of one of those proteases was identified as a dibasic amino acid motif (Adams et al., 1999). Consistent with the proteomic results, previous studies had reported the activation of the PhoPQ regulatory system by depletion of extracellular Mg^{2+} and low pH (Bearson et al., 1998).

The same proteomic strategy was performed to study the protein levels of flagellins under acidic growth conditions (Adams et al., 2001). Flagellin protein levels are repressed during the logarithmic-phase acid tolerance response (ATR) (Bearson et al., 1998; Adams et al., 2001). Acid repression of these proteins was shown to be regulated, directly or indirectly, by the PhoPQ system at the level of transcription (Adams et al., 2001). In contrast, the general stress sigma factor RpoS was found to induce flagellar gene expression (Adams et al., 2001).

A liquid chromatography-mass spectrometry-based bottom-up proteomic strategy was used to characterize S. Typhimurium proteomes in logarithmic phase, stationary phase, and an acidic and magnesium-depleted minimal medium (MgM) (Adkins et al., 2006). This analysis identified 2,343 proteins that were produced under the combined results of these three culture conditions (Adkins et al., 2006). Several proteins were uniquely detected during logarithmic growth, including a cysteine sulfinate desulfinase (DeaD), a protein related to glucose-inhibited division (GidA), and flagellar proteins (FlgD, FliA, FliZ, FliS, and FliT) (Adkins et al., 2006). Host cell invasion proteins (SipA, SipB, SipC, and SipD), a ribosome modulation factor (Rmf), and a key indicator of the stationary phase were found to be uniquely expressed under stationary growth conditions. Proteins shown to be highly and uniquely expressed in magnesium-depleted minimal media included Mg^{2+} transport proteins (MgtA, MgtB, and MgtC), propanediol utilization proteins (Pdu), ABC superfamily proteins (ArtJ, CycA, CysP, and ModF), and a sensor kinase (PhoR) (Adkins et al., 2006).

S. Typhimurium cultured in a rich medium and in a minimal nutrient medium, designed to mimic the macrophage phagosomal environment, was also studied using comparative native peptidomics (Manes et al., 2007). Comparative peptidomics is an emerging field of study and a useful tool to identify short amino acid sequences, such as defective proteins, proteins targeted for proteolysis, and bioactive peptides that are frequently missed using bottom-up proteomics. More than 5,000 peptides, originating from 682 proteins, were identified using both LC-MS/MS and LC-Fourier transform ion cyclotron resonance (FTICR)-MS (Manes et al., 2007). The analysis of the bacterial peptidome showed high levels of degraded proteins, especially from ribosomal proteins, in the phagosome-mimicking medium (Manes et al., 2007). Stress conditions and minimal media are thought to induce ribosomal protein degradation to maintain the required levels of amino acids for protein synthesis.

Two-dimensional difference gel electrophoresis (2-D DIGE) technology was applied to study the combined effect of two environmental signals found in the gut, low oxygen tension, and high osmolarity, on the proteome of S. Typhimurium

SL1344 (Sonck et al., 2009). A comparison between laboratory growth conditions and conditions that mimic in vivo growth revealed 255 differentially expressed proteins. Of the 66 proteins that were characterized, proteins involved in anaerobic fumarate respiration, FrdA and FrdB, and the utilization of 1,2-propanediol, such as the products encoded by the *pdu* operon, were upregulated under simulated in vivo conditions (Sonck et al., 2009). An arginine deiminase (ADI), located in an operon that constitutes a pathway for the catabolism of L-arginine, also showed high levels of expression under those conditions. *Salmonella* triggered expression of proteins linked to an adaptation to low oxygen concentrations and osmotic stress under in vivo-mimicking conditions (Sonck et al., 2009). The data showed a good correlation with previously reported transcriptional changes and a high percentage of proteins were also identified in previous in vivo studies (Becker et al., 2006).

7.6.4 Salmonella *Protein Expression Within Host Cells*

Proteomic methods have also been applied to study protein expression when *Salmonella* is in direct contact with the host or the host cells (Becker et al., 2006; Shi et al., 2006; Molero et al., 2009; Raghunathan et al., 2009; Shi et al., 2009). In the last few years, new strategies have emerged to solve the limitation inherent in such experiments, which include specific characterization of bacterial proteomes where the abundance of the host proteome masks them (Shi et al., 2006). Proteome profiling of the host response can also further our understanding of the host–pathogen interactions and it can provide host biomarkers for early *Salmonella* detection.

To identify proteins associated with macrophage colonization, *S.* Typhimurium strain 14028 was isolated from RAW 264.7 macrophages derived from *Salmonella*-susceptible BALB/c mice (Shi et al., 2006). An accurate mass and time (AMT) tag proteomic method was used to detect changes in protein abundance at different times after infection (Shi et al., 2006). Among the proteins identified, 315 were derived from *S.* Typhimurium and 371 were derived from mouse macrophages (Shi et al., 2006). A strong induction after infection was observed in 39 *S.* Typhimurium proteins, and 7 of these proteins are known virulence factors including IHFα, IHFβ, MgtB, OmpR, SitA, SitB, and SodCI (Shi et al., 2006). The increased protein abundance of MgtB, a Mg^{2+} transporter, and SitA/B, components of a Mn^{2+} transporter, suggested limiting divalent metal ion conditions inside the *Salmonella*-containing vacuole (SCV) (Shi et al., 2006). OmpR regulates the expression of type III secretion system (TTSS) of SPI-2 via SsrAB, confirming previous data revealing upregulation of the TTSS of SPI-2 of *S.* Typhimurium within macrophages (Cirillo et al., 1998; Hensel et al., 1998). Some bacterial proteins were found to be specifically modulated by Nramp1, including IHFα, IHFβ, and SodCI virulence factors as well as Upp, SerA, and STM3117 (Shi et al., 2006). *STM3117* mutants showed a strong decrease in the ability of *S.* Typhimurium to replicate inside macrophages that lack Nramp1 (Shi et al., 2006). STM3117, in addition to the coregulated proteins STM3118-3119, had been previously speculated to be virulence factor (Eriksson et al., 2003) and they are predicted to encode proteins involved in biosynthesis and modification of the peptidoglycan layer (Shi et al., 2006).

To investigate the macrophage response to *S.* Typhimurium infection, an LC-MS/MS-based proteomic approach was used to analyze cell lysates of RAW 264.7 macrophages infected with *S.* Typhimurium strain 14028 (Shi et al., 2009). More than 1,000 macrophage proteins and 115 *S.* Typhimurium proteins were identified that differed from the previous study (Shi et al., 2006). While 113 *S.* Typhimurium proteins were detected in previous studies, two proteins, a putative ABC transporter (STM0770) and a TTSS-1 component SopB, were newly identified (Shi et al., 2009).

A total of 1,006 macrophage proteins were identified and 244 of them (24%) were significantly affected by *S.* Typhimurium infection (Shi et al., 2009). Macrophage proteins that were altered during *Salmonella* infection were involved in a wide range of functions, including those that play a key role in the production of antibacterial NO (iNOS), production of prostaglandin H2 (COX-2), and regulation of intracellular traffic (SNX5, SNX6, and SNX9) (Shi et al., 2009). An increase in the protein levels of the mitochondrial SOD2 was revealed for the first time. The data indicated the likely involvement of *S.* Typhimurium SopB in regulating the abundance of SNX6. The results showed a global macrophage response to *S.* Typhimurium infection (Shi et al., 2009).

Recently, a reverse-phase protein array (RPA)-base strategy was used to reveal signaling events in epithelial cells in culture after *S.* Typhimurium strain SL1344 infection (Molero et al., 2009), as another approach to understanding the host cell response to *Salmonella* infection. The TTSS encoded by *Salmonella* pathogenicity island-1 (SPI-1) is intimately involved in *Salmonella* invasion of host enterocytes during acute enteric infection (Clark et al., 1998). Host cell pathways activated at different stages during the invasion of wild-type *Salmonella* were compared to those activated after infecting with mutants in the TTSS-1 (Molero et al., 2009). Two effectors translocated by this system were studied: the GTPase modulator SopE/E2 and the phosphoinositide phosphatase SigD, also known as SopB. Upon invasion, SigD plays a role in the activation of Akt, which phosphorylates its targets FoxO and GSK-3β (Molero et al., 2009). SopE/E2 promotes *Salmonella* internalization via activation of Cdc42 and Rac1. The RPA results detected a SopE/E2-dependent activation of ERK, JNK, and p38 mitogen-activated protein kinases (MAPK) pathways, as well as a minor contribution to that activation by SigD (Molero et al., 2009).

Salmonella enzyme expression was investigated during infection of mice (Becker et al., 2006). Most earlier studies were performed in vitro, either mimicking host conditions or in cell culture (Coldham and Woodward, 2004; Adkins et al., 2006; Shi et al., 2006). Identification of proteins during infection was difficult due to the presence of large amounts of host proteins and very low levels of bacterial protein. A new approach using flow cytometry and MS-based proteomics permitted the purification of *Salmonella* expressing a green fluorescent protein from infected mouse tissues and identification of *Salmonella* enzymes (Becker et al., 2006). Samples that were isolated from spleens, in a mouse typhoid fever model, and from cecum, in an enteritis model, contained 228 and 539 metabolic enzymes, respectively (Becker et al., 2006). To increase the coverage of *Salmonella* metabolic pathways, a metabolic network model of *S.* Typhimurium was generated from previously published mutant phenotypes and genome comparisons. The combined data of the proteomes and from the metabolic network model revealed in vivo information that was then analyzed. The majority of metabolic enzymes were predicted to be non-essential for *Salmonella* virulence, in part due to network redundancy and the apparent nutrient-rich host conditions (Becker et al., 2006). Among the 155 essential *Salmonella* enzymes identified, only 64 were conserved in other major human pathogens, and most of them, unfortunately, are involved in pathways that do not constitute new targets for broad-spectrum antibiotics (Becker et al., 2006).

Another genome-scale metabolic network, including 1,083 genes, was developed and validated using gene expression profiles of *S.* Typhimurium LT2 isolated from macrophage cell lines (Raghunathan et al., 2009). This model integrates high-throughput genomic, proteomic, and phenotypic experimental data and it demonstrated an accuracy of 80% for growth and virulence phenotypes (Raghunathan et al., 2009).

7.6.5 Comparative Proteomics Among Different Salmonella Serovars

S. Typhimurium LT2 has a mutation in *rpoS* that attenuates its virulence (Swords et al., 1997). The proteomes of a virulent *S.* Typhimurium strain (ATCC 14028) and LT2 were obtained using a liquid chromatography-mass spectrometry-based bottom-up proteomic approach and compared. The resulting data identified 2,121 proteins from the LT2 strain and 2,296 from the ATCC 14028 strain, and a 90% overlap of these proteins. Interestingly, the products of the *pdu* operon, involved in propanediol utilization, showed higher levels of expression in strain ATCC 14028 and when strain LT2 was grown in an acidic and magnesium-depleted minimal medium (MgM) known to mimic the phagosomal conditions (Adkins et al., 2006). The propanediol utilization operon had been associated with virulence and a correlation between the presence of the *pdu* genes and food poisoning had been suggested (Korbel et al., 2005).

More recently, the proteomes of *S.* Typhi strain Ty2 and *S.* Typhimurium LT2 were compared. Using an LC-MS/MS-based proteomics approach, 2,066 *S.* Typhi proteins were identified under logarithmic and stationary growth conditions and an acidic and low magnesium medium (MgM) (Ansong et al., 2008). Among a subset of proteins highly and uniquely expressed in logarithmic cultures, there were host cell invasion proteins (OrgA, InvH, SirA, SopE, SpaN, SpaT, SipA, SipB, and SipD), Vi polysaccharide biosynthesis proteins (TviE and TviD), and flagellar biosynthesis proteins (FlgD, FliZ, and FliT). A ribosome modulation factor (Rmf) and proteins involved in ethalonamine utilization (EutB, EutL, and EutN) were included in those proteins highly and almost uniquely expressed under stationary growth conditions. Proteins highly and almost exclusively expressed in the

MgM culture included products of the biotin operon (BioA, BioB, BioD, and BioF), hemolysin E (HlyE), a toxin-like protein (CdtB), Mg^{2+} transport proteins (MgtA, MgtB, and MgtC), products of SPI-2 virulence genes (SseA, SseB, SsaU, and SsaV), PagC, PqaB, and PgtE. *S.* Typhimurium infects a wide range of host species and causes gastroenteritis in humans, while *S.* Typhi is host specific and provokes a systemic infection only in humans (Pang et al., 1995). Analysis of protein expression of *S.* Typhi was compared to previously published data for *S.* Typhimurium LT2 proteins obtained under identical growth conditions (Adkins et al., 2006). Under MgM growth conditions, several highly expressed proteins were common to both serotypes, including Mg^{2+} transport proteins (MgtA, MgtB, and MgtC), SPI-2 virulence proteins (SsaV and SseB), phosphate regulon sensor protein (PhoR), and an outer membrane protease (PgtE). Unlike the avirulent *S.* Typhimurium LT2 strain, biotin synthesis proteins were highly expressed under conditions that mimic the macrophage phagosomal environment in *S.* Typhi. Interestingly, these products also showed an increased expression in the virulent strain *S.* Typhimurium 14028 (Adkins et al., 2006), suggesting that biotin synthesis proteins may be involved in *Salmonella* virulence. Proteins that may play a key role in human host specificity and *S.* Typhi pathogenesis were those shown to be exclusively expressed in this serotype, and these included (Vi) polysaccharide biosynthesis and export proteins, CdtB, HlyE, and putative bacteriophage proteins (Ansong et al., 2008).

A two-dimensional gel electrophoresis (2-DGE) and tandem mass spectrometry (LC-MS/MS) strategy was used to compare cytosolic proteins expression on Columbia blood agar of isolates of *S. enterica* serovar Pullorum and serovar Typhimurium (Encheva et al., 2005). Serovar Pullorum is host specific and infects only poultry, causing pullorum disease. Several proteins were found to show elevated levels of expression in the host-adapted serovar Pullorum, such as proteins involved in sulfate transport and synthesis of cysteine (Encheva et al., 2005). That work was extended to include serovars Enteritidis, Choleraesuis, and Dublin in a more recent study (Encheva et al., 2007). Serovar Enteritidis, like the host-generalist serovar Typhimurium, infects a broad range of animals (including wild rodents, poultry, pigs, and cattle) and causes gastroenteritis in humans, systemic infection in mice, and asymptomatic chronic infection in chickens (Rabsch et al., 2001). Serovars Choleraesuis and Dublin infect only a few species and are thus considered host adapted to swine and cattle, respectively. Comparison of the proteome profile of serovar Typhimurium to the proteomes of these other serotypes revealed only one protein isoform, SodA isoform II (Mn^{2+}-binding superoxide dismutase), to be specifically overexpressed in this serovar (Encheva et al., 2007). Serovars Enteritidis and Pullorum were found to express high levels of D-galactose-binding protein precursor (MglB) isoform II (Encheva et al., 2007). Serovar Typhimurium overexpressed the isoform I, while serovar Choleraesuis did not express detectable levels of neither isoform. The proteomic expression profile of serovar Choleraesuis detected one enzyme, succinate semialdehyde dehydrogenase I (GabD), that was absent from all other serovars of that study (Encheva et al., 2007). Serovar Pullorum showed expression of an additional isoform of the lysine arginine ornithine (LAO)-binding transport protein, while serovar Typhimurium only expressed two isoforms.

Additionally, the proteomic profile of serovar Dublin did not reveal any specific protein of this serovar (Encheva et al., 2007).

The standard proteomic approach combining 2-DGE and MS was applied to comparatively analyze the cytosolic proteins of *S. enterica* serovar Gallinarum and serovar Enteritidis isolated from poultry (Osman et al., 2009). The host-adapted serovar Gallinarum is the causative agent of fowl typhoid, a severe systemic disease in poultry. Twenty-two proteins of serovar Gallinarum were found to be overexpressed in this study, and some of these proteins had previously reported to be involved in virulence, including an SPI-1 effector protein, a protein similar to T-cell inhibitor protein, a response regulator protein, a protein similar to chaperone heat shock protein 90, and a paratose synthetase protein (RfbS) (Osman et al., 2009).

Proteomics is revealing critical information that cannot be determined by transcriptomics, because the rate of synthesis and degradation of proteins are a vital part of their steady state. One barrier for the highest throughput proteomic methods is the accuracy of measuring the relative abundance of proteins. Other goals to improve proteomic analysis are to continue to reduce in the amount of protein needed and to increase the complexity of mixtures that can be analyzed with confidence. As these aspects of proteomic analysis continue to be improved, biological questions that have been too difficult to address to date, such as monitoring the proteomes of bacteria and host simultaneously, will become experimentally approachable.

7.6.6 Current Limitations and Future Prospects

Overall a combination of the tools that have been developed in comparative genomics, high-throughput fitness assays, promoter capture, transcriptomics, and additional methods that are not reviewed here, including chromatin immunoprecipitation (Navarre et al., 2006) and the development of interactomes (Typas et al., 2008), will strengthen our functional genomic description of *Salmonella* biology. Such functional genomic description is likely to reveal vulnerabilities that can be exploited to minimize the impact of the bacterium in food animals, as well as in food preparation.

References

Adams P, Fowler R et al. (1999) Defining protease specificity with proteomics: a protease with a dibasic amino acid recognition motif is regulated by a two-component signal transduction system in *Salmonella*. Electrophoresis 20(11):2241–2247

Adams P, Fowler R et al. (2001) Proteomic detection of PhoPQ-and acid-mediated repression of *Salmonella* motility. Proteomics 1(4):597–607

Adkins JN, Mottaz HM et al. (2006) Analysis of the *Salmonella typhimurium* proteome through environmental response toward infectious conditions. Mol Cell Proteomics 5(8):1450–1461

Aebersold R, Mann M (2003) Mass spectrometry-based proteomics. Nature 422(6928):198–207

Agudo D, Mendoza MT et al. (2004) A proteomic approach to study *Salmonella typhi* periplasmic proteins altered by a lack of the DsbA thiol: disulfide isomerase. Proteomics 4(2):355–363

Alausa KO, Montefiore D et al. (1977) Septicaemia in the tropics. A prospective epidemiological study of 146 patients with a high case fatality rate. Scand J Infect Dis 9(3):181–185

Alpuche-Aranda CM, Racoosin EL et al. (1994) *Salmonella* stimulate macrophage macropinocytosis and persist within spacious phagosomes. J Exp Med 179:601–608

Alpuche Aranda CM, Swanson JA et al. (1992) *Salmonella typhimurium* activates virulence gene transcription within acidified macrophage phagosomes. Proc Natl Acad Sci USA 89(21):10079–10083

Althouse C, Patterson S et al. (2003) Type 1 fimbriae of *Salmonella enterica* serovar Typhimurium bind to enterocytes and contribute to colonization of swine in vivo. Infect Immun 71(11): 6446–6452

Altier C, Suyemoto M (1999) A recombinase-based selection of differentially expressed bacterial genes. Gene 240(1):99–106

Altier C, Suyemoto M et al. (2000) Regulation of *Salmonella enterica* serovar typhimurium invasion genes by *csrA*. Infect Immun 68(12):6790–6797

Andrews-Polymenis HL, Rabsch W et al. (2004) Host restriction of *Salmonella enterica* serotype Typhimurium pigeon isolates does not correlate with loss of discrete genes. J Bacteriol 186(9):2619–2628

Andrews-Polymenis HL, Santiviago C et al. (2009) Novel genetic tools for studying food borne Salmonella. Curr Opin Biotechnol 20:1–9

Ansong C, Yoon H et al. (2008) Proteomics analysis of the causative agent of typhoid fever. J Proteome Res 7(2):546–557

Ansong C, Yoon H et al. (2009) Global systems-level analysis of Hfq and SmpB deletion mutants in *Salmonella*: implications for virulence and global protein translation. PLoS One 4(3):e4809

Ansorge WJ (2009) Next-generation DNA sequencing techniques. N Biotechnol 25(4):195–203

Baba T, Ara T et al. (2006) Construction of *Escherichia coli* K-12 in-frame, single-gene knockout mutants: the Keio collection. Mol Syst Biol 2:2006.0008

Badarinarayana V, Estep PW 3rd et al. (2001) Selection analyses of insertional mutants using subgenic-resolution arrays. Nat Biotechnol 19(11):1060–1065

Bajaj V, Lucas RL et al. (1996) Co-ordinate regulation of *Salmonella typhimurium* invasion genes by environmental and regulatory factors is mediated by control of *hilA* expression. Mol Microbiol 22(4):703–714

Baldwin DN, Salama NR (2007) Using genomic microarrays to study insertional/transposon mutant libraries. Methods Enzymol 421:90–110

Bar-Meir M, Raveh D et al. (2005) Non-Typhi *Salmonella* gastroenteritis in children presenting to the emergency department: characteristics of patients with associated bacteraemia. Clin Microbiol Infect 11:651–655

Bauer-Garland J, Frye JG et al. (2006) Transmission of *Salmonella enterica* serotype Typhimurium in poultry with and without antimicrobial selective pressure. J Appl Microbiol 101(6): 1301–1308

Bäumler AJ, Kusters JG et al. (1994) *Salmonella typhimurium* loci involved in survival within macrophages. Infect Immun 62:1623–1630

Bearson SM, Bearson BL et al. (2006) Identification of *Salmonella enterica* serovar Typhimurium genes important for survival in the swine gastric environment. Appl Environ Microbiol 72(4):2829–2836

Bearson BL, Wilson L et al. (1998) A low pH-inducible, PhoPQ-dependent acid tolerance response protects *Salmonella typhimurium* against inorganic acid stress. J Bacteriol 180(9):2409–2417

Becker D, Selbach M et al. (2006) Robust *Salmonella* metabolism limits possibilities for new antimicrobials. Nature 440(7082):303–307

Bispham J, Tripathi BN et al. (2001) *Salmonella* pathogenicity island 2 influences both systemic salmonellosis and *Salmonella*-induced enteritis in calves. Infect Immun 69(1):367–377

Boddicker JD, Ledeboer NA et al. (2002) Differential binding to and biofilm formation on, HEp-2 cells by *Salmonella enterica* serovar Typhimurium is dependent upon allelic variation in the *fimH* gene of the *fim* gene cluster. Mol Microbiol 45(5):1255–1265

Bogomolnaya LM, Santiviago CA et al. (2008) 'Form variation' of the O12 antigen is critical for persistence of *Salmonella* Typhimurium in the murine intestine. Mol Microbiol 70(5): 1105–1119

Bowe F, Lipps CJ et al. (1998) At least four percent of the *Salmonella typhimurium* genome is required for fatal infection of mice. Infect Immun 66:3372–3377

Brawn LC, Hayward RD et al. (2007) *Salmonella* SPI1 effector SipA persists after entry and cooperates with a SPI2 effector to regulate phagosome maturation and intracellular replication. Cell Host Microbe 1(1):63–75

Brent A, Oundo J et al. (2006) *Salmonella* bacteremia in Kenyan children. Pediatr Infect Dis J 25(3):230–236

Brown NF, Vallance BA et al. (2005) *Salmonella* pathogenicity island 2 is expressed prior to penetrating the intestine. PLoS Pathog 1(3):e32

Bumann D (2002) Examination of *Salmonella* gene expression in an infected mammalian host using the green fluorescent protein and two-colour flow cytometry. Mol Microbiol 43(5): 1269–1283

Camilli A, Mekalanos JJ (1995) Use of recombinase gene fusions to identify *Vibrio cholerae* genes induced during infection. Mol Microbiol 18(4):671–683

Carnell SC, Bowen AJ et al. (2007) Role in virulence and protective efficacy in pigs of *Salmonella enterica* serovar Typhimurium secreted components identified by signature-tagged mutagenesis. Microbiology 153:1940–1952

Carrol ME, Jackett PS et al. (1979) Phagolysosome formation, cyclic adenosine 3′:5′-monophosphate and the fate of *Salmonella typhimurium* within mouse peritoneal macrophages. J Gen Microbiol 110(2):421–429

CDC (2005) *Salmonella* annual summary 2005. Department of Health and Human Services, Centers for Disease Control and Prevention, National Institute of Infectious Diseases

CDC (2008) Division of Bacterial and Mycotic Diseases, Disease listing: Salmonellosis. http://www.cdc.gov/salmonella

Chan K, Baker S et al. (2003) Genomic comparison of *Salmonella enterica* serovars and *Salmonella bongori* by use of an *S. enterica* serovar typhimurium DNA microarray. J Bacteriol 185(2):553–563

Chan K, Kim CC et al. (2005) Microarray-based detection of *Salmonella enterica* serovar Typhimurium transposon mutants that cannot survive in macrophages and mice. Infect Immun 73(9):5438–5449

Charles IG, Maskell DJ (2001) Transposon mediated differential hybridisation. International Patent Number WO2001/007651

Chaudhuri RR, Peters SE et al. (2009) Comprehensive identification of *Salmonella enterica* serovar typhimurium genes required for infection of BALB/c mice. PLoS Pathog 5(7):e1000529

Cheesbrough JS, Taxman BC et al. (1997) Clinical definition for invasive *Salmonella* infection in African children. Pediatr Infect Dis J 16(3):277–283

Chiu CH, Tang P et al. (2005) The genome sequence of *Salmonella enterica* serovar Choleraesuis, a highly invasive and resistant zoonotic pathogen. Nucleic Acids Res 33(5):1690–1698

Chowdhury SM, Shi L et al. (2009) A method for investigating protein-protein interactions related to *Salmonella typhimurium* pathogenesis. J Proteome Res 8(3):1504–1514

Cirillo DM, Valdivia RH et al. (1998) Macrophage-dependent induction of the *Salmonella* pathogenicity island 2 type III secretion system and its role in intracellular survival. Mol Microbiol 30(1):175–188

Clark MA, Hirst BH et al. (1998) Inoculum composition and *Salmonella* pathogenicity island 1 regulate M-cell invasion and epithelial destruction by *Salmonella typhimurium*. Infect Immun 66(2):724–731

Coldham NG, Woodward MJ (2004) Characterization of the *Salmonella typhimurium* proteome by semi-automated two-dimensional HPLC-mass spectrometry: detection of proteins implicated in multiple antibiotic resistance. J Proteome Res 3(3):595–603

Crawford RW, Rosales-Reyes R et al. (2010) Gallstones play a significant role in *Salmonella* ssp. gallbladder colonization and carriage. Proc Natl Acad Sci USA 107(9):4353–4358

Croucher NJ, Fookes MC et al. (2009) A simple method for directional transcriptome sequencing using Illumina technology. Nucleic Acids Res 37(22):e148

Crump JA, Luby SP et al. (2004) The global burden of typhoid fever. Bull World Health Organ 82(5):346–353

Datsenko KA, Wanner BL (2000) One-step inactivation of chromosomal genes in *Escherichia coli* K-12 using PCR products. Proc Natl Acad Sci USA 97(12):6640–6645

Deiwick J, Rappl C et al. (2002) Proteomic approaches to *Salmonella* Pathogenicity island 2 encoded proteins and the SsrAB regulon. Proteomics 2(6):792–799

Deng W, Liou SR et al. (2003) Comparative genomics of *Salmonella enterica* serovar Typhi strains Ty2 and CT18. J Bacteriol 185(7):2330–2337

Drecktrah D, Knodler LA et al. (2005) The *Salmonella* SPI1 effector SopB stimulates nitric oxide production long after invasion. Cell Microbiol 7(1):105–113

Drecktrah D, Knodler LA et al. (2007) *Salmonella* trafficking is defined by continuous dynamic interactions with the endolysosomal system. Traffic 8(3):212–225

Ellis HM, Yu D et al. (2001) High efficiency mutagenesis, repair, and engineering of chromosomal DNA using single-stranded oligonucleotides. Proc Natl Acad Sci USA 98(12):6742–6746

Encheva V, Wait R et al. (2005) Proteome analysis of serovars Typhimurium and Pullorum of *Salmonella enterica* subspecies I. BMC Microbiol 5:42

Encheva V, Wait R et al. (2007) Protein expression diversity amongst serovars of *Salmonella enterica*. Microbiology 153(Pt 12):4183–4193

Eriksson S, Lucchini S et al. (2003) Unravelling the biology of macrophage infection by gene expression profiling of intracellular *Salmonella enterica*. Mol Microbiol 47(1):103–118

Faucher SP, Porwollik S et al. (2006) Transcriptome of *Salmonella enterica* serovar Typhi within macrophages revealed through the selective capture of transcribed sequences. Proc Natl Acad Sci USA 103(6):1906–1911

Fields PI, Swanson RV, Haidaris CG, Heffron F (1986) Mutants of *Salmonella typhimurium* that cannot survive within the macrophage are avirulent. Proc Natl Acad Sci USA 83: 5189–5193

Fink RC, Evans MR et al. (2007) FNR is a global regulator of virulence and anaerobic metabolism in *Salmonella enterica* serovar Typhimurium (ATCC 14028s). J Bacteriol 189(6): 2262–2273

Flierl MA, Rittirsch D et al. (2007) Phagocyte-derived catecholamines enhance acute inflammatory injury. Nature 449(7163):721–725

Frye J, Karlinsey JE et al. (2006) Identification of new flagellar genes of *Salmonella enterica* serovar Typhimurium. J Bacteriol 188(6):2233–2243

Galán JE, Curtiss R 3rd (1989) Cloning and molecular characterization of genes whose products allow *Salmonella typhimurium* to penetrate tissue culture cells. Proc Natl Acad Sci USA 86:6383–6387

Galán JE, Ginocchio C et al. (1992) Molecular and functional characterization of the *Salmonella* invasion gene *invA*: homology of InvA to members of a new protein family. J Bacteriol 174:4338–4349

Garcia-Calderon CB, Casadesus J et al. (2007) Rcs and PhoPQ regulatory overlap in the control of *Salmonella enterica* virulence. J Bacteriol 189(18):6635–6644

Garcia Vescovi E, Soncini FC et al. (1996) Mg^{2+} as an extracellular signal: environmental regulation of *Salmonella* virulence. Cell 84(1):165–174

Geddes K, Cruz F et al. (2007) Analysis of cells targeted by *Salmonella* type III secretion in vivo. PLoS Pathog 3(12):e196

Geddes K, Worley M et al. (2005) Identification of new secreted effectors in *Salmonella enterica* serovar Typhimurium. Infect Immun 73(10):6260–6271

Gordon MA, Banda HT et al. (2002) Non-typhoidal *Salmonella* bacteraemia among HIV-infected Malawian adults: high mortality and frequent recrudescence. AIDS 16(12):1633–1641

Gordon MA, Walsh AL et al. (2001) Bacteraemia and mortality among adult medical admissions in Malawi – predominance of non-typhi salmonellae and *Streptococcus pneumoniae*. J Infect 42(1):44–49

Guina T, Yi EC et al. (2000) A PhoP-regulated outer membrane protease of *Salmonella enterica* serovar typhimurium promotes resistance to alpha-helical antimicrobial peptides. J Bacteriol 182(14):4077–4086

Gulig PA, Curtiss R (1987) Plasmid-associated virulence of *Salmonella typhimurium*. Infect Immun 1987:2891–2901

Gulig PA, Curtiss R 3rd (1988) Cloning and transposon insertion mutagenesis of virulence genes of the 100-kilobase plasmid of *Salmonella typhimurium*. Infect Immun 56(12):3262–3271

Gulig PA, Caldwell AL, Chiodo VA (1992) Identification, genetic analysis and DNA sequence of a 7.8-kb virulence region of the *Salmonella typhimurium* virulence plasmid. Mol Microbiol 6:1395–1411

Gulig PA, Doyle TJ (1993) The *Salmonella typhimurium* virulence plasmid increases the growth rate of Salmonellae in mice. Infect Immun 61:504–511

Gunn JS, Miller SI (1996) PhoP-PhoQ activates transcription of *pmrAB*, encoding a two-component regulatory system involved in *Salmonella typhimurium* antimicrobial peptide resistance. J Bacteriol 178(23):6857–6864

Gupta PK (2008) Single-molecule DNA sequencing technologies for future genomics research. Trends Biotechnol 26(11):602–611

Haneda T, Ishii Y et al. (2009) Genome-wide identification of novel genomic islands that contribute to *Salmonella* virulence in mouse systemic infection. FEMS Microbiol Lett 297(2):241–249

Haraga A, Ohlson MB et al. (2008) Salmonellae interplay with host cells. Nat Rev Microbiol 6:53–66

Hardt WD, Chen LM et al. (1998) *S. typhimurium* encodes an activator of Rho GTPases that induces membrane ruffling and nuclear responses in host cells. Cell 93(5):815–826

Harshey RM (2003) Bacterial motility on a surface: many ways to a common goal. Annu Rev Microbiol 57:249–273

Harshey RM, Matsuyama T (1994) Dimorphic transition in *Escherichia coli* and *Salmonella typhimurium*: surface-induced differentiation into hyperflagellate swarmer cells. Proc Natl Acad Sci USA 91(18):8631–8635

Heithoff DM, Conner CP et al. (1997) Bacterial infection as assessed by in vivo gene expression. Proc Natl Acad Sci USA 94(3):934–939

Hensel M, Shea JE et al. (1995) Simultaneous identification of bacterial virulence genes by negative selection. Science 269:400–403

Hensel M, Shea JE et al. (1997) Analysis of the boundaries of *Salmonella* pathogenicity island 2 and the corresponding chromosomal region of *Escherichia coli* K-12. J Bacteriol 179:1105–1111

Hensel M, Shea JE et al. (1998) Genes encoding putative effector proteins of the type III secretion system of *Salmonella* pathogenicity island 2 are required for bacterial virulence and proliferation in macrophages. Mol Microbiol 30(1):163–174

Hernandez LD, Hueffer K et al. (2004) *Salmonella* modulates vesicular traffic by altering phosphoinositide metabolism. Science 304(5678):1805–1807

Hinton JC, Hautefort I et al. (2004) Benefits and pitfalls of using microarrays to monitor bacterial gene expression during infection. Curr Opin Microbiol 7(3):277–282

Hisert KB, MacCoss M et al. (2005) A glutamate-alanine-leucine (EAL) domain protein of *Salmonella* controls bacterial survival in mice, antioxidant defence and killing of macrophages: role of cyclic diGMP. Mol Microbiol 56(5):1234–1245

Holt KE, Parkhill J et al. (2008) High-throughput sequencing provides insights into genome variation and evolution in *Salmonella* Typhi. Nat Genet 40(8):987–993

Holt KE, Thomson NR et al. (2009) Pseudogene accumulation in the evolutionary histories of *Salmonella enterica* serovars Paratyphi A and Typhi. BMC Genomics 10:36

Hornick RB, Greisman SE et al. (1970a) Typhoid fever: pathogenesis and immunologic control. N Engl J Med 283(13):686–691

Hornick RB, Greisman SE et al. (1970b) Typhoid fever: pathogenesis and immunologic control. 2. N Engl J Med 283(14):739–746

Huang Y, Leming CL et al. (2007) Genome-wide screen of *Salmonella* genes expressed during infection in pigs, using in vivo expression technology. Appl Environ Microbiol 73(23): 7522–7530

Jarvik T, Smillie C et al. (2010) Short-term signatures of evolutionary change in the *Salmonella enterica* serovar typhimurium 14028 genome. J Bacteriol 192(2):560–567

Jiang X, Rossanese OW et al. (2004) The related effector proteins SopD and SopD2 from *Salmonella enterica* serovar Typhimurium contribute to virulence during systemic infection of mice. Mol Microbiol 54(5):1186–1198

Jones BD, Ghori N et al. (1994) *Salmonella typhimurium* initiates murine infection by penetrating and destroying the specialized epithelial M cells of the Peyer's patches. J Exp Med 180:15–23

Joseph B, Otta SK et al. (2001) Biofilm formation by *Salmonella* spp. on food contact surfaces and their sensitivity to sanitizers. Int J Food Microbiol 64(3):367–372

Julio SM, Heithoff DM et al. (2000) *ssrA* (tmRNA) plays a role in *Salmonella enterica* serovar Typhimurium pathogenesis. J Bacteriol 182(6):1558–1563

Kang MS, Besser TE et al. (2006) Identification of specific gene sequences conserved in contemporary epidemic strains of *Salmonella enterica*. Appl Environ Microbiol 72(11): 6938–6947

Kankwatira AM, Mwafulirwa GA et al. (2004) Non-typhoidal *Salmonella* bacteraemia – an underrecognized feature of AIDS in African adults. Trop Doct 34(4):198–200

Karatzas KA, Randall LP et al. (2008) Phenotypic and proteomic characterization of multiply antibiotic-resistant variants of *Salmonella enterica* serovar Typhimurium selected following exposure to disinfectants. Appl Environ Microbiol 74(5):1508–1516

Karavolos MH, Spencer H et al. (2008) Adrenaline modulates the global transcriptional profile of *Salmonella* revealing a role in the antimicrobial peptide and oxidative stress resistance responses. BMC Genomics 9:458

Kariuki S, Revathi G et al. (2006) Characterisation of community acquired non-typhoidal *Salmonella* from bacteraemia and diarrhoeal infections in children admitted to hospital in Nairobi, Kenya. BMC Microbiol 6:101

Karlinsey JE (2007) Lambda-red genetic engineering in *Salmonella enterica* serovar Typhimurium. Methods Enzymol 421:199–209

Karzai AW, Susskind MM et al. (1999) SmpB, a unique RNA-binding protein essential for the peptide-tagging activity of SsrA (tmRNA). Embo J 18(13):3793–3799

Kim W, Killam T et al. (2003) Swarm-cell differentiation in *Salmonella enterica* serovar typhimurium results in elevated resistance to multiple antibiotics. J Bacteriol 185(10): 3111–3117

Kingsley RA, Msefula CL et al. (2009) Epidemic multiple drug resistant *Salmonella* Typhimurium causing invasive disease in sub-Saharan Africa have a distinct genotype. Genome Res 19(12):2279–2287

Klein JR, Fahlen TF et al. (2000) Transcriptional organization and function of invasion genes within *Salmonella enterica* serovar Typhimurium pathogenicity island 1, including the *prgH*, *prgI*, *prgJ*, *prgK*, *orgA*, *orgB*, and *orgC* genes. Infect Immun 68(6):3368–3376

Klumpp J, Fuchs TM (2007) Identification of novel genes in genomic islands that contribute to *Salmonella typhimurium* replication in macrophages. Microbiology 153(Pt 4):1207–1220

Korbel JO, Doerks T et al. (2005) Systematic association of genes to phenotypes by genome and literature mining. PLoS Biol 3(5):e134

Ku YW, McDonough SP et al. (2005) Novel attenuated *Salmonella enterica* serovar Choleraesuis strains as live vaccine candidates generated by signature-tagged mutagenesis. Infect Immun 73(12):8194–8203

Kukral AM, Strauch KL et al. (1987) Genetic analysis in *Salmonella typhimurium* with a small collection of randomly spaced insertions of transposon Tn*10* delta 16 delta 17. J Bacteriol 169(5):1787–1793

Langridge GC, Phan MD et al. (2009) Simultaneous assay of every *Salmonella* Typhi gene using one million transposon mutants. Genome Res 19(12):2308–2316

Lawley TD, Chan K et al. (2006) Genome-wide screen for *Salmonella* genes required for long-term systemic infection of the mouse. PLoS Pathog 2(2):e11

Ledeboer NA, Frye JG et al. (2006) *Salmonella enterica* serovar Typhimurium requires the Lpf, Pef, and Tafi fimbriae for biofilm formation on HEp-2 tissue culture cells and chicken intestinal epithelium. Infect Immun 74(6):3156–3169

Ledeboer NA, Jones BD (2005) Exopolysaccharide sugars contribute to biofilm formation by *Salmonella enterica* serovar typhimurium on HEp-2 cells and chicken intestinal epithelium. J Bacteriol 187(9):3214–3226

Lewis C, Skovierova H et al. (2009) *Salmonella enterica* Serovar Typhimurium HtrA: regulation of expression and role of the chaperone and protease activities during infection. Microbiology 155(Pt 3):873–881

Lichtensteiger CA, Vimr ER (2003) Systemic and enteric colonization of pigs by a *hilA* signature-tagged mutant of *Salmonella choleraesuis*. Microb Pathog 34(3):149–154

Liu W-Q, Feng Y, Wang Y, Zou Q-H, Chen F, Guo J-T, Peng Y-H, Jin Y, Li Y-G, Hu S-N, Johnston RN, Liu G-R, Liu S-L (2009) *Salmonella paratyphi* C: Genetic Divergence from *Salmonella choleraesuis* and Pathogenic Convergence with *Salmonella typhi*. PLoS ONE 4(2):e4510

Lodge J, Douce GR et al. (1995) Biological and genetic characterization of Tn*phoA* mutants of *Salmonella typhimurium* TML in the context of gastroenteritis. Infect Immun 63(3):762–769

MacLean D, Jones JD et al. (2009) Application of 'next-generation' sequencing technologies to microbial genetics. Nat Rev Microbiol 7(4):287–296

Mahan MJ, Slauch JM et al. (1993) Selection of bacterial virulence genes that are specifically induced in host tissues. Science 259(5095):686–688

Mahan MJ, Tobias JW et al. (1995) Antibiotic-based selection for bacterial genes that are specifically induced during infection of a host. Proc Natl Acad Sci USA 92(3):669–673

Manes NP, Gustin JK et al. (2007) Targeted protein degradation by *Salmonella* under phagosome-mimicking culture conditions investigated using comparative peptidomics. Mol Cell Proteomics 6(4):717–727

Mardis ER (2008) Next-generation DNA sequencing methods. Annu Rev Genomics Hum Genet 9:387–402

Mariconda S, Wang Q et al. (2006) A mechanical role for the chemotaxis system in swarming motility. Mol Microbiol 60(6):1590–1602

McClelland M, Sanderson KE et al. (2001) Complete genome sequence of *Salmonella enterica* serovar Typhimurium LT2. Nature 413(6858):852–856

McClelland M, Sanderson KE et al. (2004) Comparison of genome degradation in Paratyphi A and Typhi, human-restricted serovars of *Salmonella enterica* that cause typhoid. Nat Genet 36:1268–1274

Mead PS, Slutsker L et al. (1999) Food-related illness and death in the United States. Emerg Infect Dis 5(5):607–625

Merighi M, Ellermeier CD et al. (2005) Resolvase-in vivo expression technology analysis of the *Salmonella enterica* serovar Typhimurium PhoP and PmrA regulons in BALB/c mice. J Bacteriol 187(21):7407–7416

Miao EA, Miller SI (2000) A conserved amino acid sequence directing intracellular type III secretion by *Salmonella typhimurium*. Proc Natl Acad Sci USA 97(13):7539–7544

Miller SI (1991) PhoP/PhoQ: macrophage-specific modulators of *Salmonella* virulence? Mol Microbiol 5(9):2073–2078

Miller SI, Kukral AM et al. (1989a) A two-component regulatory system (phoP phoQ) controls *Salmonella typhimurium* virulence. Proc Natl Acad Sci USA 86(13):5054–5058

Miller I, Maskell D et al. (1989b) Isolation of orally attenuated *Salmonella typhimurium* following Tn*phoA* mutagenesis. Infect Immun 57(9):2758–2763

Molero C, Rodriguez-Escudero I et al. (2009) Addressing the effects of *Salmonella* internalization in host cell signaling on a reverse-phase protein array. Proteomics 9(14):3652–3665

Monack DM, Bouley DM et al. (2004) *Salmonella typhimurium* persists within macrophages in the mesenteric lymph nodes of chronically infected Nramp1$^{+/+}$ mice and can be reactivated by IFNgamma neutralization. J Exp Med 199(2):231–241

Morgan E, Campbell JD et al. (2004) Identification of host-specific colonization factors of *Salmonella enterica* serovar Typhimurium. Mol Microbiol 54(6):994–1010

Morozova O, Marra MA (2008) Applications of next-generation sequencing technologies in functional genomics. Genomics 92(5):255–264

Mottaz-Brewer HM, Norbeck AD et al. (2008) Optimization of proteomic sample preparation procedures for comprehensive protein characterization of pathogenic systems. J Biomol Tech 19(5):285–295

Nagalakshmi U, Wang Z et al. (2008) The transcriptional landscape of the yeast genome defined by RNA sequencing. Science 320(5881):1344–1349

Navarre WW, Porwollik S et al. (2006) Selective silencing of foreign DNA with low GC content by the H-NS protein in *Salmonella*. Science 313(5784):236–238

Nunn BL, Shaffer SA et al. (2006) Comparison of a *Salmonella typhimurium* proteome defined by shotgun proteomics directly on an LTQ-FT and by proteome pre-fractionation on an LCQ-DUO. Brief Funct Genomic Proteomic 5(2):154–168

Osman KM, Ali MM et al. (2009) Comparative proteomic analysis on *Salmonella* Gallinarum and *Salmonella* Enteritidis exploring proteins that may incorporate host adaptation in poultry. J Proteomics 72(5):815–821

Padalon-Brauch G, Hershberg R et al. (2008) Small RNAs encoded within genetic islands of *Salmonella typhimurium* show host-induced expression and role in virulence. Nucleic Acids Res 36(6):1913–1927

Pang T, Bhutta ZA et al. (1995) Typhoid fever and other salmonellosis: a continuing challenge. Trends Microbiol 3(7):253–255

Parkhill J, Dougan G et al. (2001) Complete genome sequence of a multiple drug resistant *Salmonella enterica* serovar Typhi CT18. Nature 413(6858):848–852

Patel JC, Galan JE (2006) Differential activation and function of Rho GTPases during *Salmonella*-host cell interactions. J Cell Biol 175(3):453–463

Patel JC, Galan JE (2008) Investigating the function of Rho family GTPases during *Salmonella*/host cell interactions. Methods Enzymol 439:145–158

Perkins TT, Kingsley RA et al. (2009) A strand-specific RNA-Seq analysis of the transcriptome of the typhoid bacillus *Salmonella typhi*. PLoS Genet 5(7):e1000569

Porwollik S, Boyd EF et al. (2004) Characterization of *Salmonella enterica* subspecies I genovars by use of microarrays. J Bacteriol 186(17):5883–5898

Porwollik S, Santiviago CA et al. (2005) Differences in gene content between *Salmonella enterica* serovar Enteritidis isolates and comparison to closely related serovars Gallinarum and Dublin. J Bacteriol 187(18):6545–6555

Prouty AM, Gunn JS (2003) Comparative analysis of *Salmonella enterica* serovar Typhimurium biofilm formation on gallstones and on glass. Infect Immun 71(12):7154–7158

Prouty AM, Schwesinger WH et al. (2002) Biofilm formation and interaction with the surfaces of gallstones by *Salmonella* spp. Infect Immun 70(5):2640–2649

Rabsch W, Tschape H et al. (2001) Non-typhoidal salmonellosis: emerging problems. Microbes Infect 3(3):237–247

Raffatellu M, Santos RL et al. (2008) Simian immunodeficiency virus-induced mucosal interleukin-17 deficiency promotes *Salmonella* dissemination from the gut. Nat Med 14:4

Raghunathan A, Reed J et al. (2009) Constraint-based analysis of metabolic capacity of *Salmonella typhimurium* during host-pathogen interaction. BMC Syst Biol 3:38

Rappl C, Deiwick J et al. (2003) Acidic pH is required for the functional assembly of the type III secretion system encoded by *Salmonella* pathogenicity island 2. FEMS Microbiol Lett 226(2):363–372

Rathman M, Sjaastad MD et al. (1996) Acidification of phagosomes containing *Salmonella typhimurium* in murine macrophages. Infect Immun 64(7):2765–2773

Rech EL, De Bem AR et al. (1996) Biolistic-mediated gene expression in guinea pigs and cattle tissues in vivo. J Med Biol Res 29(10):1265–1267

Rediers H, Rainey PB et al. (2005) Unraveling the secret lives of bacteria: use of in vivo expression technology and differential fluorescence induction promoter traps as tools for exploring niche-specific gene expression. Microbiol Mol Biol Rev 69(2):217–261

Reen FJ, Boyd EF et al. (2005) Genomic comparisons of *Salmonella enterica* serovar Dublin, Agona, and Typhimurium strains recently isolated from milk filters and bovine samples from Ireland, using a *Salmonella* microarray. Appl Environ Microbiol 71(3):1616–1625

Richter-Dahlfors A, Buchan AMJ et al. (1997) Murine salmonellosis studied by confocal microscopy: *Salmonella typhimurium* resides intracellularly inside macrophages and exerts a cytotoxic effect on phagocytes in vivo. J Exp Med 186(4):569–580

Rodland KD, Adkins JN et al. (2008) Use of high-throughput mass spectrometry to elucidate host-pathogen interactions in *Salmonella*. Future Microbiol 3(6):625–634

Rogers LD, Kristensen AR et al. (2008) Identification of cognate host targets and specific ubiquitylation sites on the *Salmonella* SPI-1 effector SopB/SigD. J Proteomics 71(1):97–108

Roumagnac P, Weill FX et al. (2006) Evolutionary history of *Salmonella typhi*. Science 314(5803):1301–1304

Santiviago C, Reynolds MM et al. (2009) Array-based analysis of pools of Salmonella targeted deletion mutants identifies novel genes under selection during infection. PLoS Pathog 5(7):e1000477

Sassetti CM, Boyd DH et al. (2001) Comprehensive identification of conditionally essential genes in mycobacteria. Proc Natl Acad Sci USA 98(22):12712–12717

Sawitzke JA, Thomason LC et al. (2007) Recombineering: in vivo genetic engineering in *E. coli*, *S. enterica*, and beyond. Methods Enzymol 421:171–199

Schmidt F, Schmid M et al. (2009) Assembling proteomics data as a prerequisite for the analysis of large scale experiments. Chem Cent J 3(1):2

Shah DH, Lee MJ et al. (2005) Identification of *Salmonella gallinarum* virulence genes in a chicken infection model using PCR-based signature-tagged mutagenesis. Microbiology 151(Pt 12):3957–3968

Shea JE, Hensel M et al. (1996) Identification of a virulence locus encoding a second type III secretion system in *Salmonella typhimurium*. Proc Natl Acad Sci USA 93:2593–2597

Shelobolina ES, Sullivan SA et al. (2004) Isolation, characterization, and U(VI)-reducing potential of a facultatively anaerobic, acid-resistant bacterium from low-pH, nitrate- and U(VI)-contaminated subsurface sediment and description of *Salmonella subterranea* sp. nov. Appl Environ Microbiol 70(5):2959–2965

Shi L, Adkins JN et al. (2006) Proteomic analysis of *Salmonella enterica* serovar typhimurium isolated from RAW 264.7 macrophages: identification of a novel protein that contributes to the replication of serovar typhimurium inside macrophages. J Biol Chem 281(39):29131–29140

Shi L, Chowdhury SM et al. (2009) Proteomic investigation of the time course responses of RAW 264.7 macrophages to infection with *Salmonella enterica*. Infect Immun 77(8):3227–3233

Sittka A, Lucchini S et al. (2008) Deep sequencing analysis of small noncoding RNA and mRNA targets of the global post-transcriptional regulator, Hfq. PLoS Genet 4(8):e1000163

Sittka A, Pfeiffer V et al. (2007) The RNA chaperone Hfq is essential for the virulence of *Salmonella typhimurium*. Mol Microbiol 63(1):193–217

Sonck KA, Kint G et al. (2009) The proteome of *Salmonella* Typhimurium grown under in vivo-mimicking conditions. Proteomics 9(3):565–579

Stanley TL, Ellermeier CD et al. (2000) Tissue-specific gene expression identifies a gene in the lysogenic phage Gifsy-1 that affects *Salmonella enterica* serovar typhimurium survival in Peyer's patches. J Bacteriol 182(16):4406–4413

Steele-Mortimer O, Meresse S et al. (1999) Biogenesis of *Salmonella typhimurium*-containing vacuoles in epithelial cells involves interactions with the early endocytic pathway. Cell Microbiol 1(1):33–49

Surette MG, Miller MB et al. (1999) Quorum sensing in *Escherichia coli*, *Salmonella typhimurium*, and *Vibrio harveyi*: a new family of genes responsible for autoinducer production. Proc Natl Acad Sci USA 96(4):1639–1644

Swords WE, Cannon BM et al. (1997) Avirulence of LT2 strains of *Salmonella typhimurium* results from a defective *rpoS* gene. Infect Immun 65(6):2451–2453

Thompson A, Rolfe MD et al. (2006) The bacterial signal molecule, ppGpp, mediates the environmental regulation of both the invasion and intracellular virulence gene programs of *Salmonella*. J Biol Chem 281(40):30112–30121

Thomson NR, Clayton DJ et al. (2008) Comparative genome analysis of *Salmonella* Enteritidis PT4 and *Salmonella* Gallinarum 287/91 provides insights into evolutionary and host adaptation pathways. Genome Res 18(10):1624–1637

Toguchi A, Siano M et al. (2000) Genetics of swarming motility in *Salmonella enterica* serovar typhimurium: critical role for lipopolysaccharide. J Bacteriol 182(22):6308–6321

Tsolis RM, Townsend SM et al. (1999) Identification of a putative *Salmonella enterica* serotype typhimurium host range factor with homology to IpaH and YopM by signature-tagged mutagenesis. Infect Immun 67(12):6385–6393

Turner AK, Lovell A et al. (1998) Identification of *Salmonella typhimurium* genes required for colonization of the chicken alimentary tract and for virulence in newly hatched chicks. Infect Immun 66:2099–2106

Typas A, Nichols RJ et al. (2008) High-throughput, quantitative analyses of genetic interactions in *E. coli*. Nat Methods 5(9):781–787

Valdivia RH, Falkow S (1996) Bacterial genetics by flow cytometry: rapid isolation of *Salmonella typhimurium* acid-inducible promoters by differential fluorescence induction. Mol Microbiol 22(2):367–378

Valdivia RH, Falkow S (1997) Fluorescence-based isolation of bacterial genes expressed within host cells. Science 277(5334):2007–2011

Valentin-Hansen P, Eriksen M et al. (2004) The bacterial Sm-like protein Hfq: a key player in RNA transactions. Mol Microbiol 51(6):1525–1533

Vernikos GS, Parkhill J (2006) Interpolated variable order motifs for identification of horizontally acquired DNA: revisiting the *Salmonella* pathogenicity islands. Bioinformatics 22(18): 2196–2203

Voetsch AC, Van Gilder TJ et al. (2004) FoodNet estimate of the burden of illness caused by nontyphoidal *Salmonella* infections in the United States. Clin Infect Dis 38(Suppl 3): S127–S134

Wang Q, Frye JG et al. (2004) Gene expression patterns during swarming in *Salmonella typhimurium*: genes specific to surface growth and putative new motility and pathogenicity genes. Mol Microbiol 52(1):169–187

Wang Q, Mariconda S et al. (2006) Uncovering a large set of genes that affect surface motility in *Salmonella enterica* serovar Typhimurium. J Bacteriol 188(22):7981–7984

Waterman SR, Holden DW (2003) Functions and effectors of the *Salmonella* pathogenicity island 2 type III secretion system. Cell Microbiol 5(8):501–511

Wilhelm BT, Marguerat S et al. (2008) Dynamic repertoire of a eukaryotic transcriptome surveyed at single-nucleotide resolution. Nature 453(7199):1239–1243

Winterberg KM, Reznikoff WS (2007) Screening transposon mutant libraries using full-genome oligonucleotide microarrays. Methods Enzymol 421:110–125

Wolters DA, Washburn MP et al. (2001) An automated multidimensional protein identification technology for shotgun proteomics. Anal Chem 73(23):5683–5690

Worley MJ, Ching KH et al. (2000) *Salmonella* SsrB activates a global regulon of horizontally acquired genes. Mol Microbiol 36(3):749–761

Worley MJ, Nieman GS et al. (2006) *Salmonella typhimurium* disseminates within its host by manipulating the motility of infected cells. Proc Natl Acad Sci USA 103(47):17915–17920

Yu D, Ellis HM et al. (2000) An efficient recombination system for chromosome engineering in *Escherichia coli*. Proc Natl Acad Sci USA 97(11):5978–5983

Zhang S, Adams LG et al. (2003a) Secreted effector proteins of *Salmonella enterica* serotype Typhimurium elicit host-specific chemokine profiles in animal models of typhoid fever and enterocolitis. Infect Immun 71:4795–4803

Zhang CG, Chromy BA et al. (2005) Host-pathogen interactions: a proteomic view. Expert Rev Proteomics 2(2):187–202

Zhang S, Kingsley RA et al. (2003b) Molecular pathogenesis of *Salmonella enterica* serotype typhimurium-induced diarrhea. Infect Immun 71(1):1–12

Zhang A, Wassarman KM et al. (2003c) Global analysis of small RNA and mRNA targets of Hfq. Mol Microbiol 50(4):1111–1124

Zhou Z, Ribeiro AA et al. (2001) Lipid A modifications in polymyxin-resistant *Salmonella typhimurium*: PMRA-dependent 4-amino-4-deoxy-L-arabinose, and phosphoethanolamine incorporation. J Biol Chem 276(46):43111–43121

Chapter 8
Genomics of *Staphylococcus*

Jodi A. Lindsay

8.1 Introduction

The staphylococci are Gram-positive cocci that divide to form clusters that look like grapes. By 16S ribosomal sequencing, they are most closely related to the Gram-positive, low G+C content *Bacillus–Lactobacillus–Staphylococcus* genera (Woese, 1987). There are over 30 species of staphylococci identified, and they are typically found on the skin and mucous membranes of mammals. About a dozen species are frequently carried on humans, including *Staphylococcus aureus, Staphylococcus epidermidis, Staphylococcus haemolyticus, Staphylococcus capitis, Staphylococcus hominis, Staphylococcus cohnii, Staphylococcus lugdunensis, Staphylococcus schleiferi, Staphylococcus saprophyticus, Staphylococcus simulans, Staphylococcus warneri* and *Staphylococcus xylosus*. All of these species are capable of causing disease in humans, but *S. aureus* is by far the most aggressive pathogen and is the species most commonly implicated in food poisoning. Therefore, most of this chapter will focus on *S. aureus*.

Staphylococcus aureus are differentiated from other staphylococci in the diagnostic laboratory by their ability to coagulate plasma. *Staphylococcus aureus* also encode and produce a wide range of toxins, immune escape mechanisms and other virulence factors and are becoming increasingly resistant to antibiotics. They are a major cause of hospital-acquired infection and an increasing cause of veterinary infection as well as a common cause of mastitis in dairy cows. *Staphylococcus aureus* is also a common cause of food poisoning, due to bacterial staphylococcal enterotoxins (SEs) that are heat resistant. Because symptoms usually resolve within 24 h without treatment, it is likely that most cases remain undiagnosed or unreported. Carriage of SE genes varies widely between *S. aureus* strains, and some enterotoxins are more commonly associated with food poisoning than others.

J.A. Lindsay (✉)
Department of Cellular and Molecular Medicine, St George's University of London,
London SW17 0RE, UK
e-mail: jlindsay@sgul.ac.uk

M. Wiedmann, W. Zhang (eds.), *Genomics of Foodborne Bacterial Pathogens*,
Food Microbiology and Food Safety, DOI 10.1007/978-1-4419-7686-4_8,
© Springer Science+Business Media, LLC 2011

This chapter will firstly cover the habitat and pathogenesis of *S. aureus*, including food poisoning and animal infections. There are currently 19 SEs implicated in food poisoning and their known activity, incidence and expression are described. The distribution of the SE toxin genes in *S. aureus* genomes is complex, and the sequencing projects and genetic studies have revealed much. Finally, methods for the detection of SEs, *S. aureus* and toxin genes in food are described.

8.2 Habitat and Pathogenesis

8.2.1 Nasal Colonisation

Carriage of *S. aureus* is typically in the vestibulum nasi of the human nares (nostrils). This is likely the key colonisation site, though *S. aureus* can also be found in the throat, axillae (armpit), groin and intestinal tract. About 25% of humans carry *S. aureus* in their nose all of the time, while another 50% are intermittent carriers and 25% seem to not carry *S. aureus* at all (Kluytmans et al., 1997; Peacock et al., 2001; Graham et al., 2006). Most carriers have only one type of *S. aureus* in their nose and are preferentially colonised by that type, suggesting that there is a strong host element to colonisation (Nouwen et al., 2004). Colonisation seems to be non-detrimental to the host, although some studies have suggested that a minor inflammatory response is associated and that non-colonisers produce specific antimicrobial peptides that prevent colonisation (Cole et al., 1999; Quinn and Cole, 2007). Colonisation is a risk factor for developing invasive disease (Peacock et al., 2001).

Methicillin-resistant *S. aureus* (MRSA) are *S. aureus* that carry the resistance gene *mecA*. Methicillin class antibiotics are the preferred treatment for *S. aureus*, and so MRSA are a serious threat to hospitalised patients who rely on antibiotics for decolonisation, prophylaxis (antibiotic treatment for the prevention of disease) and treatment. MRSA nasal colonisation is high in patients that have been exposed to hospitals, but is now increasing in other groups, such as those in nursing homes and veterinarians. In the USA, a few clones such as USA400 and more recently USA300 have spread quickly and colonise about 1.5% of the healthy population (Gorwitz et al., 2008). It could be that new strains are regularly introduced into human populations and they are then spread quickly, or it could be that MRSA strains have spread because of antibiotic and/or some other unknown pressure.

Decolonisation of *S. aureus*, particularly MRSA, is desirable in patients at high risk of developing infection such as those undergoing surgery or the immunocompromised. It may also be considered if a food handler is thought responsible for a *S. aureus* food poisoning outbreak. Chlorhexidine is widely used in the form of liquid and powder scrubs. Mupirocin, a topical antibiotic, is also used in regions where *S. aureus* resistance to this agent is low. No decolonisation strategy works particularly well, and re-colonisation, presumably from environmental sources and contact with other colonised humans, is common (Kluytmans and Wertheim, 2005).

Sneezing, nose picking and nose wiping are all actions that can disperse *S. aureus* onto hands and/or the environment, including into food. Many nasal carriers are also positively colonised in the throat, armpits, groin and/or rectum. *Staphylococcus aureus* survive on skin, but for shorter time periods as they are inhibited by fatty acids (Kenny et al., 2008). Another major reservoir is colonisation of the intestinal tract, particularly in young children and hospitalised patients, although *S. aureus* must compete with other gut microbes (Vesterlund et al., 2006). Gut colonisation does not lead to the symptoms of food poisoning, presumably because environmental conditions are not ideal for enterotoxin production.

8.2.2 Pathogenesis

Staphylococcus aureus is a common cause of minor skin and wound infections. These infections do not require antibiotics, and in healthy patients a normal polymorphonuclear cell response (innate immunity) will clear up the infection. Patients who are immunocompromised, such as the elderly, those on immunosuppressive therapies, diabetics and the very young, may not mount a sufficient immune response. The bacterium cannot penetrate skin on its own and can begin causing infection wherever it was introduced. Once in the bloodstream, it can seed to other sites. Symptoms vary widely, but fever, pus formation and elevated white cell counts are common. Symptoms depend on the site of infection and include bacteraemia (blood), pneumonia (lungs), endocarditis (heart), abscess (muscle), osteomyelitis (bone), arthritis (joint) and conjunctivitis (eye). Infections range from very minor and requiring no treatment through to serious and even fatal. The condition of the host and the management of the infection are probably the major determinants in outcome.

Staphylococci are the major cause of hospital-acquired (nosocomial) infections. This is because patients in hospital are often immunocompromised, and they often have open wounds, surgical sites, catheters, drips and needle injection sites allowing *S. aureus* to penetrate into tissue. Infections are prevented by using prophylactic antibiotics to control bacteria when the patients are most vulnerable such as during surgery. Infection control measures, including hand washing, using sterile instruments and maintaining a clean environment, are also important. As patient bacterial flora is a likely source of infection, personal hygiene probably also plays a role.

Staphylococcus aureus infections are treated by drainage of pus, removal of any devices associated with infection (e.g. catheters) and antibiotics. Penicillins were highly successful for prophylaxis and treatment of *S. aureus* infections, until *S. aureus* producing penicillinase were selected for, and these drugs are now useless. The next generation of penicillins, the methicillin family, were resistant to penicillinase, and these are still widely used. The first methicillin-resistant *S. aureus* (MRSA) were described in the 1960s, but did not become common in hospitals for another 30 years. Currently, MRSA are endemic in hospitals in many developed and developing countries (Lowy, 2003), although a few have managed to keep rates low by strict policies of screening and infection control (Wertheim et al., 2004). More recently, MRSA that have evolved in the community, such as the notorious USA300 isolates, have become more widespread, and these strains cause severe skin and soft tissue infections in healthy people (Moran et al., 2006). Resistance to a range of other antibiotics is found in *S. aureus*, though few strains are resistant to everything useful. However, this makes 'empirical therapy' difficult. In other words, if a physician has to prescribe antibiotics prior to knowing the cause of an infection, choosing suitable antibiotics is complicated by the range of agents staphylococci are potentially susceptible to. For this reason, glycopeptides such as vancomycin have been very popular as resistance was exceedingly rare. The first cases of fully vancomycin-resistant *S. aureus* were described in 2002, and the numbers of cases are steadily increasing (Zhu et al., 2008). New antibiotics have been released, such as linezolid, daptomycin, tigecycline and quinupristin/dalfopristin, but they are more expensive, have been licensed for only some conditions, may be toxic and resistance to them all has already been described. There is limited financial incentive for industry to develop new antibiotics (Projan, 2003).

8.2.3 Food Poisoning

Staphylococcus aureus food poisoning is due to ingestion of *S. aureus* enterotoxins. Symptoms include abdominal cramps, nausea, vomiting (emesis) and diarrhoea. Onset of symptoms is rapid, usually between 0.5 and 8 h after ingestion of the contaminated food, and symptoms usually resolve without intervention within 24 h. Severe symptoms requiring hospitalisation can occur in a proportion of patients, and intravenous hydration and electrolyte support are given. Deaths can also occur although at low frequency (Do Carmo et al., 2004). Onset of symptoms depends on individual susceptibility (e.g. elderly, very young) and the amount of enterotoxin ingested (Do Carmo et al., 2004). It was estimated that only 144 ng of enterotoxin A (SEA) in chocolate milk was sufficient to cause human food poisoning (Evenson et al., 1988), and a similar amount of SEA in powdered milk affected over 13,000 people (Asao et al., 2003). An outbreak affecting 4,000 people and causing 16 deaths in Brazil likely involved ingestion of milligrams of SEA toxin (Do Carmo et al., 2004).

Staphylococcus aureus is one of the most common causes of food poisoning and SEA is the toxin most commonly implicated (Holmberg and Blake, 1984; Wieneke et al., 1993). In France, staphylococci caused 25% of all reported food poisoning cases between 1999 and 2000, which is 1,651 cases (Le Loir et al., 2003). In Taiwan between 1991 and 2005, 17.5% of bacterial food poisoning cases were due to *S. aureus* (Chiang et al., 2008). If we consider that most cases resolve quickly and are not identified by health-care workers or public health departments, we have to assume that these rates are significantly underestimating the burden of food poisoning due to *S. aureus*. Mead et al. (1999) have taken this into account and estimated that staphylococci cause 185,000 cases of food poisoning in the USA every year.

The food types involved in *S. aureus* food poisoning include meat, poultry, milk and milk products, fish and shellfish, eggs and egg products. Actual products vary in different countries. In the UK from 1969 to 1990, meat and poultry, especially ham and chicken, were the major sources, followed by fish and shellfish, and milk and milk products (Wieneke et al., 1993). In France between 1999 and 2000, cheese was the major source, followed by meat, sausages, pies, fish and seafood, eggs and egg products then poultry (Le Loir et al., 2003). In the USA from 1975 to 1982, the major source was red meat, but also salads (mayonnaise), poultry and pastries (Genigeorgis, 1989). In Japan, rice balls, omelette in lunch boxes, and milk and yoghurt drinks are common sources (Asao et al., 2003).

Contamination is usually from humans to food. Open wounds are not necessary, and asymptomatic nasal colonisation may be sufficient to contaminate food products. Contamination often occurs in the home, but also in restaurants, shops and during the manufacturing process (Wieneke et al., 1993). Heating, such as pasteurisation, is sufficient to kill *S. aureus* and this organism does not produce spores. However, enterotoxins are relatively thermostable, and heating is not useful for destroying them (Asoe et al., 2003; Pepe et al., 2006). *Staphylococcus aureus* grow well in a range of conditions (Stewart et al., 2002). Growth is slowed in low temperatures and dry conditions, but is resistant to high salt and can survive fluctuations in pH. Fermented and dried meat products have lower levels of staphylococci (Ingham et al., 2005) perhaps due to low moisture content. Therefore, any prolonged period of storage of potentially contaminated food under conditions ideal for *S. aureus* growth can lead to production of the toxins (see Section 8.3.3). Education of consumers, processors and food handlers about hand washing and food storage is important.

8.2.4 Infections of Food-Producing Animals

Staphylococcus aureus is a common pathogen of dairy cows and one of the major causes of mastitis. Mastitis causes white cells to infiltrate into the milk, and this milk fetches a lower price for the dairy farmer. Treatment of infected animals with antibiotics has a modest success rate, and so some infected animals will be culled. If treated, a dairy cow will be required to undergo a 'dry' period and not be milked (or not milked for consumption) while the animal recovers. Antibiotics in milk are not tolerated, so the dry period will be extended until antibiotic levels have decreased. Mastitis also occurs in sheep and goats where it has similar economic consequences (Miles et al., 1992; Tollersrud et al., 2006).

Staphylococcus aureus is also capable of causing other types of infections in food-producing animals, such as wound and skin infections, and occasionally more serious infections such as those found in humans. Other animals raised for food consumption such as poultry and pigs (McNamee et al., 1999; van Belkum et al., 2008) can also be affected. However, an animal with a severe infection is more likely to be culled than treated and will not be fit for human consumption.

Carriage of staphylococci in the nose, gut or other locations of food-producing animals is likely, but there are relatively few recent studies (Pepe et al., 2006). Although contaminated meat is a common source of staphylococci, it is widely believed that the bacteria are probably introduced at the time of culling or handling, rather than because the animal has been colonised. It could be that evidence to prove whether contaminated meat is a source has simply not been explored sufficiently.

The strains of *S. aureus* that cause mastitis or infection of animals are presumably the same ones that colonise animals and are generally different from the human strains. They vary in the lineages they belong to (see Section 8.4), and this means they have unique combinations of surface proteins that interact directly with host (Sung et al., 2008). However, some cases of mastitis are due to 'human' isolates, often multi-drug-resistant strains that clearly come from human sources. It is also known that occasionally humans suffer from infections due to animal strains. It is possible that *S. aureus* is simply being opportunistic, and disease is usually due to carriage isolates (Sung et al., 2008).

An interesting situation has recently developed in the Netherlands, Denmark, Belgium and surrounding countries where pig farming is an important industry. Carriage of a novel MRSA lineage, ST398, has been discovered at high levels in pigs and pig farmers (van Belkum et al., 2008). Infection in pigs is rare, but occasionally infections in pig farmers or those with contact with pigs have occurred. Because the Netherlands has a low level of MRSA in humans in their hospitals, they have a rigorous nasal screening policy for MRSA in all patients admitted to hospital, and this is how the ST398 MRSA was discovered. The impact of this on human health is still debated, and as the strain is endemic in many pig farms and their environment, eradication to protect human health is unlikely to be practical or cost effective.

Animal strains of *S. aureus* are capable of producing SEs. Sung et al. (2008) have investigated 37 mastitis strains from cows in the UK by multistrain microarray, and the data show that the common lineage ST151 carries genes for *seg, seli, selm, seln* and *selo*, while *selk* was found in a few ST771 isolates. Scherrer et al. (2004) identified *S. aureus* that were positive for SE toxins in raw bulk tank milk from goats and sheep in Switzerland. Therefore, animal *S. aureus* strains are a potential source of food poisoning toxin. Pasteurisation of cow, sheep or goat milk will kill *S. aureus* and prevent further toxin production. Some milk or milk products are not pasteurised, and for example, staphylococci are tolerated in French soft cheese. The staphylococci in this case could come from the milk or from human handlers and the cheese maturing environment. Food poisoning due to contamination of unpasteurised cheese in France is common. New methods for detecting *S. aureus* lineages from animals may be useful in identifying the source of staphylococci in the future (see Section 8.5).

8.2.5 Other Staphylococcal Species

Staphylococcus intermedius, a common pathogen of dogs, can carry multiple SE genes, and SEA, SEB, SEC and SED production can be detected (Hendricks et al., 2002). Strains can also produce proteins that are cross-reactive with antibodies to SEC and cause emesis (Becker et al., 2001; Edwards et al., 1997), and at least one food poisoning outbreak has been attributed to *S. intermedius* (Khambaty et al., 1994). Coagulase-negative staphylococci rarely carry enterotoxin genes, although a few isolates of *S. epidermidis, S. cohnii, S. haemolyticus, S. xylosus* and *S. saprophyticus* have been reported (Bautista et al., 1988; Veras et al., 2008), and a few of these isolates have been implicated in food poisoning. Enterotoxin genes have not been identified in the whole genome sequences of *S. saprophyticus* ATCC15305, *S. epidermidis* ATCC12228, *S. epidermidis* RP62A or *S. haemolyticus* JCSC1435.

8.3 Staphylococcal Enterotoxins (SEs)

The *S. aureus* enterotoxins (SEs) are capable of causing emesis (vomiting). The most well studied are staphylococcal enterotoxin A (SEA) encoded by the gene *sea* (also known as *entA*), SEB, SEC, SED and SEE. SEA is the toxin most commonly associated with food poisoning outbreaks (Holmberg and Blake, 1984; Wieneke et al., 1993). The sequencing projects have revealed a range of genes with sequence homology to these, which have also been called enterotoxins SEG through SEV, although some of these new genes encode proteins that are not emetic. The distribution of toxin genes in *S. aureus* strains is highly variable, with some carried on stable regions of the chromosome associated with particular lineages and others carried on mobile genetic elements (MGEs). Expression of toxin genes is necessary for food poisoning, and this is controlled by environmental conditions in some food products.

8.3.1 Emesis and Mechanism of Action

The emetic activity of enterotoxins is measured using an animal model. The monkey emetic assay (Bergdoll, 1988; McCormick et al., 2003) involves oral administration of purified toxin in fruit juice or similar and watching the monkeys for evidence of vomiting for up to 24 h. Most enterotoxins that are emetic cause symptoms with an approximately 1 μg/kg dose. Alternatively, toxin injected intraperitoneally can elicit the same response using a much lower concentration.

Some enterotoxins are less active than others and require higher concentrations to cause emesis. For example, in monkeys, 5–20 μg per animal of SEA, SEB, SEC2, SED, SEE, SEG is sufficient, while a higher doses of SEII was required to cause emesis in only one of four monkeys (Bergdoll, 1988; Munson et al., 1998). Some putative enterotoxins have failed to produce emesis in the monkey, notably SEIL (Orwin et al., 2003). The monkey model is relatively expensive, and many putative SEs have not been tested. Those with no emetic activity in the monkey or those that have not been tested are now called 'staphylococcal enterotoxin-like (SEl) superantigens' (Lina et al., 2004).

Alternative animal models have also been investigated. The house musk shrew model is useful, and shrews can be injected intraperitoneally (Hu et al., 2003).

In house musk shrews, SEA, SEE and SEII have good activity, but higher concentrations of SEB, SEC2, SED, SEG, SEH and SEIP are needed to cause vomiting, so there are some differences compared to monkeys (Hu et al., 1999, 2003; Omoe et al., 2005a). Weanling pigs have also been used, and 20 μg of SEA caused symptoms of food poisoning (Taylor et al., 1982). SEB causes emesis in ferrets (Wright et al., 2000).

Early studies in monkeys (Sugiyama and Hayama, 1965) and recent studies in house musk shrew (Hu et al., 2007) have identified how enterotoxins cause vomiting. SEA causes increased secretion of serotonin (5-hydroxytryptamine, 5-HT) in the intestine of house musk shrews. The serotonin release stimulates serotonin receptors on vagus afferent nerve terminals and is considered to be the trigger for the emetic reflex of reverse peristalsis. Interestingly, cannabinoid receptor agonists decrease serotonin release and inhibit emesis.

All enterotoxins belong to a larger family of toxins known as superantigens and have superantigenic activity. Superantigens can bind non-specifically to MHC class II proteins interacting with T-cell receptors, triggering T-cell expansion (mitogenesis), massive cytokine release and leading to shock, such as toxic shock syndrome (Dinges et al., 2000). SEB and SEC in particular are thought to be common causes of surgical toxic shock, a relatively rare but extremely serious complication of surgery. The role of superantigens in normal invasive *S. aureus* infection is not clear. However, superantigens have been implicated in other diseases (but without definitive proof) including inflammatory bowel disease (McKay, 2001), nosocomial- or antibiotic-associated diarrhoea (Flemming and Ackermann, 2007), chronic hyperplastic sinusitis and nasal polyps (Bachert et al., 2003; Bernstein et al., 2003), atopic eczema (Mempel et al., 2003), asthma and other allergies (Herz et al., 1999; Gould et al., 2007) and chronic obstructive pulmonary disease (COPD; Rohde et al., 2004).

A cysteine loop structure is present in some SEs, and there is evidence it is necessary for emetic activity (Hovde et al., 1994; Hoffman et al., 1996). The crystal structure of several SEs has been elucidated, and regions binding specifically to the MHC-II-binding region and to the T-cell receptor, necessary for mitogenic activity, identified (Dinges et al., 2000). Interestingly, different parts of the proteins are responsible for emetic and mitogenic activities (Spero and Morlock, 1978; Harris et al., 1993). In SEA, mutations of histidine 225 are necessary for MHC binding and emesis, while mutations in histidine 61 of SEA are sufficient to prevent emetic activity but not stimulation of murine T-cell proliferation (Hoffman et al., 1996). A further peptide region (147–156 in SEA) is involved in crossing gut epithelium via an active receptor-mediated mechanism in vitro (Shupp et al., 2002).

8.3.2 Types of Enterotoxins and Distribution

There are currently 19 named SE proteins and corresponding genes, from SEA through to SEIU (Table 8.1). SEF is not used as this was originally assigned to the toxin now called toxic shock syndrome toxin 1 (*tst*), which is not emetic. SEs are identified by sequence similarity, but many do not actually cause emesis (or

Table 8.1 Properties of the staphylococcal enterotoxins

Enterotoxin	Emetic in monkey	Group[a]	GenBank accession number	Element
SEA	Yes	III	L22565	Phage
SEB	Yes	II	M11118	SaPI
SEC	Yes	II	X05815 DQ192646 M28364	SaPI
SED	Yes	III	M28521	Plasmid
SEE	Yes	III	M21319	?SaPI[b]
SEG	Yes	II	AF285760	egc[c]
SHE	Yes	III	BA000033	Transposon
SElI	Weak	V	AF285760	egc
SElJ	n.d.	III	AB075606	Plasmid
SElK	n.d.	V	U93688	SaPI
SElL	No	V	BA000018	SaPI
SElM	n.d.	V	AF285760	egc
SElN	n.d.	III	AF285760	egc
SElO	n.d.	III	AF285760	egc
SElP	n.d.	III	BA000018	Phage
SElQ	n.d.	V	U93688	SaPI
SElR	n.d.	II	AB075606	plasmid
SElU	n.d.	II	AY205307 EF030428	egc
SElV	n.d.	V	EF030427	egc

GenBank accession numbers are examples and can be used at www.ncbi.nlm.nih.gov/sites/entrez?db=pubmed, search 'Nucleotide'

n.d., not done

[a]Group based on amino acid sequence homology (McCormick et al., 2001; Jarraud et al., 2001; Lindsay, unpublished)

[b]Unconfirmed

[c]enterotoxin gene cluster

experiments have not been performed yet) and are now called SEl (staphylococcal enterotoxin-like), e.g. *selk* and toxin SElK (Lina et al., 2004). In previous nomenclature, some were referred to as enterotoxin A (*entA*), etc., but the form *sea* is now more common.

The superantigens, including the enterotoxins, have been clustered into five families based on whole gene similarity (McCormick et al., 2001; Jarraud et al., 2001; Lindsay, unpublished). The SEs cluster into three of these groups, II, III and V (Table 8.1). Only groups II and III appear to be emetic, and this may be due to the presence of the cysteine loop. The group II genes are *seb, sec* and variants, *seg, selr* and *selu*, and they each have cysteine loop separated by 10–19 amino acids that is predicted to be involved in emesis. This group also includes *ssa* and *speA* found in streptococci. Group III are *sea, sed, see, seh, selj, seln, selo* and *selp*, and each has a cysteine loop separated by nine amino acids. The remainder of the enterotoxin genes cluster into group V and are *seli, selk, sell, selm, selq* and *selv*.

Here the emetic activity of each toxin is described, along with general and sequence features, the type of genetic element it is encoded on and frequency of

Table 8.2 Percentage distribution of SE genes in carriage and food poisoning strains, detected by PCR

Gene	Carriage		Food poisoning				
	UK[a] ($n = 161$)	Japan[b] ($n = 97$)	Japan[b] ($n = 69$)	Taiwan[c] ($n = 147$)	Brazil[d] ($n = 22$)	France[e] ($n = 33$)	UK[f] ($n = 39$)
sea	30	8	58	29	45	67	23
seb	10	19	44	20	50	3	13
sec	11	7	0	7	0	3	8
sed	4	9	16	2	0	30	5
see	0	0	0	0		0	3
seg	58	52	35	2		24	44
seh	12	3	43	8		15	21
seli	57	49	35	30			23
selj	12	9	12	2			
selk	19	10	41	16			
sell	14	7	0	7			
selm	64	44	32	12			
seln		52	35	11			
selo	44	43	26	14			
selp		25	17	28			
selq	31	7	45	11			
selr		9	12	5			
selu				14			

Blank squares, not done.

[a] 161 human carriage and invasive isolates in the UK (Peacock et al., 2002; Lindsay et al., 2006, Lindsay, unpublished), includes microarray data

[b] 97 carriage isolates and 69 food poisoning isolates from 30 food poisoning outbreaks in Japan (Omoe et al., 2005b)

[c] 147 food poisoning isolates from Taiwan (Chiang et al., 2008)

[d] 22 food poisoning isolates from Brazil (Veras et al., 2008)

[e] 33 food poisoning isolates from France (Kérouanton et al., 2007)

[f] 39 food poisoning isolates from the UK (McLauchlin et al., 2000)

carriage in *S. aureus* populations. More general details about genetic elements they are encoded on, such as egc, bacteriophage, SaPIs and plasmids, and their distribution will be found in the Section 8.4. Note that many *S. aureus* strains carry no SE genes, suggesting there might be more to be identified. Other strains may carry eight or more SE genes, and it is possible multiple toxins contribute to disease (Chiang et al., 2008). The distribution of SE genes in *S. aureus* strains varies substantially (Table 8.2). It is interesting to note that only some SE toxin genes are more prevalent in food poisoning isolates than in typical *S. aureus* isolates, and there are major geographic differences. Since MRSA distribution also varies geographically, this may be a general feature of *S. aureus* (Cockfield et al., 2007). McLauchlin et al. (2000) have suggested the distribution of toxins in food poisoning *S. aureus* has varied over time in the UK.

SEA. SEA is the most common cause of staphylococcal food poisoning (Holmberg and Blake, 1984; Wieneke et al., 1993) and is the most well characterised. SEA is emetic in monkeys, and important residues for activity have been identified (Hoffman et al., 1996). *sea* is encoded on a bacteriophage (Betley and Mekalanos, 1985) of the phi3 family. Gene expression is not regulated by *agr* (Tremaine et al., 1993). Serum antibody to SEA is common in healthy people (Holtfreter et al., 2004).

SEB. SEB is emetic in monkeys (Silverman et al., 1969) and ferrets (Wright et al., 2000). It is carried on a SaPI (Yarwood et al., 2002).

SEC. SEC is often split into minor variant types called SEC1, SEC2, SEC3. The amino acid similarity between these variants is very high, in the region of 91–99% (Lina et al., 2004). Deringer et al. (1996) showed that these minor variations are enough to alter the ability to bind to T-cell receptors, although they seem unlikely to have an effect on emetic activity. A variant of *sec* has been identified in *S. intermedius*, a species of staphylococci that is coagulase positive and causes infection in dogs. SEC$_{canine}$ is emetic in monkeys and has superantigen activity (Edwards et al., 1997). A variant of *sec1* has been identified in a bovine *S. aureus* (Fitzgerald et al., 2001). *sec* genes are carried on SaPIs (Kuroda et al., 2001; Fitzgerald et al., 2001).

SED. SED is detected in *S. aureus* supernatants that are emetic in monkey feeding tests (Kokan and Bergdoll, 1987). It has been detected in milk during intramammary infection (Tollersrud et al., 2006). *sed* is carried on a large penicillinase plasmid, which often carries other SEs such as *selj* and *selr*. *sed* expression is regulated by *agr* (Bayles and Iandolo, 1989).

SEE. SEE purified from staphylococcal supernatants is emetic in monkeys (Borja et al., 1972). *see* is 84% homologous to *sea* (Couch et al., 1988). The element *see* is encoded on has not been characterised, but is UV inducible (Couch et al., 1988) and likely to be a phage or a SaPI.

SEG. SEG is emetic in monkeys (Munson et al., 1998). There are slight genetic variants in some strains, and variations may be associated with lineage (Blaiotta et al., 2006). *seg* is carried on egc (Jarraud et al., 2001).

SEH. SEH purified from culture supernatant has emetic activity (Su and Wong, 1995). Outbreaks of food poisoning due to SEH have been described (Jørgensen et al., 2005; Ikeda et al., 2005). *seh* is found next to a transposase gene in two sequenced CC1 strains (Baba et al., 2002; Holden et al., 2004) and therefore may be mobilised on a transposon. The *seh* element is downstream of SCC*mec* type IV, and it has been proposed that insertion of the *seh* transposon stabilises SCCmec and prevents excision (Noto and Archer, 2006).

SElI. SElI are lacking cysteine loop and are poorly emetic or not emetic in monkeys (Munson et al., 1998; Orwin et al., 2002), but are emetic in house musk shrews (Hu et al., 2003). *seli* genes with slight variation have been described and these are probably associated with different lineage (Blaiotta et al., 2006). *seli* is encoded on egc (Jarraud et al., 2001).

SElJ. It is not known if SElJ is emetic. *selj* is encoded on the same plasmid that *sed* was discovered on, but can also be found on other plasmids. Expression is not regulated by *agr* (Zhang et al., 1998).

SElK. SElK has superantigen activity (Orwin et al., 2001; Fitzgerald et al., 2001). It is not known if it is emetic, but it is lacking a cysteine loop. *selk* is carried on a

SaPI (Lindsay et al., 1998; Yarwood et al., 2002), and these SaPIs can also carry *selq*, *seb* or *tst*. *selk* and *selq* have similar activities but *selq* is not lethal in a rabbit miniosmotic pump model.

SElL. SElL is not emetic in monkeys, and this is probably because it is missing a cysteine loop (Orwin et al., 2003). SElL does have superantigenic activity (Fitzgerald et al., 2001). *sell* is encoded on a SaPI along with a *sec1* variant and *tst* from a bovine *S. aureus* isolate RF122 (Fitzgerald et al., 2001).

SElM. SElM has superantigen activity (Pan et al., 2007) but its emetic activity is unknown. *selm* was described by Jarraud et al. (2001) on the egc.

SElN. SElN has superantigen activity (Pan et al., 2007) but its emetic activity is unknown. *seln* was described by Jarraud et al. (2001) on the egc; it was originally called *sek*, but this has now changed (Lina et al., 2004).

SElO. The emetic and mitogenic activity of SElO is unknown. *selo* was described by Jarraud et al. (2001) on the egc; it was originally called *sel*, but this has now changed (Lina et al., 2004).

SElP. SElP is emetic in a house musk shrew assay, but only at high dose of 50–150 µg per animal. SElP has not been tested in monkeys yet (Omoe et al., 2005a). It was originally sequenced in the N315 strain and is encoded on a phage of the type 3 family, similar to *sea* (Kuroda et al., 2001). It has high sequence homology to *sea* (71%).

SElQ. SElQ has superantigen activity, but lacks a cysteine loop and is predicted to be poorly emetic. *selq* is located just next to *selk* in some SaPIs and may have resulted from a gene duplication. These SaPIs can also contain *seb* or *tst* (Lindsay et al., 1998; Orwin et al., 2002; Yarwood et al., 2002).

SElR. SElR has superantigenic activity, but its emetic activity has not been tested (Omoe et al., 2004). It is similar in sequence to *seg* (66%). *selr* is encoded on a plasmid, which can also carry *selj* and/or *sed* (Omoe et al., 2003).

SElU. The SElU protein has not been purified or its activity proven. The gene *selu* is encoded on the egc in some isolates and could be an intact copy of a gene that is typically mutated in most egc regions to form two pseudogenes, ψent1 and ψent2 (Letertre et al., 2003a). More recently, a variant of *selu2* has been described which has a different mutation leading to an intact gene formed from ψent1 and ψent2. This protein has been purified and shown to have superantigen activity (Thomas et al., 2006). The incidence of intact *selu1* and *selu2* genes is rare (Thomas et al., 2006).

SElV. SElV has superantigen activity, but its emetic activity is unknown. *selv* is a result of a recombination between *selm* and *seli* on the egc, but is rare (Thomas et al., 2006).

set and ssl. The staphylococcal enterotoxin-like (*set*) group of genes are related to the enterotoxin genes by sequence similarity (Williams et al., 2000). They are mostly located on GIα, while a small group is also found on a putative islet next to an insertion sequence. The GIα gene element shows substantial sequence variation that correlates with lineage. However, the *set* genes are clearly distinct from the SEs, and there is no evidence that they cause emesis. Recent studies have shown these genes often encode proteins that have immunostimulatory properties, so they have been renamed staphylococcal superantigen-like (*ssl*) proteins (Lina et al., 2004).

8.3.3 Expression of SE Toxin Genes

Staphylococcus aureus need to survive and grow in food in order to produce SE toxin. *Staphylococcus aureus* grow well in a wide variety of conditions, particularly those with high water activity, neutral pH, high salt and aeration. They require only a limited range of nutrients for growth and can grow anaerobically. Growth is limited in dry or low water activity conditions, in low temperatures and by competing bacteria, and *S. aureus* can be killed by heating (Genigeorgis, 1989; Stewart et al., 2002).

When grown in rich laboratory media, most *S. aureus* encoding SE genes produce SE, although *seg* and *seli* may be poorly expressed (Omoe et al., 2002; Chiang et al., 2008). However, growth of *S. aureus* in all conditions may not be sufficient for toxin production, and there are many cases where *S. aureus* survive and do not produce SEs, such as in the human gut or in some food stuffs. Expression of *S. aureus* toxins is known to be highly regulated by growth phase, environmental conditions and *S. aureus* gene regulatory proteins and small RNAs (Horsburgh, 2008; Fournier, 2008). Under growth conditions that are not ideal, toxin production by *S. aureus* may be a waste of energy and cause selection of bacteria that can best control toxin expression.

Storage of food contaminated with *S. aureus* at 25°C for as little as 24 h can result in sufficient enterotoxin production to cause food poisoning (Pereira et al., 1991). Experimental studies (Soejima et al., 2007) have estimated that milk kept above 25°C should not be consumed after 8 h. In ideal laboratory conditions, SEC production starts in log phase and is detectable as early as 7 h after inoculation (Otero et al., 1990), while more sensitive detection methods suggest that SEB might be produced as early as 4 h (Rajkovic et al., 2006). This is typical of many *S. aureus* exotoxins, which are regulated by growth phase and produced in late log phase or stationary phase.

One of the major regulators of toxin production in *S. aureus* is the accessory gene regulator (*agr*) system. Agr is activated by a quorum sensing system and is thus expressed when the cell culture reaches a critical mass (Novick et al., 1993); in the laboratory this is at late log and stationary phases. *agr* activates *seb* (Tseng and Stewart, 2005), *sec* (Regassa and Betley, 1993) and *sed* (Bayles and Iandolo, 1989) expression, but not *sea* (Tremaine et al., 1993) or *sej* (Zhang et al., 1998). *agr* is inhibited at low pH (Regassa et al., 1992), and this can be triggered by glucose fermentation. Glucose also inhibits *sec* expression independently of *agr* and pH (Regassa et al., 1991, 1992). High salt (1.2 M) inhibits the production of *sec* (Regassa and Betley, 1993) independently of *agr*. Aconitase mutations that prevent the tricarboxylic acid cycle functioning normally lead to early stationary phase and reduced *sec* expression (Somerville et al., 2002), which might suggest that toxin production requires an extended period of rapid growth for toxin production.

SarA is a gene regulator that can upregulate *seb* (Chan and Foster, 1998) and *sec* (Chien et al., 1999) expression. *sarA* expression is partly dependent on sigmaB activity, although sigmaB represses *seb* (Schmidt et al., 2004). Rot, a regulator with a structure similar to SarA, represses *seb* expression, and *agr* decreases *rot* activity (Tseng and Stewart, 2005). Regulatory pathways are complex and may vary between strains.

SE toxin expression may also be controlled post-transcriptionally, as Omoe et al. (2002) showed *seg* and *sei* mRNA production did not necessarily lead to toxin production. This could be due to small RNA interference, poor secretion or degradation.

SEs in food are not destroyed by heat, including pasteurisation, frying or baking (Pepe et al., 2006), although this depends on the food type (Schwabe et al., 1990). Nor are SEs liable to digestion with stomach acids such as trypsin (Spero and Morlock, 1978). Therefore, it is important to monitor food storage conditions at all stages of production to ensure SE is not produced in the first place.

8.4 *S. aureus* Genomes

8.4.1 Sequences

There are currently 14 fully annotated whole *S. aureus* genome sequences deposited in public databases (img.jgi.doe.gov). The first two sequences, MRSA strain N315 and VISA strain Mu50, both related and from Japanese hospitals (Kuroda et al., 2001), began the revolution. Whole genome sequences allowed the first complete look at the capabilities of *S. aureus*, in terms of metabolism, virulence factors, regulators and host–pathogen interactions. Subsequent genome sequences include US community MRSA strain MW2 (Baba et al., 2002), UK hospital epidemic MRSA252 (Holden et al., 2004), UK community invasive MSSA (Holden et al., 2004), laboratory MSSA NCTC8325 (Gillaspy et al., 2006), laboratory MRSA COL (Gill et al., 2005), hospital MSSA with mecA gene Mu3 (Ohta et al., 2004), US community USA300 MRSA strain FPR3757 (Diep et al., 2006), hospital MRSA JH1 and its VISA derivative JH9 (Mwangi et al., 2007), laboratory strain Newman (Baba et al., 2008) and the bovine *S. aureus* strain RF122 (Herron-Olson et al., 2007), and USA300 MRSA strain TCH1516 (Highlander et al., 2007). The new sequences revealed a huge diversity between the different strains, predominantly in lineage-determining surface protein genes and mobile genetic elements (see below).

Staphylococcus aureus genomes are all approximately 2.9 Mb and encode around 3,000 predicted genes or coding sequences (CDS). Most genes have a function assigned to them based on similarity to genes in other organisms of known function. These are predominantly genes involved in cell division and replication, regulation, transport, molecule biosynthesis and degradation, metabolism, cell wall constituents, ribosomes, toxins, immune evasion and adaptation. Approximately 35% of CDS have an unknown function (Lindsay and Holden 2004; Holden and Lindsay, 2008).

Staphylococcus aureus encode dozens of known or putative toxins, immune evasion proteins or proteins that interact with host. Many, like the SEs, occur in multiple variants, are carried on a variety of genetic elements and are differentially distributed amongst strains (Table 8.1). The next section will describe how genomes vary, evolve and are distributed.

8.4.2 Genome Variation and Evolution

There is substantial variation between genomes of different isolates of *S. aureus*. Variation falls into two main types. First, *S. aureus* is polyclonal and each clonal cluster or lineage carries a unique combination of hundreds of genes that are present or absent or differ substantially. Second, mobile genetic elements (MGEs) are chunks of DNA that move into and out of *S. aureus* horizontally and often encode virulence (including SE) and resistance genes.

8.4.3 Lineages

Multilocus sequence typing (MLST) was the first genetic-based typing method successfully applied to *S. aureus* that showed the species was polyclonal (Enright et al., 2000; Feil et al., 2003). Essentially, most human isolates could be clustered into about 10 major clonal complexes or lineages, with some 'orphans'. This was based on the sequence of seven housekeeping genes. Subsequently whole genome sequences showed that strains belonging to the same lineage were remarkably similar despite enormous differences in geography and time (Holden et al., 2004; Diep et al., 2006). The major difference between strains of the same lineage was their MGE content (see below). The next development was population studies of hundreds of strains using multistrain whole genome microarrays (Lindsay et al., 2006). These studies showed that each lineage was quite distinct from the other lineages and varied in hundreds of core variable (CV) genes. Surprisingly, there were few cases of a CV gene being found in only one lineage, suggesting a lot of recombination between lineages and survival and expansion of the fittest lineages (Lindsay et al., 2006). Many CV genes encode proteins expressed on the cell surface that interact with host and their regulators. They include variants of fibronectin-binding proteins, capsule, coagulase, *sasG*, *ebh*, *sasA* and regulators *agr* and *sarT* (Lindsay et al., 2006; Kuhn et al., 2006). Common human lineages are CC1, CC5, CC8, CC12, CC15, CC22, CC25, CC30, CC45, CC51, with CC30 comprising about a third of isolates (Feil et al., 2003). MRSA include CC1, CC5, CC8, CC22, CC30 (including the subgroup ST36) and CC45, as well as some rarer *S. aureus* lineages such as ST239, ST59, ST80 that have become more successful with resistance. Animal *S. aureus*, especially those from cattle, belong to unique lineages including CC151, CC771, CC188, CC97, CC130 (Sung et al., 2008). A new type of MRSA from pigs is ST398.

Lineages are evolving independently of each other. We have proposed that restriction modification system Sau1 has contributed to the independent evolution

(Waldron and Lindsay, 2006). Restriction enzymes bind to specific DNA sequences via a specificity subunit and digest the DNA. This is a mechanism for bacteria to protect themselves from foreign DNA that may be detrimental. The same specific DNA sequences on the bacteria's own DNA are protected by a modification enzyme that carries the same specificity subunit. The Sau1 system in *S. aureus* consists of one restriction enzyme subunit, *hsdR*, and two different copies of the modification and specificity subunits, *hsdM* and *hsdS*. The Sau1 system prevents the uptake of plasmids from *Escherichia coli* or enterococci to *S. aureus*. Differences in *hsdS* genes correlate strongly with lineage. DNA from different lineages is digested by the Sau1 system. Therefore, we think that SauI contributes to the independent evolution of lineages, as well as controlling the horizontal transfer of MGEs encoding virulence and resistance genes.

8.4.4 *GI and* egc

Jarraud et al. (2001) identified a cluster of genes showing homology to the enterotoxins. They were called *sel, sem, sei,* pseud*ent1,* pseud*ent2, sek* and *seg,* but have now been renamed as *selo, selm, seli, ψent1, ψent2, seln* and *seg* (Lina et al., 2004). The cluster of genes was called the enterotoxin gene cluster (*egc*), and it was speculated they formed an enterotoxin nursery where genomic rearrangements would lead to the production of new toxin genes. This has now been confirmed (Letertre et al., 2003a; Thomas et al., 2006). SEG is known to be emetic, SElI is only very weakly emetic in monkeys (Munson et al., 1998) and the remaining toxins have not been tested.

The *S. aureus* sequencing projects have identified two large regions called GIα and GIβ, and the egc sits at the end of some variants of GIβ. Both GIs are highly variable between isolates, but are found in all strains and there is no evidence that they are mobile. Microarray studies have shown that variability correlates strongly with lineage (Lindsay et al., 2006). GIα encodes Sau1 restriction modification *hsdM* and *hsdS* genes, multiple copies of related lipoproteins and multiple staphylococcal superantigen-like (*ssl*) gene variants (previously known as *set*) now known to control immune modulation (Bestebroer et al., 2007; Langley et al., 2005). GIβ encodes the second variant of Sau1 restriction modification *hsdM* and *hsdS* genes, multiple copies of related serine proteases and occasionally hyaluronate lyase, a two-component leucocidin DE, lantibiotic biosynthesis genes and/or the egc locus.

If the GIβ of multiple sequenced genomes is compared, differences can be identified and these correlate with lineage (a useful tool for viewing sequences is http://www.webact.org/WebACT/home). This shows that egc is carried at the end of the GIβ element and is only in lineages 5, 30 and 151, and not present in lineages 1 and 8. Similarly, differences can be detected using the multistrain *S. aureus* microarrays (Lindsay et al., 2006). This data also shows that the egc locus is carried only in lineages 5, 30, 151, 22, 45, 51 and 25. It is not carried in lineages 1, 8, 15, 12, 97, 239, 188, 771, 398 or ST39 (Lindsay et al., 2006; Sung et al., 2008; Lindsay, unpublished). egc loci in unsequenced genomes may have different combinations of pseudogenes or important variations that affect toxin activity that cannot be detected by microarray, and these are likely to be lineage related (Blaiotta et al., 2006). Essentially, this means that intact *seg*, an SE with known emetic activity, is probably carried by approximately half of all human *S. aureus* isolates, including MRSA, and a third of animal isolates.

Despite its prevalence, Holtfreter et al. (2004) showed that normal humans carried less antibodies in plasma to egc SEs compared to rarer SEs like SEA. This suggests egc SEs may not be expressed during colonisation or infection, although much experimental work remains to be done. Interestingly, Chiang et al. (2008) report that *seg* is rare in *S. aureus* associated with food poisoning in Taiwan (Table 8.2). This could be because the strains in Taiwan belong to different lineages, as *seli, selm, seln* and *selo* were more prevalent.

8.4.5 Mobile Genetic Elements (MGEs)

MGEs are pieces of DNA with discrete ends that encode genes that contribute to their ability to move horizontally between strains. In *S. aureus*, the major MGEs are bacteriophage, *S. aureus* pathogenicity islands (SaPIs), plasmids, transposons and staphylococcal cassette chromosomes (SCCs) (Lindsay and Holden, 2006; Lindsay, 2008). All have been reported to carry SE genes, except SCC which typically carries antibiotic resistance genes, including *mecA*. Most MGEs can move at high frequency between *S. aureus* isolates, including during the course of infection (Moore and Lindsay, 2001; Goerke et al., 2004; Lindsay, 2008).

sea is carried on a lysogenic bacteriophage (Betley and Mekalanos, 1985). The bacteriophage genome sits in the bacterial genome at a specific phage integration site and can be induced by stress to excise from the chromosome. The bacteriophage genome is replicated multiple times, and genes carried on the phage genome are expressed to form phage head and tail structures. The bacteriophage genome is packaged into the phage structures, the bacterial cell is lysed and the virulent phage are released to infect other bacteria. Phage bind to bacteria and inject their DNA directly into the bacterial cell, where it repeats the lytic pathway of replication and phage packaging, or alternatively will enter the lysogenic pathway where the phage genome integrates into the bacterial chromosome at the specific site and sits there dormantly.

The *sea* bacteriophage belong to the phi3 family (Lindsay and Holden, 2004), which always integrates into the β-haemolysin (*hlb*) gene. This phage is particularly widespread in human isolates, and phi3 can also carry genes that modulate the human immune response (van Wamel et al., 2006) including staphylokinase, *scn* and *chips*. However, not all phages of this type carry *sea*. The phage is also found in bovine isolates, including the *sea* gene (Sung et al., 2008). During stress induction, such as that caused by antibiotic exposure, the *sea* gene is expressed along with other phage genes, and this can lead to increased toxin production (Sumby and Waldor, 2003; Goerke et al., 2006). *selp* are also found in phi3 phage and presumably can be induced and transferred in the same way. Most strains of *S. aureus* carry bacteriophage in their genome, typically between one and four different types, each approximately 45 kb.

Staphylococcus aureus pathogenicity islands (SaPIs) can carry *seb, sec, selk* and *selp* genes (Lindsay et al., 1998; Fitzgerald et al., 2001; Yarwood et al., 2002). SaPIs are approximately 16 kb in size and integrate into the chromosome much like bacteriophage. However, they lack the bacteriophage genes for making phage particles. In the presence of a helper phage, they can be induced by stress to excise from the chromosome, replicate and package themselves into miniature bacteriophage phage particles (Lindsay et al., 1998; Ruzin et al., 2001). These particles can then infect other *S. aureus* and inject the SaPI DNA which then integrates into a specific location on the chromosome (Lindsay et al., 1998). SaPIs are found in most strains of *S. aureus* with many carrying two or more (Lindsay and Holden, 2006). SaPIs encoding *sec* have been identified in bovine *S. aureus* (Fitzgerald et al., 2001). When SaPIs are induced by stress and replicated, this can lead to enhanced toxin production (Ubeda et al., 2005). Early reports of the element encoding *see* suggested it is inducible by UV light, but did not function as a phage; it is therefore possible this was a SaPI (Couch et al., 1988).

sed was first identified on a 27.6-kb penicillinase-positive plasmid (Bayles and Iandolo, 1989) called pIB485. Subsequently, this plasmid was found to also carry *selj* (Zhang et al., 1998). *selr* was identified on similar plasmids that also carried *selj* and/or *sed* (Omoe et al., 2003). Large penicillinase plasmids in *S. aureus* are relatively common and presumably selected for by the widespread use of penicillin antibiotics. Penicillinase is a protein that degrades the β-lactam ring of penicillin. These plasmids are highly mosaic and are capable of carrying other resistance genes, toxins, transposons and other factors. Some also carry conjugative *tra* genes that encode a transfer mechanism allowing the plasmid to transfer horizontally between strains of *S. aureus*. Even plasmids without *tra* genes can transfer horizontally to other strains via generalised transduction (Lindsay, 2008).

seh is speculated to be on a transposon, as it sits next to a transposase gene. Transposases catalyse the excision and integration of DNA segments, and this may include the neighbouring *seh* gene. There is an association between *seh* and the lineage CC1 (Lindsay, unpublished). *seh* sits just downstream of the SCC*mec* region in some MRSA, and it has been suggested that it plays a role in the stability of the SCC*mec* region (Noto and Archer, 2006).

8.5 Diagnosis and Detection of SEs

Diagnosis of food poisoning by SEs may be performed to exclude other diagnoses or to determine the source and scope of a food poisoning outbreak in order to take preventative measures. Since there is currently debate about the role of some toxins in *S. aureus* food poisoning, diagnosis is also carried out for research purposes. Diagnosis of food poisoning involves identification of SE toxin in food and symptoms in patients that consumed the food. SE toxin was originally detected using specific antibody and gel diffusion tests, but is now usually detected in food using commercially available antibody-based kits, either ELISA or reverse passive latex agglutination (RPLA) (Thompson et al., 1986; Kokan and Bergdoll, 1987). These kits are widely available for SEA, SEB, SEC and SED. Detection kits are not available for the newer SEs. Detection sensitivity is relatively low (McLauchlin et al., 2000).

A newer approach to improve sensitivity has been described. It involves the same specific antibodies to SEs, but they are attached to a specific DNA tag (or the secondary antibody is attached to the DNA tag), and the tag is detected by real-time PCR amplification. This makes the assay 100–1,000 times more sensitive, while still enabling the test to be performed relatively quickly (Rajkovic et al., 2006; Fischer et al., 2007). Preliminary studies suggest it can be used to test food directly.

For the newer SEs, specific PCR to detect genes in food has been developed (Letertre et al., 2003b; Omoe et al., 2005b; Chiang et al., 2008). However, PCR of food is not particularly sensitive as shown in experiments using spiked foods. Some foods inhibit the PCR reaction, especially dairy products (McLauchlin et al., 2000). It is also worth considering that the presence of the toxin gene is not evidence that the toxin was expressed.

Identification of live *S. aureus* from contaminated food is not necessarily useful. *Staphylococcus aureus* is common in the environment, and on food, and usually does not cause disease. Indeed, in some countries *S. aureus* is tolerated in food, such as in France, which allows 10^3 cfu/g in raw milk cheese. Alternatively, a food substance capable of causing food poisoning may have active toxin but the bacteria may have been killed by pasteurisation (Asao et al., 2003). Nevertheless, screening for live *S. aureus* in food can support identification of the causative agent during an outbreak investigation or evaluation of whether storage and handling conditions are adequate. *Staphylococcus aureus* can be detected in food by growth on Baird Parker agar, where it can detect 10^2 cfu/g in solids or 10 cfu/g in liquids (Baird and Lee, 1995; Le Loir et al., 2003). A high concentration of *S. aureus* in food associated with acute-onset food poisoning would indicate *S. aureus* was the causative agent.

Live *S. aureus* can be cultured and then used to detect the presence of toxin by ELISA, or toxin genes by PCR (Klotz et al., 2003), which might be suggestive of the causative SE involved. This approach has been used to identify the distribution of the newer toxins in food poisoning cases (Table 8.2). In most studies so far, some isolates of *S. aureus* suspected to be involved in food poisoning have been toxin negative, and this suggests other unknown toxins may be involved or the *S. aureus* isolated was not the cause of toxin production. Since many SE toxin genes are widespread in *S. aureus*, detection of one of these genes is not diagnostic. Knowing the local epidemiology may be helpful for interpreting data. For example, in Japan, sea-positive isolates are rare, so identifying this toxin in a food poisoning-associated strain suggests this toxin played a role (Omoe et al., 2005b).

Detection of lineages can be a useful tool for classifying strains, including predicting the presence of egc enterotoxins (see Section 8.4.4). Lineages can be detected reliably using MLST or *spa* typing (Enright et al., 2000; Harmsen et al., 2003), which involves PCR and sequencing of seven or one gene, respectively. The sequence is then compared to a database available on the Internet. Microarrays can also be used to identify lineages and SE gene carriage (Sergeev et al., 2004; Lindsay et al., 2006), although this method is expensive and technically demanding. Inexpensive and rapid microarray technology is available (Anjum et al., 2007) but at the time of writing has not been developed for *S. aureus* food poisoning applications. A very simple and cheap PCR test to detect lineage by screening for *hsdS* gene variants of the Sau1 RM system has been developed (Cockfield et al., 2007).

If investigating a possible outbreak of food poisoning, it may be useful to prove that the strain of *S. aureus* in the food is the same as a strain isolated from a food handler, environment or related outbreak. *Staphylococcus aureus* strain typing is performed by a variety of methods in different reference laboratories, and experience with any method as well as an understanding of local epidemiology is always useful (Hallin et al., 2007). Ideally a typing method should identify lineage (using MLST, spa typing, microarray or the RM test) as well as the presence of a range of MGEs. SE toxin profiles (of non-egc genes) can be useful for the latter and can also be supplemented with antibiotic resistance profiles. Many MGEs are unstable, and this should be considered when interpreting results.

8.6 Future

It is clear there is still much that is not known about *S. aureus* food poisoning. The role of the newer SEs is unclear, and relatively little is known about the factors that control toxin expression in food. Better diagnostic tests are being developed, and it remains to be seen whether they help us to better understand and prevent *S. aureus* food poisoning. The role of animal strains of *S. aureus* in food poisoning is unknown, yet they could be a common source of disease. It is suspected that there are geographic and temporal changes in *S. aureus* populations and this may also impact on disease incidence, especially as there is a lot of variation in SE gene carriage between isolates. Evolution of *S. aureus* could lead to new strains carrying more or different combinations of SE genes which could impact on disease. Whether some hosts are more susceptible to developing disease is currently unknown.

8.7 Conclusions

Staphylococcus aureus enterotoxins are a very common cause of food poisoning, and *S. aureus* is so ubiquitous that food storage and handling must be carefully controlled to prevent *S. aureus* growth and toxin production. A wide range of SE toxins are produced by *S. aureus* that are capable of causing vomiting, and each is distributed amongst *S. aureus* strains differently, is regulated differently and can transfer between strains on different types of genetic elements. Detection methods are not very sensitive, and live *S. aureus* are not necessarily found in contaminated food, but improved methods are being developed. In the future, new toxins may be discovered, detection methods should become more sensitive and hopefully more may become known about factors that control *S. aureus* spread and toxin production in food.

Acknowledgements I am grateful to Pat Schlievert and Jim McLaughlin for early helpful discussions. I am especially grateful to previous members of my group and collaborators.

References

Anjum MF, Mafura M, Slickers P, Ballmer K, Kuhnert P, Woodward MJ, Ehricht R (2007) Pathotyping *Escherichia coli* by using miniaturized DNA microarrays. Appl Environ Microbiol 73:5692–5697

Asao T, Kumeda Y, Kawai T, Shibata T, Oda H, Haruki K, Nakazawa H, Kozaki S (2003) An extensive outbreak of staphylococcal food poisoning due to low-fat milk in Japan: estimation of enterotoxin A in the incriminated milk and powdered skim milk. Epidemiol Infect 130:33–40

Baba T, Bae T, Schneewind O, Takeuchi F, Hiramatsu K (2008) Genome sequence of *Staphylococcus aureus* strain Newman and comparative analysis of staphylococcal genomes: polymorphism and evolution of two major pathogenicity islands. J Bacteriol 190:300–310

Baba T, Takeuchi F, Kuroda M, Yuzawa H, Aoki K, Oguchi A, Nagai Y, Iwama N, Asano K, Naimi T, Kuroda H, Cui L, Yamamoto K, Hiramatsu K (2002) Genome and virulence determinants of high virulence community-acquired MRSA. Lancet 359:1819–1827

Bachert C, van Zele T, Gevaert P, De Schrijver L, Van Cauwenberge P (2003) Superantigens and nasal polyps. Curr Allergy Asthma Rep 3:523–531

Baird RM, Lee WH (1995) Media used in the detection and enumeration of *Staphylococcus aureus*. Int J Food Microbiol 26:15–24

Bautista L, Gaya P, Medina M, Nuñez M (1988) A quantitative study of enterotoxin production by sheep milk staphylococci. Appl Environ Microbiol 54:566–569

Bayles KW, Iandolo JJ (1989) Genetic and molecular analyses of the gene encoding staphylococcal enterotoxin D. J Bacteriol 171:4799–4806

Becker K, Keller B, von Eiff C, Brück M, Lubritz G, Etienne J, Peters G (2001) Enterotoxigenic potential of *Staphylococcus intermedius*. Appl Environ Microbiol 67:5551–5557

Bergdoll MS (1988) Monkey feeding test for staphylococcal enterotoxin. Methods Enzymol 165:324–333

Bernstein JM, Ballow M, Schlievert PM, Rich G, Allen C, Dryja D (2003) A superantigen hypothesis for the pathogenesis of chronic hyperplastic sinusitis with massive nasal polyposis. Am J Rhinol 17:321–326

Bestebroer J, Poppelier MJ, Ulfman LH, Lenting PJ, Denis CV, van Kessel KP, van Strijp JA, de Haas CJ (2007) Staphylococcal superantigen-like 5 binds PSGL-1 and inhibits P-selectin-mediated neutrophil rolling. Blood 109:2936–2943

Betley MJ, Mekalanos JJ (1985) Staphylococcal enterotoxin A is encoded by phage. Science 229:185–187

Blaiotta G, Fusco V, von Eiff C, Villani F, Becker K (2006) Biotyping of enterotoxigenic *Staphylococcus aureus* by enterotoxin gene cluster (egc) polymorphism and spa typing analyses. Appl Environ Microbiol 72:6117–6123

Borja CR, Fanning E, Huang IY, Bergdoll MS (1972) Purification and some physicochemical properties of staphylococcal enterotoxin E. J Biol Chem 247:2456–2463

Chan PF, Foster SJ (1998) Role of SarA in virulence determinant production and environmental signal transduction in *Staphylococcus aureus*. J Bacteriol 180:6232–6241

Chiang YC, Liao WW, Fan CM, Pai WY, Chiou CS, Tsen HY (2008) PCR detection of staphylococcal enterotoxins (SEs) N, O, P, Q, R, U, and survey of SE types in *Staphylococcus aureus* isolates from food-poisoning cases in Taiwan. Int J Food Microbiol 121:66–73

Chien Y, Manna AC, Projan SJ, Cheung AL (1999) SarA, a global regulator of virulence determinants in *Staphylococcus aureus*, binds to a conserved motif essential for sar-dependent gene regulation. J Biol Chem 274:37169–37176

Cockfield JD, Pathak S, Edgeworth JD, Lindsay JA (2007) Rapid determination of hospital-acquired methicillin-resistant *Staphylococcus aureus* lineages. J Med Microbiol 56:614–619

Cole AM, Dewan P, Ganz T (1999) Innate antimicrobial activity of nasal secretions. Infect Immun 67:3267–3275

Couch JL, Soltis MT, Betley MJ (1988) Cloning and nucleotide sequence of the type E staphylococcal enterotoxin gene. J Bacteriol 170:2954–2960

Deringer JR, Ely RJ, Stauffacher CV, Bohach GA (1996) Subtype-specific interactions of type C staphylococcal enterotoxins with the T-cell receptor. Mol Microbiol 22:523–534

Diep BA, Gill SR, Chang RF, Phan TH, Chen JH, Davidson MG, Lin F, Lin J, Carleton HA, Mongodin EF, Sensabaugh GF, Perdreau-Remington F (2006) Complete genome sequence of USA300, an epidemic clone of community-acquired methicillin-resistant *Staphylococcus aureus*. Lancet 367:731–739

Dinges MM, Orwin PM, Schlievert PM (2000) Exotoxins of *Staphylococcus aureus*. Clin Microbiol Rev 13:16–34

Do Carmo LS, Cummings C, Linardi VR, Dias RS, De Souza JM, De Sena MJ, Dos Santos DA, Shupp JW, Pereira RK, Jett M (2004) A case study of a massive staphylococcal food poisoning incident. Foodborne Pathog Dis 1:241–246

Edwards VM, Deringer JR, Callantine SD, Deobald CF, Berger PH, Kapur V, Stauffacher CV, Bohach GA (1997) Characterization of the canine type C enterotoxin produced by *Staphylococcus intermedius* pyoderma isolates. Infect Immun 65:2346–2352

Enright MC, Day NP, Davies CE, Peacock SJ, Spratt BG (2000) Multilocus sequence typing for characterization of methicillin-resistant and methicillin-susceptible clones of *Staphylococcus aureus*. J Clin Microbiol 38:1008–1015

Evenson ML, Hinds MW, Bernstein RS, Bergdoll MS (1988) Estimation of human dose of staphylococcal enterotoxin A from a large outbreak of staphylococcal food poisoning involving chocolate milk. Int J Food Microbiol 7:311–316

Feil EJ, Cooper JE, Grundmann H, Robinson DA, Enright MC, Berendt T, Peacock SJ, Smith JM, Murphy M, Spratt BG, Moore CE, Day NP (2003) How clonal is *Staphylococcus aureus*? J Bacteriol 185:3307–3316

Fischer A, von Eiff C, Kuczius T, Omoe K, Peters G, Becker K (2007) A quantitative real-time immuno-PCR approach for detection of staphylococcal enterotoxins. J Mol Med 85:461–469

Fitzgerald JR, Monday SR, Foster TJ, Bohach GA, Hartigan PJ, Meaney WJ, Smyth CJ (2001) Characterization of a putative pathogenicity island from bovine *Staphylococcus aureus* encoding multiple superantigens. J Bacteriol 183:63–70

Flemming K, Ackermann G (2007) Prevalence of enterotoxin producing *Staphylococcus aureus* in stools of patients with nosocomial diarrhea. Infection 35:356–358

Fournier B (2008) Global regulators of *Staphylococcus aureus* virulence genes. In: Lindsay JA (ed) Staphylococcus: molecular genetics. Caister Academic Press, Norfolk, UK, pp 131–183

Genigeorgis CA (1989) Present state of knowledge on staphylococcal intoxication. Int J Food Microbiol 9:327–360

Gill SR, Fouts DE, Archer GL, Mongodin EF, DeBoy RT, Ravel J, Paulsen IT, Kolonay JF, Brinkac L, Beanan M, Dodson RJ, Daugherty SC, Madupu R, Angiuoli SV, Durkin AS, Haft DH, Vamathevan J, Khouri H, Utterback T, Lee C, Dimitrov G, Jiang LX, Qin HY, Weidman J, Tran K, Kang K, Hance IR, Nelson KE, Fraser CM (2005) Insights on evolution of virulence and resistance from the complete genome analysis of an early methicillin-resistant *Staphylococcus aureus* strain and a biofilm-producing methicillin-resistant *Staphylococcus epidermidis* strain. J Bacteriol 187:2426–2438

Gillaspy AF, Worrell V, Orvis J, Roe BA, Dyer DW, Iandolo JJ (2006) The *Staphylococcus aureus* NCTC 8325 genome. In: Fischetti V, Novick R, Ferretti J, Portnoy D, Rood J (eds) Gram positive pathogens. ASM Press, Washington, DC, pp 381–412

Goerke C, Köller J, Wolz C (2006) Ciprofloxacin and trimethoprim cause phage induction and virulence modulation in *Staphylococcus aureus*. Antimicrob Agents Chemother 50:171–177

Goerke C, Matias y Papenberg S, Dasbach S, Dietz K, Ziebach R, Kahl BC, Wolz C (2004) Increased frequency of genomic alterations in *Staphylococcus aureus* during chronic infection is in part due to phage mobilisation. J Infect Dis 189:724–734

Gorwitz RJ, Kruszon-Moran D, McAllister SK, McQuillan G, McDougal LK, Fosheim GE, Jensen BJ, Killgore G, Tenover FC, Kuehnert MJ (2008) Changes in the prevalence of nasal colonization with *Staphylococcus aureus* in the United States, 2001–2004. J Infect Dis 197:1226–1234

Gould HJ, Takhar P, Harries HE, Chevretton E, Sutton BJ (2007) The allergic march from *Staphylococcus aureus* superantigens to immunoglobulin E. Chem Immunol Allergy 93: 106–136

Graham PL 3rd, Lin SX, Larson EL (2006) A U.S. population-based survey of *Staphylococcus aureus* colonization. Ann Intern Med 144:318–325

Hallin M, Deplano A, Denis O, De Mendonça R, De Ryck R, Struelens MJ (2007) Validation of pulsed-field gel electrophoresis and spa typing for long-term, nationwide epidemiological surveillance studies of *Staphylococcus aureus* infections. J Clin Microbiol 45:127–133

Harmsen D, Claus H, Witte W, Rothgänger J, Claus H, Turnwald D, Vogel U (2003) Typing of methicillin-resistant *Staphylococcus aureus* in a university hospital setting by using novel software for spa repeat determination and database management. J Clin Microbiol 41:5442–5448

Harris TO, Grossman D, Kappler JW, Marrack P, Rich RR, Betley MJ (1993) Lack of complete correlation between emetic and T-cell-stimulatory activities of staphylococcal enterotoxins. Infect Immun 61:3175–3183

Hendricks A, Schuberth HJ, Schueler K, Lloyd DH (2002) Frequency of superantigen-producing *Staphylococcus intermedius* isolates from canine pyoderma and proliferation-inducing potential of superantigens in dogs. Res Vet Sci 73:273–277

Herron-Olson L, Fitzgerald JR, Musser JM, Kapur V (2007) Molecular correlates of host specialization in *Staphylococcus aureus*. PLoS ONE 2:e1120

Herz U, Rückert R, Wollenhaupt K, Tschernig T, Neuhaus-Steinmetz U, Pabst R, Renz H (1999) Airway exposure to bacterial superantigen (SEB) induces lymphocyte-dependent airway inflammation associated with increased airway responsiveness – a model for non-allergic asthma. Eur J Immunol 29:1021–1031

Highlander SK, Hultén KG, Qin X, Jiang H, Yerrapragada S, Mason EO Jr, Shang Y, Williams TM, Fortunov RM, Liu Y, Igboeli O, Petrosino J, Tirumalai M, Uzman A, Fox GE, Cardenas AM, Muzny DM, Hemphill L, Ding Y, Dugan S, Blyth PR, Buhay CJ, Dinh HH, Hawes AC, Holder M, Kovar CL, Lee SL, Liu W, Nazareth LV, Wang Q, Zhou J, Kaplan SL, Weinstock GM (2007) Subtle genetic changes enhance virulence of methicillin resistant and sensitive *Staphylococcus aureus*. BMC Microbiol 7:99

Hoffman M, Tremaine M, Mansfield J, Betley M (1996) Biochemical and mutational analysis of the histidine residues of staphylococcal enterotoxin A. Infect Immun 64:885–890

Holden MTG, Feil EJ, Lindsay JA, Peacock SJ, Day NPJ, Enright MC, Foster TJ, Moore CE, Hurst L, Atkin R, Barron A, Bason N, Bentley SD, Chillingworth C, Chillingworth T, Churcher C, Clark L, Corton C, Cronin A, Doggett J, Dowd L, Feltwell T, Hance Z, Harris B, Hauser H, Holroyd S, Jagels K, James KD, Lennard N, Line A, Mayes R, Moule S, Mungall K, Ormond D, Quail MA, Rabbinowitsch E, Rutherford K, Sanders M, Sharp S, Simmonds M, Stevens K, Whitehead S, Barrell BG, Spratt BG, Parkhill J (2004) Complete genomes of two clinical *Staphylococcus aureus* strains: evidence for the rapid evolution of virulence and drug resistance. Proc Natl Acad Sci USA 101:9786–9791

Holden MTG, Lindsay JA (2008) Whole genomes: sequence, microarray and systems biology. In: Lindsay JA (ed) Staphylococcus: molecular genetics. Caister Academic Press, Norfolk, UK, pp 1–28

Holmberg SD, Blake PA (1984) Staphylococcal food poisoning in the United States. New facts and old misconceptions. JAMA 251:487–489

Holtfreter S, Bauer K, Thomas D, Feig C, Lorenz V, Roschack K, Friebe E, Selleng K, Lövenich S, Greve T, Greinacher A, Panzig B, Engelmann S, Lina G, Bröker BM (2004) egc-Encoded superantigens from *Staphylococcus aureus* are neutralized by human sera much less efficiently than are classical staphylococcal enterotoxins or toxic shock syndrome toxin. Infect Immun 72:4061–4071

Horsburgh MJ (2008) The response of *S. aureus* to environmental stimuli. In: Lindsay JA (ed) Staphylococcus: molecular genetics. Caister Academic Press, Norfolk, UK, pp 185–206

Hovde CJ, Marr JC, Hoffmann ML, Hackett SP, Chi YI, Crum KK, Stevens DL, Stauffacher CV, Bohach GA (1994) Investigation of the role of the disulphide bond in the activity and structure of staphylococcal enterotoxin C1. Mol Microbiol 13:897–909

Hu DL, Omoe K, Shimoda Y, Nakane A, Shinagawa K (2003) Induction of emetic response to staphylococcal enterotoxins in the house musk shrew (*Suncus murinus*). Infect Immun 71: 567–570

Hu DL, Omoe K, Shimura H, Ono K, Sugii S, Shinagawa K (1999) Emesis in the shrew mouse (*Suncus murinus*) induced by peroral and intraperitoneal administration of staphylococcal enterotoxin A. J Food Prot 62:1350–1353

Hu DL, Zhu G, Mori F, Omoe K, Okada M, Wakabayashi K, Kaneko S, Shinagawa K, Nakane A (2007) Staphylococcal enterotoxin induces emesis through increasing serotonin release in intestine and it is downregulated by cannabinoid receptor 1. Cell Microbiol 9:2267–2277

Ikeda T, Tamate N, Yamaguchi K, Makino S (2005) Mass outbreak of food poisoning disease caused by small amounts of staphylococcal enterotoxins A and H. Appl Environ Microbiol 71:2793–2795

Ingham SC, Engel RA, Fanslau MA, Schoeller EL, Searls G, Buege DR, Zhu J (2005) Fate of *Staphylococcus aureus* on vacuum-packaged ready-to-eat meat products stored at 21 degrees C. J Food Prot 68:327–360

Jarraud S, Peyrat MA, Lim A, Tristan A, Bes M, Mougel C, Etienne J, Vandenesch F, Bonneville M, Lina G (2001) egc, a highly prevalent operon of enterotoxin gene, forms a putative nursery of superantigens in *Staphylococcus aureus*. J Immunol 166:669–677

Jørgensen HJ, Mathisen T, Løvseth A, Omoe K, Qvale KS, Loncarevic S (2005) An outbreak of staphylococcal food poisoning caused by enterotoxin H in mashed potato made with raw milk. FEMS Microbiol Lett 252:267–272

Kenny JG, Ward D, Josefsson E, Jonsson I-M, Hinds J, Rees HH, Lindsay JA, Tarkowski A, Horsburgh MJ (2008) The *Staphylococcus aureus* response to unsaturated long chain free fatty acids: survival mechanisms and virulence implications. PloS ONE 4:e4344

Khambaty FM, Bennett RW, Shah DB (1994) Application of pulsed-field gel electrophoresis to the epidemiological characterization of *Staphylococcus intermedius* implicated in a food-related outbreak. Epidemiol Infect 113:75–81

Klotz M, Opper S, Heeg K, Zimmermann S (2003) Detection of *Staphylococcus aureus* enterotoxins A to D by real-time fluorescence PCR assay. J Clin Microbiol 41:4683–4687

Kluytmans J, van Belkum A, Verbrugh H (1997) Nasal carriage of *Staphylococcus aureus*: epidemiology, underlying mechanisms, and associated risks. Clin Microbiol Rev 10:505–520

Kluytmans JA, Wertheim HF (2005) Nasal carriage of *Staphylococcus aureus* and prevention of nosocomial infections. Infection 33:3–8

Kokan NP, Bergdoll MS (1987) Detection of low-enterotoxin-producing *Staphylococcus aureus* strains. Appl Environ Microbiol 53:2675–2676

Kuhn G, Francioli P, Blanc DS (2006) Evidence for clonal evolution among highly polymorphic genes in methicillin-resistant *Staphylococcus aureus*. J Bacteriol 188:169–178

Kuroda M, Ohta T, Uchiyama I, Baba T, Yuzawa H, Kobayashi I, Cui LZ, Oguchi A, Aoki K, Nagai Y, Lian JQ, Ito T, Kanamori M, Matsumaru H, Maruyama A, Murakami H, Hosoyama A, Mizutani-Ui Y, Takahashi NK, Sawano T, Inoue R, Kaito C, Sekimizu K, Hirakawa H, Kuhara S, Goto S, Yabuzaki J, Kanehisa M, Yamashita A, Oshima K, Furuya K, Yoshino C, Shiba T, Hattori M, Ogasawara N, Hayashi H, Hiramatsu K (2001) Whole genome sequencing of meticillin-resistant *Staphylococcus aureus*. Lancet 357:1225–1240

Kérouanton A, Hennekinne JA, Letertre C, Petit L, Chesneau O, Brisabois A, De Buyser ML (2007) Characterization of *Staphylococcus aureus* strains associated with food poisoning outbreaks in France. Int J Food Microbiol 115:369–375

Langley R, Wines B, Willoughby N, Basu I, Proft T, Fraser JD (2005) The staphylococcal superantigen-like protein 7 binds IgA and complement C5 and inhibits IgA-Fc alpha RI binding and serum killing of bacteria. J Immunol 174:2926–2933

Le Loir Y, Baron F, Gautier M (2003) *Staphylococcus aureus* and food poisoning. Genet Mol Res 2:63–76

Letertre C, Perelle S, Dilasser F, Fach P (2003a) Identification of a new putative enterotoxin SEU encoded by the egc cluster of *Staphylococcus aureus*. J Appl Microbiol 95:38–43

Letertre C, Perelle S, Dilasser F, Fach P (2003b) Detection and genotyping by real-time PCR of the staphylococcal enterotoxin genes sea to sej. Mol Cell Probes 17:139–147

Lina G, Bohach GA, Nair SP, Hiramatsu K, Jouvin-Marche E, Mariuzza R and International Nomenclature Committee for Staphylococcal Superantigens (2004) Standard nomenclature for the superantigens expressed *by Staphylococcus*. J Infect Dis 189:2334–2336

Lindsay JA (2008) *S. aureus* evolution: lineages and mobile genetic elements (MGE). In: Lindsay JA (ed) Staphylococcus: molecular genetics. Caister Academic Press, Norfolk, UK, pp 45–69

Lindsay JA, Holden MTG (2004) *Staphylococcus aureus*: superbug, supergenome? Trends Microbiol 12:378–385

Lindsay JA, Holden MTG (2006) Understanding the rise of the superbug: investigation of the evolution and genomic variation of *Staphylococcus aureus*. Funct Integr Genomics 6:186–201

Lindsay JA, Moore CE, Day NP, Peacock SJ, Witney AA, Stabler RA, Husain SE, Butcher PD, Hinds J (2006) Microarrays reveal that each of the ten dominant lineages of Staphylococcus aureus has a unique combination of surface-associated and regulatory genes. J Bacteriol 188:669–676

Lindsay JA, Ruzin A, Ross HF, Kurepina N, Novick RP (1998) The gene for toxic shock toxin is carried by a family of mobile pathogenicity islands in *Staphylococcus aureus*. Mol Microbiol 29:527–543

Lowy FD (2003) Antimicrobial resistance: the examples of *Staphylococcus aureus*. J Clin Invest 111:1265–1273

McCormick JK, Bohach GA, Schlievert PM (2003) Pyrogenic, lethal, and emetic properties of superantigens in rabbits and primates. In: Krakauer T (ed) Methods in molecular biology, vol 214: Superantigen protocols. Humana Press, Inc., Totowa, NJ, pp 245–253

McCormick JK, Yarwood JM, Schlievert PM (2001) Toxic shock syndrome and bacterial super-antigens: an update. Annu Rev Microbiol 55:77–104

McKay DM (2001) Bacterial superantigens: provocateurs of gut dysfunction and inflammation? Trends Immunol 22:497–501

McLauchlin J, Narayanan GL, Mithani V, O'Neill G (2000) The detection of enterotoxins and toxic shock syndrome toxin genes in *Staphylococcus aureus* by polymerase chain reaction. J Food Prot 63:479–488

McNamee PT, McCullagh JJ, Rodgers JD, Thorpe BH, Ball HJ, Connor TJ, McConaghy D, Smyth JA (1999) Development of an experimental model of bacterial chondronecrosis with osteomyelitis in broilers following exposure to *Staphylococcus aureus* aerosol, and inoculation with chicken anaemia and infectious bursal disease virus. Avian Pathol 28:26–35

Mead PS, Slutsker L, Dietz V, McCaig LF, Bresee JS, Shapiro C, Griffin PM, Tauxe RV (1999) Food-related illness and death in the United States. Emerg Infect Dis 5:607–625

Mempel M, Lina G, Hojka M, Schnopp C, Seidl HP, Schäfer T, Ring J, Vandenesch F, Abeck D (2003) High prevalence of superantigens associated with the egc locus in *Staphylococcus aureus* isolates from patients with atopic eczema. Eur J Clin Microbiol Infect Dis 22: 306–309

Miles H, Lesser W, Sears P (1992) The economic implications of bioengineered mastitis control. J Dairy Sci 75:596–605

Moore PC, Lindsay JA (2001) Genetic variation among hospital isolates of *Staphylococcus aureus*: evidence for horizontal transfer of virulence genes. J Clin Microbiol 39: 2760–2767

Moran GJ, Krishnadasan A, Gorwitz RJ, Rosheim GE, McDougal LK, Carey RB, Talan DA (2006) Methicillin resistant *S. aureus* infections among patients in the emergency department. N Engl J Med 355:666–674

Munson SH, Tremaine MT, Betley MJ, Welch RA (1998) Identification and characterization of staphylococcal enterotoxin types G and I from *Staphylococcus aureus*. Infect Immun 66: 3337–3348

Mwangi MM, Wu SW, Zhou Y, Sieradzki K, de Lencastre H, Richardson P, Bruce D, Rubin E, Myers E, Siggia ED, Tomasz A (2007) Tracking the in vivo evolution of multidrug resistance in *Staphylococcus aureus* by whole-genome sequencing. Proc Natl Acad Sci USA 104: 9451–9456

Noto MJ, Archer GL (2006) A subset of *Staphylococcus aureus* strains harboring staphylococcal cassette chromosome mec (SCCmec) type IV is deficient in CcrAB-mediated SCCmec excision. Antimicrob Agents Chemother 50:2782–2788

Nouwen J, Boelens H, van Belkum A, Verbrugh H (2004) Human factor in *Staphylococcus aureus* nasal carriage. Infect Immun 72:6685–6688

Novick RP, Ross HF, Projan SJ, Kornblum J, Kreiswirth B, Moghazeh S (1993) Synthesis of staphylococcal virulence factors is controlled by a regulatory RNA molecule. EMBO J 12:3967–3975

Ohta T, Hirakawa H, Morikawa K, Maruyama A, Inose Y, Yamashita A, Oshima K, Kuroda M, Hattori M, Hiramatsu K, Kuhara S, Hayashi H (2004) Nucleotide substitutions in *Staphylococcus aureus* strains, Mu50, Mu3, and N315. DNA Res 11:51–56

Omoe K, Hu DL, Takahashi-Omoe H, Nakane A, Shinagawa K (2003) Identification and characterization of a new staphylococcal enterotoxin-related putative toxin encoded by two kinds of plasmids. Infect Immun 71:6088–6094

Omoe K, Imanishi K, Hu DL, Kato H, Fugane Y, Abe Y, Hamaoka S, Watanabe Y, Nakane A, Uchiyama T, Shinagawa K (2005a) Characterization of novel staphylococcal enterotoxin-like toxin type P. Infect Immun 73:5540–5546

Omoe K, Hu DL, Takahashi-Omoe H, Nakane A, Shinagawa K (2005b) Comprehensive analysis of classical and newly described staphylococcal superantigenic toxin genes in *Staphylococcus aureus* isolates. FEMS Microbiol Lett 246:191–198

Omoe K, Imanishi K, Hu DL, Kato H, Takahashi-Omoe H, Nakane A, Uchiyama T, Shinagawa K (2004) Biological properties of staphylococcal enterotoxin-like toxin type R. Infect Immun 72:3664–3667

Omoe K, Ishikawa M, Shimoda Y, Hu DL, Ueda S, Shinagawa K (2002) Detection of seg, seh, and sei genes in *Staphylococcus aureus* isolates and determination of the enterotoxin productivities of *S. aureus* isolates harboring seg, seh, or sei genes. J Clin Microbiol 40:857–862

Orwin PM, Fitzgerald JR, Leung DY, Gutierrez JA, Bohach GA, Schlievert PM (2003) Characterization of *Staphylococcus aureus* enterotoxin L. Infect Immun 71:2916–2919

Orwin PM, Leung DY, Donahue HL, Novick RP, Schlievert PM (2001) Biochemical and biological properties of staphylococcal enterotoxin K. Infect Immun 69:360–366

Orwin PM, Leung DY, Tripp TJ, Bohach GA, Earhart CA, Ohlendorf DH, Schlievert PM (2002) Characterization of a novel staphylococcal enterotoxin-like superantigen, a member of the group V subfamily of pyrogenic toxins. Biochemistry 41:14033–14040

Otero A, García ML, García MC, Moreno B, Bergdoll MS (1990) Production of staphylococcal enterotoxins C1 and C2 and thermonuclease throughout the growth cycle. Appl Environ Microbiol 56:555–559

Pan YQ, Ding D, Li DX, Chen SQ (2007) Expression and bioactivity analysis of staphylococcal enterotoxin M and N. Protein Expr Purif 56:286–292

Peacock SJ, Moore CE, Justice A, Kantzanou M, Story L, Mackie K, O'Neill G, Day NP (2002) Virulent combinations of adhesin and toxin genes in natural populations of *Staphylococcus aureus*. Infect Immun 70:4987–4996

Peacock SJ, de Silva I, Lowy FD (2001) What determines nasal carriage of *Staphylococcus aureus*? Trends Microbiol 9:605–610

Pepe O, Blaiotta G, Bucci F, Anastasio M, Aponte M, Villani F (2006) *Staphylococcus aureus* and staphylococcal enterotoxin A in breaded chicken products: detection and behavior during the cooking process. Appl Environ Microbiol 72:7057–7062

Pereira JL, Salzberg SP, Bergdoll MS (1991) Production of staphylococcal enterotoxin D in foods by low-enterotoxin-producing staphylococci. Int J Food Microbiol 14:19–25

Projan SJ (2003) Why is big pharma getting out of antibacterial drug discovery? Curr Opin Microbiol 6:427–430

Quinn GA, Cole AM (2007) Suppression of innate immunity by a nasal carriage strain of *Staphylococcus aureus* increases its colonization on nasal epithelium. Immunology 122: 80–89

Rajkovic A, El-Moualij B, Uyttendaele M, Brolet P, Zorzi W, Heinen E, Foubert E, Debevere J (2006) Immunoquantitative real-time PCR for detection and quantification of Staphylococcus aureus enterotoxin B in foods. Appl Environ Microbiol 72:6593–6599

Regassa LB, Betley MJ (1993) High sodium chloride concentrations inhibit staphylococcal enterotoxin C gene (sec) expression at the level of sec mRNA. Infect Immun 61: 1581–1585

Regassa LB, Couch JL, Betley MJ (1991) Steady-state staphylococcal enterotoxin type C mRNA is affected by a product of the accessory gene regulator (agr) and by glucose. Infect Immun 59:955–962

Regassa LB, Novick RP, Betley MJ (1992) Glucose and nonmaintained pH decrease expression of the accessory gene regulator (agr) in *Staphylococcus aureus*. Infect Immun 60:3381–3388

Rohde G, Gevaert P, Holtappels G, Borg I, Wiethege A, Arinir U, Schultze-Werninghaus G, Bachert C (2004) Increased IgE-antibodies to *Staphylococcus aureus* enterotoxins in patients with COPD. Respir Med 98:858–864

Ruzin A, Lindsay J, Novick RP (2001) Molecular genetics of SaPI1 – a mobile pathogenicity island in *Staphylococcus aureus*. Mol Microbiol 41:365–377

Scherrer D, Corti S, Muehlherr JE, Zweifel C, Stephan R (2004) Phenotypic and genotypic characteristics of *Staphylococcus aureus* isolates from raw bulk-tank milk samples of goats and sheep. Vet Microbiol 101:101–107

Schmidt KA, Donegan NP, Kwan WA Jr, Cheung A (2004) Influences of sigmaB and agr on expression of staphylococcal enterotoxin B (seb) in *Staphylococcus aureus*. Can J Microbiol 50:351–360

Schwabe M, Notermans S, Boot R, Tatini SR, Krämer J (1990) Inactivation of staphylococcal enterotoxins by heat and reactivation by high pH treatment. Int J Food Microbiol 10:33–42

Sergeev N, Volokhov D, Chizhikov V, Rasooly A (2004) Simultaneous analysis of multiple staphylococcal enterotoxin genes by an oligonucleotide microarray assay. J Clin Microbiol 42:2134–2143

Shupp JW, Jett M, Pontzer CH (2002) Identification of a transcytosis epitope on staphylococcal enterotoxins. Infect Immun 70:2178–2186

Silverman SJ, Espeseth DA, Schantz EJ (1969) Effect of formaldehyde on the immunochemical and biological activity of staphylococcal enterotoxin B. J Bacteriol 98:437–442

Soejima T, Nagao E, Yano Y, Yamagata H, Kagi H, Shinagawa K (2007) Risk evaluation for staphylococcal food poisoning in processed milk produced with skim milk powder. Int J Food Microbiol 115:29–34

Somerville GA, Chaussee MS, Morgan CI, Fitzgerald JR, Dorward DW, Reitzer LJ, Musser JM (2002) *Staphylococcus aureus* aconitase inactivation unexpectedly inhibits post-exponential-phase growth and enhances stationary-phase survival. Infect Immun 70:6373–6382

Spero L, Morlock BA (1978) Biological activities of the peptides of staphylococcal enterotoxin C formed by limited tryptic hydrolysis. J Biol Chem 253:8787–8791

Stewart CM, Cole MB, Legan JD, Slade L, Vandeven MH, Schaffner DW (2002) *Staphylococcus aureus* growth boundaries: moving towards mechanistic predictive models based on solute-specific effects. Appl Environ Microbiol 68:1864–1871

Su YC, Wong AC (1995) Identification and purification of a new staphylococcal enterotoxin, H. Appl Environ Microbiol 61:1438–1443

Sugiyama H, Hayama T (1965) Abdominal viscera as a site of emetic action for staphylococcal enterotoxin in the monkey. J Infect Dis 115:330–336

Sumby P, Waldor MK (2003) Transcription of the toxin genes present within the staphylococcal phage phiSa3ms is intimately linked with the phage's life cycle. J Bacteriol 185:6841–6851

Sung JM, Lloyd DH, Lindsay JA (2008) *Staphylococcus aureus* host specificity: comparative genomics of human versus animal isolates by multi-strain microarray. Microbiology 154:1949–1959

Taylor SL, Schlunz LR, Beery JT, Cliver DO, Bergdoll MS (1982) Emetic action of staphylococcal enterotoxin A on weanling pigs. Infect Immun 36:1263–1266

Thomas DY, Jarraud S, Lemercier B, Cozon G, Echasserieau K, Etienne J, Gougeon ML, Lina G, Vandenesch F (2006) Staphylococcal enterotoxin-like toxins U2 and V, two new staphylococcal superantigens arising from recombination within the enterotoxin gene cluster. Infect Immun 74:4724–4734

Thompson NE, Razdan M, Kuntsmann G, Aschenbach JM, Evenson ML, Bergdoll MS (1986) Detection of staphylococcal enterotoxins by enzyme-linked immunosorbent assays and radioimmunoassays: comparison of monoclonal and polyclonal antibody systems. Appl Environ Microbiol 51:885–890

Tollersrud T, Kampen AH, Kenny K (2006) *Staphylococcus aureus* enterotoxin D is secreted in milk and stimulates specific antibody responses in cows in the course of experimental intramammary infection. Infect Immun 74:3507–3512

Tremaine MT, Brockman DK, Betley MJ (1993) Staphylococcal enterotoxin A gene (sea) expression is not affected by the accessory gene regulator (agr). Infect Immun 61:356–359

Tseng CW, Stewart GC (2005) Rot repression of enterotoxin B expression in *Staphylococcus aureus*. J Bacteriol 187:5301–5309

Ubeda C, Maiques E, Knecht E, Lasa I, Novick RP, Penadés JR (2005) Antibiotic-induced SOS response promotes horizontal dissemination of pathogenicity island-encoded virulence factors in staphylococci. Mol Microbiol 56:836–844

van Belkum A, Melles DC, Peeters JK, van Leeuwen WB, van Duijkeren E, Huijsdens XW, Spalburg E, de Neeling AJ, Verbrugh HA, Dutch Working Party on Surveillance and Research of MRSA-SOM (2008) Methicillin-resistant and -susceptible *Staphylococcus aureus* sequence type 398 in pigs and humans. Emerg Infect Dis 14:479–483

van Wamel WJ, Rooijakkers SH, Ruyken M, van Kessel KP, van Strijp JA (2006) The innate immune modulators staphylococcal complement inhibitor and chemotaxis inhibitory protein of *Staphylococcus aureus* are located on beta-hemolysin-converting bacteriophages. J Bacteriol 188:1310–1315

Veras JF, Do Carmo LS, Tong LC, Shupp JW, Cummings C, Dos Santos DA, Cerqueira MM, Cantini A, Nicoli JR, Jett M (2008) A study of the enterotoxigenicity of coagulase-negative and coagulase-positive staphylococcal isolates from food poisoning outbreaks in Minas Gerais, Brazil. Int J Infect Dis 12:410–415

Vesterlund S, Karp M, Salminen S, Ouwehand AC (2006) *Staphylococcus aureus* adheres to human intestinal mucus but can be displaced by certain lactic acid bacteria. Microbiology 152:1819–1826

Waldron DE, Lindsay JA (2006) Sau1: a novel lineage-specific type I restriction-modification system that blocks horizontal gene transfer into *Staphylococcus aureus* and between *S. aureus* isolates of different lineages. J Bacteriol 188:5578–5585

Wertheim HF, Vos MC, Boelens HA, Voss A, Vandenbroucke-Grauls CM, Meester MH, Kluytmans JA, van Keulen PH, Verbrugh HA (2004) Low prevalence of methicillin-resistant *Staphylococcus aureus* (MRSA) at hospital admission in the Netherlands: the value of search and destroy and restrictive antibiotic use. J Hosp Infect 56:321–325

Wieneke AA, Roberts D, Gilbert RJ (1993) Staphylococcal food poisoning in the United Kingdom, 1969–90. Epidemiol Infect 110:519–531

Williams RJ, Ward JM, Henderson B, Poole S, O'Hara BP, Wilson M, Nair SP (2000) Identification of a novel gene cluster encoding staphylococcal exotoxin-like proteins: characterization of the prototypic gene and its protein product, SET1. Infect Immun 68:4407–4415

Woese CR (1987) Bacterial evolution. Microbiol Rev 51:221–271

Wright A, Andrews PL, Titball RW (2000) Induction of emetic, pyrexic, and behavioral effects of *Staphylococcus aureus* enterotoxin B in the ferret. Infect Immun 68:2386–2389

Yarwood JM, McCormick JK, Paustian ML, Orwin PM, Kapur V, Schlievert PM (2002) Characterization and expression analysis of *Staphylococcus aureus* pathogenicity island 3. Implications for the evolution of staphylococcal pathogenicity islands. J Biol Chem 277: 13138–13147

Zhang S, Iandolo JJ, Stewart GC (1998) The enterotoxin D plasmid of *Staphylococcus aureus* encodes a second enterotoxin determinant (sej). FEMS Microbiol Lett 168:227–233

Zhu W, Clark NC, McDougal LK, Hageman J, McDonald LC, Patel JB (2008) Vancomycin-resistant *Staphylococcus aureus* isolates associated with Inc18-like vanA plasmids in Michigan. Antimicrob Agents Chemother 52:452–457

Chapter 9
Genomics of Pathogenic *Vibrio* Species

Michelle Dziejman and Fitnat H. Yildiz

9.1 Introduction

Members of the heterotrophic bacterial family *Vibrionaceae* are native inhabitants of aquatic environments worldwide, constituting a diverse and abundant component of marine microbial organisms. Over 60 species of the genus *Vibrio* have been identified (Thompson et al., 2004) and their phenotypic heterogeneity is well documented. The ecology of the genus remains less well understood, however, despite reports that vibrios are the dominant microorganisms inhabiting the superficial water layer and colonizing the chitinous exoskeleton of zooplankton (e.g., copepods, Thompson et al., 2004). Although some species were originally isolated from seawater as free living organisms, most were isolated in association with marine life such as bivalves, fish, eels, or shrimp. The colonization of the light organ of the *Euprymna scolopes* squid by *V. fischeri* is an example of a symbiotic *Vibrio* relationship, and the association of vibrios with zooplankton may reflect commensal relationships. However, several species are pathogenic for marine life. *V. fortis*, *V. campbellii*, *V. shiloi*, and *V. harvei* have been associated with coral reef bleaching, and *V. anguillarum*, *V. alginolyticus*, *V. penaeicida*, *V. salmonicida*, *V. vulnificus*, and *V. harveyi* are among the many species pathogenic for populations of fish, mollusk, and eel and as such can severely impact the economics of aquaculture.

Eleven *Vibrio* species have been documented as medically relevant to humans (Igbinosa and Okoh, 2008). *V. alginolyticus*, *V. cincinnatiensis*, *V. furnissii*, *V. metchnikovii*, and *V. mimicus* have each been associated with limited gastroenteritis and extra-intestinal infections such as wound infections and meningitis. *V. fluvialis* has been associated with gastrointestinal infections in several countries and may represent an underreported cause of gastroenteritis. Two species have each recently been reclassified as a different genus, *Grimontia hollisae* and *Photobacterium damselae*

M. Dziejman (✉)
Department of Microbiology and Immunology, University of Rochester School of Medicine and Dentistry, Rochester, NY 14625, USA
e-mail: michelle_dziejman@urmc.rochester.edu

M. Wiedmann, W. Zhang (eds.), *Genomics of Foodborne Bacterial Pathogens*, Food Microbiology and Food Safety, DOI 10.1007/978-1-4419-7686-4_9, © Springer Science+Business Media, LLC 2011

and appear to be only rarely identified as the causative agent of disease. Several factors render *V. cholerae*, *V. parahaemolyticus*, and *V. vulnificus* of considerable importance in terms of food-borne disease. *V. parahaemolyticus* and *V. cholerae* are most often associated with acute diarrhea in humans. *V. vulnificus*, which is commonly found in association with oysters, can also cause gastroenteritis. In addition, *V. vulnificus* is responsible for primary septicemia and wound infections in humans as well as disease in other marine populations (Gulig et al., 2005). All three species are globally distributed and found abundantly either in surface waters that serve as drinking reservoirs in developing nations or in association with consumable marine life (especially in its raw form). Together, *V. cholerae*, *V. parahaemolyticus, and V. vulnificus* contribute significantly to worldwide morbidity as a result of food-borne illnesses.

   V. cholerae was recognized as the causative agent of cholera over 100 years ago and has a rich history associated with epidemiology and bacteriology. *V. parahaemolyticus* was not associated with human gastroenteritis until 1951 and *V. vulnificus* was not recognized as a pathogen until the late 1970s (Blake et al., 1979). Consequently, there is a significant body of literature dedicated to understanding various facets of *V. cholerae* pathogenesis, whereas comparatively less is understood with regard to *V. parahaemolyticus-* and *V. vulnificus*-induced disease. In addition, much remains unanswered about the shared properties of the pathogenic *Vibrio* species, the potential for horizontal gene transfer among species, and the molecular mechanisms at work that drive the evolution and diversity of strains within each species.

9.2 Disease Characteristics and Susceptibility

V. cholerae-associated illness is characterized by a profuse, watery stool (classically described as rice water) and vomiting that can manifest within several hours to 5 days after ingestion of contaminated food or water (Sack et al., 2004). Cholera is not only endemic in developing nations, but also arises as a prominent disease in situations where overcrowding, war, and/or environmental crises lead to a collapse of infrastructure and failed sanitation. Infection is typically self-limiting, and although many infections are asymptomatic or result in mild illness, some cases can progress to severe diarrhea, electrolyte imbalance, circulatory collapse, and shock, resulting in death. Simple rehydration can reduce the mortality rate to less than 1%, but the severity of the disease should not be underestimated since in the absence of treatment, mortality rates can reach greater than 50%. The factors contributing to epidemic disease have recently been explored, and although little is understood about strain-specific factors that confer epidemic potential, reports indicate that both environmental- and bacteriophage-related interactions appear to contribute to the rise and collapse of epidemics (Lipp et al., 2003; Faruque et al., 2005a, b). All ages are susceptible to infection, although serogroup-specific immunity is conferred following recovery from infection. As a result, the susceptible population is represented mainly by children and the elderly.

The infectious dose has historically been reported to be on the order of 10^{7-8} bacteria, but food buffers the acidity of the stomach and increases the susceptibility to infection. The inoculum size in endemic areas is not well defined, but reports aimed at isolating bacteria from drinking water suggest that it may be as low as 10^3 organisms. Because of the acid-sensitive nature of *V. cholerae*, individuals with achlorhydria are at an increased risk for infection. Interestingly, individuals having blood group type O are at an increased risk for more severe infection, although the molecular basis for this observation is unclear (Barua and Greenough, 1992; Harris et al., 2005).

V. parahaemolyticus gastroenteritis is similarly associated with watery diarrhea and is often accompanied by abdominal cramps, nausea, vomiting, and fever. Symptoms typically appear within 24 h, typically after consumption of raw or undercooked seafood (both fish and shellfish). The self-limiting infection usually resolves in 3–7 days. In severe cases, the watery stool includes blood and mucus. There is no particular limitation to the susceptible population, other than that imposed by the ability or desire to consume raw seafood. *V. parahaemolyticus* infections are of considerable significance in Japan, where the consumption of raw seafood is especially common, and as predicted, *V. parahaemolyticus*-associated disease occurs less frequently in countries where seafood is eaten only after cooking. Reports of infectious dose vary, but are generally considered to be $\sim10^5$ organisms (Yeung and Boor, 2004).

V. vulnificus is a common cause of wound infections following injury in or exposure to contaminated seawater. It is not clear whether *V. vulnificus* truly causes a gastroenteritis type of illness, since disease is thought to be self-limiting and rarely severe enough to warrant hospitalization (Gulig et al., 2005). Typical symptoms of gastroenteritis, such as diarrhea, abdominal pain, and vomiting, are reported in less than half the cases. Yet, in the United States, *V. vulnificus* has emerged as an important food safety issue since primary septicemia resulting from consumption of *V. vulnificus*-contaminated seafood accounts for the highest death rate among food-borne infections. Worldwide, *V. vulnificus*-associated illness has one of the highest mortality rates of any food-borne disease. Symptoms appear several hours to several days after infection and most typically include fever, chills, nausea, and hypotension. Secondary lesions occur on extremities in more than half the cases and frequently progress to necrotizing fasciitis or vasculitis. Progression to septicemia is accompanied by a grave prognosis, and the mortality rate is estimated to be 50% even with antibiotic treatment. Although septicemia typically requires a predisposing medical condition, such as a liver- or blood-related disorder that increases serum iron levels, nearly half of the cases of *V. vulnificus* caused disease present as primary septicemia following consumption of contaminated seafood (Gulig et al., 2005; Jones and Oliver, 2009). Therefore, *V. vulnificus* is a food-borne pathogen of considerable importance.

The incidence of *V. vulnificus*-associated disease appears to be restricted to countries where consumption of raw oysters is common, including Europe, Japan, New Zealand, Republic of Korea, and the United States. The infectious dose is thought to be quite low, on the order of less than 100 organisms, and bacteria are known to multiply within host tissues extremely rapidly. Environmental factors, such as warm water and moderate salinity, can increase the number of *V. vulnificus* organisms in shellfish. Infections are seasonal; over 85% of infections occur between May and October and correlate with the *V. vulnificus* oyster-associated populations (Gulig et al., 2005).

9.3 Strain Classification

Vibrio species classification is based largely on 16S rRNA sequence. The results of phylogenetic analyses of concatenated 16S rRNA, *recA* and *rpoA* gene sequences demonstrate that *V. cholerae*, *V. parahaemolyticus*, and *V. vulnificus* fall into closely related clades within the family *Vibrionaceae*, distinct from the *V. fischeri* group that now encompasses the *Photobacteriaceae* and *Enterovibrionaceae* families. DNA–DNA hybridization studies have suggested that *V. cholerae* is most closely related to *V. mimicus* and less closely related to all other species (Thompson et al., 2004).

Each of the species known as *V. cholerae*, *V. parahaemolyticus*, and *V. vulnificus* is represented by an array of strains. Within each species, both pathogenic and non-pathogenic strains occupy the same environmental niches, and phenotypic similarities along with pathogenic potential have been used as markers for strain classification schemes. Because strain variation within different phenotypic categories has been observed, strict strain classification has been somewhat difficult and most frequently represents a single aspect or a few related phenotypes. However, serotyping has served as a useful marker for recording the correlation of disease with specific strains and establishing the epidemiological patterns associated with strain pathogenicity. Serogroup specificity has been correlated with pathogenic potential for *V. cholerae* and *V. parahaemolyticus*, and serogroup association, multilocus sequence analysis and comparative genomics have been used to identify evolutionary relationships and construct models predicting events leading to the recent emergence of pathogenic strains (Faruque and Mekalanos, 2003; Reen et al., 2006).

9.3.1 Vibrio cholerae

While *V. cholerae* strains can be classified into over 250 serogroups based on the structure of the O somatic antigen, only strains of the O1 and O139 serogroups have caused epidemic and pandemic disease (Chatterjee and Chaudhuri, 2003). The O1 serogroup can be further subdivided into the classical and El Tor biotypes, based on biochemical, immunological, and phage-related properties. It is important to note that the fifth and sixth cholera pandemics were caused by O1 serogroup classical biotype strains, whereas the seventh pandemic, which began in 1961 in Indonesia, was caused by O1 serogroup strains of the El Tor biotype (Wachsmuth et al., 1994). More recently, O139 serogroup strains emerged as a cause of epidemic cholera. Multiple lines of evidence indicate that the horizontal transfer of serogroup-specific genes resulted in the emergence of O139 serogroup strains in 1991 (CholeraWorkingGroup, 1993; Nair et al., 1994; Faruque et al., 2003). O139 serogroup strains continue to cause significant disease and in some geographic locations, co-exist along with O1 El Tor strains. Today, classical strains cannot be isolated from the environment and appear to have been completely replaced by O1 El Tor and O139 strains as the cause of disease. Although O1 El Tor strains have historically been thought of as more "environmentally fit," the molecular and ecological basis for this is not well understood.

Both pathogenic and non-pathogenic strains of O1 and O139 serogroups are found in environmental reservoirs along with strains of other serogroups (non-O1/non-O1319 strains). As a group, non-O1/non-O139 serogroup strains display more phenotypic and genetic diversity than O1 and O139 serogroup strains, which appear to be clonal in origin (Faruque et al., 2003). Some non-O1/non-O139 strains are pathogenic, but cause limited or sporadic disease year round rather than seasonal, epidemic cholera.

9.3.2 Vibrio parahaemolyticus

V. parahaemolyticus strains can also be classified based on O-antigen, but the description of serovars typically includes the flagellar antigen (K) as well (Twedt et al., 1972). There are 13 O serogroups and 71 K antigenic differences, and as many as 75 different serovars resulting from different combinations have been identified. In contrast to *V. cholerae*, a wide variety of serovars have been recognized as the causative agent of disease. However, an O3:K6 serovar strain emerged as a clonal population of clinically associated strains beginning in 1996 with large numbers of patients in Calcutta and Japan, and the incidence of disease attributed to strains of the O3:K6 serovar has continued to rise (Nair et al., 2007). Many suggest that 1996 marked the beginning of the first *V. parahaemolyticus* pandemic, and although the serovar is indeed globally distributed, it is important to note that the overall incidence of *V. parahaemolyticus*-associated disease is small in comparison to that caused by *V. cholerae*. Interestingly, the clonal association of disease-causing strains has recently been disrupted by frequent recombination events, promoting clonal diversification and suggesting that the O3:K6 subset of strains continues to evolve (Gonzalez-Escalona et al., 2008). The resulting groups of related O3:K6 clonal strains and the O3:K6 clone in particular have now been globally disseminated to Asia, North and South America, Africa, and Europe. Importantly, this clone and its serovariants have the ability to become the dominant strains once introduced into a specific geographic locale (Nair et al., 2007).

9.3.3 *Vibrio vulnificus*

The genetic and biochemical heterogeneity among *V. vulnificus* strains is at least as large as that observed for *V. cholerae* and *V. parahaemolyticus*, and perhaps more so since the species characteristically resists typical classification schemes (Jones and Oliver, 2009). Three different biotypes have been identified based on biochemical characteristics and the specific populations in which they cause disease. Human infections are caused predominantly by biotype 1, whereas biotype 2 causes infections in eels. Biotype 3 was identified in 1999 as associated with human disease in individuals handling fish in Israel, and it was recently suggested that biotype 3 is a hybrid of biotypes 1 and 2 (Bisharat et al., 2005, 2007). Categorization based on LPS structure has been complicated by the association of a single type of LPS with biotype 2 (designated serogroup E), whereas the LPS of biotype 1 strains is antigenically much more complex. Five LPS groups have been associated with biotype 1, but more than 40% of strains could not be group associated when a larger panel of strains was considered (Gulig et al., 2005). Similarly, biochemical composition analysis of the *V. vulnificus* capsule identified 94 distinguishable patterns among 120 strains (Bush et al., 1997). Ribotyping differentiated four clusters (A–D) among strains having a clinical vs. an environmental origin, but this interpretation is complicated by the ability of environmental strains and genetically diverse strains to demonstrate virulence in a mouse model (Hor et al., 1995). More recently, however, a 33 kb genomic island (discussed below) was identified as uniquely associated with biotype 1 strains, suggesting that it may encode functions that increase the pathogenic potential for this subset of strains (Cohen et al., 2007).

9.4 *Vibrio* spp. Genomic Advances

The completed genome sequence of seventh pandemic *V. cholerae* O1 El Tor strain N16961 was published in 2000, and genomic sequencing of *Vibrio* species has progressed rapidly since (Heidelberg et al., 2000). Academic efforts and a joint effort of the NIAID and microbial sequencing centers have resulted in the complete or partial genome sequencing of at least 10 additional *V. cholerae* (http://msc.jcvi.org/vibrio/index.shtml). Two different clinical isolates each of *V. parahaemolyticus* and *V. vulnificus* strains have also been sequenced (Chen et al., 2003; Makino et al., 2003; Boyd et al., 2008). The general genome information for the completely sequenced strains was compiled from data presented on the web site for the Comprehensive Microbial Resource of the J. Craig Venter Institute (http://cmr.jcvi.org/cgi-bin/CMR/CmrHomePage.cgi) and is shown in Tables 9.1 and 9.2. In addition, the genomes of multiple other *Vibrio* species (e.g., *V. fischeri*, *V. alginolyticus*, *V. splendidus*, and *V. harveyi*) and different strains within species have been sequenced by public and private institutes (Ruby et al., 2005). In progress and unpublished genomes can be viewed at http://genomesonline.org/index2.htm and at http://www.ncbi.nlm.nih.gov/sites/entrez?db=genomeprj by searching for "*Vibrio.*"

Table 9.1 Whole-genome statistics

	V. cholerae N16961		V. cholerae O395		V. parahaemolyticus RIMD 2210633		V. vulnificus YJO16		V. vulnificus CMCP6	
Sequencing center	J. Craig Venter Institute	Percentage of genome	Nankai University & JCVI:MSC	Percentage of genome	Japanese Consortium RIKEN	Percentage of genome	Yang-Ming University Taiwan	Percentage of genome	Chonnam National University	Percentage of genome
Total number of all DNA molecules	2	100	2	100	2	100	3	100	2	100
Total size of all DNA molecules	4,033,464 bp	100	4,132,319 bp	100	5,165,770 bp	100	5,260,086 bp	100	5,126,798 bp	100
Number of primary annotation coding bases	3,562,101 bp	88	3,621,410 bp	87	4,475,448 bp	86	4,635,540 bp	88	4,356,550 bp	85
Number of G+C bases	1,915,376 bp	47	1,964,499 bp	47	2,343,877 bp	45	2,455,465 bp	47	2,393,740 bp	47
	Primary annotation summary		Primary annotation summary		JCVI automated annotation summary		JCVI automated annotation summary		JCVI automated annotation summary	
Total genes	4,009	100	3,998	100	4,708	100	4,897	100	4,796	100
Protein-coding genes	3,887	96	3,878	97	4,548	97	4,758	97	4,796	Na
Genes assigned a role category	2,313	59	2,492	64	3,520	77	2,856	60	3,604	75

Table 9.1 (continued)

Sequencing center	V. cholerae N16961		V. cholerae O395		V. parahaemolyticus RIMD 2210633		V. vulnificus YJ016		V. vulnificus CMCP6	
	J. Craig Venter Institute		Nankai University & JCVI:MSC		Japanese Consortium RIKEN		Yang-Ming University Taiwan		Chonnam National University	
Genes not assigned a role category	2	0.05	13	0.33	228	5	608	13	417	9
Conserved hypothetical genes	628	16	671	17	692	15	634	13	456	10
Hypothetical genes	944	24	702	18	108	2	660	14	319	7
tRNA genes	98	2	96	2	126	3	112	2	na	na
rRNA genes	24	0.6	24	0.6	34	0.7	27	0.6	Na	na

Table 9.2 Chromosome statistics

	V. cholerae N16961		V. cholerae O395		V. parahaemolyticus RIMD 2210633		V. vulnificus YJ016		V. vulnificus CMCP6	
Chromosome 1										
GenBank accession version	AE003852.1		CP000626.1		BA000031.2		BA000037.2		AE016795.1	
Sequence length	2,961,149 bp	100%	1,108,250 bp	100%	3,288,558 bp	100%	3,354,505 bp	100%	3,281,945 bp	100%
Coding regions	2,632,214 bp	89%	957,430 bp	86%	2,807,649 bp	85%	2,873,292 bp	86%	2,813,179 bp	86%
Intergenic regions	328,935 bp	11%	150,820 bp	14%	480,909 bp	15%	481,213 bp	14%	468,766 bp	14%
Number of genes	2,773	100%	1,133	100%	2,899	100%	3,079	100%	3,119	100%
Number of genes not assigned to role ids	2	0.07%	1	0.08%	123	4%	392	13%	251	8%
Conserved hypothetical genes	460	17%	209	18%	415	14%	407	13%	326	10%
Hypothetical genes	525	19%	331	29%	65	2%	358	12%	182	6%
tRNA	na		na		112	4%	100	3%	0	0%
rRNA	na		na		31	1%	24	0.74%	0	0%
Chromosome 2										
GenBank accession version	AE003853.1		CP000627.1		BA000032.2		BA000038.2		AE016796.1	
Sequence length	1,072,315 bp	100%	3,024,069 bp	100%	1,877,212 bp	100%	1,857,073 bp	100%	1,844,853 bp	100%
Coding regions	929,887 bp	87%	2,663,980 bp	88%	1,602,982 bp	85%	1,632,244 bp	88%	1,611,879 bp	87%
Intergenic regions	142,428 bp	13%	360,089 bp	12%	274,230 bp	15%	224,829 bp	12%	232,974 bp	13%
Number of genes	1,114	100%	2,745	100%	1,649	100%	1,629	100%	1,677	100%
Number of genes not assigned to role ids	0	0%	12	0.43%	105	6%	205	13%	166	10%
Conserved hypothetical genes	168	15%	462	17%	277	17%	223	14%	130	8%
Hypothetical genes	419	38%	371	14%	43	3%	274	17%	137	8%
tRNA	na		na		14	0.84%	12	0.72%	0	0%
rRNA	na		na		3	0.18%	3	0.18%	0	0%

M. Dziejman and F.H. Yildiz

Table 9.2 (continued)

	V. cholerae N16961	V. cholerae O395	V. parahaemolyticus RIMD 2210633	V. vulnificus YJ016	V. vulnificus CMCP6
Plasmid					
GenBank accession Version				AP005352.1	
Sequence length				48,508 bp	100%
Coding regions				38,059 bp	78%
Intergenic regions				10,449 bp	21%
Number of genes				50	100%
Number of genes assigned to role ids				7	14%
Number of genes not assigned to role ids				11	22%
Conserved hypothetical genes				4	8%
Hypothetical genes				28	56%
tRNA				0	0%
rRNA				0	0%

9.5 Common Genomic Features of *V. cholerae*, *V. parahaemolyticus*, and *V. vulnificus*

9.5.1 Two Chromosomes

Genomic sequence analysis of *V. cholerae* strain N16961 confirmed that *Vibrio* spp. carry two circular chromosomes of unequal size and revealed a number of interesting features (Trucksis et al., 1998; Heidelberg et al., 2000). The *V. cholerae* N16961 large chromosome is 2.9 Mb in size, whereas the small chromosome is 1.1 Mb. The combined genome is predicted to encode 3,885 open reading frames (ORFs). The small chromosome carries an integron island, a gene capture system ancestrally derived from antibiotic resistance plasmids (Mazel et al., 1998). Approximately 40% of the predicted proteins encoded by the *V. cholerae* N16961 genome are classified as hypothetical proteins of no known function, a disproportionate number of which lie within the integron island or elsewhere on the small chromosome. The complete genome of an O1 serogroup, classical biotype strain from the sixth pandemic, was sequenced by two groups. Although differences exist in terms of gene content and sequence polymorphisms, the coding capacity and general characteristics of the genome are similar to that of N16961.

The genome sequence of *V. parahaemolyticus* strain RIMD2210633 encodes 4,832 genes, split between a large chromosome of 3.29 Mbp and a small chromosome of 1.88 Mbp (Makino et al., 2003). The genome of *V. vulnificus* strain YJ016 is approximately 900 kb larger and is predicted to encode 4,897 ORFs (Chen et al., 2003). The large chromosome is similar in size to that of *V. cholerae* (3.4 Mb) but the small chromosome is nearly double in size at 1.9 Mb. Assembly of the genomic sequence of strain YJ016 revealed a ~48 kb plasmid (pYJ016) in addition to two circular chromosomes (Chen et al., 2003). The plasmid encodes 69 ORFs, including genes related to plasmid conjugation that are oriented and arranged similarly to genes found on the F plasmid. Although pYJ016 does not encode the full complement of proteins thought to be necessary for conjugation, the plasmid has been reported to be transferred between bacteria by conjugation in the laboratory (Chen et al., 2003). A second *V. vulnificus* strain, CMPC6, has also been sequenced, encoding 4,537 genes in a total genome size of 3.3 and 1.8 Mbp. The *V. vulnificus* genome is therefore more similar in size to the *V. parahaemolyticus* genome. All species share a similar G+C% composition between 46 and 47%, similar to that of other enteric pathogens. Detailed genome statistics for each chromosome are shown in Table 9.2.

Several lines of evidence suggest that essential genes are located mainly on the large chromosome (Judson and Mekalanos, 2000; Chen et al., 2003; Makino et al., 2003). However, exceptions exist (e.g., genes encoding proteins necessary for glycolysis, ribosomal proteins), suggesting that the smaller chromosome has ensured its retention by encoding functions essential for growth and viability. In all three species, chromosome II carries genes whose products expand the range of substrates transported or utilized, or the ability of the organism to survive under specific conditions. This observation is consistent with the view that chromosome II encodes proteins necessary for adaptation. It has been noted that for all three species, the smaller chromosome has a higher proportion of genes unique to each *Vibrio* and a higher than expected number of genes predicted to encode hypothetical proteins.

9.5.2 Genetic Fluidity

Significant differences in the chromosomal organization of the *Vibrio* pathogens have been identified by performing in silico whole-genome comparisons of *V. parahaemolyticus* strain RIMD2210633 to the genomes of *V. cholerae* N16961, *V. vulnificus* YJ016, and *V. vulnificus* CMCP6. Analyses have provided evidence that multiple intra- and inter-chromosomal rearrangements have occurred in the *Vibrio* genomes. The data suggest that genome plasticity not only plays an important role in species evolution, but also indicates that a "core" set of genes has been retained by *Vibrio* spp. Chen et al. (2003) conducted pair-wise, chromosome-by-chromosome analysis of the *V. vulnificus* YJ016 sequence with the *V. cholerae* El Tor N16961 sequence and the *V. parahaemolyticus* RIMD 2210633 sequence. The results provided a comparison of the relative positions of genes that are conserved between two species (*V. vulnificus* and either *V. cholerae* or *V. parahaemolyticus*). While there was significant alignment of the *V. vulnificus* and *V. parahaemolyticus* large chromosomes, a greater number of recombination events that resulted in reordered regions was found when the large chromosomes from *V. vulnificus* and *V. cholerae* were aligned. The linear gene organization of chromosome II appears better preserved between *V. vulnificus* and *V. parahaemolyticus* as well, consistent with the *V. cholerae* chromosome II having a smaller genome size and encoding different proteins relative to both *V. parahaemolyticus* and *V. vulnificus*. Evidence of gene duplications and transposition events was found more frequently in the *V. vulnificus* genome compared to *V. cholerae*; such events could contribute to the expanded genome size of *V. vulnificus*.

Further evidence supporting the occurrence of multiple intra- and inter- chromosomal rearrangements in *Vibrio* evolution is derived from the results of additional in silico analyses using the genomes of *V. cholerae*, *V. vulnificus*, and *V. parahaemolyticus* (Rogozin et al., 2004; Reen et al., 2006). Genome diagonal plots indicating orthologous genes for *Vibrio* spp. identified regions of synteny across each chromosome as well as chromosomal segments that have retained gene order but that have been exchanged between chromosome I and chromosome II. Genetic rearrangements are found both between species and within strains of the same species, however, it is important to note that true colinearity of genomes can only be addressed when genomic sequence data are compared for multiple strains within a given species.

Comparative genomic studies of *V. cholerae* strains using microarrays and in silico comparisons identified two regions that appear to be unique to seventh pandemic strains, *Vibrio* seventh pandemic islands I and II (VSP-1 and VSP-2) (Dziejman et al., 2002; O'Shea et al., 2004). VSP-I is a ~11 kb region extending from VC0175-VC0185, and VSP-II is a ~27 kb region encompassing VC0495-VC0516. Although many of the functions encoded within these regions remain to be identified, additional studies suggest that these islands may have been derived from mobile elements and remain variable segments of the genome (Murphy and Boyd, 2008; Nusrin et al., 2009). Interestingly, VSP-II sequences share significant sequence similarity with a region of the *V. vulnificus* chromosome; VSP-I and VSP-II are otherwise unique and approximately half the predicted proteins have no known homology (O'Shea et al., 2004).

Boyd and coworkers have performed extensive comparative studies using multiple *Vibrio* spp. genomic sequence data, and the results have contributed to our understanding of the evolution of strains and unique coding sequences contributing to pathogenicity. Genome BLAST atlases were constructed that identified regions present in the *V. parahaemolyticus* strain, but absent from other *Vibrio* genomes (Boyd et al., 2008). Twenty-four regions, each greater than 10 kb, were identified as unique to RIMD2210633. Fourteen of the regions had been previously identified and included LPS and capsule polysaccharide gene clusters, a class 1 integron, two f237 phage regions, and the seven VpaI regions denoting genomic island that were previously identified as sequences acquired by horizontal transfer (including the two T3SSs discussed below; Hurley et al., 2006). It is interesting to note that when a larger panel of strains was evaluated, five of the seven genomic islands were found mainly in post-1995 pandemic strains. The functions encoded by one of these (VPaI-1) are thought to include cold adaptation and swarming ability (Nishioka et al., 2008). These findings are consistent with the hypothesis that unique DNA regions may encode functions necessary for adaptation or survival under specific conditions, as well as having the potential to encode virulence factors.

Ten of the 24 regions were newly identified. Two were located on the large chromosome and eight were found on the small chromosome and included genes predicted to encode proteins involved in osmotic shock protection, pili, degradation processes, and proteins related to a type VI secretion system (T6SSs, discussed below; Boyd et al., 2008). When the analyses were extended to 28 additional *Vibrionaceae* genomes, including other species as well as multiple strains of *V. cholerae*, a small number of additional species were found to encode some of the same regions. For example, the Type III Secretion System 2 (T3SS2, discussed below) is also present in a subset of *V. cholerae* non-O1/non-O139 strains, and a type I pilin, T3SS1, and osmotolerance genes were identified in *V. alginolyticus*, *V. harveyi*, and *Vibrio* spp. Ex25.

Fourteen of the identified regions have the characteristics of genomic islands, providing further evidence that the *V. vulnificus* genome is extremely variable and plastic (Quirke et al., 2006). As for all unique regions associated with a specific strain, it is not clear whether in general, *V. vulnificus* strains carry large regions of strain-specific DNA or whether YJ016 is especially rich in unique sequences. However, it has been suggested that a high recombination rate in *V. vulnificus* may account for the generation of genetic diversity (Bisharat et al., 2007). A 33 kb region present on chromosome II of both YJ016 and CMPC6, but absent in other sequenced *Vibrio* species, was found more commonly associated with a more virulent lineage of *V. vulnificus*, a division established by multilocus sequence typing data and 16S rRNA sequence (Nilsson et al., 2003; Cohen et al., 2007). The region is predicted to encode proteins important for sulfur accumulation and metabolism function and is predicted to impart an advantage for survival in the human host (Cohen et al., 2007).

9.5.3 The Superintegron

Superintegrons provide a mechanism for bacteria to incorporate exogenous open reading frames (ORFs) by site-specific recombination into the bacterial genome and have been referred to as assembly platforms (reviewed by Mazel, 2006). The integrated ORFs are converted to functional genes using expression mechanisms that are a part of the superintegron structure. The integron genes therefore provide unique functions as well as a pool of genes that serve as the basis for genetic diversity through duplication and distribution. Superintegrons are present in many γ-proteobacterial genomes, including those of the *Vibrionaeceae*. The superintegrons of different *Vibrio* species all contain a large number of gene cassettes, ranging from 72 associated with the superintegron of *V. parahaemolyticus* to more than 200 in *V. vulnificus* (Mazel et al., 1998; Chen et al., 2003; Makino et al., 2003). In *V. cholerae*, the SI is present on the small chromosome, whereas in *V. parahaemolyticus* and *V. vulnificus*, it is found on the large chromosome.

Reports of the genetic content for superintegrons for different species and strains indicate little conservation of gene cassettes. The *V. parahaemolyticus* SI is completely different than that of *V. cholerae* and *V. vulnificus*. Indeed, for the two sequenced *V. vulnificus* strains, only 29 of the 129 unique SI ORFs from the YJ016 strain are conserved >80% with SI ORFs in the CMPC6 strain; only 6 ORFs are completely identical between the two strains (Chen et al., 2003). Comparison of the SI genes distributed among different *V. cholerae* strains reveals a similar diversity in the region, especially within the serogroups comprising non-O1/non-O139 strains (Dziejman et al., 2002, 2005; Labbate et al., 2007; Pang et al., 2007). The findings are consistent with the concept of the superintegron as a significant contributor to the expanding repertoire of genes that can be associated with the supragenome of the *Vibrio* species.

9.5.4 Motility

Vibrio species are recognized as highly motile, and *V. cholerae*, *V. parahaemolyticus*, and *V. vulnificus* each produce a monotrichous, sheathed, polar flagella (McCarter, 2001). *V. parahaemolyticus* also produces unsheathed, peritrichous flagella. More than 40 proteins comprise the flagellar apparatus of *V. cholerae*, and the majority of flagellar genes reside within three large clusters on the large chromosome, including five genes encoding flagellins (*flaA-E*). The genes encoding the motor proteins are found in three additional locations (Heidelberg et al., 2000). Based on the functions encoded and genetic epistasis experiments, the genes can be classified into four different classes that also incorporate a hierarchy of expression, utilizing both sigma54-dependent and sigma28-dependent promoters (Prouty et al., 2001). The roles of motility and the more than 40 chemotaxis proteins encoded by strain N16961 in the pathogenesis and the aquatic lifestyle of *V. cholerae* are active areas of investigation (Butler and Camilli, 2005). The role of flagella in the host immune response has also been an area of investigation and it was recently shown that purified flagella and secreted flagellin proteins (FlaC and FlaD) can induce the release of IL-8 form epithelial cells via Toll-like receptor 5 (Xicohtencatl-Cortes et al., 2006). Other studies have demonstrated that *V. cholerae* flagella significantly reduced NF-kappa B activation in epithelial cells when compared to levels induced by flagella from *Salmonella*. One possible explanation is that the *V. cholerae* flagella is less likely to be shed as dissociated monomers due to its sheath, thereby repressing the host response to the presence of flagella (Yoon and Mekalanos, 2008). A large, sequence-defined transposon insertion library was recently constructed and used to identify all genes required for motility (Cameron et al., 2008). The screen identified VC2208 (*flgT*) as essential for motility, which is predicted to encode a protein similar to TolB. Since TolB functions in orchestrating outer membrane architecture, it was hypothesized that FlgT may function in coordinating use of the cell's outer membrane in sheath formation. Considering that strains lacking motility were critical for eliminating the reactogenicity of candidate vaccine strains, these studies are of considerable importance in contributing to our understanding of the role of motility and the host immune response to *Vibrio* infection (Kenner et al., 1995; Qadri et al., 2007).

V. parahaemolyticus produces both polar and lateral flagella. Polar flagella are produced when bacteria are grown planktonically (in liquid media), whereas lateral flagella are produced in response to growth in viscous media or on a sold surface (McCarter, 2004). Therefore, conditions exist where *V. parahaemolyticus* produces both types of flagella, although the flagella are genetically distinct. The ~50 genes encoding the polar flagella reside on the large chromosome, and the ~40 genes necessary for lateral flagella production are found on the small chromosome. Gene expression is controlled by sigma54 and sigma28-dependent factors that respond temporally to the synthesis and assembly of the flagellar components. There appear to be more than 60 genes involved in chemotaxis for *V. parahaemolyticus*, and it is thought that the signal transduction mechanisms are shared between the two flagella types. The role of motility in virulence and other phenotypes has been explored, and lateral flagella have been shown to play an important role in chitin-associated adherence and colonization in the aquatic reservoir (Yeung et al., 2002; Merino et al., 2006). Lateral flagella have also been shown to play a role in adhesion to HeLa cells and in biofilm formation.

The role of motility in *V. vulnificus* infection or oyster colonization is less well studied, although whole cell lysates were used to immunize rabbits to generate anti-sera used for immunoscreening of a *V. vulnificus* gene expression library. Clones that interacted with the anti-serum were identified, and the results indicated that the product of the *flgE* gene was recognized by the rabbit immune system. Subsequent studies using a *flgE* deletion strain demonstrated that a non-motile mutant was decreased in its ability to form a biofilm, to adhere to INT-407 cells, and showed attenuation in a mouse model of infection (Lee et al., 2004). The results are consistent with data suggesting that other *V. vulnificus* flagellar components have a role in adherence and virulence (Jones and Oliver, 2009)

9.5.5 Quick Replication – A Hallmark of Vibrio Species

Vibrios are reported to have some of the fastest replication times of all bacteria, and it is thought that this may contribute to difficulty in assessing tolerance levels of organisms associated with food. A recent study by Dryselius and coworkers reported doubling times in standard rich laboratory media of 12–14 min for *V. parahaemolyticus*, 16–20 min for *V. cholerae*, and 18–22 min for *V. vulnificus* (Dryselius et al., 2008). Others have reported replication times of 8–9 min for *V. parahaemolyticus* and other *Vibrio* spp. (Daniels et al., 2000; Aiyar et al., 2002). Numerous studies have focused on the replication, partitioning, and maintenance of two chromosomes. It has been speculated that the different replication requirements of each chromosome have sustained the bipartite genome (Egan et al., 2005). An alternative explanation suggests that replication of two smaller chromosomes vs. one large chromosome provides an evolutionary advantage by facilitating faster replication (Egan and Waldor, 2003; Egan et al., 2004; Reen et al., 2006).

9.5.6 Role of Phages in Vibrio Genome Evolution

Vibriophages are abundantly found in marine waters that are occupied by *Vibrio* spp. and have long been used as tools used for typing of strains. For each species, phages can be readily isolated from environmental waters alone or from bacteria found in association with other organisms such as oysters (DePaola et al., 1998; Comeau et al., 2006; Miller et al., 2003). The genomic era has facilitated both the identification and sequence analysis of phage genomes carried by different strains, some of which have a broad host range (e.g., KVP40; Makino et al., 2003; Miller et al., 2003; Seguritan et al., 2003; Lan et al., 2009). The CTX phage is the best known and studied bacteriophage infecting *V. cholerae* (Waldor and Friedman, 2005). Accessory phage has also been shown to assist in the horizontal transfer of cholera toxin genes, and *V. cholerae* carries several genomic islands bearing features or remnants of bacteriophage genomes (Faruque and Mekalanos, 2003; Murphy and Boyd, 2008). The role phages play in the different aspects of *Vibrio* infection, transmissibility, and environmental survival are beginning to be addressed, and results suggest that phage titers influence the infectivity of *V. cholerae* as well as the growth of strains in the environmental reservoir (Faruque et al., 2005a, b; Nelson et al., 2008; Zahid et al., 2008). However, many questions remain about the relationship and interplay of the numerous phages in the epidemiology and ecology of *Vibrio* spp.

9.6 Genes for Environmental Fitness

9.6.1 Chitin Utilization and Natural Competence

Chitin, a polymer of *N*-acetylglucosamine (GlcNAc) and one of the most abundant biopolymers in nature, can be used as a carbon and energy source by vibrios. Chitin degradation is an important part of the marine ecosystem; consequently, vibrios provide an essential function in the environmental reservoir. The metabolic pathways and relevant chitin-associated phenotypes are best described for *V. cholerae*, although genes necessary for the chitin utilization pathway have been identified from a number of *Vibrio* species and the "core chitin utilization pathway" is conserved (Hunt et al., 2008).

In *V. cholerae*, growth on chitinous surfaces or on chitin oligosaccharides (GlcNAc)$_n$ ($n>1$) induces expression of the genes necessary for chitin colonization, degradation, and assimilation (Meibom et al., 2004). Chitin utilization begins with degradation of chitin polymer, through the action of extracellular chitinases, into chitin oligosaccharides. A chitoporin transports (GlcNAc)$_n$ into the periplasm, and nonspecific porins transport the monomer (GlcNAc). In the periplasm, chitin oligomers are degraded by chitinodextrinases and *N*-acetylglucosaminidases. Some are modified by deacetylases, leading to conversion of GlcN–GlcNAc to GlcN–GlcN. An ABC-type transporter system transports (GlcNAc)$_2$, while (GlcNAc) and (GlcN)$_2$ are transported into the cytosol by their respective PTS systems. In the cytosol or during transport by a PTS system (GlcNAc)$_2$, (GlcNAc), and (GlcN)$_2$ can be converted into phosphorylated intermediates and eventually degraded into fructose-6-phosphate, acetate, and NH$_3$. The degradation products are then used as a carbon and nitrogen source leading to proliferation of the organism.

Growth on chitin surfaces induces expression not only of genes necessary for chitin utilization but also of genes involved in competence development, thereby facilitating horizontal gene transfer (HGT; Meibom et al., 2005). Chitin-induced competence requires genes encoding a type IV pilus termed chitin-regulated pili (ChiRP, also known as the *pilA* gene cluster): VC0857-VC0861, VC2423-VC2426, VC2630-VC2634, VC1612, *tfoX*^{VC} (VC1153), and the *tfoX*^{VC} regulated transcriptional regulator VC1917 (Meibom et al., 2004, 2005). The quorum sensing end-point regulator HapR and the stress/starvation sigma factor RpoS positively regulates chitin-induced competence, indicating that the process is stimulated in environments of high cell density and nutrient limitation (Meibom et al., 2005). *V. vulnificus* and *V. parahaemolyticus* also harbor the genes required for natural competence in their genome, suggesting that this processes is a common property of *Vibrio* species (Hunt et al., 2008).

The importance of chitin-induced natural transformation in the evolution of pathogenic *V. cholerae* species was revealed in two recent studies. In the first study, Blokesch and Schoolnik reported that in a "mixed serogroup biofilm" containing *V. cholerae* O1 and O139 serogroups formed on a chitin surface, "serogroup-specific gene clusters" such as the DNA cluster encoding the lipopolysaccharide (LPS) O-antigen can be transferred between the serogroups (Blokesch and Schoolnik,

2007). Thus, natural transformation can lead to serogroup conversion, which when coupled with phage selection pressure can lead to emergence of new pathogenic variants of *V. cholerae*. Indeed, the emergence of *V. cholerae* O139 from *V. cholerae* O1 El Tor in 1992 could be due in part to chitin-induced natural transformation. In the second study, Udden et al. reported that the classical biotype CTX prophage could be acquired by El Tor biotype in a process involving lytic phages and chitin-induced natural transformation (Udden et al., 2008). Pathogenic *V. cholerae* strains harbor the genes encoding cholera toxin (CT), which are contained in the genome of a filamentous bacteriophage, CTXΦ, that exists as a lysogen. CTXΦ genomes of classical (CTX^{class}Φ) and El Tor (CTX^{ET}Φ) biotypes are polymorphic in the repressor gene *rtsR*, thereby allowing their differentiation (Davis et al., 2000; Kimsey and Waldor, 2004). Recent epidemiological studies revealed the presence of "hybrid variants" with characteristics of the El Tor biotype but harboring CTX^{class}Φ. However, the mechanism by which "hybrid variants" emerged remained elusive (Nair et al., 2002). Transfer of the CTX^{class}Φ to an El Tor biotype strain was markedly enhanced in a microcosm containing *V. cholerae* O141 harboring CTX^{class}Φ (donor strain), *V. cholerae* O1 El Tor (recipient strain), lytic phage for *V. cholerae* O141, and a chitin source (required for natural transformation. This finding indicated that vibriophages can also contribute to the evolution of new pathogenic variants by increasing the abundance of environmental DNA through lysis of *V. cholerae* cells (Udden et al., 2008). Clearly, the life cycles of *V. cholerae* and vibriophages are interwoven and this relationship may be particularly relevant to the emergence of new pathogenic strains. In contrast, relatively little is known about chitin-induced natural transformation in the evolution of pathogenic *V. vulnificus* and *V. parahaemolyticus* strains, although it can be speculated that similar mechanisms may play a role across the *Vibrio* species.

9.6.2 Biofilm Formation

Biofilms, surface attached microbial communities surrounded by extra-polymeric substances composed of exopolysaccharide(s) (EPS), proteins, and DNA, are important for environmental survival and transmission of pathogenic *Vibrio* species (Yildiz and Visick, 2009). Biofilm formation begins with transport and attachment of the bacterium to a surface. Bacterial surface properties (outer membrane proteins, LPS, EPS, flagella, pili) can greatly influence attachment of a bacterium to a surface. Several gene products are critical for *V. cholerae* biofilm development. Mannose-sensitive hemagglutinin type IV pilus (MSHA) and flagellum have been found to facilitate attachment to surfaces (Watnick et al., 1999; Watnick and Kolter, 1999; Watnick et al., 2001). Similarly, a polar flagellum and pili facilitate attachment to the surfaces in both *V. parahaemolyticus* (Enos-Berlage et al., 2005; Shime-Hattori et al., 2006) and *V. vulnificus* (Paranjpye and Strom, 2005). After the initial attachment, movement and growth of attached bacteria lead to surface colonization and formation of microcolonies. Growth of attached bacteria and continued production of EPS lead to development of mature biofilm structures that are characterized by pillars of cells and channels. Development of these structures depends on cell growth rate, motility, signaling molecules, and exopolysaccharide production.

Polysaccharides are a major component of biofilms and are required for formation of mature biofilms. *Vibrio* exopolysaccharide (VPS) production is required for the development of mature biofilms of *V. cholerae* (Watnick and Kolter, 1999; Yildiz et al., 2004; Yildiz and Visick, 2009). While the chemical structure of the VPS is not yet known, initial chemical analysis of VPS revealed that the VPS contains nearly equal amounts of glucose and galactose, with smaller amounts of *N*-acetylglucosamine and mannose (Yildiz and Schoolnik, 1999). The genes involved in VPS production are clustered in two regions on the large chromosome of *V. cholerae* (Fig. 9.1a): *vps*U (VC0916), *vpsA-K*, VC0917-27 (*vps*-I cluster) and *vpsL-Q*, VC0934-9 (*vps*-II cluster) (Yildiz and Schoolnik, 1999; Yildiz and Visick, 2009). Based on predicted amino acid sequences of the VPS proteins, we can make limited predictions about the possible functions of these genes. The proteins encoded by VC0917, VC0918, and VC0927 appear to be required for production of nucleotide sugar precursors. VC0920, VC0925, and VC0934 encode glycosyltransferases. These enzymes catalyze the transfer of sugar moieties to specific acceptor molecules, forming glycosidic bonds that are predicted to be involved in initiation and/or elongation of possible VPS subunits (repeat units). VC0921, VC0924, and VC0937 are predicted to encode proteins required for polymerization and export functions. VC0919 and VC0923 encode proteins predicted to have acyltransferase activity and could be involved in addition of an acetyl group to VPS. VC0922, VC0926, VC0935, VC0936, VC0938, and VC0939 encode hypothetical proteins. Whether all of the *vps* cluster genes are required for VPS production has not yet been studied, nor has their individual role in biofilm formation been examined.

In addition to the *vps* gene clusters, other loci contribute to biofilm formation in *V. cholerae*. Some of these loci encode proteins required for the biosynthesis of the O-antigen of LPS and/or capsule. For example *V. cholerae* O139 contains a locus required for CPS and LPS (O-antigen) production. Both the O-antigen and CPS are composed of LPS-O-antigen polysaccharide. Although the role of CPS and LPS in biofilm formation has not been extensively investigated, initial studies revealed that in *V. cholerae* O139, O-antigen capsule interferes with biofilm formation when tested under specific conditions, in the absence of VPS (Kierek and Watnick, 2003b). A mutation in *wbfF*, which prevents export of capsular precursors and leads to CPS production but not O-antigen production, results in an increase in formation of *vps*-independent biofilms (Kierek and Watnick, 2003a). In contrast, a mutation in *wbfR*, which eliminates CPS and LPS-O-antigen production, markedly decreases biofilm formation. Interestingly, mutation in *waaL*, which eliminates LPS-O-antigen production but not CPS production, did not significantly alter biofilm formation, suggesting that the O139 capsule itself is able to mediate *vps*-independent biofilm development (Kierek and Watnick, 2003b). Structural constraints resulting from interaction of LPS-O-antigen with calcium and exposure of other surface antigens mediating surface attachment in the absence of CPS are thought to be responsible for enhanced biofilm formation capacity of a CPS-deficient mutant.

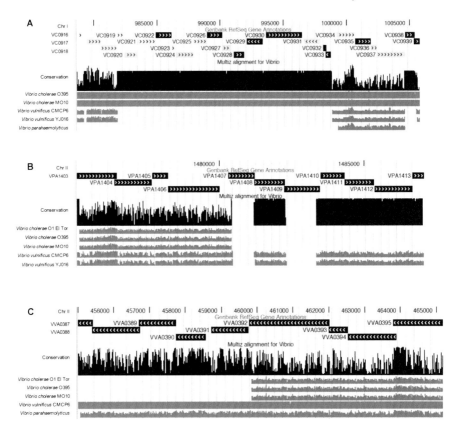

Fig. 9.1 Conservation analysis of exopolysaccharide loci from pathogenic *Vibrio* spp. (**a**) *vps* locus of *V. cholerae* O1 El Tor, (**b**) *cps* locus of *V. parahaemolyticus*, and (**c**) *wcr* locus of *V. vulnificus*. All GenBank annotated genes within this region are represented by *blocks* with *arrowheads* indicating the direction of transcription. The conservation track shows a measure of evolutionary conservation based on a phylogenetic hidden Markov model (phylo-HMM). The conservation score is calculated based on a multiple genome alignment of *V. cholerae* El Tor, *V. cholerae* O395, *V. cholerae* MO10, *V. vulnificus* CMCP6, *V. vulnificus* YJ016, and *V. parahaemolyticus*. A general conservation score is listed for the entire multiple alignment (*black bars*) while pair-wise conservation between *V. cholerae* El Tor and the related species is represented by *gray bars*. The height of the *bars* indicates the degree of conservation for that region. Missing *gray bars* reflect genomic regions absent in the related species genome. Screen shots taken from the UCSC Bacterial Genome Browser (http://microbes.ucsc.edu)

In *V. parahaemolyticus*, the *cps* locus (*cpsA-K*, VPA1403-VPA1413) is required for production of a capsular polysaccharide (CPSA; Fig. 9.1b; Enos-Berlage and McCarter, 2000). CPSA is mainly composed of fucose, galactose, glucose, and *N*-acetylglucosamine and plays a key role in biofilm formation. In addition to the CPSA, two other cell surface polysaccharides, the LPS-O-antigen and the O-antigen capsule, influence biofilm formation in *V. parahaemolyticus*. Mutations in the genes VP0235, a putative nucleoside-diphosphate sugar epimerase involved in production of O-antigen capsule (VP0214–VP0237), and VP2713, a putative lipid A a core:O-antigen ligase/polymerase, markedly changed biofilm architecture even in the presence of CPSA (Enos-Berlage et al., 2005). Thus, in contrast to *V. cholerae*, each type of extracellular polysaccharide contributes positively to development of biofilm architecture in *V. parahaemolyticus*. It should be noted that little is known about biofilm fitness, and a correlation has not been established between biofilm survival capacity and the structure of the biofilms. Thus, it is yet to be determined how CPS and LPS contribute to biofilm structure and function.

Similarly, multiple gene clusters involved in polysaccharide production are found on the *V. vulnificus* genomes. One such cluster, located on the small chromosome of the sequenced *V. vulnificus* strains YJ016 (VVA0395–VVA0387) and CMCP6 (VV21574–VV21582), is *wcrA-K* locus. It is associated with the synthesis of capsular and rugose polysaccharide and hence formation of both OP and rugose colonies (Grau et al., 2008). The *wcr* locus most resembles the *V. parahaemolyticus cps* locus and bears some similarity to the *vps* loci (Fig. 9.1c). As this locus is involved in formation of rugose colonies, which are associated with enhanced biofilm formation in other *Vibrios*, *wcr* most likely contributes to biofilm formation in *V. vulnificus*. Understanding of the role of the *wcr* locus in biofilm formation awaits further investigation.

In addition, *V. vulnificus* harbors a group 1 capsule locus that contain the *wza*, *wzb*, and *wzc* genes (involved in polysaccharide assembly and export) and polymorphic genes (which vary significantly among *V. vulnificus* strains) with predicted functions in CPS or LPS biosynthesis (Wright et al., 2001; Chatzidaki-Livanis et al., 2006; Nakhamchik et al., 2007). In *V. vulnificus* strain MO6-24/O, the group 1 CPS locus is responsible for the production of a capsular polymer composed of repeating units of three *N*-acetyl-quinovosamine and one *N*-acetyl-galactosamine uronic acid residues. Group 1 CPS inhibits biofilm formation as a mutation in the *wza* gene resulting in increased biofilm formation (Joseph and Wright, 2004). Taken together, *Vibrio* species are capable of producing LPS, CPS, and EPS, and the contribution of each of these polysaccharides to biofilm formation varies between the species.

9.7 Adaptive Responses

As facultative human pathogens with an aquatic life cycle, pathogenic *Vibrio* species encounter fluctuations in physical, chemical, and biological factors in their natural habitats and during the infectious process. The ability of *Vibrio* spp. to sense, respond, and adapt to fluctuations in environmental parameters is critical for both in vivo and ex vivo fitness of these organisms (Faruque et al., 2004). For example, it has been well documented that environmental survival/abundance and distribution of *Vibrio* species are influenced by salinity and temperature. In general, epidemiological analyses indicate that numbers of pathogenic *Vibrio* spp. are most abundant in warm waters of moderate salinity.

9.7.1 Salinity/Osmolarity

The molecular mechanism by which *Vibrio* spp. mediate osmolarity and salinity adaptation is currently emerging. A recent study in *V. cholerae* reported changes in the whole-genome expression pattern upon exposure of this pathogen to varying degrees of osmolarity and salinity stress (Shikuma and Yildiz, 2009). In response to high osmolarity and high-salinity stress, the transcription of genes encoding Na^+/H^+ antiporters and compatible solute biosynthesis/transport genes were increased. Na^+/H^+ antiporters are critical for generating a Na^+ gradient across the cytoplasmic membrane for Na-dependent transport and flagellar motility functions and for Na^+/H^+ homeostasis to prevent Na^+ toxicity and to modulate intracellular pH under alkaline conditions. In *V. cholerae*, three genes encoding Na^+/H^+ antiporters (VC2037, VCA0193, and *nhaA* (VC1627)) exhibited greater expression in cells grown in medium of high NaCl concentrations. Genes encoding Na^+/H^+ antiporters are also present on *V. vulnificus* and *V. parahaemolyticus* genomes. *V. vulnificus* possesses homologous genes for *nhaA* (VV1434), *nhaB* (VV2123), and three predicted *nhaC* (VV1170, VV1347, and VV1181) genes; and *V. parahaemolyticus* possesses *nhaA* (VP1228), *nhaB* (VP2072), six predicted *nhaC* homologs (VP0618, VP0632, VP1134, VP1982, VP2125, and VP2115), and *nhaD* (VPA0051).

Under high osmolarity conditions, to control their internal water activity, microorganisms produce or transport compatible solutes and low molecular weight organic compounds that are compatible with the metabolism. Common compatible solutes are sugars, amino acids, and their derivatives such as betaine, glycinebetaine, and ectoines. In *V. cholerae*, expression of ectoine biosynthesis genes, *ectABC* (VCA0825–VCA0823), and a putative aspartokinase (VCA0822), compatible solute transporters, *opuD* (VC1279) and *putP* (VCA1071) increased in cells grown in high-salinity medium (Shikuma and Yildiz, 2009). These responses are critical to surviving environmental stress. Indeed, *V. cholerae* mutants lacking *ectA* are impaired in their ability to survive in high osmolarity media (Pflughoeft et al., 2003).

Analyses of the *V. parahaemolyticus* genome revealed the presence of ectoine biosynthesis genes, *ectABC*, as well as betaine biosynthesis genes *betABI*, suggesting that this organism has the ability to synthesize two different compatible solutes, ectoine and betaine (Naughton et al., 2009). It is unknown if *V. parahaemolyticus* prefers one compatible solute over the other. In addition, there are genes encoding compatible solute transporters: four genes encoding betaine/carnitine/coline transporters (BCCT) and two genes encoding ProU, which shows a preference for glycine betaine and proline betaine. Mutational and physiological analysis showed that *V. parahaemolyticus* can synthesize ectoine at high salinity and that a *proU1* mutation does not lead to a growth defect, indicating a redundancy in compatible solute transporters (Naughton et al., 2009). In the *V. vulnificus* genome, while betaine biosynthesis genes *betABI* are present, ectoine biosynthesis genes, *ectABC*, are missing (Naughton et al., 2009). With respect to compatible solute transporters, a ProU transporter and a BCCT are predicted to be present. The role of these loci in *V. vulnificus* osmoadaptation has yet to be investigated.

A comparison of genomes of pathogenic *Vibrio* species with respect to "osmoadaptation" genes showed that the number of genes predicted to be involved osmoadaptation is greater in *V. parahaemolyticus* than in the genomes of either *V. cholerae* or *V. vulnificus*. Comparative growth analysis of these species at different salinity and temperature conditions showed that *V. parahaemolyticus* had a growth advantage under all conditions tested, demonstrating the physiological relevance of "osmoadaptation" gene content (Naughton et al., 2009).

Relatively little is known about low-osmolarity and low-salinity stress in *Vibrio* spp. In *V. cholerae* expression of genes involved in biosynthesis of membrane-derived oligosaccharides *mdoG* (VC1288) and *mdoH* (VC1287) and in biosynthesis of spermidine (small organic polycations commonly known as polyamines), *speA* and *speB* (VCA0815 and VCA0814) are induced in cells grown in low-salinity medium (Shikuma and Yildiz, 2009). These genes are also present in genomes of *V. parahaemolyticus* and *V. vulnificus*. The role of MDOs and polyamines in salinity/osmolarity adaptation in *Vibrio* spp. has yet to be examined.

9.7.2 Cold Shock

As facultative human pathogens inhabiting aquatic environments, vibrios are subjected to fluctuations in water temperature. Thus, one important factor for environmental survival and transmission of these pathogens is their ability to exhibit cold shock response and adaptation. It is documented that in response to cold shock all of the *Vibrio* spp. enter into a state of dormancy known as "viable but culturable state," the molecular basis of which is unknown and will not be further discussed (the reader is referred to a review by Oliver, 2005). We will limit our discussion mainly to cold shock proteins.

The cold shock response and adaptation studies in *Escherichia coli* have shown that upon a temperature downshift two classes of genes are induced (Graumann and Marahiel, 1996). The first class, class I, represents genes whose expression is low at 37°C but markedly induced upon cold shock. These include cold shock proteins (CspA, CspB, CspG, and CspI) predicted to function as RNA/DNA chaperones that function at transcriptional and translational levels, proteins with RNA helicase activity (CsdA), a ribosomal-binding factor (RbfA), proteins involved in RNA termination, antitermination (NusA), and a ribonuclease (PNP). Class II proteins are expressed at 37°C but their expression is moderately induced after cold shock. Class II proteins include recombination protein (RecA), a nucleoid-associated protein H-NS, and DNA supercoiling protein DNA gyrase. It is evident from the list of cold shock-induced genes that the cold shock response involves changes in mRNA stability, cold adaptation of ribosomes, and changes in DNA topology (negative supercoiling increases transient upon cold shock). It is important to note that cold shock also induces changes to membrane lipid composition. The proportion of unsaturated fatty acids in membrane lipids increases through cold activation of specific enzymes.

The molecular basis of cold adaptation in *Vibrio* spp. is being investigated. Upon a temperature downshift from 37°C, *V. cholerae* can acclimate and grow at temperatures as low as 15°C, below which viability is lost (Carroll et al., 2001; Datta and Bhadra, 2003). *V. cholerae* does not exhibit adaptive cold shock response, since pre-adaptation of *V. cholerae* to cold stress by first growing at 15°C and then exposing to 5°C does not allow cells to grow at low temperatures (Carroll et al., 2001; Datta and Bhadra, 2003). A survey of the *V. cholerae* genome revealed four genes encoding Csp family of proteins. Three of these genes VCA0166 (*cspA*), VCA0184, and VCA0933 (*cpsV*) are in the small chromosome, while the fourth one, VC1142 (*cspD*), is in the large chromosome. It has been documented that the *V. cholerae* cold shock response involves the induction of *cpsA* and *cpsV* expression; however, the physiological consequences of *cpsA* and *cpsV* mutations in *V. cholerae* cold shock response is yet to be determined (Carroll et al., 2001).

There are five genes encoding Csp family of proteins in *V. vulnificus* genome: VV1 2119 (CspD), VV1 2532, VV1 2757, VV2 0503, and VV2 0519. An analysis of expression of these genes embedded in oysters under shellstock refrigeration conditions revealed that the expression of many of these genes was induced upon cold shock and the response was strain dependent (Limthammahisorn et al., 2009). In the *V. parahaemolyticus* genome, there are four genes encoding the Csp family of proteins VP1012 (CspD), VP1889 (CspA), VPA0552, and VPA1289. However, the protein products and functions encoded by these genes have not been characterized.

Considering the wide range of temperatures in environmental waters, it seems likely that all the pathogenic *Vibrio* species experience cold stress before being transmitted to humans. Understanding the molecular basis of the processes critical for the pathogens' environmental survival may provide clues for how best to reduce their ability to interact with human populations.

9.7.3 Acid Tolerance

One of the first stresses vibrios encounter during infection is exposure to the acidic environment of the stomach. In vitro, bacteria are sensitive even to mildly acidic pH. However, *V. cholerae*, *V. vulnificus*, and *V. parahaemolyticus* are each capable of mounting an acid-tolerant response (ATR), similar to that observed for *Salmonella*. The CadA protein, (a lysine decarboxylase) uses lysine and a proton to produce carbon dioxide and cadaverine in the cytoplasm, thereby mediating an ATR (Merrell and Camilli, 1999). *cadA* is co-transcribed with *cadB*, which encodes an inner membrane proteins responsible for lysine import and cadaverine export. *cadBA* expression is controlled by CadC, a ToxR-like transcriptional regulatory proteins that spans the inner membrane. *cadA* expression is induced during infection in experimental animal models and there is some evidence to suggest that an ATR can contribute to the ability of the bacteria to survive subsequent acid challenges and establish growth in the small intestine more rapidly than bacteria that are not acid adapted (Merrell and Camilli, 1999).

V. vulnificus and *V. parahaemolyticus* also encode CadAB and CadC, and although the ATR is likely to function similarly, there is some evidence that the molecular details may differ (Rhee et al., 2006, 2008; Tanaka et al., 2008). In both *V. cholerae* and *V. vulnificus*, low pH stimulates *cadAB* expression in a CadC-dependent manner, although in *V. vulnificus*, *cadC* expression is dependent on the activity of a LysR family transcriptional regulator, AphB, which functions in *V. cholerae* as a component of virulence gene regulation (Skorupski and Taylor, 1999; Kovacikova et al., 2004; Rhee et al., 2006). In *V. parahaemolyticus*, *cadB* expression is constitutive, but *cadA* expression is acid inducible (Tanaka et al., 2008).

9.7.4 Interaction with Protozoa

Bacterial pathogens in the environment become part of the microbial food web and their abundance can be modulated by protozoa, which are ubiquitous in natural environments. Protozoan grazing was shown to reduce the planktonic *V. cholerae* population in coastal marine environments (Worden et al., 2006). Recent studies showed that *V. cholerae* has the capacity to defend itself against protozoan predation using a novel secretion system designated as type VI secretion (T6SS; Pukatzki et al., 2006). Many of the genes encoding proteins of T6SS apparatus are clustered on the small chromosome virulence-associated secretion (VAS) (VCA0107-VCA123). *V. cholerae* T6SS is required for secretion of three related proteins VgrG-1 (VC1416), VgrG-2 (VCA0018), and VgrG-3 (VCA0123), and Hcp (VCA0017), all lacking N-terminal leader sequences (Pukatzki et al., 2006, 2007). Besides being T6SS effectors, VgrG-2 and Hcp are also necessary for secretion of other T6SS substrates, indicating that they are part of the T6SS apparatus. The molecular basis of formation of T6SS apparatus is emerging. Vrg proteins exhibit structural similarities to trimeric tail-spike proteins of bacteriophage T4, suggesting that the T6SS structural apparatus may be similar (Pukatzki et al., 2007). The Vrg proteins could form a complex that is involved in puncturing host cell membranes and/or constitute part of a channel that functions to transport effector proteins out of the cell. Besides the Vrg proteins, Hcp and ClpB proteins are also predicted to form oligomers with a central channel. Precisely how components of the T6SS interact and form a translocation apparatus spanning the cytoplasmic membrane, periplasm, and outer membrane remains unknown.

The lack of T6SS genes resulted in loss of cytotoxicity toward amoebae and macrophages (Pukatzki et al., 2006, 2007; Ma et al., 2009); and recent studies have provided an insight into mechanisms by which effectors of T6SS function. One of the *V. cholerae* T6SS effector proteins VgrG-1 harbors a C-terminal actin crosslinking domain and its activity is required for cytotoxicity (Pukatzki et al., 2007). Interestingly, T6SS appears to be activated after *V. cholerae* are taken up by endocytosis in response to an unknown stimulus (Ma et al., 2009). Following endocytosis, the effector protein VgrG-1 is transferred into the cytosol, where it can covalently crosslink host actin and, hence, prevent actin polymerization and phagocytosis (Ma et al., 2009). It is speculated that this process ensures survival of the pathogen population. Both *V. vulnificus* and *V. parahaemolyticus* harbor putative T6SS genes (Boyd et al., 2008; Shrivastava and Mande, 2008); however, a more complete understanding of the roles of T6SS in the biology of *Vibrio* spp. awaits further investigation.

9.7.5 Iron Acquisition

Iron is an essential element required for survival of *Vibrio* spp. both in aquatic ecosystems and in the host environment. *Vibrio* species produce siderophores (high-affinity, low molecular weight, iron-binding compounds) under iron deplete conditions. *V. cholerae* has the capacity to synthesize the catechol siderophore vibriobactin (Wyckoff et al., 2007). Accordingly, the genes required for vibriobactin biosynthesis (VC0771–VC0775) and transport (VC0776–VC0779) and receptors (VC2211) are found on the large chromosome of *V. cholerae*. Mutants unable to produce or transport vibriobactin exhibit a small defect in host growth and colonization, implying that either vibriobactin plays a modest role during in vivo iron acquisition, other iron transport systems play a more significant role, or that there is functional redundancy (Sigel et al., 1985).

In addition to vibriobactin genes, the presence of *irgA* (VC0475) and *vctA* (VCA0232) encoding enterobactin receptors indicates that *V. cholerae* takes advantage of siderophores produced by other organisms. Indeed, *V. cholerae* can transport enterobactin, and an *irgA* and *vctA* double mutant lacks the capacity to transport enterobactin (Mey et al., 2002). Mutation in *irgA*, however, did not result in attenuation for virulence (Mey et al., 2002). It is suggested that such a capacity could be important for the environmental persistence of *V. cholerae*, as this pathogen most likely encounters enterobactin-producing organisms in sewage-contaminated waters. In addition to enterobactin, *V. cholerae* appears to transport agrobactin and fluvibactin, most likely via ViuA (VC2211), a vibriobactin transporter (Wyckoff et al., 2007).

V. vulnificus produces two siderophores for iron acquisition: a catechol and a hydroxymate siderophore (Simpson and Oliver, 1983). Several genes predicted to be involved in catechol siderophore vulnibactin biosynthesis (VVA1294–VVA1310) in *V. vulnificus* and the vulnibactin receptor is encoded by *vuuA* (Webster and Litwin, 2000). Vulnibactin production is critical for virulence (Litwin et al., 1996), but the role of the second hydroxymate siderophore in iron uptake remains to be studied. *V. vulnificus* can also utilize the iron from hydroxymate siderophores (Kim et al., 2007) and aerobactin produced by other organisms for growth (Tanabe et al., 2005).

V. parahaemolyticus produces a new class of siderophore, vibrioferrin (Yamamoto et al., 1994). The genes required for vibrioferrin biosynthesis (*pvs*ABCDE *V. parahaemolyticus* vibrioferrin synthesis) and transport (*pvu*BCDE *V. parahaemolyticus* vibrioferrin utilization) have been identified (Tanabe et al., 2003, 2006). Similar to *V. vulnificus*, *V. parahaemolyticus* can also utilize deferoxamine and aerobactin for growth (Funahashi et al., 2009).

Heme is another potential iron source for pathogenic microorganisms. There are three heme transporters, *hutA* (VCA0576), *hutR* (VCA0064), and *hasR* (VCA0625), in the *V. cholerae* genome. HutA can transport both heme and hemoglobin, and a mutant lacking *hutA* exhibits a marked reduction in hemoglobin utilization capacity. A mutant lacking all three transporters is unable to utilize heme as an iron source (Mey and Payne, 2001). *V. vulnificus* was shown to utilize non-transferrin-bound iron through the heme receptor HupA, encoded by *hupA* (VVA0781).

Besides outer membrane receptors, the transport of siderophores or heme-bound iron across the outer membrane requires energy-transducing TonB–ExbBD complexes. In *V. cholerae*, two TonB–ExbBD systems, TonB–ExbBD 1 and TonB2–ExbBD, function together with vibriobactin, enterobactin, and heme outer membrane receptors for transport of iron (Mey and Payne, 2003; Wyckoff et al., 2007). TonB1 and TonB2 have specific and redundant functions in iron transport. While the vibriobactin receptor and heme receptors HutA and HutR function with both Ton systems, enterobactin receptor and heme receptor HasR use the TonB2 system. Mutants unable to produce both TonB systems have reduced colonization capacity, indicating that TonB-dependent iron transport is utilized for in vivo iron acquisition. Three different TonB systems are predicted to be encoded by *V. vulnificus*, although the specific relationship between the different TonB systems and the vulnibactin receptor has not been determined.

To acquire iron, *Vibrio* spp. also utilizes TonB-independent iron transporters. In *V. cholerae* these include the *fbpABC* (VC0608–VC0610) ABC transporter involved in ferric iron transport and *feoABC* involved in ferrous iron transport. A *vibB*, *feoB*, and *fbpA* triple mutant, which is unable to produce vibriobactin or transport ferrous and ferric iron, is still able to grow in the presence of the iron chelator EDDA. The results implies the presence of a yet to be identified high-affinity iron transport system (Wyckoff et al., 2006). Genome sequence analysis revealed that there are two additional loci predicted to encode ferric iron ABC transporters: VCA0685–VCA0687 and VCA0601–VCA0603. However, initial studies performed in *Shigella flexneri* do not support this annotation (Wyckoff et al., 2006). Similarly, there are multiple loci on the *V. vulnificus* and *V. parahaemolyticus* genomes that are predicted to be involved in ferrous and ferric iron transport. Altogether, these results and analyses show that *Vibrio* species possess multiple means of iron acquisition, further confirming the importance of iron for growth and survival of these pathogens.

9.8 Pathogenic Determinants

9.8.1 Adherence and Colonization

V. cholerae encodes the toxin co-regulated pilus (TCP), a type IV bundle forming pilus that is essential for colonization in the murine model and also necessary for human colonization (Taylor et al., 1987; Rhine and Taylor, 1994). TcpA also serves as the receptor for the cholera toxin phage and has been shown to be important for adherence to chitin (Waldor and Mekalanos, 1996; Reguera and Kolter, 2005).

In *V. cholerae*, the accessory colonization factors (*acfA-C*) were identified as co-regulated genes within the TCP island and their deletion results in a colonization defect in the infant mouse model (Peterson and Mekalanos, 1988). Although the human receptor for TCP has not yet been identified, antibody to TcpA was identified in the sera of convalescent patients using an in vivo-induced antigen screening technology (IVIAT) (Hang et al., 2003). The study also identified PilA as an in vivo-induced, immunogenic protein. PilA (ChiRP) is typeIV pilin having homology to pili involved in twitching motility, although this phenotype has not yet been reported for *V. cholerae*. PilA has been reported to play a role in chitin-induced competence, as mentioned earlier.

The role of other *V. cholerae* pili, such as the mannose-sensitive hemagglutinin (MSHA), has also been recognized as multifunctional (Thelin and Taylor, 1996; Tacket et al., 1998; Watnick et al., 1999). MSHA is essential for biofilm formation and importantly, its expression has been shown to be downregulated during infection (Hsiao et al., 2009). MSHA has been dubbed the anti-colonization factor, since constitutive MSHA expression renders strains attenuated for colonization in the infant mouse model (Hsiao et al., 2006). This finding is consistent with the hypothesis that downregulation of certain genes is as critical as upregulation of virulence factors during the infectious process (Hsiao et al., 2006).

In vitro studies using *V. parahaemolyticus* and cell lines or animal model intestinal tracts have provided a survey of conditions and adhesive properties for the organism (Nakasone and Iwanaga, 1990; Falcioni et al., 2005; Ferreira et al., 2008). MSHA as well as pili with different carbohydrate-binding specificities were identified as candidates that promote adherence (Nasu et al., 2000; Yeung et al., 2002).

In *V. vulnificus*, few specific adhesions or pili have been associated with human colonization, although it is generally accepted that contact with host cells is necessary for cytotoxicity and pathogenesis (Jones and Oliver, 2009). Mutations in *pilA* result in a 1–2 log increase in LD50 and decreased attachment to epithelial cells. In addition, *pilA* mutants also have decreased persistent association with oysters, suggesting that as for *V. cholerae*, PilA plays a role both in human colonization/disease and in the aquatic environment (Paranjpye and Strom, 2005). IVIAT studies performed using sera from *V. vulnificus*-infected patients did not identify proteins with obvious similarities to known colonization factors, but did identify proteins involved in signaling, metabolism, and secretion (Kim et al., 2003).

9.8.2 Cholera Toxin

Cholera toxin (CT) is a classic AB structure toxin and is unique to *V. cholerae*. The enzymatic moiety has an ADP-ribosylating activity that targets the stimulatory subunit of G alpha proteins that comprise the adenylate cyclase complex. The sustained activation of adenylate cyclase promotes increased cAMP levels, which in turn promote chloride secretion through the CFTR, accompanied by fluid loss. Cholera toxin is nearly entirely responsible for the severe osmotic diarrhea characteristic of the disease (Mekalanos et al., 1983). The *ctxAB* genes are carried on the filamentous phage CTXφ, which uses the toxin co-regulated pilus (TCP) as its receptor to enter cells and establish lysogeny (Waldor and Mekalanos, 1996). A single copy of the toxin genes and associated phage sequences are found on chromosome I of strain N16961, flanked by 17-bp *attRS1* integration sites. Duplicate copies of the *ctxAB* genes lie within the genome of classical O1 isolates, one on each chromosome, consistent with data showing that classical strains can produce more toxin than El Tor strains (Mekalanos, 1983; Mekalanos et al., 1983). The enzymatic function of CT, its retrograde transport through the host cell, and its transcriptional control by ToxR regulon proteins have been well studied and have served as a paradigm for understanding toxin transport and virulence gene regulation (Sanchez and Holmgren, 2008).

9.8.3 Type III Secretion Systems

Type III secretion systems (T3SSs) are found in a wide variety of pathogenic bacteria and are a contact-dependent secretion apparatus that translocates bacterial virulence proteins directly into eukaryotic host cells. Although previously associated with many other pathogenic bacteria, T3SSs were first identified in *Vibrio* spp. when the genomic sequence of *V. parahaemolyticus* revealed the presence of two T3SSs: T3SS1, found on the large chromosome and having a genetic organization similar to the T3SS of *Yersinia* spp., and T3SS2, located on the small chromosome but unrelated to any known T3SSs (Makino et al., 2003). Initial surveys of strains indicate that the T3SS1 genes are ubiquitously present in all strains of *V. parahaemolyticus*, as well as strains of *V. alginolyticus* and the mollusk pathogen, *V. tubiashii*. In contrast, T3SS2 is found almost exclusively in pathogenic *V. parahaemolyticus* strains that are also positive for the Kanagawa phenomenon (encoded by the *tdh* gene, described below), although exceptions have been identified (Meador et al., 2007; Nair et al., 2007). Although T3SS1 is found in all *V. parahaemolyticus* strains, mutational analyses have shown that T3SS1 genes are associated with a cytotoxic activity when cultured in vitro with HeLa cells (Park et al., 2004). The T3SS1 effector VopS (VP1686) was shown to promote cell rounding by covalently attaching an adenosine 5′-monophosphate (AMP) moiety (termed AMPylation) to a conserved threonine residue on Rho, Rac, and Cdc42 proteins (Casselli et al., 2008; Yarbrough et al., 2009). The interaction of Rho GTPases with downstream effectors is affected by the AMPylation, thereby disrupting actin assembly in the host cell.

A T3SS island, similar to the T3SS2 of *V. parahaemolyticus*, was identified in a non-O1/non-O139 clinical isolate by genomic sequence analysis (Dziejman et al., 2005). Southern analysis and genomic sequence analysis of additional strains identified identical, or divergent but related, T3SSs in other non-O1/non-O139 isolates (Dziejman et al., 2005; Chen et al., 2007). Recent work describes the activity of one *V. cholerae* effector protein, VopF, which has amino acid similarity to the *V. parahaemolyticus* T3SS2 VopL effector (Liverman et al., 2007; Tam et al., 2007). VopF contains both formin and WH2 domains and can bind and polymerize actin in vitro. VopL also functions in actin assembly, although the mechanisms and functional protein domains differ. The identification of additional effector proteins, their combined effect on host cell signaling and physiology, and their collective effort resulting in T3SS2-mediated pathogenesis remain active areas of investigation.

In *V. parahaemolyticus*, deletion of genes comprising T3SS2 decreased the enterotoxicity of bacterial cells in a rabbit model of fluid accumulation (Park et al., 2004). Earlier reports suggested that *V. parahaemolyticus* strains could invade mammalian cells in vitro, and an O3:K6 serotype strain possesses an activity that alters host cell cytoskeletal organization and is invasive when co-cultured with Caco-2 cells (Akeda et al., 1997, 2002). Similar observations have been reported for non-O1/non-O139 strains, but prior to the identification of the T3SS island. The bacterial factors responsible for promoting invasion and/or aspects of possible T3SS involvement in invasion have not yet been identified (Panigrahi et al., 1990; Russell et al., 1992; Dalsgaard et al., 1995).

As mentioned earlier, strain-specific genome sequence analysis has revealed that both *V. parahaemolyticus* and *V. cholerae* strains can encode T3SS variants. A T3SS variant was identified in *trh*-positive *V. parahaemolyticus* strains (producing a thermostable-related hemolysin), indicating that at least two lineages contribute to the T3SS positive-phenotype associated with *V. parahaemolyticus* (Okada et al., 2009). T3SS2 association with pandemic stains has also been addressed, and screens performed using a wide panel of strains suggest that T3SS2 is occasionally found in non-pandemic strains (Meador et al., 2007). Differences have also been observed for *V. cholerae* strains and indicate that the T3SS mechanism has been diversified in *V. cholerae* as well (Chen et al., 2007).

9.8.4 Rtx Toxins

Originally named RtxA, the repeat-in-toxin gene of *V. cholerae* was initially associated with an actin-binding, cell rounding activity during in vitro co-culture with mammalian cells (Fullner and Mekalanos, 1999, 2000). Genomic sequence annotation confirmed that the *rtxA* gene is the largest ORF encoded by the N16961 genome (13,677 bp) and identified a related protein that is encoded by *V. vulnificus* (Heidelberg et al., 2000; Chen et al., 2003). Now referred to as multifunctional autoprocessing RTX toxin (MARTX), the *Vibrio* proteins are recognized not only as possessing sequences similar to those found in other RTX toxins, but also as displaying unique features that distinguish them from the RTX toxins of other species. For example, the *V. cholerae* MARTX causes cell rounding via actin depolymerization rather than pore formation (reviewed in Satchell, 2007).

In both *V. cholerae* and *V. vulnificus*, the genes that encode the MARTX toxins are organized within a cluster such that *rtxA* (encoding the MARTX protein) is transcribed in a single operon along with genes encoding functions necessary for toxin processing and export. In *V. cholerae*, the RTX gene cluster (VC1446–VC1451) is linked to the cholera toxin gene cluster, which includes the CTX structural genes and bacteriophage elements, TLC and *rstR* genes. In *V. vulnificus* strain YJ016, the homologous RTX gene cluster (VVA1030, VVA1032, VVA1034, VVA1035, VVA1036) is found on the small chromosome. *V. vulnificus* also appears to encode paralogous MARTX proteins, one on the large chromosome (VV1546) and another found on the small chromosome (VVA0331). An RTX gene cluster has not been identified in *V. parahaemolyticus*, although an *rtxA* gene that encodes an activity promoting virulence in a juvenile Atlantic salmon model was recently identified in the fish pathogen *V. anguillarum* (Lee et al., 2008; Li et al., 2008).

As its name suggests, the *V. cholerae* MARTEX protein is multifunctional. In addition to possessing three domains of characteristic glycine-rich repeat sequences (two at the amino terminus and one domain at the carboxy terminus), RtxA encodes an actin crosslinking domain, a domain promoting RhoGTPase inactivation, an uncharacterized domain that is similar to members of the alpha–beta hydrolase protein family, and an autocatalytic cysteine protease domain (CPD) whose activity is essential to the toxic action of MARTX. The *V. vulnificus* MARTEX protein is composed of five distinct "activity" domains, which possess structural and/or sequence similarities to domains found in a diverse array of bacterial toxins. A domain associated with actin crosslinking is absent, but present are domains associated with Rho GTPase inactivation, the alpha–beta hydrolase motif, and three regions encoding sequences similar to toxins from *Xenorhabus*, *Pasteurella*, and *Photorhabdus*. Both *V. vulnificus* and *V. cholerae* MARTEX proteins carry domains of unknown function as well (Satchell, 2007).

Deletion of *V. cholerae* MARTEX does not result in a detectable phenotype in the infant mouse model, but does decrease pathogenicity in a streptomycin-treated adult mouse model of infection (Olivier et al., 2007). The *V. vulnificus* RtxA was shown to contribute to virulence in the mouse model, and RtxA was reported to induce apoptotic death through a mitochondria-dependent pathway in human intestinal epithelial cells exposed to *V. vulnificus* (Lee et al., 2007, 2008).

9.8.5 Hemolysins

All three *Vibrio* species encode numerous genes whose products are annotated as hemolysins. Annotation of the *V. cholerae* genome identified multiple putative proteases genes and at least seven ORFs whose products are predicted to encode hemolysins, but the true functions of the proteins are not well defined (Heidelberg et al., 2000). Other hemolysins have received more attention. The HlyA hemolysin has enterotoxic activity, whereas the *hap* gene product is a metalloprotease that disrupts intestinal epithelial cell tight junctions (Alm et al., 1988; Wu et al., 1996; Coelho et al., 2000; Mitra et al., 2000). Both *hlyA* and *hapA* are found on chromosome II in *V. cholerae*. Whether these predicted hemolysins function to disrupt cellular membranes in the human host or have alternative functions important for the aquatic lifestyle or association with marine organisms remains to be determined. N16961 does not encode ST, a heat stable toxin that is produced by some pathogenic TCP-CT- non-O1/non-O139 strains (Pal et al., 1992; Dalsgaard et al., 1995; Heidelberg et al., 2000; Rahman et al., 2008).

In *V. vulnificus*, a number of extracellular enzymes and toxins, including metalloprotease, phospholipase, and cytolysin, have been implicated in causing tissue destruction and subsequent bacterial invasion into the blood stream (Strom and Paranjpye, 2000; Jones and Oliver, 2009). A cytolysin gene, *vvhA* (VVA0965), is present on the small chromosome and appears to be unique. A metalloprotease gene, *vvp* (VVA1465), and a phospholipase gene, *vpl* (VVA0303), both of which are highly homologous to those in *V. cholerae*, are located on the small chromosome (Chen et al., 2003). Although purified metalloprotease and cytolysin produced pathologic effects in multiple animal models (Strom and Paranjpye, 2000; Jones and Oliver, 2009), a mutant deficient in both factors is only slightly reduced in its virulence in mice (Fan et al., 2001). It seems likely that still unidentified cytotoxin(s) contribute significantly to the pathogenesis of *V. vulnificus*. Indeed, more than 10 homologs of known cytotoxins and hemolysins were identified in the genome sequence of *V. vulnificus* (Chen et al., 2003).

9.8.6 Thermostable Direct Hemolysin (TDH)

Pathogenic *V. parahaemolyticus* strains have typically been differentiated from non-pathogenic strains based on the ability to cause hemolysis on a specialized blood agar media known as Wagatsuma agar (Miyamoto et al., 1969). This activity is referred to as the Kanagawa phenomenon and is mediated by a thermostable direct hemolysin (TDH) that is encoded by the *tdhA* (*tdh2*) gene (Nishibuchi and Kaper, 1995). Genomic sequence analysis revealed that the *tdhA* gene and the related *tdhS* (*tdh1*) gene lie within TTSS2 (Makino et al., 2003). Previous studies identified five variants of the *tdh* sequence, which share >97% identity at the amino acid level, encode similar functions, and can be present in varying combination in different strains (Nakaguchi et al., 2004). The transcription of *tdh2* is significantly higher than that of the others and therefore, TDH activity is usually the result of only *tdh2* expression, even if strains carry multiple *tdh* genes (Nakaguchi and Nishibuchi, 2005). Some pathogenic strains carry *trh* genes that encode a thermostable-related hemolysin (TRH), and these nucleotide sequences are ~65% identical to the *tdh2* gene (Nishibuchi et al., 1989; Kishishita et al., 1992). While the biological activities of TRH are related to those observed for TDH, it appears that *trh* expression is much lower than that of *tdh2*.

TDH is thought to function as a major virulence factor. Several studies suggest that its activity mediates an increase in intracellular calcium levels and may affect PKC phosphorylation (either directly or indirectly), resulting in calcium-mediated fluid secretion by intestinal epithelial cells (Raimondi et al., 1995; Takahashi et al., 2000a, b; Takahashi et al., 2001). Both the *tdh* and *trh* genes have been considered a marker for *V. parahaemolyticus* pathogenic strains. A strain deleted for *tdh* showed decreased enterotoxicity in a rabbit model, but conflicting results have been reported regarding the ability of similar deletions to alter the cytotoxicity of HeLa cells (Nishibuchi et al., 1992; Park et al., 2004). As discussed earlier, it is generally thought that proteins in addition to TDH contribute to *V. parahaemolyticus* virulence.

The T3SS island of non-O1/non-O139 *V. cholerae* strains carries a single *tdh* gene, although it is sufficiently divergent from the *V. parahaemolyticus tdh* gene to question whether it is more similar to the TRH that is produced by some *V. para-haemolyticus* strains (Dziejman et al., 2005). Reports of *V. cholerae* carrying a plasmid or insertion sequence mediated *tdh* gene, (and expressing a related activity) further suggest that the gene can be mobilized and distributed among strains. The role of the TDH in *V. cholerae* virulence has not been fully investigated (Honda et al., 1986; Baba et al., 1991; Terai et al., 1991). No homologue has yet been identified in *V. vulnificus*.

9.8.7 Capsule

The ability of microorganisms to evade phagocytosis and serum killing is commonly attributed to the presence of a capsule, and this is the case for *V. vulnificus* as well. However, it has been suggested that the capsule may also contribute to the highly invasive behavior, although *V. vulnificus* has not been shown to be invasive in mammalian cell culture. As stated earlier, the capsule of different *V. vulnificus* strains is heterogeneous among clinical and environmental isolates, and minor genetic variations in the metabolic pathway for CPS biosynthesis suggest a genetic basis for antigenic diversity displayed by the CPS. CPS is therefore recognized as an important contributor to *V. vulnificus* virulence (Wright et al., 1999; Jones and Oliver, 2009).

V. parahaemolyticus can produce more than 70 different K or capsular antigens, and recent studies suggest that capsule may play roles in adherence to target cells, as well as the formation of biofilms (mentioned earlier) (Guvener and McCarter, 2003; Hsieh et al., 2003). The capsule of the O139 serogroup of *V. cholerae* has been studied for both its importance during infection and its role in the environmental niche has recently been addressed (Waldor et al., 1994; Sengupta et al., 1996; Abd et al., 2009). In addition, some non-O1/non-O139 strains are also reported to synthesize capsule (Johnson et al., 1992; Chen et al., 2007).

9.8.8 Summary: Molecular Mechanisms of Gastroenteritis

How do the diverse virulence factors in the three *Vibrio* species function to dysregulate the homeostasis of human intestinal epithelial cells, resulting in clinically similar or related manifestations of diarrheal disease? The cAMP-mediated mechanism of action for cholera toxin is well understood to increase electrolyte and water transport through the CFTR channel, and *V. parahaemolyticus* TDH is thought to elicit its toxic effects via the secondary messenger calcium. However, the identification and elucidation of the functions of the effector proteins carried on the T3SS island(s) represent new and exciting areas of investigation for both *V. parahaemolyticus* and the non-O1/non-O139 *V. cholerae* strains that employ this conserved mechanism of virulence. The mechanisms of *V. vulnificus*-associated gastroenteritis (and disease, in general) are also less clear, and *V. vulnificus* has been an especially difficult *Vibrio* to study, in part due to the difficulty in translating in vivo phenotypes to in vitro model systems. For example, it is well established that invasion is an important attribute contributing to *V. vulnificus* virulence and the rapid progression to septicemia, yet genomic sequence analysis has not yet identified a protein whose function promotes an invasive phenotype. Nonetheless, recent efforts have increased our understanding of the diverse genetic repertoire carried by this species and the importance of the host component in infection. For all three *Vibrio* species, our collective view of the supragenome has been significantly expanded by genomic advances, not only widening our understanding of the molecular mechanisms used by pathogenic *Vibrios* to cause disease, but also illustrating the diversity of targets available for development of diagnostics, therapeutics, and vaccine components.

Acknowledgments We would like to acknowledge the work of the many *Vibrio* labs that contributed to the information provided in this chapter. We have cited original publications when possible and referred the reader to several excellent review articles when appropriate, as they provide in depth information on specific topics that were beyond the scope of this chapter. Work in our labs is supported by grants from the NIH R01AI055987 to FHY and R01AI073785 to MD. We thank members of our laboratories for their valuable comments on the manuscript and especially Elaine Hamilton (MD) for assistance in preparing the tables.

References

Abd H, Saeed A, Weintraub A, Sandstrom G (2009) *Vibrio cholerae* O139 requires neither capsule nor LPS O side chain to grow inside Acanthamoeba castellanii. J Med Microbiol 58(1):125–131

Aiyar SE, Gaal T, Gourse RL (2002) rRNA promoter activity in the fast-growing bacterium *Vibrio natriegens*. J Bacteriol 184(5):1349–1358

Akeda Y, Kodama T, Kashimoto T, Cantarelli V, Horiguchi Y, Nagayama K, Iida T, Honda T (2002) Dominant-negative Rho, Rac, and Cdc42 facilitate the invasion process of *Vibrio parahaemolyticus* into Caco-2 cells. Infect Immun 70(2):970–973

Akeda Y, Nagayama K, Yamamoto K, Honda T (1997) Invasive phenotype of *Vibrio parahaemolyticus*. J Infect Dis 176(3):822–824

Alm RA, Stroeher UH, Manning PA (1988) Extracellular proteins of *Vibrio cholerae*: nucleotide sequence of the structural gene (*hlyA*) for the haemolysin of the haemolytic El Tor strain O17 and characterization of the *hlyA* mutation in the non-heamolytic classical strain 569B. Mol Microbiol 2:481–488

Baba K, Shirai H, Terai A, Kumagai K, Takeda Y, Nishibuchi M (1991) Similarity of the tdh gene-bearing plasmids of *Vibrio cholerae* non-O1 and *Vibrio parahaemolyticus*. Microb Pathog 10(1):61–70

Barua D, Greenough WB (1992) Cholera. Plenum Medical Book Co, New York, NY

Bisharat N, Amaro C, Fouz B, Llorens A, Cohen DI (2007) Serological and molecular characteristics of *Vibrio vulnificus* biotype 3: evidence for high clonality. Microbiology 153(Pt 3):847–856

Bisharat N, Cohen DI, Harding RM, Falush D, Crook DW, Peto T, Maiden MC (2005) Hybrid *Vibrio vulnificus*. Emerg Infect Dis 11(1):30–35

Bisharat N, Cohen DI, Maiden MC, Crook DW, Peto T, Harding RM (2007) The evolution of genetic structure in the marine pathogen, *Vibrio vulnificus*. Infect Genet Evol 7(6): 685–693

Blake PA, Merson MH, Weaver RE, Hollis DG, Heublein PC (1979) Disease caused by a marine *Vibrio*. Clinical characteristics and epidemiology. N Engl J Med 300(1):1–5

Blokesch M, Schoolnik GK (2007) Serogroup conversion of *Vibrio cholerae* in aquatic reservoirs. PLoS Pathog 3(6):e81

Boyd EF, Cohen AL, Naughton LM, Ussery DW, Binnewies TT, Stine OC, Parent MA (2008) Molecular analysis of the emergence of pandemic *Vibrio parahaemolyticus*. BMC Microbiol 8:110

Bush CA, Patel P, Gunawardena S, Powell J, Joseph A, Johnson JA, Morris JG (1997) Classification of *Vibrio vulnificus* strains by the carbohydrate composition of their capsular polysaccharides. Anal Biochem 250(2):186–195

Butler SM, Camilli A (2005) Going against the grain: chemotaxis and infection in *Vibrio cholerae*. Nat Rev Microbiol 3(8):611–620

Cameron DE, Urbach JM, Mekalanos JJ (2008) A defined transposon mutant library and its use in identifying motility genes in *Vibrio cholerae*. Proc Natl Acad Sci USA 105(25):8736–8741

Carroll JW, Mateescu MC, Chava K, Colwell RR, Bej AK (2001) Response and tolerance of toxigenic *Vibrio cholerae* O1 to cold temperatures. Antonie Van Leeuwenhoek 79(3–4): 377–384

Casselli T, Lynch T, Southward CM, Jones BW, DeVinney R (2008) *Vibrio parahaemolyticus* inhibition of Rho family GTPase activation requires a functional chromosome I type III secretion system. Infect Immun 76(5):2202–2211

Chatterjee SN, Chaudhuri K (2003) Lipopolysaccharides of *Vibrio cholerae*. I. Physical and chemical characterization. Biochim Biophys Acta 1639(2):65–79

Chatzidaki-Livanis M, Jones MK, Wright AC (2006) Genetic variation in the *Vibrio vulnificus* group 1 capsular polysaccharide operon. J Bacteriol 188(5):1987–1998

Chen Y, Bystricky P, Adeyeye J, Panigrahi P, Ali A, Johnson JA, Bush CA, Morris JG Jr, Stine OC (2007) The capsule polysaccharide structure and biogenesis for non-O1 *Vibrio cholerae* NRT36S: genes are embedded in the LPS region. BMC Microbiol 7:20

Chen Y, Johnson JA, Pusch GD, Morris JG Jr, Stine OC (2007) The genome of non-O1 *Vibrio cholerae* NRT36S demonstrates the presence of pathogenic mechanisms that are distinct from those of O1 *Vibrio cholerae*. Infect Immun 75(5):2645–2647

Chen C-Y, Wu K-M, Chang Y-C, Chang C-H, Tsai H-C, Liao T-L, Liu Y-M, Chen H-J, Shen AB-T, Li J-C, Su T-L, Shao C-P, Lee C-T, Hor L-I, Tsai S-F (2003) Comparative Genome analysis of *Vibrio vulnificus*, a marine pathogen. Genome Res 13(12):2577

CholeraWorkingGroup (1993) Large epidemic of cholera-like disease in Bangladesh caused by *Vibrio cholerae* O139 synonym Bengal. Lancet 342:387–390

Coelho A, Andrade JR, Vicente AC, Dirita VJ (2000) Cytotoxic cell vacuolating activity from *Vibrio cholerae* hemolysin. Infect Immun 68(3):1700–1705

Cohen AL, Oliver JD, DePaola A, Feil EJ, Boyd EF (2007) Emergence of a virulent clade of *Vibrio vulnificus* and correlation with the presence of a 33-kilobase genomic island. Appl Environ Microbiol 73(17):5553–5565

Comeau AM, Chan AM, Suttle CA (2006) Genetic richness of vibriophages isolated in a coastal environment. Environ Microbiol 8(7):1164–1176

Dalsgaard A, Albert MJ, Taylor DN, Shimada T, Meza R, Serichantalergs O, Echeverria P (1995) Characterization of *Vibrio cholerae* non-O1 serogroups obtained from an outbreak of diarrhea in Lima, Peru. J Clin Microbiol 33(10):2715–2722

Dalsgaard A, Serichantalergs O, Shimada T, Sethabutr O, Echeverria P (1995) Prevalence of *Vibrio cholerae* with heat-stable enterotoxin (NAG-ST) and cholera toxin genes; restriction fragment length polymorphisms of NAG-ST genes among *V. cholerae* O serogroups from a major shrimp production area in Thailand. J Med Microbiol 43(3):216–220

Daniels NA, MacKinnon L, Bishop R, Altekruse S, Ray B, Hammond RM, Thompson S, Wilson S, Bean NH, Griffin PM, Slutsker L (2000) *Vibrio parahaemolyticus* infections in the United States, 1973–1998. J Infect Dis 181(5):1661–1666

Datta PP, Bhadra RK (2003) Cold shock response and major cold shock proteins of *Vibrio cholerae*. Appl Environ Microbiol 69(11):6361–6369

Davis BM, Moyer KE, Boyd EF, Waldor MK (2000) CTX prophages in classical biotype *Vibrio cholerae*: functional phage genes but dysfunctional phage genomes. J Bacteriol 182(24): 6992–6998

DePaola A, Motes ML, Chan AM, Suttle CA (1998) Phages infecting *Vibrio vulnificus* are abundant and diverse in oysters (*Crassostrea virginica*) collected from the Gulf of Mexico. Appl Environ Microbiol 64(1):346–351

Dryselius R, Izutsu K, Honda T, Iida T (2008) Differential replication dynamics for large and small *Vibrio* chromosomes affect gene dosage, expression and location. BMC Genomics 9:559

Dziejman M, Balon E, Boyd D, Fraser CM, Heidelberg JF, Mekalanos JJ (2002) Comparative genomic analysis of *Vibrio cholerae*: genes that correlate with cholera endemic and pandemic disease. Proc Nat Acad Sci USA 99(3):1556–1561

Dziejman M, Serruto D, Tam VC, Sturtevant D, Diraphat P, Faruque SM, Rahman MH, Heidelberg JF, Decker J, Li L, Montgomery KT, Grills G, Kucherlapati R, Mekalanos JJ (2005) Genomic characterization of non-O1, non-O139 *Vibrio cholerae* reveals genes for a type III secretion system. Proc Natl Acad Sci USA 102(9):3465–3470

Egan ES, Fogel MA, Waldor MK (2005) Divided genomes: negotiating the cell cycle in prokaryotes with multiple chromosomes. Mol Microbiol 56(5):1129

Egan ES, Lobner-Olesen A, Waldor MK (2004) Synchronous replication initiation of the two *Vibrio cholerae* chromosomes. Curr Biol 14(13):R501–502

Egan ES, Waldor MK (2003) Distinct replication requirements for the two *Vibrio cholerae* chromosomes. Cell 114(4):521–530

Enos-Berlage JL, Guvener ZT, Keenan CE, McCarter LL (2005) Genetic determinants of biofilm development of opaque and translucent *Vibrio parahaemolyticus*. Mol Microbiol 55(4): 1160–1182

Enos-Berlage JL, McCarter LL (2000) Relation of capsular polysaccharide production and colonial cell organization to colony morphology in *Vibrio parahaemolyticus*. J Bacteriol 182(19): 5513–5520

Falcioni T, Papa S, Campana R, Mannello F, Casaroli A, Burattini S, Baffone W (2005) Flow cytometric evaluation of *Vibrio parahaemolyticus* adhesion inhibition to human epithelial cells. Cytometry B Clin Cytom 66B(1):25–35

Fan JJ, Shao CP, Ho YC, Yu CK, Hor LI (2001) Isolation and characterization of a *Vibrio vulnificus* mutant deficient in both extracellular metalloprotease and cytolysin. Infect Immun 69(9): 5943–5948

Faruque SM, Islam MJ, Ahmad QS, Faruque AS, Sack DA, Nair GB, Mekalanos JJ (2005a) Self-limiting nature of seasonal cholera epidemics: role of host-mediated amplification of phage. Proc Natl Acad Sci USA 102(17):6119–6124

Faruque SM, Mekalanos JJ (2003) Pathogenicity islands and phages in *Vibrio cholerae* evolution. Trends Microbiol 11(11):505–510

Faruque SM, Nair GB, Mekalanos JJ (2004) Genetics of stress adaptation and virulence in toxigenic *Vibrio cholerae*. DNA Cell Biol 23(11):723–741

Faruque SM, Naser IB, Islam MJ, Faruque AS, Ghosh AN, Nair GB, Sack DA, Mekalanos JJ (2005b) Seasonal epidemics of cholera inversely correlate with the prevalence of environmental cholera phages. Proc Natl Acad Sci USA 102(5):1702–1707

Faruque SM, Sack DA, Sack RB, Colwell RR, Takeda Y, Nair GB (2003) Emergence and evolution of *Vibrio cholerae* O139. Proc Natl Acad Sci USA 100(3):1304–1309

Ferreira RBR, Antunes LCM, Greenberg EP, McCarter LL (2008) *Vibrio parahaemolyticus* ScrC modulates cyclic dimeric GMP regulation of gene expression relevant to growth on surfaces. J Bacteriol 190(3):851–860

Fullner KJ, Mekalanos JJ (1999) Genetic characterization of a new Type IV-A pilus gene cluster found in both Classical and El Tor biotypes of *Vibrio cholerae*. Infect Immun 67(3): 1393–1404

Fullner KJ, Mekalanos JJ (2000) In vivo covalent cross-linking of cellular actin by the *Vibrio cholerae* RTX toxin. EMBO J 19(20):5315–5323

Funahashi T, Tanabe T, Shiuchi K, Nakao H, Yamamoto S (2009) Identification and characterization of genes required for utilization of desferri-ferrichrome and aerobactin in *Vibrio parahaemolyticus*. Biol Pharm Bull 32(3):359–365

Gonzalez-Escalona N, Martinez-Urtaza J, Romero J, Espejo RT, Jaykus L-A, DePaola A (2008) Determination of molecular phylogenetics of *Vibrio parahaemolyticus* strains by multilocus sequence typing. J Bacteriol 190(8):2831–2840

Grau BL, Henk MC, Garrison KL, Olivier BJ, Schulz RM, O'Reilly KL, Pettis GS (2008) Further characterization of *Vibrio vulnificus* rugose variants and identification of a capsular and rugose exopolysaccharide gene cluster. Infect Immun 76(4):1485–1497

Graumann P, Marahiel MA (1996) Some like it cold: response of microorganisms to cold shock. Arch Microbiol 166(5):293–300

Gulig PA, Bourdage KL, Starks AM (2005) Molecular Pathogenesis of *Vibrio vulnificus*. J Microbiol 43(Spec No):118–131

Guvener ZT, McCarter LL (2003) Multiple regulators control capsular polysaccharide production in *Vibrio parahaemolyticus*. J Bacteriol 185(18):5431–5441

Hang L, John M, Asaduzzaman M, Bridges EA, Vanderspurt C, Kirn TJ, Taylor RK, Hillman JD, Progulske-Fox A, Handfield M, Ryan ET, Calderwood SB (2003) Use of in vivo-induced antigen technology (IVIAT) to identify genes uniquely expressed during human infection with *Vibrio cholerae*. Proc Natl Acad Sci USA 100(14):8508–8513

Harris JB, Khan AI, LaRocque RC, Dorer DJ, Chowdhury F, Faruque ASG, Sack DA, Ryan ET, Qadri F, Calderwood SB (2005) Blood group, immunity, and risk of infection with *Vibrio cholerae* in an area of endemicity. Infect Immun 73(11):7422–7427

Heidelberg JF, Elsen JA, Nelson WC, Clayton RJ, Gwinn ML, Dodson RJ, Haft DH, Hickey EK, Peterson JD, Umayam L, Gill SR, Nelson K, Read TD, Tettelin H, Richardson D, Ermolaeva MD, Vamathevan J, Bass S, Qin H, Dragoi I, Sellers P, McDonald L, Utterback T, Fleishmann RD, Nierman WC, White O, Salzberg S, Smith. HO, Colwell RR, Mekalanos JJ, Venter JC, Fraser C (2000) DNA sequence of both chromosomes of the cholera pathogen *Vibrio cholerae*. Nature 406:477–484

Honda T, Nishibuchi M, Miwatani T, Kaper JB (1986) Demonstration of a plasmid-borne gene encoding a thermostable direct hemolysin in *Vibrio cholerae* non-O1 strains. Appl Environ Microbiol 52(5):1218–1220

Hor LI, Gao CT, Wan L (1995) Isolation and characterization of *Vibrio vulnificus* inhabiting the marine environment of the Southwestern area of Taiwan. J Biomed Sci 2(4):384–389

Hsiao A, Liu Z, Joelsson A, Zhu J (2006) Vibrio cholerae virulence regulator-coordinated evasion of host immunity. Proc Natl Acad Sci USA 103(39):14542–14547

Hsiao A, Xu X, Kan B, Kulkarni RV, Zhu J (2009) Direct regulation by the Vibrio *cholerae* regulator ToxT to modulate colonization and anticolonization pilus expression. Infect Immun 77(4):1383–1388

Hsieh Y-C, Liang S-M, Tsai W-L, Chen Y-H, Liu T-Y, Liang C-M (2003) Study of capsular polysaccharide from *Vibrio parahaemolyticus*. Infect Immun 71(6):3329–3336

Hunt DE, Gevers D, Vahora NM, Polz MF (2008) Conservation of the chitin utilization pathway in the *Vibrionaceae*. Appl Environ Microbiol 74(1):44–51

Hurley CC, Quirke A, Reen FJ, Boyd EF (2006) Four genomic islands that mark post-1995 pandemic *Vibrio parahaemolyticus* isolates. BMC Genomics 7:104

Igbinosa EO, Okoh AI (2008) Emerging *Vibrio* species: an unending threat to public health in developing countries. Res Microbiol 159(7–8):495–506

Johnson J, Panigrahi P, Morris JG (1992) Non-O1 *Vibrio cholerae* NRT36S produces a polysaccharide capsule that determines colony morphology, serum resistance, and virulence in mice. Infect Immun 60:864–869

Jones MK, Oliver JD (2009) *Vibrio vulnificus*: disease and pathogenesis. Infect Immun 77(5):1723–1733

Joseph LA, Wright AC (2004) Expression of *Vibrio vulnificus* capsular polysaccharide inhibits biofilm formation. J Bacteriol 186(3):889–893

Judson N, Mekalanos JJ (2000) TnAraOut, a transposon-based approach to identify and characterize essential bacterial genes. Nat Biotechnol 18:740–745

Kenner J, Coster T, Trofa A, Taylor D, Barrera-Oro M, Hyman T, Adams J, Beattie D, Killeen K, Mekalanos JJ, Sadoff JC (1995) Peru-15, a live, attenuated oral vaccine candidate for *Vibrio cholerae* O1 El Tor. J Infect Dis 172:1126–1129

Kierek K, Watnick PI (2003a) Environmental determinants of *Vibrio cholerae* biofilm development. Appl Environ Microbiol 69(9):5079–5088

Kierek K, Watnick PI (2003b) The Vibrio cholerae O139 O-antigen polysaccharide is essential for Ca^{2+}-dependent biofilm development in sea water. Proc Natl Acad Sci USA 100(24): 14357–14362

Kim YR, Lee SE, Kim CM, Kim SY, Shin EK, Shin DH, Chung SS, Choy HE, Progulske-Fox A, Hillman JD, Handfield M, Rhee JH (2003) Characterization and pathogenic significance of *Vibrio vulnificus* antigens preferentially expressed in septicemic patients. Infect Immun 71(10):5461–5471

Kim CM, Park YJ, Shin SH (2007) A widespread deferoxamine-mediated iron-uptake system in *Vibrio vulnificus*. J Infect Dis 196(10):1537–1545

Kimsey HH, Waldor MK (2004) The CTXphi repressor RstR binds DNA cooperatively to form tetrameric repressor-operator complexes. J Biol Chem 279(4):2640–2647

Kishishita M, Matsuoka N, Kumagai K, Yamasaki S, Takeda Y, Nishibuchi M (1992) Sequence variation in the thermostable direct hemolysin-related hemolysin (*trh*) gene of *Vibrio parahaemolyticus*. Appl Environ Microbiol 58(8):2449–2457

Kovacikova G, Lin W, Skorupski K (2004) Vibrio cholerae AphA uses a novel mechanism for virulence gene activation that involves interaction with the LysR-type regulator AphB at the *tcpPH* promoter. Mol Microbiol 53(1):129–142

Labbate M, Boucher Y, Joss MJ, Michael CA, Gillings MR, Stokes HW (2007) Use of chromosomal integron arrays as a phylogenetic typing system for *Vibrio cholerae* pandemic strains. Microbiology 153(Pt 5):1488–1498

Lan S-F, Huang C-H, Chang C-H, Liao W-C, Lin IH, Jian W-N, Wu Y-G, Chen S-Y, Wong H-c (2009) Characterization of a new plasmid-like prophage in a pandemic *Vibrio parahaemolyticus* O3:K6 strain. Appl Environ Microbiol 75(9):2659–2667

Lee BC, Choi SH, Kim TS (2008) *Vibrio vulnificus* RTX toxin plays an important role in the apoptotic death of human intestinal epithelial cells exposed to *Vibrio vulnificus*. Microbes Infect 10(14–15):1504–1513

Lee JH, Kim MW, Kim BS, Kim SM, Lee BC, Kim TS, Choi SH (2007) Identification and characterization of the *Vibrio vulnificus* rtxA essential for cytotoxicity in vitro and virulence in mice. J Microbiol 45(2):146–152

Lee BC, Lee JH, Kim MW, Kim BS, Oh MH, Kim K-S, Kim TS, Choi SH (2008) *Vibrio vulnificus* rtxE is important for virulence, and its expression is induced by exposure to host cells. Infect Immun 76(4):1509–1517

Lee JH, Rho JB, Park KJ, Kim CB, Han YS, Choi SH, Lee KH, Park SJ (2004) Role of flagellum and motility in pathogenesis of *Vibrio vulnificus*. Infect Immun 72(8): 4905–4910

Li L, Rock JL, Nelson DR (2008) Identification and characterization of a repeat-in-toxin gene cluster in *Vibrio anguillarum*. Infect Immun 76(6):2620–2632

Limthammahisorn S, Brady YJ, Arias CR (2009) In vivo gene expression of cold shock and other stress-related genes in *Vibrio vulnificus* during shellstock temperature control conditions in oysters. J Appl Microbiol 106(2):642–650

Lipp EK, Rivera IN, Gil AI, Espeland EM, Choopun N, Louis VR, Russek-Cohen E, Huq A, Colwell RR (2003) Direct detection of *Vibrio cholerae* and *ctxA* in Peruvian coastal water and plankton by PCR. Appl Environ Microbiol 69(6):3676–3680

Litwin CM, Rayback TW, Skinner J (1996) Role of catechol siderophore synthesis in *Vibrio vulnificus* virulence. Infect Immun 64(7):2834–2838

Liverman AD, Cheng HC, Trosky JE, Leung DW, Yarbrough ML, Burdette DL, Rosen MK, Orth K (2007) Arp2/3-independent assembly of actin by *Vibrio* type III effector VopL. Proc Natl Acad Sci USA 104(43):17117–17122

Ma AT, McAuley S, Pukatzki S, Mekalanos JJ (2009) Translocation of a *Vibrio cholerae* type VI secretion effector requires bacterial endocytosis by host cells. Cell Host Microbe 5(3):234–243

Makino K, Oshima K, Kurokawa K, Yokoyama K, Uda T, Tagomori K, Iijima Y, Najima M, Nakano M, Yamashita A, Kubota Y, Kimura S, Yasunaga T, Honda T, Shinagawa H, Hattori M, Iida T (2003) Genome sequence of *Vibrio parahaemolyticus*: a pathogenic mechanism distinct from that of *V. cholerae*. Lancet 361(9359):743–749

Mazel D (2006) Integrons: agents of bacterial evolution. Nat Rev Microbiol 4(8):608–620

Mazel D, Dychinco B, Webb VA, Davies J (1998) A distinctive class of integron in the *Vibrio cholerae* genome. Science 280(5363):605–608

McCarter LL (2001) Polar flagellar motility of the *Vibrionaceae*. Microbiol Mol Biol Rev 65(3):445–462

McCarter LL (2004) Dual flagellar systems enable motility under different circumstances. J Mol Microbiol Biotechnol 7(1–2):18–29

Meador CE, Parsons MM, Bopp CA, Gerner-Smidt P, Painter JA, Vora GJ (2007) Virulence gene- and pandemic group-specific marker profiling of clinical *Vibrio parahaemolyticus* isolates. J Clin Microbiol 45(4):1133–1139

Meibom KL, Blokesch M, Dolganov NA, Wu CY, Schoolnik GK (2005) Chitin induces natural competence in *Vibrio cholerae*. Science 310(5755):1824–1827

Meibom KL, Li XB, Nielsen AT, Wu CY, Roseman S, Schoolnik GK (2004) The *Vibrio cholerae* chitin utilization program. Proc Natl Acad Sci USA 101(8):2524–2529

Mekalanos JJ (1983) Duplication and amplification of toxin genes in *Vibrio Cholerae*. Cell 35(1):253–263

Mekalanos JJ, Swartz DJ, Pearson GD, Harford N, Groyne F, deWilde M (1983) Cholera toxin genes: nucleotide sequence, deletion analysis and vaccine development. Nature 306(5943):551–557

Merino S, Shaw JG, Tomás JM (2006) Bacterial lateral flagella: an inducible flagella system. FEMS Microbiol Lett 263(2):127–135

Merrell DS, Camilli A (1999) The *cadA* gene of *Vibrio cholerae* is induced during infection and plays a role in acid tolerance. Mol Microbiol 34(4):836–849

Mey AR, Payne SM (2001) Haem utilization in *Vibrio cholerae* involves multiple TonB-dependent haem receptors. Mol Microbiol 42(3):835–849

Mey AR, Payne SM (2003) Analysis of residues determining specificity of *Vibrio cholerae* TonB1 for its receptors. J Bacteriol 185(4):1195–1207

Mey AR, Wyckoff EE, Oglesby AG, Rab E, Taylor RK, Payne SM (2002) Identification of the *Vibrio cholerae* enterobactin receptors VctA and IrgA: Irga is not required for virulence. Infect Immun 70(7):3419–3426

Miller ES, Heidelberg JF, Eisen JA, Nelson WC, Durkin AS, Ciecko A, Feldblyum TV, White O, Paulsen IT, Nierman WC, Lee J, Szczypinski B, Fraser CM (2003) Complete genome sequence of the broad-host-range vibriophage KVP40: comparative genomics of a T4-related bacteriophage. J Bacteriol 185(17):5220–5233

Mitra R, Figueroa P, Mukhopadhyay AK, Shimada T, Takeda Y, Berg D, Nair GB (2000) Cell vacuolation, a manifestation of the El Tor hemolysin of *Vibrio cholerae*. Infect Immun 68(4):1928–1933

Miyamoto Y, Kato T, Obara Y, Akiyama S, Takizawa K, Yamai S (1969) In vitro hemolytic characteristic of *Vibrio parahaemolyticus*: its close correlation with human pathogenicity. J Bacteriol 100(2):1147–1149

Murphy RA, Boyd EF (2008) Three pathogenicity islands of *Vibrio cholerae* can excise from the chromosome and form circular intermediates. J Bacteriol 190(2):636–647

Nair GB, Faruque SM, Bhuiyan NA, Kamruzzaman M, Siddique AK, Sack DA (2002) New variants of *Vibrio cholerae* O1 biotype El Tor with attributes of the classical biotype from hospitalized patients with acute diarrhea in Bangladesh. J Clin Microbiol 40(9):3296–3299

Nair GB, Ramamurthy T, Bhattacharya SK, Dutta B, Takeda Y, Sack DA (2007) Global dissemination of *Vibrio parahaemolyticus* serotype O3:K6 and its serovariants. Clin Microbiol Rev 20(1):39–48

Nair GB, Shimada T, Kurazono H, Okuda J, Pal A, Karasawa T, Mihara T, Uesaka Y, Shirai H, Garg S, Saha P, Mukhopadhyay A, Ohashi T, Tada J, Nakayama T, Fukushima S, Takeda T, Takeda Y (1994) Characterization of phenotypic, serological, and toxigenic traits of *Vibrio cholerae* O139 Bengal. J Clin Microbiol 32:2775–2779

Nakaguchi Y, Ishizuka T, Ohnaka S, Hayashi T, Yasukawa K, Ishiguro T, Nishibuchi M (2004) Rapid and specific detection of *tdh*, *trh1*, and *trh2* mRNA of *Vibrio parahaemolyticus* by transcription-reverse transcription concerted reaction with an automated system. J Clin Microbiol 42(9):4284–4292

Nakaguchi Y, Nishibuchi M (2005) The promoter region rather than its downstream inverted repeat sequence is responsible for low-level transcription of the thermostable direct hemolysin-related hemolysin (*trh*) gene of *Vibrio parahaemolyticus*. J Bacteriol 187(5):1849–1855

Nakasone N, Iwanaga M (1990) Pili of a *Vibrio parahaemolyticus* strain as a possible colonization factor. Infect Immun 58(1):61–69

Nakhamchik A, Wilde C, Rowe-Magnus DA (2007) Identification of a Wzy polymerase required for group IV capsular polysaccharide and lipopolysaccharide biosynthesis in *Vibrio vulnificus*. Infect Immun 75(12):5550–5558

Nasu H, Iida T, Sugahara T, Yamaichi Y, Park KS, Yokoyama K, Makino K, Shinagawa H, Honda T (2000) A filamentous phage associated with recent pandemic *Vibrio parahaemolyticus* O3:K6 strains. J Clin Microbiol 38(6):2156–2161

Naughton LM, Blumerman SL, Carlberg M, Boyd EF (2009) Osmoadaptation among Vibrio species and unique genomic features and physiological responses of *Vibrio parahaemolyticus*. Appl Environ Microbiol 75(9):2802–2810

Nelson EJ, Chowdhury A, Flynn J, Schild S, Bourassa L, Shao Y, LaRocque RC, Calderwood SB, Qadri F, Camilli A (2008) Transmission of *Vibrio cholerae* is antagonized by lytic phage and entry into the aquatic environment. PLoS Pathog 4(10):e1000187

Nilsson WB, Paranjype RN, DePaola A, Strom MS (2003) Sequence polymorphism of the 16S rRNA gene of *Vibrio vulnificus* is a possible indicator of strain virulence. J Clin Microbiol 41(1):442–446

Nishibuchi M, Fasano A, Russell RG, Kaper JB (1992) Enterotoxigenicity of *Vibrio parahaemolyticus* with and without genes encoding thermostable direct hemolysin. Infect Immun 60(9):3539–3545

Nishibuchi M, Kaper JB (1995) Thermostable direct hemolysin gene of *Vibrio parahaemolyticus*: a virulence gene acquired by a marine bacterium. Infect Immun 63(6):2093–2099

Nishibuchi M, Taniguchi T, Misawa T, Khaeomanee-Iam V, Honda T, Miwatani T (1989) Cloning and nucleotide sequence of the gene (*trh*) encoding the hemolysin related to the thermostable direct hemolysin of *Vibrio parahaemolyticus*. Infect Immun 57(9):2691–2697

Nishioka T, Kamruzzaman M, Nishibuchi M, Satta Y (2008) On the origin and function of an insertion element VPaI-1 specific to post-1995 pandemic *Vibrio parahaemolyticus* strains. Genes Genet Syst 83(2):101–110

Nusrin S, Gil AI, Bhuiyan NA, Safa A, Asakura M, Lanata CF, Hall E, Miranda H, Huapaya B, Vargas GC, Luna MA, Sack, DA, Yamasaki S, Nair GB (2009) Peruvian Vibrio cholerae O1 El Tor strains possess a distinct region in the *Vibrio* seventh pandemic island-II that differentiates them from the prototype seventh pandemic El Tor strains. J Med Microbiol 58(3):342–354

Okada N, Iida T, Park K-S, Goto N, Yasunaga T, Hiyoshi H, Matsuda S, Kodama T, Honda T (2009) Identification and characterization of a novel type III secretion system in *trh*-positive *Vibrio parahaemolyticus* strain TH3996 reveal genetic lineage and diversity of pathogenic machinery beyond the species level. Infect Immun 77(2):904–913

Oliver JD (2005) The viable but nonculturable state in bacteria. J Microbiol 43(Spec No):93–100

Olivier V, Salzman NH, Satchell KJ (2007) Prolonged colonization of mice by Vibrio cholerae El Tor O1 depends on accessory toxins. Infect Immun 75(10):5043–5051

O'Shea YA, Finnan S, Reen FJ, Morrissey JP, O'Gara F, Boyd EF (2004) The *Vibrio* seventh pandemic island-II is a 26.9 kb genomic island present in *Vibrio cholerae* El Tor and

O139 serogroup isolates that shows homology to a 43.4 kb genomic island in *V. vulnificus*. Microbiology 150(Pt 12):4053–4063

Pal A, Ramamurthy T, Bhadra RK, Takeda T, Shimada T, Takeda Y, Nair GB, Pal SC, Chakrabarti S (1992) Reassessment of the prevalence of heat-stable enterotoxin (NAG-ST) among environmental Vibrio cholerae non-O1 strains isolated from Calcutta, India, by using a NAG-ST DNA probe. Appl Environ Microbiol 58(8):2485–2489

Pang B, Yan M, Cui Z, Ye X, Diao B, Ren Y, Gao S, Zhang L, Kan B (2007) Genetic diversity of toxigenic and nontoxigenic Vibrio cholerae Serogroups O1 and O139 revealed by array-based comparative genomic hybridization. J Bacteriol 189(13):4837–4849

Panigrahi P, Tall BD, Russell RG, Detolla LJ, Morris JG Jr (1990) Development of an in vitro model for study of non-O1 Vibrio cholerae virulence using Caco-2 cells. Infect Immun 58(10):3415–3424

Paranjpye RN, Strom MS (2005) A *Vibrio vulnificus* type IV pilin contributes to biofilm formation, adherence to epithelial cells, and virulence. Infect Immun 73(3):1411–1422

Park KS, Ono T, Rokuda M, Jang MH, Iida T, Honda T (2004) Cytotoxicity and enterotoxicity of the thermostable direct hemolysin-deletion mutants of *Vibrio parahaemolyticus*. Microbiol Immunol 48(4):313–318

Park KS, Ono T, Rokuda M, Jang MH, Okada K, Iida T, Honda T (2004) Functional characterization of two type III secretion systems of *Vibrio parahaemolyticus*. Infect Immun 72(11):6659–6665

Peterson KM, Mekalanos JJ (1988) Characterization of the *Vibrio cholerae* ToxR regulon: identification of novel genes involved in intestinal colonization. Infect Immun 56(11):2822–2829

Pflughoeft KJ, Kierek K, Watnick PI (2003) Role of ectoine in *Vibrio cholerae* osmoadaptation. Appl Environ Microbiol 69(10):5919–5927

Prouty MG, Correa NE, Klose KE (2001) The novel sigma54- and sigma28-dependent flagellar gene transcription hierarchy of *Vibrio cholerae*. Mol Microbiol 39(6):1595–1609

Pukatzki S, Ma AT, Revel AT, Sturtevant D, Mekalanos JJ (2007) Type VI secretion system translocates a phage tail spike-like protein into target cells where it cross-links actin. Proc Natl Acad Sci USA 104(39):15508–15513

Pukatzki S, Ma AT, Sturtevant D, Krastins B, Sarracino D, Nelson WC, Heidelberg JF, Mekalanos JJ (2006) Identification of a conserved bacterial protein secretion system in *Vibrio cholerae* using the *Dictyostelium* host model system. Proc Natl Acad Sci USA 103(5): 1528–1533

Qadri F, Chowdhury MI, Faruque SM, Salam MA, Ahmed T, Begum YA, Saha A, Al Tarique A, Seidlein LV, Park E, Killeen KP, Mekalanos JJ, Clemens JD, Sack DA (2007) Peru-15, a live attenuated oral cholera vaccine, is safe and immunogenic in Bangladeshi toddlers and infants. Vaccine 25(2):231–238

Quirke AM, Reen FJ, Claesson MJ, Boyd EF (2006) Genomic island identification in *Vibrio vulnificus* reveals significant genome plasticity in this human pathogen. Bioinformatics 22(8):905

Rahman MH, Biswas K, Hossain MA, Sack RB, Mekalanos JJ, Faruque SM (2008) Distribution of genes for virulence and ecological fitness among diverse *Vibrio cholerae* population in a cholera endemic area: tracking the evolution of pathogenic strains. DNA Cell Biol 27(7):347–355

Raimondi F, Kao JP, Kaper JB, Guandalini S, Fasano A (1995) Calcium-dependent intestinal chloride secretion by *Vibrio parahaemolyticus* thermostable direct hemolysin in a rabbit model. Gastroenterology 109(2):381–386

Reen FJ, Almagro-Moreno S, Ussery D, Boyd EF (2006) The genomic code: inferring *Vibrionaceae* niche specialization. Nat Rev Micro 4(9):697–704

Reguera G, Kolter R (2005) Virulence and the environment: a novel role for *Vibrio cholerae* toxin-coregulated pili in biofilm formation on chitin. J Bacteriol 187(10):3551–3555

Rhee JE, Jeong HG, Lee JH, Choi SH (2006) AphB influences acid tolerance of *Vibrio vulnificus* by activating expression of the positive regulator CadC. J Bacteriol 188(18):6490–6497

Rhee JE, Kim KS, Choi SH (2008) Activation of the *Vibrio vulnificus* cadBA Operon by Leucine-responsive regulatory protein is mediated by CadC. J Microbiol Biotechnol 18(11):1755–1761

Rhine JA, Taylor RK (1994) TcpA pilin sequences and colonization requirements for O1 and O139 *Vibrio cholerae*. Mol Microbiol 13(6):1013–1020

Rogozin IB, Makarova KS, Wolf YI, Koonin EV (2004) Computational approaches for the analysis of gene neighbourhoods in prokaryotic genomes. Brief Bioinform 5(2):131–149

Ruby EG, Urbanowski M, Campbell J, Dunn A, Faini M, Gunsalus R, Lostroh P, Lupp C, McCann J, Millikan D, Schaefer A, Stabb E, Stevens A, Visick K, Whistler C, Greenberg EP (2005) Complete genome sequence of *Vibrio fischeri*: a symbiotic bacterium with pathogenic congeners. Proc Natl Acad Sci USA 102(8):3004–3009

Russell RG, Tall BD, Morris JG Jr (1992) Non-O1 *Vibrio cholerae* intestinal pathology and invasion in the removable intestinal tie adult rabbit diarrhea model. Infect Immun 60(2):435–442

Sack DA, Sack RB, Nair GB, Siddique AK (2004) Cholera. Lancet 363(9404):223–233

Sanchez J, Holmgren J (2008) Cholera toxin structure, gene regulation and pathophysiological and immunological aspects. Cell Mol Life Sci 65(9):1347–1360

Satchell KJ (2007) MARTX, multifunctional autoprocessing repeats-in-toxin toxins. Infect Immun 75(11):5079–5084

Seguritan V, Feng IW, Rohwer F, Swift M, Segall AM (2003) Genome sequences of two closely related *Vibrio parahaemolyticus* phages, VP16T and VP16C. J Bacteriol 185(21): 6434–6447

Sengupta DK, Boesman-Finkelstein M, Finkelstein RA (1996) Antibody against the capsule of *Vibrio cholerae* O139 protects against experimental challenge. Infect Immun 64(1):343–345

Shikuma NJ, Yildiz FH (2009) Identification and characterization of OscR, a transcriptional regulator involved in osmolarity adaptation in *Vibrio cholerae*. J Bacteriol 191(13):4082–4096

Shime-Hattori A, Iida T, Arita M, Park KS, Kodama T, Honda T (2006) Two type IV pili of *Vibrio parahaemolyticus* play different roles in biofilm formation. FEMS Microbiol Lett 264(1):89–97

Shrivastava S, Mande SS (2008) Identification and functional characterization of gene components of Type VI Secretion system in bacterial genomes. PLoS ONE 3(8):e2955

Sigel SP, Stoebner JA, Payne SM (1985) Iron-vibriobactin transport system is not required for virulence of *Vibrio cholerae*. Infect Immun 47(2):360–362

Simpson LM, Oliver JD (1983) Siderophore production by *Vibrio vulnificus*. Infect Immun 41(2):644–649

Skorupski K, Taylor RK (1999) A new level in the *Vibrio cholerae* ToxR virulence cascade: aphA is required for transcriptional activation of the *tcpPH* operon. Mol Microbiol 31(3): 763–771

Strom MS, Paranjpye RN (2000) Epidemiology and pathogenesis of *Vibrio vulnificus*. Microbes Infect 2(2):177–188

Tacket CO, Taylor RK, Losonsky G, Lim Y, Nataro JP, Kaper JB, Levine MM (1998) Investigation of the roles of toxin-coregulated pili and mannose-sensitive hemagglutinin pili in the pathogenesis of *Vibrio cholerae* O139 infection. Infect Immun 66(2):692–695

Takahashi A, Iida T, Naim R, Naykaya Y, Honda T (2001) Chloride secretion induced by thermostable direct haemolysin of *Vibrio parahaemolyticus* depends on colonic cell maturation. J Med Microbiol 50(10):870–878

Takahashi A, Kenjyo N, Imura K, Myonsun Y, Honda T (2000a) Cl(-) secretion in colonic epithelial cells induced by the *Vibrio parahaemolyticus* hemolytic toxin related to thermostable direct hemolysin. Infect Immun 68(9):5435–5438

Takahashi A, Sato Y, Shiomi Y, Cantarelli VV, Iida T, Lee M, Honda T (2000b) Mechanisms of chloride secretion induced by thermostable direct haemolysin of *Vibrio parahaemolyticus* in human colonic tissue and a human intestinal epithelial cell line. J Med Microbiol 49(9):801–810

Tam VC, Serruto D, Dziejman M, Brieher W, Mekalanos JJ (2007) A type III secretion system in Vibrio cholerae translocates a formin/spire hybrid-like actin nucleator to promote intestinal colonization. Cell Host Microbe 1(2):95–107

Tanabe T, Funahashi T, Nakao H, Miyoshi S, Shinoda S, Yamamoto S (2003) Identification and characterization of genes required for biosynthesis and transport of the siderophore vibrioferrin in *Vibrio parahaemolyticus*. J Bacteriol 185(23):6938–6949

Tanabe T, Naka A, Aso H, Nakao H, Narimatsu S, Inoue Y, Ono T, Yamamoto S (2005) A novel aerobactin utilization cluster in *Vibrio vulnificus* with a gene involved in the transcription regulation of the *iutA* homologue. Microbiol Immunol 49(9):823–834

Tanabe T, Nakao H, Kuroda T, Tsuchiya T, Yamamoto S (2006) Involvement of the *Vibrio parahaemolyticus pvsC* gene in export of the siderophore vibrioferrin. Microbiol Immunol 50(11):871–876

Tanaka Y, Kimura B, Takahashi H, Watanabe T, Obata H, Kai A, Morozumi S, Fujii T (2008) Lysine decarboxylase of *Vibrio parahaemolyticus*: kinetics of transcription and role in acid resistance. J Appl Microbiol 104(5):1283–1293

Taylor RK, Miller VL, Furlong DB, Mekalanos JJ (1987) Use of *phoA* gene fusions to identify a pilus colonization factor coordinately regulated with cholera toxin. Proc Natl Acad Sci USA 84(9):2833–2837

Terai A, Baba K, Shirai H, Yoshida O, Takeda Y, Nishibuchi M (1991) Evidence for insertion sequence-mediated spread of the thermostable direct hemolysin gene among *Vibrio* species. J Bacteriol 173(16):5036–5046

Thelin KH, Taylor RK (1996) Toxin-coregulated pilus, but not mannose-sensitive hemagglutinin, is required for colonization by *Vibrio cholerae* O1 El Tor biotype and O139 strains. Infect Immun 64(7):2853–2856

Thompson FL, Iida T, Swings J (2004) Biodiversity of vibrios. Microbiol Mol Biol Rev 68(3):403–431

Trucksis M, Michalski J, Deng YK, Kaper JB (1998) The *Vibrio cholerae* genome contains two unique circular chromosomes. Proc Natl Acad Sci USA 95(24):14464–14469

Twedt RM, Spaulding PL, Johnson HM (1972) Antigenic relationships among strains of *Vibrio parahaemolyticus*. Appl Microbiol 23(5):966–971

Udden SM, Zahid MS, Biswas K, Ahmad QS, Cravioto A, Nair GB, Mekalanos JJ, Faruque SM (2008) Acquisition of classical CTX prophage from *Vibrio cholerae* O141 by El Tor strains aided by lytic phages and chitin-induced competence. Proc Natl Acad Sci USA 105(33):11951–11956

Wachsmuth IK, Olsvik Ø, Evins GM, Popovic T (1994) Molecular epidemiology of cholera. In: Wachsmuth IK, Blake PA, Olsvik Ø Vibrio cholerae and cholera: molecular to global perspectives. American Society for Microbiology, Washington, DC, pp 357–370

Waldor MK, Colwell R, Mekalanos JJ (1994) The *Vibrio cholerae* O139 serogroup antigen includes an O-antigen capsule and lipopolysaccharide virulence determinants. Proc Natl Acad Sci USA 91(24):11388–11392

Waldor MK, Friedman DI (2005) Phage regulatory circuits and virulence gene expression. Curr Opin Microbiol 8(4):459–465

Waldor MK, Mekalanos JJ (1996) Lysogenic conversion by a filamentous phage encoding cholera toxin. Science 272:1910–1914

Watnick PI, Fullner KJ, Kolter R (1999) A role for the mannose-sensitive hemagglutinin in biofilm formation by *Vibrio cholerae* El Tor. J Bacteriol 181(11):3606–3609

Watnick PI, Kolter R (1999) Steps in the development of a *Vibrio cholerae* El Tor biofilm. Mol Microbiol 34(3):586–595

Watnick PI, Lauriano CM, Klose KE, Croal L, Kolter R (2001) The absence of a flagellum leads to altered colony morphology, biofilm development and virulence in *Vibrio cholerae* O139. Mol Microbiol 39(2):223–235

Webster AC, Litwin CM (2000) Cloning and characterization of vuuA, a gene encoding the *Vibrio vulnificus* ferric vulnibactin receptor. Infect Immun 68(2):526–534

Worden AZ, Seidel M, Smriga S, Wick A, Malfatti F, Bartlett D, Azam F (2006) Trophic regulation of *Vibrio cholerae* in coastal marine waters. Environ Microbiol 8(1):21–29

Wright AC, Powell JL, Kaper JB, Morris JG Jr (2001) Identification of a group 1-like capsular polysaccharide operon for *Vibrio vulnificus*. Infect Immun 69(11):6893–6901

Wright AC, Powell JL, Tanner MK, Ensor LA, Karpas AB, Morris JG Jr, Sztein MB (1999) Differential expression of *Vibrio vulnificus* capsular polysaccharide. Infect Immun 67(5):2250–2257

Wu Z, Milton D, Nybom P, Sjo A, Magnusson KE (1996) Vibrio cholerae hemagglutinin/protease (HA/protease) causes morphological changes in cultures epithelial cells and perturbs their paracellular barrier function. Microb Pathog 21(2):11–123

Wyckoff EE, Mey AR, Leimbach A, Fisher CF, Payne SM (2006) Characterization of ferric and ferrous iron transport systems in *Vibrio cholerae*. J Bacteriol 188(18):6515–6523

Wyckoff EE, Mey AR, Payne SM (2007) Iron acquisition in *Vibrio cholerae*. Biometals 20(3–4):405–416

Xicohtencatl-Cortes J, Lyons S, Chaparro AP, Hernandez DR, Saldana Z, Ledesma MA, Rendon MA, Gewirtz AT, Klose KE, Giron JA (2006) Identification of proinflammatory flagellin proteins in supernatants of *Vibrio cholerae* O1 by proteomics analysis. Mol Cell Proteomics 5(12):2374–2383

Yamamoto S, Okujo N, Yoshida T, Matsuura S, Shinoda S (1994) Structure and iron transport activity of vibrioferrin, a new siderophore of *Vibrio parahaemolyticus*. J Biochem 115(5): 868–874

Yarbrough ML, Li Y, Kinch LN, Grishin NV, Ball HL, Orth K (2009) AMPylation of Rho GTPases by *Vibrio* VopS disrupts effector binding and downstream signaling. Science 323(5911): 269–272

Yeung PS, Boor KJ (2004) Epidemiology, pathogenesis, and prevention of foodborne *Vibrio parahaemolyticus* infections. Foodborne Pathog Dis 1(2):74–88

Yeung PSM, Hayes MC, DePaola A, Kaysner CA, Kornstein L, Boor KJ (2002) Comparative phenotypic, molecular, and virulence characterization of *Vibrio parahaemolyticus* O3:K6 isolates. Appl Environ Microbiol 68(6):2901–2909

Yildiz FH, Liu XS, Heydorn A, Schoolnik GK (2004) Molecular analysis of rugosity in a *Vibrio cholerae* O1 El Tor phase variant. Mol Microbiol 53(2):497–515

Yildiz FH, Schoolnik GK (1999) Vibrio cholera O1 El Tor: identification of a gene cluster required for the rugose colony type, exopolysaccharide production, chlorine resistance, and biofilm formation. Proc Natl Acad Sci USA 96(7):4028–4033

Yildiz FH, Visick KL (2009) *Vibrio* biofilms: so much the same yet so different. Trends Microbiol 17(3):109–118

Yoon SS, Mekalanos JJ (2008) Decreased potency of the *Vibrio cholerae* sheathed flagellum to trigger host innate immunity. Infect Immun 76(3):1282–1288

Zahid MSH, Udden SMN, Faruque ASG, Calderwood SB, Mekalanos JJ, Faruque SM (2008) Effect of phage on the infectivity of *Vibrio cholerae* and emergence of genetic variants. Infect Immun 76(11):5266–5273

Chapter 10
Genomic and Transcriptomic Analyses of Foodborne Bacterial Pathogens

Wei Zhang, Edward G. Dudley, and Joseph T. Wade

10.1 Introduction

DNA microarrays (often interchangeably called DNA chips or DNA arrays) are among the most popular analytical tools for high-throughput comparative genomic and transcriptomic analyses of foodborne bacterial pathogens. A typical DNA microarray contains hundreds to millions of small DNA probes that are chemically attached (or "printed") onto the surface of a microscopic glass slide. Depending on the specific "printing" and probe synthesis technologies for different microarray platforms, such DNA probes can be PCR amplicons or in situ synthesized short oligonucleotides. DNA microarray technologies have revolutionized the way that we investigate the biology of foodborne bacterial pathogens. The major advantage of these technologies is that DNA microarrays allow comparison of subtle genomic or transcriptomic variations between two bacterial samples, such as genomic variations between two different bacterial strains or transcriptomic alterations of same bacterial strain under two different treatments. Some applications of comparative genomic hybridization microarrays and global gene expression microarrays have been covered in previous chapters of this book. To avoid duplications, we focus this chapter on introducing several newer high-throughput experimental approaches representing the "next generation" technologies for bacterial comparative genomic and transcriptomic analyses. These technologies include (i) array-based comparative genome sequencing (CGS) which permits detection of small genomic variations down to the single nucleotide levels; (ii) array-based pan-genomic comparative genomic hybridization (pan-CGH) analysis which, in theory, can be used to estimate the core genome for any given bacterial species; (iii) transcriptome sequencing or RNA sequencing (RNA-seq) that may soon replace DNA microarrays for profiling global gene expressions; and (v) chromatin immunoprecipitation-based methods that can be combined with either DNA microarrays (ChIP-chip) or DNA sequencing

W. Zhang (✉)

National Center for Food Safety and Technology, Illinois Institute of Technology, Summit, IL 60501, USA

e-mail: zhangw@iit.edu

M. Wiedmann, W. Zhang (eds.), *Genomics of Foodborne Bacterial Pathogens*, Food Microbiology and Food Safety, DOI 10.1007/978-1-4419-7686-4_10, © Springer Science+Business Media, LLC 2011

(ChIP-seq) to probe genome-wide DNA–protein interactions and to identify regulons for specific bacterial transcription factors. These newer technologies have brought tremendous new opportunities for food microbiologists to study the gene regulation, stress response, virulence, and pathogenesis, as well as ecology and evolution of foodborne bacterial pathogens.

10.2 Array-Based Comparative Genome Sequencing (CGS)

A rapid and cost-effective approach to compare closely related bacterial genomes is array-based comparative genomic sequencing (CGS), which was previously named as comparative genomic resequencing (CGR). High-density tiling arrays are used in CGS to allow precise identification of insertions and deletions (indels), copy numbers of repetitive elements, and single-nucleotide polymorphisms relative to a reference genome (Albert et al., 2005). Figure 10.1 shows the general steps of CGS analysis for identification of single-nucleotide polymorphisms in bacterial genomes.

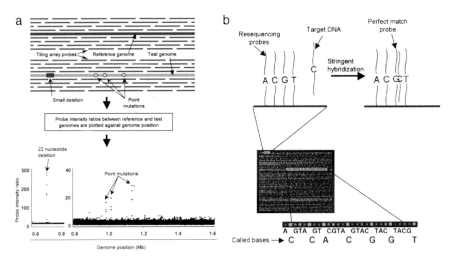

Fig. 10.1 Schematic of "comparative genomic sequencing" strategy. Step 1: Mutation mapping: Labeled bacterial genomic DNAs from a reference strain and a test strain are hybridized in parallel to custom "mapping" microarrays comprised of 29 base long oligonucleotides (*red*) tiled with 7-base spacing. Hybridization intensity ratios (reference/test) plotted vs. genome position map possible mutation sites to 29-nucleotide (probe length) windows. *Arrows* indicate positions of mutations in genes. Step 2: Array-based sequencing: Mapped mutations are next sequenced using custom sequencing arrays. Array-based sequencing exploits differential hybridization of genomic DNA to perfect-match (PM) vs. mismatch (MM) oligonucleotides. Each nucleotide queried is located centrally in an oligonucleotide (29–39 bases, depending on sequence and calculated T_m). Differences in hybridization signal intensities identify PM vs. MM oligonucleotides and thereby the correct base at each queried position

10.2.1 Methodology

10.2.1.1 Microarray Design and Probe Synthesis

To run CGS analysis, a reference genome or specific genomic regions to be compared should be first identified. Oligonucleotide tiling probes, which provide full coverage of the targeted genomic regions, can be designed using software packages such as ArrayScribe (NimbleGen System, Madison, WI) based on the reference genome sequence. The tiling oligonucleotide probes can be directly synthesized in situ using the Maskless Array Synthesis (MAS) technology (Singh-Gasson et al., 1999; Nuwaysir et al., 2002; Albert et al., 2003) on the microarray in a random layout.

10.2.1.2 Fragmentation and Labeling of Genomic DNA

For each bacterial strain to be compared, an aliquot of genomic DNA is digested with DNaseI and end-labeled with Biotin-N6-ddATP using terminal deoxynucleotidyl transferase. Terminal transferase can be inactivated by incubation at 95°C before hybridization.

10.2.1.3 Mutation Mapping

Mutation mapping arrays (Fig. 10.1a) are used to locate potential mutations (e.g., SNPs or small indels) in the genome of a test strain by tiling 29-mer probes with 7-base spacing for both strands (Albert et al., 2005). A mutation mapping array can accommodate up to 2 million probes to tile complete nucleotide sequences of the selected genomic regions. Each labeled genomic DNA sample is hybridized with a separate mutation mapping array. After hybridization, the array is washed with non-stringent buffer and stringent buffer and then stained with Cy3–streptavidin conjugation solution followed by a secondary labeling with biotinylated goat anti-streptavidin. After washes, the array is spun dry and scanned at 5 μm resolution. Pixel intensities are extracted, and probe intensity ratios between the test and reference genes are calculated and plotted against the reference genome position, providing a high-resolution map of mutation sites across the genome and localizing each mutation to a window of 29 bases (the length of the reporting probe). Probe intensity ratios significantly above the background are designated as putative mutation sites, and the corresponding probe sequences are then selected for the second step of resequencing analysis.

10.2.1.4 Resequencing

Resequencing microarrays (Fig. 10.1b) are used to identify the exact nucleotide at each potential SNP site by detecting the differential hybridization signals of sample DNA to short perfect-match (PM) and mismatch (MM) probes (Wong et al., 2004; Albert et al., 2005; Zhang et al., 2006). Each potential SNP to be queried is located near the central position of a PM oligonucleotide probe. Three additional

MM probes representing the three possible mismatch nucleotides at the same position are also synthesized to query each base position on the array. Probe length and mismatch positions may be varied based on probe melting temperature. The differences in hybridization signal intensities between sequences that bond strongly to the PM probes and those that bond poorly to the corresponding MM probes are used to determine the matched base at a given SNP position (Albert et al., 2005; Zhang et al., 2006). Resequencing microarrays are hybridized with labeled genomic DNA samples and scanned following the same protocol as described in the mutation mapping step.

10.2.2 Applications of CGS

High-density sequencing microarrays have been previously used to identify novel genetic variations in populations of viral genomes (Wong et al., 2004) and small regions of bacterial genomes (Read et al., 2002; Zwick et al., 2004). The first application of CGS microarrays in studying foodborne bacterial pathogens was reported in Zhang et al. (2006). In this study, CGS was used for high-throughput resequencing of ~1.2 Mb of nucleotides, including 1,199 chromosomal genes and 92,721 bp of the large virulence plasmid (pO157) of 11 human outbreak-associated Shiga toxin-producing *Escherichia coli* O157:H7 isolates. CGS provides a cost-effective alternative to traditional capillary sequencing or standard resequencing arrays for genome-wide SNP discovery. Using the CGS approach, Zhang et al. discovered 906 SNPs in 523 chromosomal genes and observed a high level of DNA polymorphisms among the pO157 plasmids (Zhang et al., 2006). Based on a uniform rate of synonymous substitution for *E. coli* and *Salmonella enterica* (4.7×10^{-9} per site per year), Zhang et al. estimated that the contemporary β-glucuronidase-negative, non-sorbitol-fermenting *E. coli* O157 strains diverged from their most recent common ancestor ca. 40,000 years ago (Zhang et al., 2006). Zhang et al. also compared the phylogeny of the *E. coli* O157 strains based on the informative synonymous SNPs to the maximum parsimony trees inferred from pulsed-field gel electrophoresis and multilocus variable numbers of tandem repeats analysis. The topological discrepancies indicated that, in contrast to the synonymous point mutations, different regions of *E. coli* O157 genomes may have evolved through different mechanisms with variable divergence rates (Zhang et al., 2006).

10.2.3 Major Limitations

Unlike whole-genome shotgun sequencing, CGS is a "sequencing by hybridization" approach based on available reference genomes. Therefore, it cannot be used to discover novel genomic variations (e.g., new genes) that are present in test bacterial strains but absent in reference genome (Albert et al., 2005; Zhang et al., 2006). In addition, CGS has limited capacity to resolve repetitive elements and therefore is not suitable for analyzing gene duplications or tandem repeats which are often the predominant form of genetic variations among highly clonal bacterial genomes (Zhang et al., 2006).

10.3 Pan-Genomic Comparative Genomic Hybridization (CGH) Microarrays

DNA microarrays have been quite extensively used for the comparative genomic hybridization analysis of various bacterial pathogens; however, many of the earlier CGH arrays only target partial genome of the bacterium (e.g., a small subset of the known genes). It has been increasingly recognized that a few sequenced genomes may not fully represent the entire genetic repertoire of a given organism (Medini et al., 2005; Tettelin et al., 2008; Bentley, 2009; Lapierre and Gogarten, 2009). For this reason, the pan-genome concept has triggered new investigations on the genomic diversity for several bacterial species, including *Streptococcus* spp. (Tettelin et al., 2005; Hiller et al., 2007; Lefebure and Stanhope, 2007), *Haemophilus influenzae* (Hogg et al., 2007), *Neisseria meningitidis* (Schoen et al., 2008), *E. coli* (Willenbrock et al., 2007; Rasko et al., 2008; Touchon et al., 2009), and *Lactococcus lactis* (Bayjanov et al., 2009). Pan-genome refers to the total genetic repertoire of a given species, typically composed of "core" genes plus some "dispensable" or "accessory" genes (Medini et al., 2005; Tettelin et al., 2008; Bentley, 2009; Lapierre and Gogarten, 2009). Pan-genomic CGH arrays, which can probe the full genetic repertoire and provide unbiased coverage of probes across the entire bacterial genome, have gained popularity for surveying genomic diversity in prokaryotic species, including the discovery of novel transcripts, splicing variants, protein-binding sites, and DNA polymorphisms (Mockler et al., 2005). Depending on the offset between adjacent probe locations, whole-genome tilings can be gapped, end to end, or overlapping as shown in Fig. 10.2 (Phillippy et al., 2009).

With the explosion in microarray densities, it is now possible to design pan-genome tiling arrays that contain all genomic sequences from the bacterial pan-genome. Phillippy et al. described a computational method for pan-genome tiling array design (PanArray) that both minimizes the number of probes required and guarantees that all sequences in the pan-genome are fully tiled by the array (Phillippy et al., 2009). Using this method, CGH arrays have been successfully designed for several different bacterial pan-genomes, including *Listeria monocytogenes*, *Francisella tularensis*, *Staphylococcus aureus*, *Bacillus anthracis*, *Vibrio cholerae*, *Burkholderia pseudomallei*, *E. coli*, and *Shigella* spp. (Phillippy et al., 2009). A pan-genome array makes it possible to analyze uncharacterized strains in the context of the entire pan-genome. Pan-genome tiling arrays have all the applications of single-strain tiling arrays, but with enhanced flexibility and the ability to analyze previously uncharacterized strains belonging to the same or related species. Pan-genome CGH also offers an economical alterative to sequencing for determining the genomic makeup of uncharacterized strains and explaining the causative factors of phenotypic differences between strains.

a. Tiling Density   b. Pan-genome Tiling

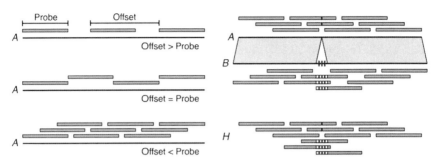

Fig. 10.2 Illustration of different tiling densities and pan-genome tiling. Genomes are represented as *horizontal lines* and probes as *colored rectangles*. The *offset* between probes is the distance between the start of one probe and the start of the next. **a** Three different tiling densities are shown for genome A. The *top figure* illustrates a gapped tiling, the *middle* an end-to-end tiling, and the *bottom* an overlapping tiling. **b** Pan-genome tiling is shown for two genomes. Genomes A and B are identical except for a small insertion in B, represented by *vertical red bars*. *Solid blue probes* are conserved in both genomes, and probes spanning the insertion event are colored by variant. Set H shows the non-redundant set of probes needed to tile the pan-genome including A and B (from Phillippy et al., 2009)

10.3.1 Methodology

10.3.1.1 Array Hybridization and Data Analysis

A detailed description of pan-CGH array hybridization and data analysis was recently reported by Deng et al. (2010). Briefly, genomic DNA labeled with Cy3 or Cy5 dye of each bacterial strain is co-hybridized with that of a reference genome on a pan-CGH array. Two dye-swap replicates are performed for each test/reference strain pair to eliminate dye bias and test the array reproducibility. A probe-based intensity classification scheme can be used to provide the most flexibility for pan-genome array data analysis, allowing any locus to be classified based on the aggregated scores of its individual probes, without reference to control hybridization (Deng et al., 2010). For instance, raw signal intensities can be transformed to log values, and log intensities for replicate hybridizations can be normalized using quantile normalization (Bolstad et al., 2003). Replicates can be combined at the probe level by taking the average of the normalized log intensities for each probe. Quantile normalization assumes similar intensity distributions, so to avoid cross-sample normalization bias. To preserve sensitivity for small polymorphisms, intensity data are not smoothed or segmented. Instead, individual probes can be each classified as present or absent using a minimum kernel density (MKD) method. The MKD method has performed well for the binary classification of both genes and segments (Willenbrock et al., 2007; Carter et al., 2008) and can be extended to the classification of individual probes. Because the array typically contains random control probes, there is expected to be a significant fraction of both present and absent probe intensities for any bacterial sample. Therefore, the distribution of probe intensities is generally bimodal, and the minima between the present and absent peaks can be used as an effective threshold for binary classification. For each sample, the probability density function of the observed intensities can be estimated using kernel density estimation, and the central minima of this function can be identified as the optimal cutoff. This method is preferred because it is non-parametric without potential normalization bias and requires no training. Therefore, each sample can be processed independently without affecting the accuracy (Deng et al., 2010). It is also extremely flexible, in that a classification for any gene can be generated by aggregating the classifications of the probes targeting that gene. For this purpose, genes can be scored by collecting all probes known to target a specific gene and computing the fraction of probes classified as present, the positive fraction (PF). A PF threshold can then be chosen by analysis of ROC curves for the hybridized reference controls to minimize the total error rate (false-positive rate + false-negative rate) vs. the tblastn 50% protein similarity threshold. PF is favored because it does not depend on cross-sample normalization, as would be necessary for an intensity threshold, and additional genomes can be analyzed independently without affecting accuracy. This makes it ideal for rapid and economical analysis of novel isolates, while maintaining comparable accuracy to alternative analysis methods (Willenbrock et al., 2007; Bayjanov et al., 2009; Deng et al., 2010).

10.3.1.2 Pan-Genomic Analysis

Pan-genomic analysis can be performed using the methods introduced by Tettelin et al. (2005), with modifications on the conservation threshold and permutation sampling. For instance, annotated proteins for each genome can be aligned to the six frame translations of all other genomes using tblastn. Query proteins are marked as present in a subject genome if the corresponding amino acid sequences align at $\geq 50\%$ similarity with an E-value $\leq 10^{-5}$, where "similarity" is defined as the number of positively scored residues divided by the length of the protein sequence. This threshold is more stringent than originally proposed (Tettelin et al., 2005), but less stringent than those used in other studies (e.g., Rasko et al., 2008). The 50% threshold is empirically selected as a compromise between tolerating draft genomes with fragmented annotations and avoiding false-positive detections due to conserved domains and distant paralogs. A PF threshold can be chosen as an analogous threshold for the CGH results, as described above. The addition of an Nth genome can be simulated by examining ordered combinations of N genomes. Due to the large number of available genomes for certain bacterial species, it may not be feasible to consider all possible permutations. Instead, a randomly selected subset of permutations can be considered for the addition of each N, and the mean (or median) values are computed from this subset (Deng et al., 2010).

10.3.1.3 Identification of Homologous Groups

Homologous groups (HGs) are used for pan-genome phylogenetic reconstruction and core genome estimation. For sequenced genomes, an HG is called present if at least one member protein of the HG aligns above say 50% similarity threshold. For CGH genomes, an HG is called present if at least one member gene of the HG hybridizes with a PF cutoff. Results based on this threshold can be converted to a unified binary matrix indicating gene presence or absence for all HGs in all genomes to be compared. These binary vectors can then be used for measuring evolutionary distance using the maximum-likelihood method (Huson and Steel, 2004; Deng et al., 2010).

10.3.2 Applications of Pan-CGH Arrays

The first application of pan-CGH arrays in studying foodborne bacterial pathogens was reported by Deng et al. (2010). In this study, Deng et al. combined in silico comparative genomic analysis and high-density CGH arrays to explore the genomic diversity of *L. monocytogenes*, as this pathogenic species displays significant intraspecific variations in host preference, ecological fitness, and virulence among the three genetic lineages (Kathariou, 2002; Freitag et al., 2009; Rasmussen et al., 1995; Wiedmann et al., 1997; Doumith et al., 2004; Zhang et al., 2004). The pan-CGH approach allowed vigorous core genome estimation and phylogenomic reconstruction, which in turn was nearly impossible for low-quality, short-read draft genome assemblies with hundreds of contigs. Using the pan-CGH approach, Deng et al. compared the genomes of 26 *L. monocytogenes* strains representing the three major genetic lineages and uncovered 86 genes and 8 small regulatory RNAs that likely make *L. monocytogenes* lineages differ in carbohydrate utilization and stress resistance during their residence in natural habitats and passage through the host gastrointestinal tract (Deng et al., 2010). Deng et al. also identified 2,330–2,456 core genes that define the *L. monocytogenes* species along with an open pan-genome pool that contains more than 4,052 genes.

Phylogenomic reconstructions based on pan-CGH results of 3,560 homologous groups showed a robust estimation of phylogenetic relatedness among *L. monocytogenes* strains (Deng et al., 2010). Based on Deng et al., a typical *L. monocytogenes* strain carries about 75% of the pan genes of this species (Deng et al., 2010). That said, experiments based on a single reference strain may not adequately sample the total genetic repertoire and, therefore, may not fully interpret the versatile biology of *L. monocytogenes*. Applications of pan-genomic CGH arrays have also been reported in studying the intraspecific genomic variations of several other bacterial species (Table 10.1). With a more defined bacterial species core genome, we may be able to supplement new genomic criterion for taxonomic classification of foodborne bacterial pathogens, as some traditional methods are often inconclusive and controversial. Such pan-genomic approaches can also be used to explore the genomic diversity in other pathogenic bacterial species, as such information would be invaluable for a better understanding of the intraspecific variations in virulence and the ecology, epidemiology, and evolution of microbial pathogens.

Table 10.1 Summary of published bacterial pan-genomic studies

Species	No. of genomes[a]	Pan-genome[b]	No. of core genes	No. of pan genes	Avg. no. of genes	References
Escherichia coli, Shigella	20	Open	1,976	>17,838	4,700	Touchon et al. (2009)
E. coli	17	Open	2,200	>13,000	5,020	Rasko et al. (2008)
E. coli	32	Open	1,563	>9,433	4,537	Willenbrock et al. (2007)
Haemophilus influenzae	13	Finite	1,461	4,425–6,052	1,970	Hogg et al. (2007)
Listeria monocytogenes	26	Open	2,350–2,450	>4,000	2,978	Deng et al. (2010)
Neisseria meningitidis	7	Open	1,333	>3,290	1,963	Schoen et al. (2008)
Streptococcus agalactiae	8	Open	1,806	>2,750	2,245	Tettelin et al. (2005)
S. agalactiae	8	Open[c]	1,472	>2,800[c]	2,198	Lefébure and Stanhope (2007)
Streptococcus pneumoniae	17	Finite	1,380	5,100	2,438	Hiller et al. (2007)
Streptococcus pyogenes	11	Closed[c]	1,376	2,500[c]	1,878	Lefébure and Stanhope (2007)

All numbers are estimates.

[a]Only studies including more than five strains are shown.

[b]Pan-genome growth behaviors as described by the authors.

[c]Estimated from figures, but not explicitly stated in the chapter.

10.3.3 Limitations of Pan-CGH Arrays

Pan-genomic CGH arrays can be used to compare bacterial genomes in pursuit of genes with a biased distribution that potentially promote the fitness in food-related environment and virulence of bacterial strains in human. At the time we wrote this chapter, array CGH approach was relatively more cost-effective compared to the sequencing and closure required to make accurate gene calls using whole-genome shotgun sequencing. Unlike whole-genome sequencing, however, the CGH approach has several inherent limitations in detecting novel genes or pseudogenes, inferring sequence-based phylogenies, and for a host of other analyses inaccessible with array data.

A particular challenge of using the pan-CGH approach is to unify the analysis of both genome sequence and CGH array data because the sensitivity of the two methods is fundamentally different. BLAST searches are capable of precisely measuring amino acid similarity and can identify orthologs and detect distant homologies. In contrast, DNA array hybridizations measure nucleotide conservation and are only capable of detecting highly conserved DNA sequences. In addition, hybridization gives no positional information and is non-specific, making it difficult to discriminate between paralogs. For this reason, Deng et al. used homologous groups for gene content comparison and permitted variant sequences to hybridize to their nearest neighbor in a group, rather than a single selected variant (Deng et al., 2010). The HG method improved the agreement between the array and BLAST detection strategies, which was critical for the phylogenetic analysis of the combined data.

10.4 RNA/Transcriptome Sequencing

Microarray technologies have been invaluable for deciphering the transcriptional networks of biological systems in response to changing environmental conditions, for identifying genes under the control of specific transcriptional regulators, and for comparative transcriptomic studies. Probes spotted onto early arrays were often limited to those from annotated open reading frames; however, the advent of high-density tiling microarrays permitted the interrogation of intergenic regions as well. Very recently, Next-Gen sequencing platforms have been used to visualize the transcriptome in unprecedented detail. This technique is designated "RNA-seq" and was first used to study the yeast transcriptome (Nagalakshmi et al., 2008). The first report of RNA-seq applied to a prokaryotic system was described in 2008 (Mao et al., 2008), and subsequently other studies using this technique for bacterial transcriptomics have been published. This section will cover the basic protocol of RNA-seq and some recently described variations, discuss how it has changed our view of bacterial transcription within two short years, and elaborate on some challenges and future directions. At the time of writing, we were only aware of three studies that had applied RNA-seq to foodborne pathogens (Liu and Camilli, 2010; Oliver et al., 2009; Perkins et al., 2009) and only a few additional studies that targeted other bacteria (Table 10.1). For the purpose of this discussion we will focus on the former papers and will include RNA-seq studies of non-foodborne pathogens when necessary to complement the discussion.

10.4.1 Methodology

The initial steps of sample preparation for RNA-seq are not unlike those performed for microarray analyses (Fig. 10.3). Bacteria are first harvested from laboratory medium, and total RNA is prepared using standard phenol/guanidine isothiocyanate methods or commercially available spin columns, and contaminating DNA is enzymatically digested. However, unlike sample preparation for microarray analysis, RNA-seq next includes a step for depleting 16S and 23S rRNA using a commercially available kit. Liu et al. (2009) have also described a custom method of removing high-abundance small RNAs from *V. cholerae* preparations. In cases where directionality of transcription is not important to the investigator, the next step involves double-stranded cDNA synthesis by random priming and reverse transcriptase treatment, followed by platform-dependent cDNA preparation for high-throughput sequencing (e.g., 454, Illumina, SOLiD). The sequencing reads obtained are mapped to a reference genome, and gene expression is reported as the gene expression index (GEI) (number of reads per 100 bp) (Oliver et al., 2009; Yoder-Himes et al., 2009) or arithmetic mean (AM) per base pair (number of reads covering each base pair) (Perkins et al., 2009). Commercial and publically available software, such as

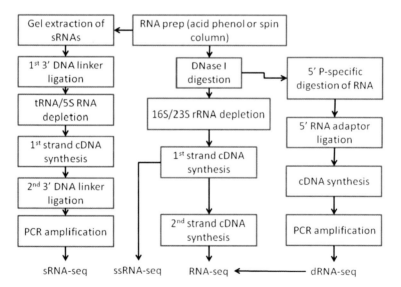

Fig. 10.3 Flowchart of the sample preparation process for RNA-seq and three variations reported in bacterial transcriptomic studies. The *arrow* connecting "dRNA-seq" and "RNA-seq" indicates that the data from both of these techniques are compared during analysis to identify transcribed genes and transcriptional start sites

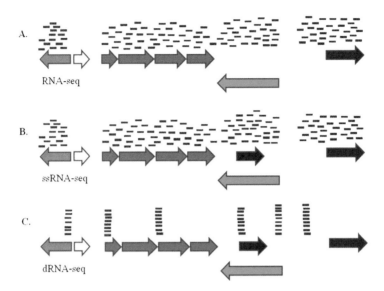

Fig. 10.4 Schematic comparing hypothetical output of three different RNA-seq protocols. *Colored arrows* represent annotated coding sequences within a genome, and *small red* and *blue rectangles* above the *arrows* depict short sequence reads that have been aligned to the genome region. **a** Output from RNA-seq experiment depicting monocistronic genes (*orange, green,* and *purple arrows*), a non-expressed gene (*white arrow*), and a putative operon (*blue arrows*). Note the aligned reads upstream of the annotated coding sequence depicted by the *purple arrow,* suggesting either the presence of a 5′ untranslated region or an inaccurate annotation of the translational start site; **b** ssRNA-seq analysis of the same genomic region shown in (**a**), with short reads aligning to the top DNA strand shown in *red* and reads aligning to the *bottom* strand in *blue.* This technique suggests the presence of a transcript (*brown arrow*) antisense and internal to the coding sequence designated by the *green arrow;* **c** dRNA-seq analysis of the same genomic region, with the majority of reads aligning to the 5′ end of predicted transcripts. This output reveals a second promoter within the operon, suggesting that these genes comprise two transcripts or the presence of an alternate, operon internal promoter. The output also confirms the presence of overlapping antisense transcripts and supports RNA-seq data that the coding sequence indicated by the *purple arrow* either contains an extended 5′ untranslated region or that the start codon is improperly annotated

that available through the Sanger Institute (Croucher et al., 2009), are available for creating graphical outputs such as that recreated in Fig. 10.4.

Additional sample preparation methods are described that permit the determination of transcriptional direction (strand-specific RNA-seq or ssRNA-seq) (Croucher et al., 2009; Perkins et al., 2009), increase the accuracy of transcriptional start site prediction (differential RNA-seq or dRNA-seq) (Albrecht et al., 2010; Sharma et al., 2010), or target small RNAs that are <200 nucleotides in length (small RNA-seq or sRNA-seq) (Liu et al., 2009). These methods provide further insights into the bacterial transcriptome and will be discussed below as well.

10.4.2 Applications of RNA-Seq

Although RNA-seq has a short history of use in the study of prokaryotes, papers published have demonstrated that this technique provides powerful insights and has a number of advantages over microarray analyses. The following is a short discussion of research questions addressed using RNA-seq and some of the improvements this technique provides over other technologies.

10.4.2.1 Increasing the Dynamic Range and Lower Limit of Detection for Gene Expression

As RNA-seq quantifies gene expression by counting sequence tags, there is theoretically no limit to the dynamic range for quantifying transcription. The analysis of the transcriptome of *L. monocytogenes* used a log GEI of 0.7 (corresponding to five reads per 100 bp) as the cutoff between expressed and non-expressed genes and also identified tmRNA as the most abundant transcript during stationary phase with a GEI of 8,566.2. This represents a dynamic range of over four log cycles, which is approximately 100-fold greater than that seen with microarrays [for a side-by-side comparison of microarrays with RNA-seq, see table 1 from Wang et al. (2009)]. Additionally, due to the ability to detect very low levels of transcription, RNA-seq studies have suggested that the majority of coding sequences (CDSs) are transcribed during growth and stationary phases in various media. For example, transcripts were detected for 83 and 66% of annotated CDS in stationary phase *L. monocytogenes* (Oliver et al., 2009) and *S. enterica* serovar Typhi (Perkins et al., 2009) grown in complex medium to OD600 of 0.6, respectively. Yoder-Himes et al. (2009) reported that Illumina reads could be assigned to 99.8% of CDS in the *Burkholderia cenocepacia* genome when bacteria were grown under conditions mimicking human sputum and the soil environment. These authors state that any CDS with one read aligned could be considered expressed, suggesting a high level of stochastic gene expression. Given the high sensitivity of these sequencing technologies, it could also be argued that low-abundance sequence tags could represent DNA that escaped DNase digestion during sample prep. The GEI that distinguishes expressed from non-expressed genes can be calculated by identifying one or more control genes that are known to be non-expressed under given conditions, as shown by Oliver et al. (2009). Similar analyses should be performed when possible to provide a rational discrimination between these two possibilities.

10.4.2.2 Identification of Intergenic Small RNAs (IGR-sRNAs)

A large number of small, typically untranslated RNAs termed sRNAs are known to regulate protein expression by base pairing with mRNAs and affecting their translation, stability, or transcription (Liu and Camilli, 2010). Tiling microarrays have been used to identify sRNAs in the past, and recent RNA-seq studies have discovered a large number of previously unknown transcripts. In the most fascinating example thus far, Liu et al. (2009) devised a method termed sRNA-seq, where RNA from *V. cholerae* was first size fractionated (ca. 14–200 nucleotide transcripts), annealed to a mixture of oligonucleotides that were complementary to tRNAs and 5S RNA and digested with RNaseH to specifically degrade DNA/RNA hybrids. The remaining RNA was used as template for cDNA synthesis, and the products were sequenced by 454. A total of 500 IGR-sRNAs were identified, including all of the previously identified transcripts. Six of seven tested IGR-sRNAs were confirmed by northern blotting, and one of these was shown to negatively regulate the mRNA for the mannitol transporter MtlA when overexpressed in *V. cholerae*. This study demonstrates the power of RNA-seq to serve as a high-throughput platform for gaining new insights into the biology of IGR-sRNAs. RNA-seq with *L. monocytogenes* also discovered seven previously unknown IGR-sRNAs; however, no further characterization of their function was reported.

10.4.2.3 Identification of Antisense sRNAs (AS-sRNAs)

RNA sample prep methods have been reported that permit the directionality of transcription to be determined. This can be accomplished by omitting second-strand cDNA synthesis (ssRNA-seq) (Croucher et al., 2009) or ligating sequence tags to RNA prior to cDNA synthesis (Liu et al., 2009; Sharma et al., 2010). These studies permit the identification of RNA molecules that are transcribed antisense and internal to CDS, termed antisense small RNAs (AS-sRNAs). While a large number of AS-sRNAs were previously reported (Lee et al., 2001; Selinger et al., 2000), RNA-seq will undoubtedly speed the process of AS-sRNA discovery as specialized microarrays are not needed. To date, RNA-seq has been used to identify AS-sRNAs in transcriptomes (Liu et al., 2009; Perkins et al., 2009), but no attempt was made to mechanistically characterize them. Reports that AS-sRNAs regulate transcription (Duhring et al., 2006) and may also be translated (Kim et al., 2009) will certainly stimulate future interest in the use of RNA-seq to catalog these in a variety of bacterial systems. 5' End mapping of RNAs has identified a lot of antisense RNAs (e.g., Shama et al. 2010, Dornenburg et al. 2010).

10.4.2.4 Transcriptional Start Sites and Processed Transcript

The massive number of reads generated by RNA-seq allows for a rough estimation of the transcriptional start site of genes (Albrecht et al., 2010; Oliver et al., 2009). Additionally, a recently described method termed dRNA-seq predicted the transcriptional start sites (TSSs) for 60 of 69 genes in *Helicobacter pylori* within two nucleotides of previously reports. The dRNA-seq protocol digests RNA prior to the synthesis of cDNA libraries with an exonuclease that has specificity for 5' monophosphorylated RNA. As RNA containing a 5' triphosphate moiety such as mRNA and sRNA is resistant to digestion, the enzymatic treatment selectively eliminates 16S and 23S rRNAs and RNA fragments 3' of nick sites that may occur during in vivo processing or RNA isolation. Therefore, the RNA pool for cDNA synthesis is enriched for fragments containing the 5' end of mRNA and sRNA, and the RNA-seq output permits high-resolution identification of the TSS. Although peer-reviewed articles applying this technique to foodborne pathogens are not yet reported, two papers focused on other human pathogens illustrate the power of this technique. dRNA-seq identified 363 and 1,907 transcription start sites in *Chlamydia trachomatis* (Albrecht et al., 2010) and *H. pylori* (Sharma et al., 2010), respectively, and the latter study provided strong confirmation of predictions made previously with a smaller number of TSSs, including the existence of an "extended" Pribnow box (−10 hexamer) and an AT-rich promoter element at position −14. This technique also permitted the identification of previously unknown promoters, posttranscriptional RNA processing, and improved identification of operons.

10.4.2.5 Identification of Misannotated Genes

Deep sequencing of transcriptomes has proven useful in identifying incorrect annotations in genome sequences. For example, transcript was only detected from the putative non-coding strand opposite the annotated hypothetical gene t2145 of *S.* Typhi (Perkins et al., 2009). The authors further argued that the conservation of DNA sequence but not of the intact reading frame in analogous genomic regions of other enteric bacteria supported the argument that t2145 is not a true CDS. RNA-seq data were similarly used to argue that some genes in the *V. cholerae* genome are incorrectly annotated as well (Liu et al., 2009). RNA-seq has also been useful for identifying CDS where the translational start site was incorrectly identified during annotation (Perkins et al., 2009; Sharma et al., 2010).

10.4.2.6 Identification of Riboswitches

RNA-seq studies have identified a large number of untranslated regions 5′ (5′ UTRs) and 3′ (3′ UTRs) of CDS (Perkins et al., 2009; Sharma et al., 2010). Many of these are known or putative riboswitches, defined as UTRs that modulate in cis the expression of genes often through the binding of metabolites. The existence of a riboswitch is often supported by sequence conservation with known riboswitches and/or secondary structure conservation. A number of putative riboswitches were identified by RNA-seq in the transcriptome of *S.* Typhi (Perkins et al., 2009). Four of these are encoded within *Salmonella* pathogenicity island 1 (SPI-1), which mediates the internalization of bacteria into host cells. Based on their location, it is predicted that these riboswitches regulate genes encoding the *araC*-like transcriptional regulators *sprA* and *sprB*, the iron transport integral membrane protein *sitD*, and the invasion protein regulator *iagA*. These predictions have yet to be supported experimentally.

10.4.2.7 Identification of Phage Cargo

Many bacterial pathogens carry lysogenic phages that encode genes contributing to virulence phenotypes such as toxin production, intracellular invasion, and intracellular survival. These phage-encoded accessory genes are also termed cargo genes. Perkins et al. (2009) demonstrated through ssRNA-seq that while most prophage genes encoded within *S.* Typhi are non-expressed or expressed at very low levels (as expected for lysogenic phage), significant activity was detected for a minority of genes. Bioinformatic analysis of these regions indicated that the genes encode the type III secretion effector SopE, hypothetical genes conserved in *E. coli* O157:H7, two putative kinases, and a putative threonine/serine kinase. The authors of this chapter make an interesting case that RNA-seq may prove useful in identifying new genes that mediate virulence in bacterial pathogens.

10.4.3 Major Limitations of RNA-Seq

RNA-seq data are only as good as the prepared cDNA libraries and the rigor of data interpretation. Most methods of sample preparation use a spin column step during the purification process; however, these kits do not effectively recover RNA molecules that are <200 nucleotides in size. Therefore, many publications are likely underreporting the number of ncRNAs expressed. Sample prep may also lead to selective purification or loss of classes of RNA transcripts. Little overlap was found between previously identified ncRNAs that interact with Hfq and ncRNAs identified by RNA-seq (Perkins et al., 2009). The authors postulate that this may be due to these transcripts precipitating with Hfq during protein purification steps or removal of ncRNA–Hfq–rRNA complexes during the 16S/23S rRNA depletion step. Lastly, as elegantly shown by Oliver et al. (2009), experimental platform and stringency of statistical analysis greatly affect results. This study reported that while RNA-seq and two previously reported microarray datasets identified 346 genes in total as being regulated by the *L. monocytogenes* σ^B, only 72 genes were common to all three studies. This highlights the need to perform careful replicates, to be familiar with the shortcomings of your statistical approach, and the need to follow up interesting observations with alternative techniques such as qRT-PCR.

10.4.4 Conclusion and Future Perspectives

In the relatively short time since its first description, RNA-seq has made significant impacts in the field of prokaryotic transcriptomics. This technique has demonstrated the great complexity of the bacterial transcriptome, including the large number of ncRNAs, AS-RNAs, and level of posttranscriptional processing. The power of deep sequencing can be used to accurately identify transcriptional start sites and operons. A few publications have also demonstrated the agreement between GEI and transcript abundance as determined by microarrays and qRT-PCR (Oliver et al., 2009; Yoder-Himes et al., 2009), indicating that RNA-seq is a robust technique for comparative transcriptomic studies as well. While most insights provided by RNA-seq can also be made with microarrays as well, the former has a few distinct advantages. First, as it is a sequence-based technique rather than hybridization-based technique, the output data are less ambiguous. For example, expression levels of two genes that are highly similar by DNA sequence would be easier to determine by RNA-seq than by microarray analysis. Second, RNA-seq can be performed on any organism for which a genome is available, which is less of a hindrance than microarray analysis that may require the construction of a new chip for each new genome sequenced. Studies can also be performed using non-closed genomes (Oliver et al., 2009), and arguably as Next-Gen sequencing reads increase in read length and number per sequence run, a genome sequence may become less important for some applications such as comparative transcriptomics. Lastly, as mentioned above the dynamic range of RNA-seq is much greater than microarrays, providing more information on the expression of weakly transcribed genes.

The application of this technology to prokaryotic systems is still in its infancy, and papers at the time of writing with some notable exceptions (Liu et al., 2009; Oliver et al., 2009) simply report general conclusions about transcriptome analysis with few mechanistic insights into specific biological processes. This will undoubtedly change very rapidly as other researchers begin applying RNA-seq to their systems. Possibly the biggest holdup at this time is that each sequencing run accumulates in the order of 10^5–10^7 short sequence reads per sample, and depending on the number of samples and replicates a very large amount of data are generated. Alignment to a reference genome and further analysis requires computational power beyond that found in most experimental laboratories. There has also been a scarcity of easy-to-use software platforms that are designed for researchers with only modest bioinformatics training. This is now changing, with freeware such as plug-ins for the genome browser Artemis that can display transcriptome mapping files graphically (Perkins et al., 2009) and Galaxy (Taylor et al., 2007) that performs analyses without the need to download anything locally. Much like genome sequencing, which too was once only accessible to a limited number of laboratories, hardware upgrades and the creation of user-friendly software will make RNA-seq approachable for most laboratories in the near future (Table 10.2).

Table 10.2 Major accomplishments in prokaryotic RNA-seq

Organism	Platform(s) used	Number of reads mapped per sample or library	Major accomplishments	References
Sinorhizobium meliloti	454	8,000–10,000	First report of RNA-seq with bacterium	Mao et al. (2008)
V. cholerae	454	4×10^5	Discovery of large number of ncRNAs	Liu et al. (2009)
B. cenocepacia	Illumina	1.7–4.5×10^6	First comparative RNA-seq study with bacteria	Yoder-Himes et al. (2009)
Salmonella enterica subsp. *enterica* serovar Typhi	Illumina	2.8–6.5×10^6	First use of ssRNA-seq to identify transcript directionality	Perkins et al. (2009)
L. monocytogenes	Illumina	3.0–4.8×10^6	First use of RNA-seq to characterize a bacterial regulon	Oliver et al. (2009)
H. pylori	454 and Illumina	8.4×10^5–3.1×10^6 454 reads	First use of dRNA-seq to identify transcriptional start sites	Sharma et al. (2010)
C. trachomatis	454	7.4–9.9×10^5	Comparative transcriptomics between elementary bodies and reticulate bodies; identification and functional prediction of ncRNAs	Albrecht et al. (2010)

10.5 ChIP-Chip and ChIP-Seq

Protein–DNA interactions play key roles in many cellular processes, including transcription, DNA replication, DNA repair, recombination, and chromosome structure. Many protein–DNA interactions are critical to the growth and virulence of bacterial pathogens that cause foodborne illness. For example, DNA-binding transcription factors are required for the virulence of species including *S. enterica* serovar Typhimurium (e.g., PhoP), *V. cholerae* (e.g., ToxT), and *L. monocytogenes* (e.g., PrfA). Many different techniques can be used to determine the position and strength of protein–DNA interactions for given proteins and/or DNAs, both in vitro and in vivo. In this section, we describe a powerful technique, chromatin immunoprecipitation (ChIP), that can be combined with either microarrays (ChIP-chip) or next generation sequencing (ChIP-seq) to determine the association of a given protein with all genomic regions simultaneously. These methods hold great promise for the study of foodborne illness and will likely help elucidate many aspects of the physiology and virulence of the bacteria that cause these infections. Bacteria are particularly amenable to study using ChIP-chip and ChIP-seq because of the large number of fully sequenced genomes, their relatively small genome size, and the multitude of DNA-binding proteins that regulate important cellular processes.

10.5.1 Methodology

10.5.1.1 Chromatin Immunoprecipitation (ChIP)

ChIP is used to determine the association of a specific protein with a specific DNA sequence in vivo. The ChIP method is summarized in Fig. 10.5. Proteins are crosslinked to DNA using formaldehyde, a small molecule that can freely diffuse into living cells. Proteins can be crosslinked directly to DNA or indirectly through protein–protein interactions. Formaldehyde treatment immediately stops cell growth and traps protein–protein and protein–DNA interactions due to the extensive crosslinking. Thus, ChIP provides a "snapshot" of the cell at the moment of formaldehyde addition. Following crosslinking, cells are lysed and DNA is fragmented by sonication. Typically, DNA is fragmented to a size of ~200–500 bp. Sonicated cell lysates are immunoprecipitated with an antibody specific to the protein of interest. This can immunoprecipitate both the protein and any DNA that is directly or indirectly crosslinked to the protein. Crosslinks are reversed using heat, and DNA is purified for further analysis.

Immunoprecipitated DNA from a ChIP experiment can be analyzed in a variety of different ways. Quantitative PCR (typically real-time PCR) is most commonly used to measure the association of a protein with specific DNA sequences. Alternatively, combining ChIP with microarrays (ChIP-chip) or next generation sequencing (ChIP-seq) provides an unbiased, comprehensive view of protein association across an entire genome.

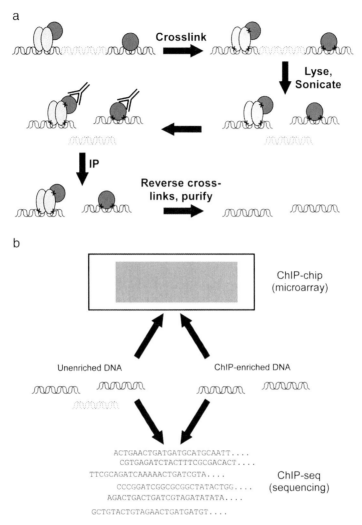

Fig. 10.5 Schematic of chromatin immunoprecipitation (ChIP), ChIP-chip, and ChIP-seq. **a** Schematic of ChIP. Proteins and DNA are crosslinked in living cells with formaldehyde. Cells are lysed and DNA is fragmented by sonication. The protein of interest is immunoprecipitated using a specific antibody. Crosslinks are reversed using heat and DNA is purified. **b** Schematic of ChIP-chip and ChIP-seq. DNA samples from a ChIP experiment and a control experiment can be amplified, labeled, and hybridized to a DNA microarray (ChIP-chip) or directly sequenced using high-throughput sequencing (ChIP-seq)

10.5.1.2 Sample Amplification for ChIP-Chip and ChIP-Seq

A typical ChIP experiment generates between 1 and 10 ng DNA. This is insufficient for ChIP-chip and most ChIP-seq platforms. Furthermore, for ChIP-chip, DNA must first be appropriately labeled, e.g., with Cy3 dye; for ChIP-seq, DNA must be modified to include flanking sequence compatible with the sequencing procedure. Several methods have been developed for sample amplification and labeling for each of ChIP-chip and ChIP-seq. For ChIP-chip, these methods must include a step for incorporation of a modified nucleotide, e.g., Cy3-dCTP, such that the sample can be visualized on the microarray scan. There are four described methods for amplification and labeling of ChIP samples for ChIP-chip: ligation-mediated PCR, random priming followed by PCR, T7 transcription followed by reverse transcription, and strand displacement amplification. All methods have been described in detail elsewhere (Buck and Lieb, 2004; Rhodius and Wade, 2009). ChIP-seq libraries are typically constructed by ligation-mediated PCR from a ChIP sample. For ChIP-seq libraries, labeling is not required.

10.5.1.3 Controls

There are several possible controls for a ChIP-chip or a ChIP-seq experiment. The most commonly used control is DNA purified from the sonicated, lysed, crosslinked cells before the immunoprecipitation step of the ChIP procedure. This is considered the "input" material, and all genomic regions are present at equal concentration. An alternative control is DNA from a "mock" ChIP experiment using an irrelevant antibody or no antibody in the immunoprecipitation step or using the relevant antibody with a bacterial strain not expressing the protein of interest. In cases where the protein of interest is epitope tagged, a mock ChIP experiment can be performed using an untagged strain. For ChIP-chip samples where Cy3/Cy5 dye is used as a label, the control sample is labeled with the opposite dye to the ChIP sample and pairs of control and experimental samples are hybridized to microarrays simultaneously. If duplicate experiments are performed, dye swaps help reduce dye-specific biases in signal intensity.

For ChIP-chip, the experimental read-out is typically a ratio of fluorescence values for the control and ChIP samples. An example of raw ChIP-chip data for *E. coli* σ^{70} (Reppas et al., 2006) is shown in Fig. 10.6. For ChIP-seq, sequencing reads are mapped back to the genome using standard software, e.g., Bowtie (Langmead et al., 2009), and the experimental read-out is the sum of all sequence reads at any given genomic position. Enriched genomic sequences can be identified using a wide

Fig. 10.6 Examples of ChIP-chip data from bacteria. Raw ChIP-chip data for *E. coli* K-12 σ^{70} (Reppas et al., 2006). Data are shown for a ~40 kb region of the genome, and the scale is indicated by a *black bar* representing 1 kb. Genes are shown as *dark gray boxes* (position above or below the *central line* indicates gene orientation). Peaks, interpreted as binding sites for σ^{70}, are indicated with *black dashes*

variety of methods that are described in detail elsewhere (Rhodius and Wade, 2009; Pepke et al., 2009).

10.5.1.4   Choice of Microarray or Sequencing Platform

Both microarray and high-throughput sequencing platforms have developed rapidly. Early microarrays were spotted with PCR products representing either genes or intergenic regions. These are not ideal for ChIP-chip because of the relatively low resolution. Nevertheless, spotted PCR product microarrays have proved extremely useful in the study of transcription factor binding (e.g., Lee et al., 2002). More recent microarray designs provide higher density, often with >100,000 probes per microarray. This allows for genome "tiling," whereby all regions of the genome are represented by at least one probe on the microarray. This is especially true for bacterial genomes, which are relatively small. For high-density, tiling microarrays, probes are typically oligonucleotides that are synthesized in situ. Tiling microarrays allow for higher resolution analysis of protein–DNA interactions and give increased quality data due to redundancy of closely spaced probes. The increased resolution permits distinction of intergenic and intragenic binding sites.

For ChIP-seq, the key parameter when selecting a sequencing platform is the number of sequence reads per run. Based on studies in yeast, we estimate that ~500,000 sequencing reads are necessary to confidently (>95%) identify genomic sequences that are enriched $\geq$twofold relative to an unbound control region (Lefrancois et al., 2009). These numbers are based on a 3-Mb genome which is typical for a bacterium. The number of nucleotides per sequence read is a far less important parameter. The relatively small size of bacterial genomes means that only a short sequence read ($\leq$20 nt) is needed to map the sequence uniquely to the genome; all commonly used high-throughput sequencing methods provide read lengths of at least 20 nt. Multiplexing of ChIP-seq libraries is a simple way to increase efficiency and reduce cost. ChIP-seq libraries can be tagged with a "barcode," a short DNA sequence that flanks the genomic DNA and is sequenced at the same time. Since only 500,000 reads are necessary for high-quality data analysis, many barcoded libraries can be sequenced simultaneously without significant loss of data quality. Methods for creating barcoded libraries have been described in detail elsewhere (Parameswaran et al., 2007; Lefrancois et al., 2009).

10.5.1.5   Antibodies and Epitope Tagging

ChIP requires an antibody specific to the protein of interest. Very few antibodies are commercially available for bacterial DNA-binding proteins. Furthermore, generating polyclonal antibody by immunization of an animal is expensive and time-consuming, and there is no guarantee that the resulting antibody will work efficiently or specifically in a ChIP experiment. An alternative to using an antibody specific to the protein of interest is to use an epitope-tagged protein. This is particularly appealing when applying ChIP to bacterial species since they are often genetically tractable and epitope tagging on the chromosome or on a plasmid is often straightforward. Commonly used epitope tags include FLAG, HA, myc, and

VSV-G. Some bacterial species are naturally competent for DNA. Others, e.g., *S. enterica*, are not, but advances in recombineering using bacteriophage recombinase genes allow for simple, rapid mutagenesis of chromosomal DNA in these species (Datsenko and Wanner, 2000). Plasmids have been created that use these recombineering methods to rapidly epitope tag proteins at their chromosomal locus, leaving only a short "scar" sequence (Uzzau et al., 2001; Cho et al., 2006). Once a protein is epitope tagged it can be immunoprecipitated using a commercially available monoclonal antibody. A potential disadvantage of epitope tagging is that addition of amino acids at the C-terminus of a protein can affect protein function. However, large-scale tagging studies in yeast and *E. coli* demonstrate that most essential genes can be tagged without loss of viability, suggesting that epitope tagging rarely affects protein function (Ghaemmaghami et al., 2003; Huh et al., 2003; Hu et al., 2009).

10.5.1.6 Detailed Analysis of ChIP-Chip and ChIP-Seq Data

Basic analysis of ChIP-chip and ChIP-seq data involves identification of genomic sequences that are significantly enriched relative to unbound sequence. This identifies "target" regions but does not identify the precise bases that are contacted by the protein of interest. ChIP resolution is limited by the size of DNA fragments following sonication and, in the case of ChIP-chip, the spacing of probes on the microarray. High-quality ChIP-chip or ChIP-seq data allow for resolution of ~50 bp (Rhodius and Wade, 2009). For sequence-specific DNA-binding proteins, e.g., a transcription factor, the precise binding site can be predicted by searching for DNA sequence patterns that are enriched in the target regions. This will identify a DNA sequence motif represented by a position weight matrix (PWM) that indicates the probability of each nucleotide at each position within the DNA site (Fig. 10.7). There are a variety of methods to perform this type of motif analysis (Rhodius and Wade, 2009). The quality of motif analysis can be improved further by including sequence information from regions with low-level enrichment, since these are likely to have poor matches to the consensus sequence for the protein of interest (Tanay, 2006).

Depending on the protein being studied, additional analyses may prove to be informative. For example, ChIP-chip studies of H-NS, a nucleoid-associated DNA-binding protein in *S. enterica* serovar Typhimurium, demonstrated that H-NS has a preference for A/T-rich sequence (Lucchini et al., 2006; Navarre et al., 2006).

Fig. 10.7 DNA sequence motif prediction. DNA sequence motif for *E. coli* K-12 LexA identified from ChIP-chip data (Wade et al., 2005) using the software MEME (Bailey and Elkan, 1994). The height of the letters indicates the likelihood of that nucleotide at that position within the DNA sequence motif. These data were visualized using WebLogo (Crooks et al., 2004)

Comparison of ChIP-chip or ChIP-seq data for different DNA-binding proteins is often informative as it can identify functionally interacting proteins, e.g., StpA and H-NS in *Salmonella* Typhimurium (Navarre et al., 2006; Lucchini et al., 2006; 2009).

10.5.2 Application of ChIP-Chip

ChIP-chip and ChIP-seq are ideally suited to bacteria for several reasons. First, hundreds of bacterial genomes have been fully sequenced. Second, bacterial genomes are relatively small, reducing problems of cross-hybridization with ChIP-chip and guaranteeing high-resolution data. Third, many bacteria are genetically tractable, allowing for straightforward epitope tagging. Fourth, there are a huge number of DNA-associated proteins that are required for the growth, survival, and virulence of different bacterial species. ChIP-chip has been applied to many bacterial species, as described below. There are currently no published ChIP-seq studies of any bacterium. This is because the technique is still in its infancy; there is no technical hurdle that prohibits the application of ChIP-seq to bacteria.

Many ChIP-chip studies have focused on the model bacteria *E. coli*, *Bacillus subtilis*, and *Caulobacter crescentus* (Wade et al., 2007). These studies have led to important advances in our understanding of the mechanisms of transcription and chromosome segregation. While these studies do not directly pertain to foodborne illness bacteria, they provide broad insight into the mechanisms of transcription and chromosome segregation that are conserved across many bacterial species.

ChIP-chip studies of transcription factors in *E. coli* have demonstrated the existence of many more binding sites than expected, even for well-studied proteins such as LexA (Wade et al., 2007). This is in large part due to the identification of many target regions for these transcription factors that adjacent to genes whose expression does not change upon deletion or overexpression of the corresponding transcription factor. Such targets may have functional importance but would be impossible to identify using other genomic approaches.

Several ChIP-chip studies have determined the association of RNA polymerase and associated σ factors with the *E. coli* genome (Grainger et al., 2005; Herring et al., 2005; Reppas et al., 2006; Wade et al., 2006). These studies have revealed the existence of many more promoters than previously known and have demonstrated a substantial degree of functional overlap between different σ factors that were previously thought to be largely distinct (Wade et al., 2006).

Two proteins involved in chromosome segregation have been studied using ChIP-chip in *B. subtilis*. RacA was identified as a protein that connects chromosomes to the cell pole to allow chromosome segregation following replication (Ben-Yehuda et al., 2003). ChIP-chip of RacA identified a specific sequence recognized by RacA that is localized proximal to the origin of replication (Ben-Yehuda et al., 2005). SpoOJ is also required for chromosome segregation. ChIP-chip of SpoOJ helped to refine the consensus sequence for this protein, identified unexpected binding sites distal to the origin of replication, and showed that SpoOJ spreads along DNA after initial binding. Together, these ChIP-chip studies have significantly advanced our understanding of chromosome segregation.

ChIP-chip and ChIP-seq have the potential to greatly expand our understanding of the physiology and virulence of bacteria that cause foodborne illness. To date, only four published ChIP-chip studies involve foodborne illness bacteria, all of which focus on *S.* Typhimurium. For many bacteria that cause foodborne illness there is no technical hurdle that blocks application of these methods. Thus, the potential for new discoveries is huge.

In *S.* Typhimurium, four DNA-binding proteins, H-NS, StpA, SsrB, and HilA, have been studied using ChIP-chip (Lucchini et al., 2009, 2006; Navarre et al., 2006; Thijs et al., 2007; Tomljenovic-Berube et al., 2010). H-NS and StpA are highly abundant nucleoid-associated proteins that bind DNA with low sequence specificity. In groundbreaking work, two groups demonstrated using ChIP-chip that H-NS and StpA in *S.* Typhimurium have a preference for binding A/T-rich DNA which results in the binding and transcriptional silencing of horizontally acquired DNA sequence (Lucchini et al., 2009, 2006; Navarre et al., 2006). This phenomenon has been termed "xenogeneic silencing" and has also been observed with H-NS in *E. coli* using ChIP-chip (Grainger et al., 2006; Oshima et al., 2006).

Survival in host cells of *S. enterica* requires the production of a type III secretion system and associated effector proteins. Transcription of the corresponding genes is activated by SsrB, a transcription factor that is part of a two-component system. ChIP-chip of SsrB identified many novel targets (Tomljenovic-Berube et al., 2010). Intriguingly, approximately half of the SsrB targets identified by ChIP-chip are located within ORFs. This phenomenon has been observed for several other transcription factors in other bacterial species (Shimada et al., 2008; Wade et al., 2007), but the significance of these sites is not yet understood. It is possible that transcription factors bound at intragenic sites regulate expression of non-coding RNAs that initiate within genes. Consistent with this hypothesis, large numbers of intragenic RNAs have been identified in *E. coli* and *H. pylori* (Sharma et al., 2010; Dornenburg et al., 2010). ChIP-chip of SsrB also allowed detailed determination of the consensus DNA site for this protein which in turn was used to identify binding sites of an SsrB homologue in another bacterial species, *Sodalis glossinidius* (Tomljenovic-Berube et al., 2010).

There are many more transcription factors in foodborne illness bacteria that are known to regulate genes involved in virulence, e.g., CRP and PhoP in *S.* Typhimurium, PrfA in *L. monocytogenes*, Rns in Enterotoxigenic *E. coli*, and ToxT in *V. cholerae*. There are also likely to be at least as many transcription factors that have not yet been demonstrated to regulate virulence genes. ChIP-chip or ChIP-seq studies of these proteins will likely reveal important insights into virulence. For example, novel targets of known virulence gene regulators are likely to themselves be virulence genes. Other transcription factors are required for normal cell growth and/or survival in the host. ChIP-chip or ChIP-seq studies of these proteins will likely provide insights into the changes that occur in foodborne illness bacteria upon entering the host. A greater understanding of virulence gene regulation and gene regulation that impacts bacterial survival in the host will promote the development of new antimicrobial therapies.

10.5.3 Related Techniques

There are several genomic techniques that are related to ChIP-chip and ChIP-seq. In many cases these approaches are complementary to ChIP-chip and ChIP-seq.

10.5.3.1 Transcription Profiling and RNA-Seq

Transcription profiling refers to microarray-based analysis of changes in RNA levels under different growth conditions or following deletion, overexpression, or mutation of a gene (Rhodius and Wade, 2009). In many cases, the mutated gene encodes a transcription factor. RNA-seq refers to the equivalent experiments performed using high-throughput sequencing (Wang et al., 2009). These approaches identify genes whose expression changes due to direct or indirect effects of a transcription factor. Indirect effects can be limited by transiently comparing RNA levels in cells lacking a transcription factor to cells transiently expressing the transcription factor (Rhodius and Wade, 2009). Nevertheless, indirect effects confound interpretation of these data whereas ChIP-chip and ChIP-seq avoid indirect effects by focusing on the protein–DNA interactions rather than downstream events. Furthermore, ChIP-chip and ChIP-seq can identify protein–DNA interactions that do not result in regulation under the conditions tested, e.g., due to redundancy of transcription factors or inappropriate growth conditions. An advantage of transcription profiling and RNA-seq is that they identify all regulated genes in an operon and they identify the type of regulation, i.e., positive or negative, whereas ChIP-chip and ChIP-seq identify only the DNA site for the transcription factor. Combining ChIP-chip and transcription profiling provides insights beyond those gained using either technique alone, since it permits identification of all direct regulatory targets and the type of regulation (Gao et al., 2004; Tachibana et al., 2005; Hu et al., 2007).

10.5.3.2 DIP-Chip/Genomic SELEX, Protein-Binding Microarray Analysis, and CSI Array Analysis

Three in vitro genomic approaches have been developed to identify the sequence specificity of a DNA-binding protein. All three methods use microarrays but could easily be adapted to use high-throughput sequencing. DIP-chip and genomic SELEX involve mixing of purified DNA-binding protein with fragmented genomic DNA, immunoprecipitation, and hybridization of the associated DNA to a microarray. This identifies all genomic regions that are bound by the protein of interest in vitro. Protein-binding microarray (PBM) analysis involves binding of a DNA-binding protein directly to a microarray and detection of protein–DNA interactions using a fluorescently labeled antibody that is specific to the protein of interest.

Depending on the microarray design, this method can be used to identify genomic DNA sequences bound in vitro, or a collection of DNA sequences that can be used to infer a DNA sequence motif. CSI array analysis involves hybridization of purified DNA-binding protein to a random library of dsDNA. This is followed by immunoprecipitation of the DNA-binding protein and hybridization of the associated DNA to a microarray. The microarray contains probes with every possible combination of bases up to a certain length (length determined by the density of the microarray). In all three cases, the method identifies the DNA sequence specificity of the protein. The major disadvantage of these techniques is that they do not identify protein–DNA interactions in vivo and so ignore possible effects of other proteins. DIP-chip/genomic SELEX, PBM analysis, and CSI array analysis could be used to compare the in vitro and in vivo binding of a DNA-binding protein which would provide insights into the control of binding to target sites in vivo.

10.5.3.3 ROMA

Run-off microarray analysis (ROMA) is a technique that identifies transcriptional changes caused by a specific transcription factor in a defined in vitro system using microarray analysis (MacLellan et al., 2009). It is analogous to transcription profiling but eliminates the possibility of indirect effects on transcription. ROMA can provide similar information to ChIP-chip since transcriptionally induced or repressed genes must be regulated by transcription factors bound at their promoters. A major disadvantage of ROMA is that it cannot detect regulatory effects that require multiple transcription factors. As with DIP-chip/genomic SELEX, protein-binding microarrays, and CSI array, ROMA could be used in combination with ChIP-chip to compare the in vitro and in vivo binding specificities of a transcription factor, although ROMA does not identify protein–DNA interactions that do not affect transcription in vitro.

10.5.3.4 RNA-Immunoprecipitation Chip

RNA-immunoprecipitation-chip (RIP-chip) is very similar to ChIP-chip but is used to study protein–RNA interactions rather than protein–DNA interactions (Baroni et al., 2008). An RNA-binding protein is immunoprecipitated from whole-cell extracts, and then RNA is reversely transcribed and hybridized to a microarray or sequenced using high-throughput sequencing. Typically RIP-chip does not involve crosslinking, but formaldehyde crosslinking has been successfully incorporated in the RIP method (Gilbert et al., 2004). Alternatively, modified nucleotides can be introduced to cells that can be photo-crosslinked after incorporation into RNA. RIP-chip using this photo-crosslinking approach is termed PAR-CLIP. Analysis by high-throughput sequencing has the advantage that the precise site of association can be detected because the labeled nucleotide is altered as a consequence of the crosslinking (Hafner et al., 2010). Thus, the resolution of PAR-CLIP is extremely high.

10.5.4 Conclusion and Future Perspectives

ChIP-chip and ChIP-seq are powerful genomic approaches that determine the genome-wide association of a DNA-binding protein in vivo. They are particularly well suited to the study of bacterial pathogens that cause foodborne illness because of the availability of many fully sequenced genomes, the relatively small genome size, and the large number of DNA-binding proteins that regulate key cellular processes in these species. Published studies have demonstrated the utility of ChIP-chip in understanding gene regulation in foodborne illness bacteria. Future studies are expected to uncover many new protein–DNA interactions that are important for the virulence of these bacteria. Lastly, ChIP-chip and ChIP-seq can be combined with other genomic approaches such as transcription profiling/RNA-seq to provide even greater insight into gene regulation.

References

Albert TJ, Dailidiene D, Dailide G, Norton JE, Kalia A, Richmond TA, Molla M, Singh J, Green RD, Berg DE (2005) Mutation discovery in bacterial genomes: metronidazole resistance in Helicobacter pylori. Nat Methods 2:951–953

Albert TJ, Norton J, Ott M, Richmond T, Nuwaysir K, Nuwaysir EF, Stengele KP, Green RD (2003) Light-directed 5′→3′ synthesis of complex oligonucleotide microarrays. Nucleic Acids Res 31:35–44

Albrecht M, Sharma CM, Reinhardt R, Vogel J, Rudel T (2010) Deep sequencing-based discovery of the Chlamydia trachomatis transcriptome. Nucleic Acids Res 38:868–877

Bailey TL, Elkan C (1994) Fitting a mixture model by expectation maximization to discover motifs in biopolymers. Proc Int Conf Intell Syst Mol Biol 2:28–36

Baroni TE, Chittur SV, George AD, Tenenbaum SA (2008) Advances in RIP-chip analysis: RNA-binding protein immunoprecipitation microarray profiling. Methods Mol Biol 419:93–108

Bayjanov JR, Wels M, Starrenburg M, van Hylckama Vlieg JE, Siezen RJ, Molenaar D (2009) PanCGH: a genotype-calling algorithm for pangenome CGH data. Bioinformatics 25(3):309–314

Bentley S (2009) Sequencing the species pan-genome. Nat Rev Microbiol 7(4):258–259

Ben-Yehuda S, Fujita M, Liu XS, Gorbatyuk B, Skoko D, Yan J, Marko JF, Liu JS, Eichenberger P, Rudner DZ, Losick R (2005) Defining a centromere-like element in Bacillus subtilis by identifying the binding sites for the chromosome-anchoring protein RacA. Mol Cell 18:773–782

Ben-Yehuda S, Rudner DZ, Losick R (2003) RacA, a bacterial protein that anchors chromosomes to the cell poles. Science 299:532–536

Bolstad BM, Irizarry RA, Astrand M, Speed TP (2003) A comparison of normalization methods for high density oligonucleotide array data based on variance and bias. Bioinformatics 19(2):185–193

Buck MJ, Lieb JD (2004) ChIP-chip: considerations for the design, analysis, and application of genome-wide chromatin immunoprecipitation experiments. Genomics 83:349–360

Carter B, Wu G, Woodward MJ, Anjum MF (2008) A process for analysis of microarray comparative genomics hybridisation studies for bacterial genomes. BMC Genomics 9:53

Cho BK, Knight EM, Palsson BO (2006) PCR-based tandem epitope tagging system for Escherichia coli genome engineering. Biotechniques 40:67–72

Crooks GE, Hon G, Chandonia JM, Brenner SE (2004) WebLogo: a sequence logo generator. Genome Res 14:1188–1190

Croucher NJ, Fookes MC, Perkins TT, Turner DJ, Marguerat SB, Keane T, Quail MA, He M, Assefa S, Bahler J, Kingsley RA, Parkhill J, Bentley SD, Dougan G, Thomson NR (2009) A simple method for directional transcriptome sequencing using Illumina technology. Nucleic Acids Res 37:e148

Datsenko KA, Wanner BL (2000) One-step inactivation of chromosomal genes in Escherichia coli K-12 using PCR products. Proc Natl Acad Sci USA 97:6640–6645

Deng X, Phillippy AM, Li Z, Salzberg SL, Zhang W (2010) Probing the pan-genome of Listeria monocytogenes: new insights into intraspecific niche expansion and genomic diversification. BMC Genomics 11:500

Dornenburg JE, DeVita AM, Palumbo MJ, Wade JT (2010). Widespread antisense transcription in Escherichia coli. mBio 1, e00024–10

Doumith M, Cazalet C, Simoes N, Frangeul L, Jacquet C, Kunst F, Martin P, Cossart P, Glaser P, Buchrieser C (2004) New aspects regarding evolution and virulence of Listeria monocytogenes revealed by comparative genomics and DNA arrays. Infect Immun 72(2): 1072–1083

Dühring U, Axmann IM, Hess WR, Wilde A (2006) An internal antisense RNA regulates expression of the photosynthesis gene isiA. Proc Natl Acad Sci USA 103:7054–7058

Freitag NE, Port GC, Miner MD (2009) Listeria monocytogenes – from saprophyte to intracellular pathogen. Nat Rev Microbiol 7(9):623–628

Gao F, Foat BC, Bussemaker HJ (2004) Defining transcriptional networks through integrative modeling of mRNA expression and transcription factor binding data. BMC Bioinformatics 5:31

Ghaemmaghami S, Huh WK, Bower K, Howson RW, Belle A, Dephoure N, O'Shea EK, Weissman JS (2003) Global analysis of protein expression in yeast. Nature 425:737–741

Gilbert C, Kristjuhan A, Winkler GS, Svejstrup JQ (2004) Elongator interactions with nascent mRNA revealed by RNA immunoprecipitation. Mol Cell 14:457–464

Grainger DC, Hurd D, Goldberg MD, Busby SJ (2006) Association of nucleoid proteins with coding and non-coding segments of the Escherichia coli genome. Nucleic Acids Res 34: 4642–4652

Grainger DC, Hurd D, Harrison M, Holdstock J, Busby SJ (2005) Studies of the distribution of Escherichia coli cAMP-receptor protein and RNA polymerase along the E. coli chromosome. Proc Natl Acad Sci USA 102:17693–17698

Hafner M, Landthaler M, Burger L, Khorsid M, Hausser J, Berninger P, Rothballer A, Ascano MJ, Jungkamp A, Munschauser M, Ulrich A, Wardle GS, Dewell S, Zavolan L, Tuschl T (2010) Transcriptome-wide identification of RNA-binding protein and microRNA target sites by PAR-CLIP. Cell 141:129–141

Herring CD, Rafaelle M, Allen TE, Kanin EI, Landick R, Ansari AZ, Palsson BO (2005) Immobilization of Escherichia coli RNA polymerase and location of binding sites by use of chromatin immunoprecipitation and microarrays. J Bacteriol 187:6166–6174

Hiller NL, Janto B, Hogg JS, Boissy R, Yu S, Powell E, Keefe R, Ehrlich NE, Shen K, Hayes J et al (2007) Comparative genomic analyses of seventeen Streptococcus pneumoniae strains: insights into the pneumococcal supragenome. J Bacteriol 189(22):8186–8195

Hogg JS, Hu FZ, Janto B, Biossy R, Hayes J, Keefe R, Post JC, Ehrlich GD (2007) Characterization and modeling of the Haemophilus influenzae core and supragenomes based on the complete genomic sequences of Rd and 12 clinical nontypeable strains. Genome Biol 8:R103

Hu P, Janga SC, Babu M, Díaz-Mejía JJ, Yang BGW, Pogoutse O, Guo X, Phanse S, Wong P, Chandran S, Christopoulos C, Nazarians-Armavil A, Nasseri NK, Musso G, Ali M, Nazemof N, Eroukova V, Golshani A, Paccanaro A, Greenblatt JF, Moreno-Hagelsieb G, Emili A (2009) Global functional atlas of Escherichia coli encompassing previously uncharacterized proteins. PLoS Biol 7:e96

Hu Z, Killion PJ, Iyer VR (2007) Genetic reconstruction of a functional transcriptional regulatory network. Nat Genet 39:683–687

Huh WK, Falvo JV, Gerke LC, Carroll AS, Howson RW, Weissman JS, O'Shea EK (2003) Global analysis of protein localization in budding yeast. Nature 425:686–691

Huson DH, Steel M (2004) Phylogenetic trees based on gene content. Bioinformatics 20(13): 2044–2049

Kathariou S (2002) Listeria monocytogenes virulence and pathogenicity, a food safety perspective. J Food Prot 65(11):1811–1829

Kim W, Silby MW, Purvine SO, Nicoll JS, Hixson KK, Monroe M, Nicora CD, Lipton MS, Levy SB (2009) Proteomic detection of non-annotated protein-coding genes in Pseudomonas fluorescens Pf0-1. PLoS ONE 4:e8455

Langmead B, Trapnell C, Pop M, Salzberg SL (2009) Ultrafast and memory-efficient alignment of short DNA sequences to the human genome. Genome Biol 10:R25

Lapierre P, Gogarten JP (2009) Estimating the size of the bacterial pan-genome. Trends Genet 25(3):107–110

Lee TI, Rinaldi NJ, Robert F, Odom DT, Bar-Joseph Z, Gerber GK, Hannett NM, Harbison CT, Thompson CM, Simon I, Zeitlinger J, Jennings EG, Murray HL, Gordon DB, Ren B, Wyrick JJ, Tagne JB, Volkert TL, Fraenkel E, Gifford DK, Young RA (2002) Transcriptional regulatory networks in Saccharomyces cerevisiae. Science 298:799–804

Lee JM, Zhang S, Saha S, Santa Anna S, Jiang C, Perkins J (2001) RNA expression analysis using an antisense Bacillus subtilis genome array. J Bacteriol 183:7371–7380

Lefebure T, Stanhope MJ (2007) Evolution of the core and pan-genome of Streptococcus: positive selection, recombination, and genome composition. Genome Biol 8(5):R71

Lefrancois P, Euskirchen GM, Auerbach RK, Rozowsky J, Gibson T, Yellman CM, Gerstein M, Snyder M (2009) Efficient yeast ChIP-Seq using multiplex short-read DNA sequencing. BMC Genomics 10:37

Liu JM, Camilli A (2010) A broadening world of bacterial small RNAs. Curr Opin Microbiol 13:18–23

Liu JM, Livny J, Lawrence MS, Kimball MD, Waldor MK, Camilli A (2009) Experimental discovery of sRNAs in Vibrio cholerae by direct cloning, 5S/tRNA depletion and parallel sequencing. Nucleic Acids Res 37:e46

Lucchini S, McDermott P, Thompson A, Hinton JC (2009) The H-NS-like protein StpA represses the RpoS (s38) regulon during exponential growth of Salmonella Typhimurium. Mol Microbiol 74:1169–1186

Lucchini S, Rowley G, Goldberg MD, Hurd D, Harrison M, Hinton JC (2006) H-NS mediates the silencing of laterally acquired genes in Bacteria. PLoS Pathog 18:2

MacLellan SR, Eiamphungporn W, Helmann JD (2009) ROMA: an in vitro approach to defining target genes for transcription regulators. Methods 47:73–77

Mao C, Evans C, Jensen R, Sobral B (2008) Identification of new genes in Sinorhizobium meliloti using the Genome Sequencer FLX system. BMC Microbiol 8:72

Medini D, Donati C, Tettelin H, Masignani V, Rappuoli R (2005) The microbial pan-genome. Curr Opin Genet Dev 15(6):589–594

Mockler TC, Chan S, Sundaresan A, Chen H, Jacobsen SE, Ecker JR (2005) Applications of DNA tiling arrays for whole-genome analysis. Genomics 85(1):1–15

Nagalakshmi U, Wang Z, Waern K, Shou C, Raha D, Gerstein M, Snyder M (2008) The transcriptional landscape of the yeast genome defined by RNA sequencing. Science 320: 1344–1349

Navarre WW, Porwollik S, Wang Y, McClelland M, Rosen H, Libby SJ, Fang FC (2006) Selective silencing of foreign DNA with low GC content by the H-NS protein in Salmonella. Science 313:236–238

Nuwaysir EF, Huang W, Albert TJ, Singh J, Nuwaysir K, Pitas A, Richmond T, Gorski T, Berg JP, Ballin J et al (2002) Gene expression analysis using oligonucleotide arrays produced by maskless photolithography. Genome Res 12:1749–1755

Oliver H, Orsi R, Ponnala L, Keich U, Wang W, Sun Q, Cartinhour S, Filiatrault M, Wiedmann M, Boor K (2009) Deep RNA sequencing of L. monocytogenes reveals overlapping and extensive stationary phase and sigma B-dependent transcriptomes, including multiple highly transcribed noncoding RNAs. BMC Genomics 10:641

Oshima T, Ishikawa S, Kurokawa K, Aiba H, Ogasawara N (2006) Escherichia coli histone-like protein H-NS preferentially binds to horizontally acquired DNA in association with RNA polymerase. DNA Res 13:141–153

Parameswaran P, Jalili R, Tao L, Shokralla S, Gharizadeh B, Ronaghi M, Fire AZ (2007) A pyrosequencing-tailored nucleotide barcode design unveils opportunities for large-scale sample multiplexing. Nucleic Acids Res 35:e130

Pepke S, Wold B, Mortazavi A (2009) Computation for ChIP-seq and RNA-seq studies. Nat Methods 6:S22–S32

Perkins TT, Kingsley RA, Fookes MC, Gardner PP, James KD, Yu L, Assefa SA, He M, Croucher NJ, Pickard DJ, Maskell DJ, Parkhill J, Choudhary J, Thomson NR, Dougan G (2009) A strand-specific RNA-Seq analysis of the transcriptome of the typhoid bacillus Salmonella Typhi. PLoS Genet 5:e1000569

Phillippy AM, Deng X, Zhang W, Salzberg SL (2009) Efficient oligonucleotide probe selection for pan genomic tiling arrays. BMC Bioinformatics 10:293

Rasko DA, Rosovitz MJ, Myers GS, Mongodin EF, Fricke WF, Gajer P, Crabtree J, Sebaihia M, Thomson NR, Chaudhuri R et al (2008) The pangenome structure of Escherichia coli: comparative genomic analysis of E. coli commensal and pathogenic isolates. J Bacteriol 190(20):6881–6893

Rasmussen OF, Skouboe P, Dons L, Rossen L, Olsen JE (1995) Listeria monocytogenes exists in at least three evolutionary lines: evidence from flagellin, invasive associated protein and listeriolysin O genes. Microbiology 141:2053–2061

Read TD, Salzberg SL, Pop M, Shumway M, Umayam L, Jiang L, Holtzapple E, Busch JD, Smith KL, Schupp JM et al (2002) Comparative genome sequencing for discovery of novel polymorphisms in Bacillus anthracis. Science 296:2028–2033

Reppas NB, Wade JT, Church G, Struhl K (2006) The transition between transcriptional initiation and elongation in E. coli is highly variable and often rate-limiting. Mol Cell 24:747–757

Rhodius VA, Wade JT (2009) Technical considerations in using DNA microarrays to define regulons. Methods 47:63–72

Schoen C, Blom J, Claus H, Schramm-Gluck A, Brandt P, Muller T, Goesmann A, Joseph B, Konietzny S, Kurzai O et al (2008) Whole-genome comparison of disease and carriage strains provides insights into virulence evolution in Neisseria meningitidis. Proc Natl Acad Sci USA 105(9):3473–3478

Selinger DW, Cheung KJ, Mei R, Johansson EM, Richmond CS, Blattner FR, Lockhart DJ, Church GM (2000) RNA expression analysis using a 30 base pair resolution Escherichia coli genome array. Nat Biotech 18:1262–1268

Sharma CM, Hoffmann S, Darfeuille F, Reignier J, Findeiss S, Sittka A, Chabas S, Reiche K, Hackermüller J, Reinhardt R, Stadler PF, Vogel J (2010) The primary transcriptome of the major human pathogen Helicobacter pylori. Nature 464:250–255

Shimada T, Ishihama A, Busby SJ, Grainger DC (2008) The Escherichia coli RutR transcription factor binds at targets within genes as well as intergenic regions. Nucleic Acids Res 36: 3950–3955

Singh-Gasson S, Green RD, Yue Y, Nelson C, Blattner F, Sussman MR, Cerrina F (1999) Maskless fabrication of light-directed oligonucleotide microarrays using a digital micromirror array. Nat Biotechnol 17:974–978

Tachibana C, Yoo JY, Tagne JB, Kacherovsky N, Lee TI, Young ET (2005) Combined global localization analysis and transcriptome data identify genes that are directly coregulated by Adr1 and Cat8. Mol Cel Biol 25:2138–2146

Tanay A (2006) Extensive low-affinity transcriptional interactions in the yeast genome. Genome Res 16:962–972

Taylor J, Schenck I, Blankenberg D, Nekrutenko A (2007) Using Galaxy to perform large-scale interactive data analyses. Curr Protoc Bioinform 19:10.5.1–10.5.25

Tettelin H, Masignani V, Cieslewicz MJ, Donati C, Medini D, Ward NL, Angiuoli SV, Crabtree J, Jones AL, Durkin AS et al (2005) Genome analysis of multiple pathogenic isolates of

Streptococcus agalactiae: implications for the microbial "pan-genome". Proc Natl Acad Sci USA 102(39):13950–13955

Tettelin H, Riley D, Cattuto C, Medini D (2008) Comparative genomics: the bacterial pan-genome. Curr Opin Microbiol 11(5):472–477

Thijs IM, De Keersmaecker SC, Fadda A, Engelen K, Zhao H, McClelland M, Marchal K, Vanderleyden J (2007) Delineation of the *Salmonella enterica* serovar Typhimurium HilA regulon through genome-wide location and transcript analysis. J Bacteriol 189:4587–4596

Tomljenovic-Berube AM, Mulder DT, Whiteside MD, Brinkman FS, Coombes BK (2010) Identification of the regulatory logic controlling Salmonella pathoadaptation by the SsrA-SsrB two-component system. PLoS Genet 6:e1000875

Touchon M, Hoede C, Tenaillon O, Barbe V, Baeriswyl S, Bidet P, Bingen E, Bonacorsi S, Bouchier C, Bouvet O et al (2009) Organised genome dynamics in the Escherichia coli species results in highly diverse adaptive paths. PLoS Genet 5(1):e1000344

Uzzau S, Figueroa-Bossi N, Rubino S, Bossi L (2001) Epitope tagging of chromosomal genes in Salmonella. Proc Natl Acad Sci USA 98:15264–15269

Wade JT, Reppas NB, Church GM, Struhl K (2005) Genomic analysis of LexA binding reveals the permissive nature of the Escherichia coli genome and identifies unconventional target sites. Genes Dev 19:2619–2630

Wade JT, Roa DC, Grainger DC, Hurd D, Busby SJW, Struhl K, Nudler E (2006) Extensive functional overlap between sigma factors in Escherichia coli. Nat Struct Mol Biol 13:806–814

Wade JT, Struhl K, Busby SJ, Grainger DC (2007) Genomic analysis of protein-DNA interactions in bacteria: insights into transcription and chromosome organization. Mol Microbiol 65:21–26

Wang Z, Gerstein M, Snyder M (2009) RNA-Seq: a revolutionary tool for transcriptomics. Nat Rev Genet 10:57–63

Wiedmann M, Bruce JL, Keating C, Johnson AE, McDonough PL, Batt CA (1997) Ribotypes and virulence gene polymorphisms suggest three distinct Listeria monocytogenes lineages with differences in pathogenic potential. Infect Immun 65(7):2707–2716

Willenbrock H, Hallin PF, Wassenaar TM, Ussery DW (2007) Characterization of probiotic Escherichia coli isolates with a novel pan-genome microarray. Genome Biol 8(12):R267

Wong CW, Albert TJ, Vega VB, Norton JE, Culter DJ, Richmond TA, Stanton LW, Liu ET, Miller LD (2004) Tracking the evolution of the SARS coronavirus using high-throughput, high-density resequencing arrays. Genome Res 14:398–405

Yoder-Himes DR, Chain PSG, Zhu Y, Wurtzel O, Rubin EM, Tiedje JM, Sorek R (2009) Mapping the Burkholderia cenocepacia niche response via high-throughput sequencing. Proc Natl Acad Sci USA 106:3976–3981

Zhang W, Jayarao BM, Knabel SJ (2004) Multi-virulence-locus sequence typing of Listeria monocytogenes. Appl Environ Microbiol 70(2):913–920

Zhang W, Qi W, Albert TJ, Motiwala AS, Alland D, Hyytia-Trees EK, Ribot EM, Fields PI, Whittam TS, Swaminathan B (2006) Probing genomic diversity and evolution of Escherichia coli O157 by single nucleotide polymorphisms. Genome Res 16:757–767

Zwick ME, Mcafee F, Culter DJ, Read TD, Ravel J, Bowman GR, Galloway DR, Mateczun A (2004) Microarray-based resequencing of multiple Bacillus anthracis isolates. Genome Biol 6:R10

Chapter 11
Proteomics of Foodborne Bacterial Pathogens

Clifton K. Fagerquist

11.1 Preface

This chapter is intended to be a relatively brief overview of proteomic techniques currently in use for the identification and analysis of microorganisms with a special emphasis on foodborne pathogens. The chapter is organized as follows. First, proteomic techniques are introduced and discussed. Second, proteomic applications are presented specifically as they relate to the identification and qualitative/quantitative analysis of foodborne pathogens.

11.2 Techniques

Before mass spectrometry-based proteomic techniques became widely available for protein identification, proteins were identified by N-terminal sequencing using Edman degradation chemistry (Edman, 1956). Since its inception, the technology and instrumentation of Edman sequencers have undergone continuous development resulting in advancements in speed and sensitivity. A major milestone in this technical development was the gas-phase sequenator which largely superceded the previous "spinning-cup" technology (Hewick et al., 1981). The sensitivities of state-of-the-art Edman sequencers allow protein spots in 2-D gels to be identified with confidence. Edman sequencers are still used for research in protein chemistry and proteomics often in conjunction with mass spectrometry-based proteomic techniques. If knowledge of the N-terminal sequence of a protein is critical, for instance in confirming the removal of a signal peptide (an indicator of protein export), N-terminal Edman sequencing can provide such information unambiguously whereas mass spectrometry-based proteomic techniques may not always provide sequence coverage of the N-terminus.

C.K. Fagerquist (✉)
Produce Safety and Microbiology Research Unit, US Department of Agriculture, Western Regional Research Center, Agricultural Research Service, Albany, CA 94710, USA
e-mail: clifton.fagerquist@ars.usda.gov

M. Wiedmann, W. Zhang (eds.), *Genomics of Foodborne Bacterial Pathogens*, Food Microbiology and Food Safety, DOI 10.1007/978-1-4419-7686-4_11, © Springer Science+Business Media, LLC 2011

11.2.1 Ionization Techniques Used in Mass Spectrometry-Based Proteomics

The revolution in mass spectrometry-based proteomics was triggered by the development of two ionization techniques developed in the late 1980s and early 1990s: *m*atrix-*a*ssisted *l*aser *d*esorption/*i*onization or MALDI (Tanaka et al., 1987; Karas et al., 1987; Karas and Hillenkamp, 1988) and *e*lectro*s*pray *i*onization or ESI (Fenn et al., 1989). It is difficult to overestimate the impact that these two key inventions had on furthering the development of mass spectrometry and its application to analysis of large biomolecules including peptides and proteins. This impact was recognized by the 2002 Nobel Prize in Chemistry.

11.2.1.1 Matrix-Assisted Laser Desorption/Ionization (MALDI)

MALDI involves embedding an analyte in a matrix of a weak organic acid. MALDI organic acids are solids at room temperature (e.g., α-*c*yano-4-*h*ydroxy*c*innamic *a*cid or CHCA). CHCA (and similar compounds) absorbs strongly in the near UV. Typically, a pulsed nitrogen laser ($\lambda = 337$ nm) or a solid-state laser (third harmonic of YAG, $\lambda = 355$ nm) is used to irradiate a dried spot of matrix containing the analyte. The strong absorption of the matrix at the laser wavelength results in an instantaneous desorption of matrix and analyte molecules into the gas phase. During the desorption process, the analyte is protonated (or deprotonated). The MALDI ionization process typically results in singly charged (protonated or deprotonated) analytes; however, doubly and triply charged states are possible (although at lower signal intensities). In the case of peptides and proteins, protonation occurs on the side chains of basic residues (arginine, lysine, histidine) and at the N-terminus. The relatively low charge states produced by MALDI mean that the mass-to-charge ratio (m/z) of proteins would be beyond the mass range of most mass analyzers. In consequence, the mass analyzer most compatible with MALDI is one of theoretically unlimited mass range: time-of-flight mass spectrometry (TOF-MS). The coupling of MALDI and TOF-MS also has the advantage of high throughpout.

11.2.1.2 Electrospray Ionization (ESI)

Another ionization technique that has played a critical role in accelerating development of mass spectrometry-based proteomics is *e*lectro*s*pray *i*onization (ESI). ESI involves the formation of electrostatically charged liquid droplets in a high-voltage electric field. The droplets are emitted from a narrow capillary. The potential difference between the capillary tip and the entrance to the mass spectrometer results in a stream of charged droplets. Droplet diameter is determined by the capillary diameter, liquid flow rate, droplet surface tension, and other parameters. A countercurrent flow of dry inert gas (usually nitrogen) facilitates desolvation of the charged droplets reducing droplet diameter and concomitantly increasing Coulomb repulsion. When Coulomb repulsion exceeds droplet surface tension, the droplet is unstable and forms two smaller charged droplets by a process referred to as Coulomb explosion (Bakhoum and Agnes, 2005). The smaller charged droplets similarly undergo a process of desolvation and Coulomb explosion until finally the analyte molecules become protonated (or deprotonated). ESI can generate multiply charged analyte ions, i.e., multiply protonated ions in positive ion mode and multiply deprotonated ions in negative ion mode, if there are sufficient number of protonation sites in the analyte molecule. For peptides and proteins, protonation occurs at the side chain of basic residues and the N-terminus (and deprotonation at the side chain of acidic residues and the C-terminus). Adjustment of solvent pH with mild acids and bases facilitates ionization in positive ion mode or negative ion mode, respectively. At very low pH, it is possible to protonate the peptide backbone when all available side chain basic residues are fully occupied by a proton. Higher charge state protein ions are more susceptible to fragmentation because of Coulomb repulsion as well as protonation of the polypeptide backbone which facilitates peptide backbone cleavage. A multiply charged protein will form a series of ion signals with an m/z corresponding to the following formula $[M + xH]^{+x}$ for positively charged ions and $[M - xH]^{-x}$ for negatively charged ions (where M is the molecular weight of the protein and x is the number of protons added (or removed) from the peptide or protein). The series of ion signal intensities in a mass spectrum is referred to as the charge state envelope.

The higher charge states attainable with ESI compared to MALDI have both advantages and disadvantages. One obvious advantage is that the higher charge reduces the m/z of the ions to within the upper mass range of most mass spectrometers ($m/z < 2,000$). In addition the charge state envelope from an ESI-MS spectrum can be deconvoluted to obtain an accurate molecular weight determination that is often more accurate than the molecular weight obtained by MALDI-TOF-MS. A disadvantage of ESI is that multiply charged ions from multiple analytes can result in peak congestion making deconvolution or peak isolation difficult. This problem can be overcome by separating a sample chromatographically reducing the complexity of a mass spectrum. In contrast, peak congestion is usually not a problem for MALDI because ions are singly or at most doubly charged.

Another disadvantage of ESI is that it is more sensitive toward salts and small molecule contaminants that may interfere with ionization of a larger biomolecules. Ionization by ESI relies upon charge competition among analyte molecules. Ionization of a peptide or protein by ESI is dependent on their gas-phase basicity. As most peptides and proteins have basic residues, ionization by ESI is usually quite highly efficient. However, a small molecule contaminant of higher basicity can interfere with ionization of peptides and proteins by sequestering all available protons in a droplet. In consequence, samples introduced by flow injection must be free of contaminants that might interfere with ionization. MALDI is somewhat more tolerant of the presence of contaminants, although sample pre-cleanup is still important.

An important advantage of ESI is its compatibility with liquid chromatography (LC). In-line LC-ESI-MS is now a standard technique for small molecule, peptide, and protein analysis. Complex samples of peptides and proteins can be chromatographically separated prior to ionization by ESI. In addition, traps can be employed to retain peptides and proteins and other important analytes, prior to analytical separation, allowing salts and buffers (and other interfering contaminants) to be diverted to waste so that the analytical column and the ESI source are not contaminated. LC is compatible with MALDI only when it is performed off-line.

11.2.2 Mass Analyzers Used in Mass Spectrometry-Based Proteomics

Along with the development of MALDI and ESI ionization techniques, mass analyzers have undergone rapid development and improvement over the past two decades. The advent of MALDI and ESI and their application to the analysis of ever more complex biological molecules has driven the need to improve existing mass analyzers as well as development of hybrid mass spectrometry instruments.

11.2.2.1 Triple Quadrupole Mass Spectrometry

The triple quadrupole mass spectrometer has been and continues to be the workhorse instrument of the pharmaceutical analytical laboratories for structural characterization and quantitation of small molecules, drug metabolites as well as peptides. In its simplest design configuration, the triple quadrupole has three sequential quadrupoles aligned in a co-linear fashion (Yost and Enke, 1978). The first quadrupole can be operated to mass select or isolate a single m/z allowing a precursor ion to pass to the second quadrupole. All other ions because of their unstable trajectories in the quadrupole do not transit to the second quadrupole. A collision or target gas (typically nitrogen or argon) is introduced into the second quadrupole. The mass-selected precursor ion collides with target gas molecules in the second quadrupole. These collisions result in a net increase in internal energy of the precursor ion leading to its subsequent fragmentation in the second quadrupole. This process is referred to as collision-induced dissociation (CID) or collision-activated dissociation (CAD). The third quadrupole is then scanned across an m/z range from m/z 10 Th to above the precursor ion m/z. The intensity of the fragment ion and precursor ion signals is then amplified by an electron multiplier. The triple quadrupole is one of the simplest kind of *tandem* mass spectrometers (MS/MS). A recent variant of the triple quadrupole has become commercially available where the third quadrupole can be operated as a linear trap allowing MS/MS/MS. For MS^N experiments (where $N = 3$ or higher) "trapping" mass analyzers, such as quadrupole ion traps or *Fourier transform ion cyclotron resonance* (FT-ICR) instruments, are required. The triple quadrupole is considered a low-resolution/low mass accuracy instrument, and collision energies do not typically exceed 100 eV (1 eV corresponds to 23 kcal/mol). The limited resolution and mass accuracy mean that the instrument is not used for determination of elemental formula of unknown compounds. However, for quantitative, targeted analysis of known analytes or structural characterization of metabolites this instrument remains a reliable and rugged technology.

11.2.2.2 Quadrupole Ion Trap Mass Spectrometry (QIT-MS)

Another tandem mass spectrometer of comparable resolution, mass accuracy, and fragmentation energy is the *q*uadrupole *i*on *t*raps or QITs (March, 1997). The most important distinction between the triple quadrupole and the QIT mass analyzers is that QIT instruments *store* as well as analyze ions whereas triple quadrupole instruments do not store ions during the analysis. The ion storage aspect of traps carries both advantages and disadvantages. The most important advantage is that it is possible to perform multiple, sequential MS/MS experiments, i.e., MS^N (where $N = 3$–10). From the standpoint of peptide sequencing this can be a significant advantage. Ion isolation is accomplished using a tailored waveform (Taylor et al., 1995) that can eject ions from the trap from all but narrow m/z region. The remaining ions within the specified m/z window are then excited into trajectories that result in energetic collision with the "bath" gas resulting in the production of fragment ions by CAD. Because the excitation radiofrequency (RF) is specific only for the m/z of the precursor ion, fragment ions generated are not excited by the RF signal because they have a different m/z. This should be contrasted with the operation of a triple quadrupole mass spectrometer where a rapidly formed fragment ion in the second quadrupole may itself undergo ion/target gas collision and subsequent fragmentation. In addition to acting as a collision gas, the "bath" gas (i.e., helium) also facilitates the trapping of ions during the "fill" cycle by dampening the translational motion of the ions as well as removing internal energy from the ions resulting in concentration of the ions to the center of the trap volume.

Another important distinction between triple quadrupoles and QIT instruments relates to the stability of fragment ion trajectories in QIT which is a function of fragment ion m/z in relation to precursor ion m/z. The Mathieu stability diagram which applies to QIT instruments indicates that there is an m/z limit on the stability of fragment ion trajectories below which fragment ions have unstable orbits in the trap and thus cannot be detected (March, 1997). This stability limit is dependent on the m/z of the precursor ion. Thus, some fragment ions generated from an MS/MS experiment may not be detected because they have unstable orbits in the QIT. This, of course, can have consequences for peptide sequencing as not all fragment ions generated will be detected.

QITs have an ion storage capacity of ∼1,000 charges. Exceeding this limit will result in a loss of resolution and mass accuracy due to space charge effects. The ion storage capacity is determined by the trap volume (about the size of a marble). In order to address the limitation of ion storage capacity, QIT instruments incorporate a pre-scan analysis of the number of ions being transferred to the trap per unit time. The pre-scan measurement is used to determine the "fill" time, i.e., the time required to inject ions into the trap and not exceed its ion storage capacity, for the analytical analysis. This feature allows QITs to obtain maximum sensitivity while at the same time maintaining acceptable resolution and mass accuracy.

It should be emphasized that it is *not* the number of ions that is the critical factor of QIT storage capacity but the number of *charges* in the trap. This distinction becomes readily apparent when one attempts to analyze a large intact protein, such as bovine serum albumin (BSA, MW = 66 kDa), using a QIT (even with pre-scan). ESI of BSA produces a wide charge state envelope with the highest charge states having in excess of 100 charges (protons) for each protein molecule. Thus, the charge limit of the trap can be quickly reached, and yet obtaining good signal to noise (S/N) requires a sufficient number of protein ions for analysis. This problem is further exacerbated by alkali metal adduction present in some large proteins which can further spread ion signal across a broader m/z range reducing the S/N to unacceptable levels.

In spite of these limitations, QIT instruments have found a niche in small molecule and bottom-up proteomics because of their low cost, ruggedness, and reliability. These instruments are used extensively in "shotgun" proteomic experiments in conjunction with multidimensional chromatography (LC/LC), e.g., *multi*dimensional *p*rotein *i*dentification *t*echnology or MudPIT (Wolters et al., 2001). MudPIT, a non-gel-based proteomic approach, involves the enzymatic digestion of the entire complement of proteins present in a cell (or subcellular compartment) of an organism under specific physiological conditions related to the question being addressed. In this approach, intact proteins are *not* separated (either by gel electrophoresis or by LC) prior to enzymatic digestion. This complex mixture of proteins becomes an even more complex mixture of peptides with a concentration range of several orders-of-magnitude. Multidimensional liquid chromatography is employed to reduce this complexity by fractionating the peptides, first by strong cation exchange and second by reverse-phase chromatography. Typically, proteins are digested with trypsin which results in peptides with C-terminal basic residues. Peptides ionized by ESI are typically singly, doubly, or triply charged with one charge (proton) conveniently located at the C-terminus of the peptide. In consequence, the ion storage capacity of the QIT is not overwhelmed by higher charge state species (>+4), and the m/z of most peptides fall below the upper m/z limit (<2,000–3,000 m/z) of most QIT instruments.

Because of the relatively low resolution and mass accuracy of QIT instruments, free, open-source software has been developed specifically to address the issue of false-positive identifications of peptides and proteins for proteomic applications (Washburn et al., 2001). The software also addresses the issue of how the QIT is operated to acquire MS/MS data of peptides. In most LC/mass spectrometry-based proteomic experiments data collection involves operation of the QIT in a cycle of repetitive scan modes. First, a wide mass range scan is acquired to determine which precursor ions are good candidates of MS/MS based on their relative intensity. Ions are ranked in order of relative intensity with the most abundant ions taking precedence for MS/MS fragmentation. It is possible, in theory and practice, to perform a very narrow mass range scan across a precursor ion before MS/MS in order to determine the charge of the ion. This information would be helpful when trying to identify the peptide during a database search. However, in practice, some laboratory operators may skip this step because it would reduce the duty cycle of the instrument for acquiring as many MS/MS spectra as possible. By this approach, the software determines identification of a species from a database with no prior knowledge of whether the ion species is +1, +2, +3, +4, etc. Invariably, this approach can lead to false-positive identifications, and a sophisticated statistical analysis is employed by the software to compute the probability of false-positive identifications. The identifications of peptides and proteins are thus accompanied by a score indicating the probability that the identification could have occurred randomly (Washburn et al., 2001).

One last note about QIT instruments: these instruments are primarily (if not exclusively) used for bottom-up proteomic experiments, such as MudPIT. Their limited resolution, mass accuracy, and ion storage capacity make them largely inappropriate for top-down identification of large proteins. However, in spite of these significant limitations, the low cost, ruggedness, and reliability of QIT instruments have made them a viable alternative for doing proteomic research compared to the more expensive higher resolution mass spectrometers.

11.2.2.3 Linear Ion Trap Mass Spectrometry (LIT-MS)

A recent addition to the arena of tandem mass spectrometers used for proteomic research is the *l*inear *i*on *t*rap or LIT (Schwartz and Senko, 2004; Douglas et al., 2005). Like the QIT and FT-ICR instruments, LITs store ions during mass analysis. The ion storage volume of LITs is larger than that of QIT (but smaller than that of FT-ICR) with an ion capacity of $\approx$10,000 charges. The LITs are low-resolution/low mass accuracy instruments. Probably the most significant design feature of LITs is that ions can be introduced from either side of the mass analyzer. This quadrupole mass analyzer is segmented into three physically distinct regions which allow ions to be introduced from opposite sides of the instrument to be stored and electrostatically partitioned from ions collected at the other side of the instrument. Ions from both ends can be combined into the center section. The center section is slotted allowing ions to be analyzed and ejected radially to detectors positioned around the center section. This configuration allows controlled ion/ion chemistry experiments to be conducted in a single mass analyzer. The most important ion/ion experiment, relevant to proteomic research, is *e*lectron *t*ransfer *d*issociation or ETD (Syka et al., 2004). ETD (which will be discussed in greater detail in the next section) involves transfer of an electron from a small molecule singly charged anion to a doubly, triply, or quadruply charged (protonated) peptide. This charge transfer process results in efficient fragmentation of the peptide backbone producing primarily c and z fragment ions. Posttranslational modifications (PTMs), such as phophorylation or glycosylation, remain attached to fragment ions facilitating their identification and localization. The efficiency of the fragmentation "ladder" is so complete as to allow the manual interpretation of spectra as relatively straightforward, in spite of the fact that LITs are low-resolution and low mass accuracy instruments. Because of this innovative dissociation technique, LITs have found a niche in proteomic research as a stand-alone mass analyzer. However, LITs have been combined with triple quadrupole, FT-ICR, and Orbitrap^{TM} instruments to generate even more powerful hybrid instruments for proteomic research.

11.2.2.4 Fourier Transform Ion Cyclotron Resonance
Mass Spectrometry (FT-ICR-MS)

At the other end of the mass spectrometry spectrum is *Fourier transform ion cyclotron resonance mass spectrometry* or FT-ICR-MS (Comisarow and Marshall, 1974). These instruments, which have exceedingly high resolution (routinely 100–500 k) and sub-ppm mass accuracy, have become the instrument of choice (when cost is not an issue) for bottom-up or top-down proteomics. Like QIT instruments, FT-ICR-MS instruments "store" ions as part of their analysis. However, FT-ICR possesses a much greater ion storage capacity due to their larger ion cell volume: ~1,000,000 charges can be stored without seriously impacting resolution and mass accuracy. In consequence, this instrument has found applications in top-down analysis of intact proteins as well as bottom-up analysis of peptides. The FT-ICR-MS instruments have also been utilized at the forefront of studies of ion/molecule chemistry as well as development of new dissociation techniques for small and large biomolecules. In addition to CID or CAD, *infrared multiphoton dissociation* or IRMPD, *sustained off-resonance irradiation collision-activated dissociation* or SORI-CAD, *blackbody infrared radiative dissociation* or BIRD, *ultraviolet radiative dissociation* or UVD, and *electron capture dissociation* or ECD either made their debut on FT-ICR or were adapted to FT-ICR instruments. The range of dissociation techniques, adapted to (and now commercially available on) FT-ICR instruments, has greatly facilitated the analysis of large intact biomolecules for top-down proteomics. Like the QIT, the FT-ICR can be used to perform multiple MS/MS experiments sequentially (MS^N where $N = 3$–10). For tandem MS experiments, ions are isolated by a technique referred to as *stored waveform inverse Fourier transform* or SWIFT (Marshall et al., 1985). SWIFT utilizes a combination of multiple radiofrequency (RF) signals to eject ions from the ICR cell whose m/z is above and below the m/z of the precursor ion that one wants to retain in the cell for subsequent ion/molecule or dissociation experiments. In addition, hybrid instrument configurations have combined other mass analyzers for ion dissociation (e.g., linear traps) with FT-ICR for its high-resolution, sub-ppm mass accuracy analysis of fragment ions.

The high resolution and sub-ppm mass accuracy of FT-ICR-MS instruments dramatically reduce the false positives in peptide and protein identification for bottom-up proteomic experiments. The reason for this is simple: the high resolution and sub-ppm mass accuracy of FT-ICR eliminate many more false-positive peptide identifications which could not be accomplished using lower resolution/lower mass accuracy instruments. In addition, the high resolution of the instrument results in unambiguous identification of the charge state of both precursor and fragment ions which facilitates correct identification of the peptide/protein.

Thus, proteomic identifications from FT-ICR MS/MS data need not undergo complicated statistical validation algorithms with respect to false-positive identification as these are effectively eliminated due to the higher quality of the MS/MS data.

The greatest disadvantage of FT-ICR-MS instruments is its cost which can range from $500 k to in excess of $1 million. As these instruments utilize superconducting magnets to generate strong magnetic fields with field strengths of 3–21 T (1 T = 10,000 G), both liquid helium and liquid nitrogen are required to cool the coils of the superconducting magnets. These high-performance instruments also require a great deal of expertise to maintain and operate, although the commercially available models have incorporated many fail-safe features to protect the instrument from operator error.

11.2.2.5 The Orbitrap™

A very recent and novel addition to the world of mass analyzers is the Orbitrap™ (Makarov, 1999, 2000; Hu et al., 2005). This mass analyzer is an ion trap; however, unlike other trapping mass analyzers that use either strong magnetic fields or electrodynamic radiofrequency (RF) for ion confinement, the Orbitrap™ uses electrostatics for confinement with ions orbiting around a central electrode. Ions are generated externally and injected into the Orbitrap™ for a brief interval when the electric field is off. Ions in an Orbitrap™ revolve around the central electrode while at the same time oscillating from pole to pole. This oscillation (which is independent of the initial position, angle, and energy of the injected ions) generates an image current in the outer, confining electrode. The frequency of the oscillation is fast Fourier transformed into m/z signal of the respective ion. Ions of different m/z will have different frequencies of oscillation in the trap. Thus, the Orbitrap™ is both an ion trap and a nondestructive ion detector. The Orbitrap™ can achieve resolutions exceeding 100,000 (FWHM) and single-digit ppm mass accuracy. Only FT-ICR instruments have higher resolutions. One of the advantages of the Orbitrap™ is that it does not have superconducting magnets that require liquid helium and nitrogen for magnet cooling. In the current commercially available platform (Makarov et al., 2006) the Orbitrap™ is part of a hybrid instrument with a linear ion trap on the front end for MS^N experiments followed by a modified linear trap (C-trap) for collisionally cooling and storing ions before their injection into the Orbitrap™ for high-resolution, high mass accuracy analysis of precursor and fragment ions. In this configuration, the Orbitrap™ is primarily utilized for MS analysis.

11.2.2.6 Time-of-Flight Mass Spectrometry (TOF-MS)

Like many analytical techniques used in the chemical and biological sciences, *time-of-flight mass spectrometry* or TOF-MS began as a tool developed by physicists (Stephens, 1946; Wiley and MacLaren, 1955; Mamyrin et al., 1973). Relatively simple in design and inexpensive to construct, TOF instruments are widely used in the fields of physics and physical chemistry for research in gas-phase molecular dynamics, photochemistry, and ion chemistry research. The first TOF analyzers were linear in design (Stephens, 1946; Wiley and MacLaren, 1955). With continuous ion extraction, ions are accelerated from the source upon their formation by a discrete ionization event, e.g., a pulsed ionizing beam of electrons or a pulsed laser beam. The "pulsed" ionizing event serves as the "start" signal for the flight of ions to the detector. As all singly charged ions have the same energy (but different velocities), their arrival time at the detector is used to calculate their mass. Wiley–MacLaren introduced to the linear TOF design a timing delay between the ionizing event (a pulsed electron beam) and the ion extraction pulse. This "time-lag focusing" improved instrument resolution (Wiley and MacLaren, 1955). With the advent of MALDI, this innovation was further elaborated by introducing a delay time between desorption/ionization laser pulse and the pulsed extraction/acceleration of ions from the source (Brown and Lennon, 1995). Improvements in TOF resolution were furthered by the development of the reflectron-TOF mass analyzer (Mamyrin et al., 1973). Reflectron-TOFs extend the ion flight distance/time (and thus resolution) by deflecting ions nearly 180° from their original trajectory from the source to a detector. This configuration allows a more compact instrument without the necessity of increasing the length of a linear flight tube to increase resolution. In addition, the reflectron provides high-order focusing of ions beyond that achievable on linear instruments resulting in an increase in resolution and mass accuracy. The success of the reflectron-TOF configuration led to the design of multiple reflectrons sequentially arranged, i.e., multi-pass reflectron configurations to extend the ion flight path and with it an increased resolution.

The success of the reflectron-TOF-MS to increase resolution and mass accuracy led to its incorporation into "hybrid" instruments that combined the reflectron-TOF with quadrupole mass analyzers (Shevchenko et al., 1997) and quadrupole ion traps (Qian et al., 1995). The reflectron-TOF served as the "back-end" mass analyzer of these instruments providing high resolution and mass accuracy of fragment ions generated by the "front-end" quadrupole mass analyzers.

The first significant impact of TOF-MS on proteomics came from the use of MALDI-TOF instruments for *peptide mass fingerprinting* or PMF (Henzel et al., 2003). By this approach, proteins are separated typically by 1-D or 2-D gel electrophoresis. Individual gel bands (or spots) are excised and subjected to in-gel digestion usually with trypsin. The resulting peptides are then ionized my MALDI and analyzed by TOF-MS. An MS spectrum is obtained ("fingerprint") consisting of primarily singly charged peptide ions. The m/z of peptide ions are then compared against a protein database that is derived from genomic sequencing. The database of protein sequences is digested in silico resulting in a list of m/z of in silico peptide ions for each protein sequence. The m/z of these in silico peptide ions are compared against the experimentally observed m/z of the MS spectrum. A scoring algorithm is used to calculate the probability that an identification occurred by chance. As one of the first mass spectrometry-based techniques developed for proteomics, PMF was very successful and continues to be used in conjunction with other proteomic techniques. Its one drawback is that a gel band (or spot) may contain multiple proteins. The peptides generated from digestion of multiple proteins can complicate the identification process when the MS spectrum is compared to in silico MS spectra of single proteins.

11.2.2.7 Tandem Time-of-Flight Mass Spectrometry (TOF-TOF-MS)

Because of the potential for high-throughput proteomic analysis by MALDI-TOF-MS, it quickly became apparent of the advantages of a MALDI-TOF-MS with *tandem* MS capabilities, specifically for MS/MS of peptides. At first this objective was attempted with MS instruments by scanning the reflectron voltages in such a way as to identify fragment ions from specific precursor ions observed in normal TOF-MS mode. As this scanning could only be done in finite steps to avoid ion defocusing, the scan "steps" had to be "stitched" together post-acquisition. The energy for ion fragmentation came from energy deposited into the molecule during the MALDI desorption/ionization process. This type of dissociation was referred to as *post-source dissociation* or PSD. This particular experimental approach was facilitated by the development of the curved-field reflectron mass analyzer (Cornish and Cotter, 1993) which allowed fragment ions produced from PSD to be more easily focused and detected.

The development of MALDI tandem TOF instruments (i.e., TOF-TOF) was very much an extension of existing TOF technology. A sequential, co-linear arrangement of two TOF mass analyzers was used. The first TOF analyzer was used for ion mass selection on the basis of its arrival time at a "gate." PSD (or high-energy CID) of the mass-selected ions resulted in fragment ions which were analyzed by the second TOF (Medzihradszky et al., 2000; Schnaible et al., 2002; Cotter et al., 2004). Several TOF–TOF instruments are now available commercially. In one specific design, precursor ions are isolated for fragmentation on the basis of their arrival time at a timed-ion selector (mass gate) at the end of the first TOF analyzer. After acceleration in the first source to ∼8 kV, ions are decelerated to 1–2 kV for fragmentation in a collision cell located after the first TOF analyzer. The collision gas is air, nitrogen, or argon. Precursor ion fragmentation can occur as a result of PSD and/or CID. Fragment ions are then reaccelerated to 16–17 kV in the second source and analyzed in the second TOF analyzer which is a reflectron-TOF. Reflectron-TOF resolution is 15,000–20,000 (FWHM) with a mass accuracy of 50 ppm (external calibration). In order to increase sample throughput for proteomic applications, the latest tandem MALDI instruments employ a pulsed solid-state YAG laser (third harmonic, $\lambda = 335$ nm) with a repetition rate of 200 Hz with very rapid sample plate robotics to avoid complete desorption of matrix from any one site of a matrix spot. Contrast this with the pulsed nitrogen laser ($\lambda = 337$ nm), still used in conventional MALDI-TOF instruments, that has a repetition rate of 20 Hz at best (and normally operated at 10 Hz or lower) (Medzihradszky et al., 2000).

One advantage of the tandem MALDI instruments (beyond generating high-throughput MS/MS data of peptide ions) is that peptide ions are fragmented at significantly higher collision energies than that obtained on quadrupole, ion trap, or FT-ICR instruments. This has certain advantages in that some of the fragmentations occur as a result of direct bond cleavage. At collision energies of ∼100 eV (or below), fragmentation occurs, not as a result of direct bond cleavage but as a result of energy deposition followed by energy randomization followed by dissociation at the lowest energy dissociation channel, i.e., a statistical dissociation. For example, leucine and isoleucine have exactly the same elemental mass but different geometric structures (structural isomers) that cannot be distinguished from one another by low-energy CID. However, at collision energies of 1–2 keV these two residues give slightly different fragment ions which can be used to distinguish these two nearly identical residues (Medzihradszky et al., 2000).

Tandem MALDI instruments have also found utilization in conjunction with off-line LC separation of peptides (Pan et al., 2008). Deposition of the peptide-containing LC eluent onto a 384-spot MALDI plate allows correlation of the MS/MS identification of a peptide to a specific LC retention time. In contrast with online LC/MS/MS analysis, one can return to a matrix spot of interest to do further interrogation. Tandem MALDI has also found applications in tissue imaging, e.g., diseased versus normal, as well as monitoring of the distribution of a drug and its metabolites in tissues (Reyzer and Caprioli, 2005).

11.2.2.8 Quadrupole-TOF Mass Spectrometry (Quadrupole-TOF-MS)

One of the most successful instruments for proteomic research is a hybrid configuration instrument: the quadrupole/time-of-flight (TOF) mass spectrometer (Morris et al., 1996; Shevchenko et al., 1997; Chernushevich et al., 2001). The quadrupole/TOF is essentially a triple quadrupole instrument where the third quadrupole has been replaced with a reflectron-TOF mass analyzer. This relatively simple design innovation resulted in an instrument with the ion selection and fragmentation characteristics of a triple quadrupole with the resolution (10^4 FWHM) and the mass accuracy (10 ppm) of a TOF instrument. Like the triple quadrupole, the quadrupole/TOF does *not* store ions (at least not for any appreciable amount of time). Ions are mass selected by the first quadrupole and fragmented in the second quadrupole, and the fragment ions are analyzed by the TOF. A critical feature of this instrument design is that, whereas the first and second quadrupoles are co-linear, ions transferred to the TOF source (from the second quadrupole) are extracted/accelerated *orthogonally* from their direction of introduction. This ensures the highest resolution and mass accuracy obtainable with the TOF analyzer. In terms of resolution and mass accuracy, the quadrupole/TOFs represent a significant improvement over the QIT instruments. Although the quadrupole/TOFs do not possess the resolution or mass accuracy of FT-ICR instruments, they are approximately one-half to one-third the cost of these instruments. The quadrupole/TOF instruments can utilize both ESI and MALDI (Loboda et al., 2003) for ion generation.

In terms of operation for bottom-up proteomic experiments, quadrupole/TOFs acquire a broad-range *m/z* scan as the first acquisition step. From this "survey" scan, the most abundant ions are selected for *m/z* isolation by the first quadrupole, followed by fragmentation in the second quadrupole and analysis of the fragment ions in the TOF. The resolution and mass accuracy of the quadrupole/TOF allow for the determination of the *m/z* and charge state of an ion "on the fly" from the survey scan. This information is useful as the collision energy can be automatically adjusted to the *m/z* and charge of the ion being fragmented, i.e., a rolling collision energy. Knowledge of the charge state prior to fragmentation is also useful when analyzing tryptic digested protein samples since it is possible to *exclude* from fragmentation those ions whose charge would indicate that they are unlikely to be peptide ions, e.g., singly charged ions. Peptides generated from tryptic digestion are typically doubly, triply, or quadruply charged when ionized by ESI. Thus, important analysis time is not wasted on isolating and fragmenting singly charged contaminant ions. Finally, database searches of MS/MS data collected on a quadrupole/TOF instrument provide 10 ppm mass accuracy of the m/z of the precursor and fragment ions in addition to the charge state of the precursor ion, thus significantly reducing false-positive peptide identifications.

11.2.2.9 Quadrupole Ion Trap-TOF Mass Spectrometry (QIT-TOF-MS or IT-TOF-MS)

Another hybrid instrument is the quadrupole ion trap-TOF developed by Lubman and coworkers (Qian et al., 1995; Qian and Lubman, 1995). This instrument combines the tandem (MS^N) capabilities of the quadrupole ion trap with the resolution and mass accuracy of a reflectron-TOF analyzer. Utilizing a variety of ionization sources, there are several commercially available versions of this instrument and are most commonly used for small molecule analysis as well as proteomic research (Wu et al., 1996).

11.2.2.10 Ion Mobility Spectrometry (IMS)

Ion mobility spectrometry (IMS) when coupled to mass spectrometry has also found a niche in proteomic research. Ion mobility spectrometry involves separation of ions in a drift tube across which there is an electrostatic potential. The tube is pressurized with helium gas at a few torr. Ions are generated by ESI, stored briefly in a quadrupole ion trap, and then pulsed into the drift tube. Ions are pulled forward by the electrostatic potential difference across the drift tube while simultaneously experiencing numerous collisions with helium atoms resulting in a viscous drag effect on the ions. The time required for an ion to traverse the drift tube (measured in milliseconds) is dependent not only on the m/z of the ion but also on its *shape*. Two ions with the same mass and m/z but with different collision cross sections (Ω) will traverse the drift tube at different velocities with the ion with the smaller Ω traversing the drift tube faster than the ion with the larger Ω. For example, a native-state protein will typically have a folded compact structure with a Ω that is smaller than the same protein in its unfolded, denatured state. IMS can separate such protein ions with the same m/z but with different conformational structures. It is still necessary to measure the m/z of ions exiting the drift tube, and this is accomplished by ion transfer to the source of an orthogonal extraction reflectron-TOF mass analyzer. The microsecond timescale of TOF-MS analysis is particularly suited for the measurement of ions separated on the millisecond timescale by ion mobility as one can obtain numerous TOF "slices" of packet of ions of a specified drift time. An ion chromatogram can then be constructed of ion drift time versus m/z. If liquid chromatography (LC) is added to the analysis prior to ESI, a reconstructed chromatogram can include three parameters: LC retention time, ion drift time, and ion m/z.

In addition to studying conformational structures of a variety of low symmetry molecules such as peptides, proteins, synthetic polymers, oligosaccharides, IMS-MS is also particularly suited for analysis of complex mixtures such as protein digests from "shotgun" proteomic experiments (Valentine et al., 1998; Hoaglund et al., 1998; Hoaglund-Hyzer et al., 2000, 2002; Lee et al., 2002). The resolving power of IMS has been increased by the development of tandem IMS (IMS-IMS and IMS-IMS-IMS) using a series of interconnected drift tubes where ions are transferred from one drift tube to another. Ion funnels are utilized to reduce ion dispersion caused by diffusion prior to exiting one drift tube and transfer to the next tube (Merenbloom et al., 2006). A modified quadrupole-TOF instrument with a drift tube located between the first quadrupole and the TOF is now commercially available and has been utilized for conformational studies of proteins (Ruotolo et al., 2007).

11.2.3 Ion Dissociation/Manipulation in Mass Spectrometry-Based Proteomics

11.2.3.1 Collision-Induced Dissociation (CID) and Collision-Activated Dissociation (CAD)

Dissociation techniques (as well as ion/molecule and ion/ion reactions) play a critical role in both mass spectrometry and proteomics. Collision-induced dissociation (or CID) also referred to as collision-activated dissociation (or CAD) is probably the most widely used dissociation technique. CID (or CAD) involves the fragmentation of a molecular ion by collision with neutral target atom or molecule. The energy deposited into the molecular ion as a result of a collision is rapidly redistributed throughout the vibrational/rotational degrees of freedom of the molecule. This redistribution of the internal energy eventually results in fragmentation of the molecular ion at its weakest bonds as these dissociation channels are the most accessible with respect to the internal energy of the molecule. Such a statistical dissociation involves collision energies in the tens to hundreds of electron volts (eV). However, collision energies in the kilovolt range may result in fragmentation by direct bond rupture from collision with the target gas.

11.2.3.2 Sustained Off-Resonance Irradiation Collision-Activated Dissociation (SORI-CAD)

A variation of CID/CAD is sustained off-resonance irradiation collision-activated dissociation or SORI-CAD (Senko et al., 1994). This technique is primarily available on FT-ICR instruments. In essence, ions trapped in an ICR cell are excited slightly from their cyclotron orbits by off-resonance excitation. The ions, thus excited, collide with background gas molecules that are pulsed into the ICR cell. Ion–molecule collisions deposit energy into the molecular ions and the ion subsequently fragments. The technique is primarily used for depositing small amounts of energy into a molecule and monitoring fragmentations over extended periods of time (seconds to minutes).

11.2.3.3 Post-source Dissociation (PSD)

Post-source dissociation or PSD is the fragmentation of a molecular ion *after* the ion has been accelerated from the source. The energy of fragmentation comes from the process of ionization or collisions with background gas in the source as the ion is accelerated. PSD thus relies on the intrinsic delay time between energy deposition into the ion and its actual fragmentation after the ion has reached its full acceleration. PSD is possible on a number of instruments, e.g., TOF-MS, triple quadrupoles as well as hybrid instruments. PSD can be deliberately propagated or it may be an unwanted artifact of instrument design. In MALDI-TOF and MALDI-TOF-TOF instruments, PSD can be deliberately increased by simply increasing the laser power (fluence) above normal operating conditions resulting in absorption of photons by the molecular ion. In instruments that use differential vacuum pumping, such as triple quadrupole and hybrid instruments, PSD can also be increased by increasing the acceleration of ions *prior* to their entry into the high-vacuum region of the mass spectrometer.

11.2.3.4 Infrared Multiphoton Dissociation (IRMPD) and Blackbody Infrared Radiative Dissociation (BIRD)

*I*nfrared *m*ulti*p*hoton *d*issociation (or IRMPD) and *b*lackbody *i*nfrared *r*adiative *d*issociation or BIRD (Price et al., 1996) are different but related techniques for dissociation of molecular ions by absorption of IR photons over brief or long periods of time. IRMPD and BIRD are primarily available on FT-ICR instruments. IRMPD has found applications in the analysis of glycosylated peptides and proteins as well as the analysis of long complex sugar and carbohydrate structures (Lancaster et al., 2006). BIRD is not widely used for proteomic studies but has found use in fundamental studies of ion chemistry. Both dissociation techniques involve statistical dissociation of molecular ions by gradual "heating" of the molecular ions in the ICR cell through absorption of IR photons either from a continuous CO_2 laser (as in the case for IRMPD) or by gradual heating of the stainless steel vacuum chamber surrounding the ICR cell (as in the case of BIRD).

11.2.3.5 Electron Capture Dissociation (ECD)

*E*lectron *c*apture *d*issociation (or ECD) and *e*lectron *t*ransfer *d*issociation (or ETD) are related dissociation techniques that are relatively new and yet have had a tremendous impact on bottom-up and top-down proteomic research. First reported in 1998, ECD involves recombination of electrons with multiply charged (protonated) peptide and protein ions (Zubarev et al., 1998). ECD was first demonstrated on an FT-ICR instrument. Unlike CID/CAD (and other ergodic dissociation techniques) that generates the peptide backbone fragmentations producing *b* and *y* ions, ECD produces primarily *c* and *z* ions. In contrast with CID/CAD which often results in dissociative loss of PTMs, ECD leaves PTMs attached to the fragment ion which facilitates their localization and identification. However, ECD suffers from low fragmentation efficiency due to the difficulty of obtaining good spatial overlap of the electron beam and the cations in the ICR cell. However, technical improvements have been made to increase the overlap between the electron beam and the ion cloud resulting in an increase in dissociation efficiency. There has been some limited success with adapting ECD to QIT instruments.

11.2.3.6 Electron Transfer Dissociation (ETD)

First reported in 2003, ETD involves an ion/ion recombination with transfer of an electron from a small singly charged molecular anion (fluoroanthracene anion) to a doubly, triply, or quadruply charged (protonated) peptide (Syka et al., 2004). This charge transfer process, which was demonstrated on a linear ion trap instrument, results in efficient fragmentation of the peptide backbone producing primarily c and z fragment ions similar to ECD. And like ECD, PTMs remain attached to the fragment ions facilitating their localization and identification. The efficiency of recombination and peptide backbone fragmentation (at nearly every residue) facilitates relatively simple interpretation of MS/MS spectra in spite of the low resolution and mass accuracy of the LIT instruments. The success of ETD as a peptide and protein sequencing tool has resulted in LIT instruments being added as an adjunct to FT-ICR and Orbitrap^{TM} instruments. In such a hybrid configuration, the LIT is used to carry out ETD fragmentation and the fragment ions are transferred to the FT-ICR^{TM} or Orbitrap instruments for high-resolution/high mass accuracy analysis.

11.2.3.7 Gas-Phase Ion/Ion Chemistry

In addition to bringing about molecular ion dissociation, ion/ion reactions have also been used to reduce the complexity of MS/MS spectra generated by CID and ETD and other dissociation techniques (Reid and McLuckey, 2002; Coon et al., 2005). In this case, recombination results in the *removal* of a proton from multiply protonated fragment ions to a small molecular anion (e.g., benzoyl anion). Deprotonation of multiply protonated fragment ions results in singly protonated fragment ions which greatly facilitates interpretation of MS/MS spectra. Deprotonation of fragment ions can also be accomplished by ion/molecule gas-phase reactions where a gas-phase neutral molecule of high basicity collides with and removes a proton from a multiply protonated fragment ion.

11.2.4 Quantitative Techniques for Proteomic Analysis

11.2.4.1 Two-Dimensional Gel Electrophoresis (2-DE)

*Two-d*imensional gel *e*lectrophoresis (2-DE) has been (and continues to be) the "classical" approach for separating and quantifying the proteome of single-cell and multicellular organisms. Even before the development of mass spectrometry-based platforms for proteomic identification, 2-DE had already established itself as a technique for the separation and analysis of large numbers of proteins by their p*I* and MW (MacGillivray and Rickwood, 1974; O'Farrell, 1975). With the wealth of biological information contained in genomic databases becoming available and the rapid improvements in the ionization and analysis of biomolecules in the field of mass spectrometry, it became feasible, within a reasonable analysis time, to identify a large number of protein spots observed in 2-D gels.

The first of these 2-D gel/mass spectrometry-based proteomic techniques combined the qualitative and quantitative analyses of proteins by 2-DE separations with in-gel digestion, peptide mass "fingerprinting" using MALDI-TOF-MS, and database searching (Rabilloud, 2002; Sinha et al., 2001). This approach has been very successful such that it continues to be used by many proteomic researchers. In this approach, protein expression (i.e., up/downregulation of a protein) is measured by comparative image analysis of the spot intensities (or "volumes") of two gels: one gel representing protein expression under "normal" conditions, i.e., the "control" sample, and the other gel representing protein expression under the "perturbed" or "stressed" state of the organism. The "perturbation" could be any of an infinite number of possible stimuli or environmental stressors on the organism, e.g., exposure to a drug, a change of pH, or osmolarity or temperature. A number of software packages have been developed for image analysis of 2-D gels with conventional gel stains, e.g., Coomassie blue or silver staining. Under this proteomic approach, mass spectrometry is used exclusively for protein identification. Even with the higher resolving power of 2-DE, it is not an unusual occurrence that a single 2-D gel spot may actually represent not one protein, but several proteins. Such an occurrence raises complexities with respect to correlating the change in gel spot intensity to the change in expression of a single protein. However, with respect to protein identification, this increased complexity has been surmounted to some degree by the use of nano-liquid chromatography (nano-LC) coupled with tandem mass spectrometry (MS/MS). Numerous tryptic peptides from multiple proteins may confound analysis by peptide mass "fingerprinting"; however, nano-LC/MS/MS can provide sequence-specific fragmentation of tryptic peptides leading to the identification of multiple proteins contained in a single gel spot.

11.2.4.2 Difference Gel Electrophoresis (2-D DIGE)

The labor-intensive nature of 2-DE gel work and its reliance on replicate analysis, as well issues related to sample-to-sample and gel-to-gel variability, have led to improvements in the quantitative aspect of 2-D gel analysis. Of particular significance is *di*fference *g*el *e*lectrophoresis or 2-D DIGE (Unlü et al., 1997; Tonge et al., 2001). In this approach, two (or three) different fluorescent dyes are used to tag proteins in two (or three) different samples, e.g., "control" versus "stressed #1" or "stressed #2." The fluorescent dye molecules covalently bind to the side chain amine groups of lysine residues. The amount of dye is adjusted to a low concentration in order to allow one dye molecule per protein molecule in a sample. The dyes fluoresce at different wavelengths; thus it is possible to correlate the amount of fluorescence in a spot to the amount of a particular protein (or proteins) in a gel spot of a particular sample. This technique has the advantage that two (or three) samples are run on a single gel. This allows the most exact comparison of changes in gel spot intensities (or volumes), i.e., protein expression, from two (or three) samples because the samples are run under exactly the same gel conditions. Contrast this with conventional 2-D gels (one sample/one gel) where image analysis requires distortion (or warping) of the gel images in order to better align gel spots from multiple gels for comparison purposes. Multiple samples on a single gel also reduce the total number of gels required to run. Finally, the dynamic range of fluorescent dyes is purported to be four orders of magnitude which is very useful for quantitative studies (Tonge et al., 2001).

11.2.4.3 "Shotgun" Proteomics and MudPIT

Even with these improvements in quantitative 2-D gel analysis there remain significant limitations. For instance, highly basic (>pI 10), highly acidic (<pI 3.0), very small (MW < 10 kDa), very large (MW > 140 kDa), and low-abundant proteins are difficult to analyze by 2-DE. In addition, membrane proteins are also difficult to analyze by 2-DE due to solubility issues. In response to these limitations and driven by the goal of increased proteomic throughput, new mass spectrometry-based techniques and chromatographic techniques were developed to analyze the peptides from direct enzymatic digestion of a proteome without any prior protein separation or fractionation. This approach, referred to as "shotgun" proteomics, involves enzymatic digestion of a complex mixture of proteins (such as that found in a proteome or subproteome of an organism) to produce an even more complex mixture of peptides (McCormack et al., 1997). These peptides are then separated by reverse-phase microcapillary HPLC, ionized by ESI, and subjected to CAD/MS/MS analysis. Sequence-specific fragmentation of these peptides allowed their identification by comparison to a database of in silico fragment ions of peptides generated from in silico enzymatic digestion of proteins in databases using a database software search engine, e.g., SEQUEST (McCormack et al., 1997). However, it quickly became apparent from the sheer complexity of the number of peptides generated by the "shotgun" approach that many peptides would not be detected or identified due to co-eluting peptides having different ionization efficiencies and abundances as well as instrument limitations of MS/MS duty cycle. Improvements in chromatographic separation could temporally extend peptide elution and resolution thus leading to increased number of peptide identifications. Multidimensional chromatographic (LC/LC) separation was introduced to address the complex mixture of peptides from "shotgun" proteomics (Opiteck et al., 1997; Wolters et al., 2001). *Multidimensional protein identification technology* or MudPIT combines *strong cation exchange* (SCX) chromatography with reverse-phase chromatography in a single biphasic analytical column interfaced to a tandem mass spectrometer (Wolters et al., 2001). A sample is loaded onto the SCX stationary phase of the column, and peptides were selectively eluted from the SCX phase using discrete salt concentration steps. Selective elution of a *portion* of the peptide mixture at a single salt concentration is followed by reverse-phase separation of those peptides on the C_{18} stationary phase part of the analytical column. Peptides are ionized by ESI and analyzed by MS/MS. This online fractionation of the "shotgun" peptide mixture allowed for more automated LC/LC/MS/MS analysis increasing the detection and identification of peptides than would be possible using a single stationary phase chromatographic separation. One drawback of the MudPIT approach is that chromatographic analysis time is increased from ~2 to ~20 h depending on the number of salt fractionation steps used. MS data acquisition for such an extended period of time resulted in the generation of thousands of MS/MS spectra from a single MudPIT experiment. As the data generated from such experiments could no longer be analyzed manually, software programs have been developed to analyze MS/MS automatically.

11.2.4.4 Stable Isotope Labeling for Quantitative Mass Spectrometry-Based Proteomics

Quantitation of "shotgun" proteomics was addressed by the invention of isotopically labeled chemical reagents. The first of these was named *i*sotope-*c*oded *a*ffinity *t*ags or ICAT (Gygi et al., 1999a). ICAT are isotopically labeled chemical reagents that have, at one end of the molecule, functional groups capable of covalently bonding to sulfhydryl groups of cysteinyl residues. The other end of the ICAT molecule has a biotin affinity tag and between is an isotopically labeled "linker" polyether chain. The "heavy" ICAT reagent possesses eight deuterium on the "linker" region of the molecule. The "light" ICAT reagent has eight hydrogen on the "linker." The isotopically "light" and "heavy" ICAT reagent is then used to label equal amounts of extracted proteins from the organism under normal conditions (the control) and a sample from the organism while under some physiological stress or perturbation. After ICAT labeling, the two samples are combined and digested. The combined digested sample is subjected to affinity binding extraction so that only peptides with the ICAT biotin tag are retained. This greatly reduces the number of peptides analyzed by LC/MS/MS as the cysteine residue is relatively rare in proteins. Only the ICAT-labeled peptides are analyzed by LC/MS/MS. As the "light" and "heavy" ICAT-labeled peptides are chemically identical their chromatographic retention times are identical (although there may be slight differences in retention time due to the chromatographic effect of eight hydrogen versus eight deuterium). Co-elution of a heavy/light pair of ICAT-labeled peptides allows simultaneous MS analysis and alternating MS/MS analysis of the two peptides. For doubly charged peptides, the *m/z* difference between the "light" and the "heavy" labeled peptides is 4 Th. MS/MS analysis is used for proteomic confirmation of each peptide. Relative quantitation of the "heavy" and "light" peptides is accomplished by integrating the area of the chromatographic peak generated from an extracted ion chromatogram of the LC/MS data for each "heavy" and "light" peptide *m/z* (Gygi et al., 1999a). An updated version of ICAT, cleavable ICAT, was used by Hardwidge et al. to identify host proteins that bind to virulence proteins of enteropathogenic *Escherichia coli* (Hardwidge et al., 2006).

Another isotopic labeling technique for quantitative proteomics that measures *absolute* concentrations of a peptide/protein is AQUA (an acronym for *a*bsolute *quan*titation) (Gerber et al., 2003). This approach involves the use of a synthetic peptide as an internal standard that is isotopically labeled at a single residue, e.g., leucine ($^{13}C_6$ and ^{15}N). The protein to be quantified generates the native version of the peptide upon enzymatic digestion. The MS/MS fragmentation pattern of the peptide internal standard is used to select two (or more) characteristic precursor ion to fragment ion transitions to be used in a *s*elected *r*eaction *m*onitoring (SRM) experiment using a triple quadrupole mass spectrometer. As the native peptide is chemically identical to the internal standard peptide, the two peptides will elute with the same retention time and ionize with the same efficiency. Monitoring of the transitions of the labeled standard and the unlabeled native peptide will allow the generation of an extracted ion chromatograms whose integrated area is used for measuring relative abundances. Knowledge of the absolute amount of internal standard added to a sample allows absolute quantitation of the native peptide and thus the native protein. This approach has the added advantage that the absolute concentration of a posttranslationally modified protein can be measured if a peptide internal standard can be synthesized with the desired covalent modification, e.g., glycosylation, phophorylation (Gerber et al., 2003).

Another technique that has become increasingly utilized for quantitative proteomic studies is metabolic incorporation of stable isotopes (nonradioactive) into proteins during cell growth and division. The approach involves using an isotopically enriched media (e.g., ^{15}N) for cell growth. A parallel experiment is also conducted with naturally abundant isotope media. As the cells grow and divide under the specific conditions of the experiment, the stable isotope is incorporated into the organism's biosynthesized proteins. Equal amounts of cells from both the control sample (cells grown on naturally abundant stable isotope media) and the cells grown on isotopically enriched media are combined, lysed, separated (e.g., PAGE), digested, and identified by mass spectrometry, e.g., PMF by MALDI-TOF-MS (Oda et al., 1999). Mass spectrometry not only identifies the proteins by PMF but also measures the relative abundance of peptides and thus the up/downregulation of proteins. As the peptides are isotopically distinguishable but chemically identical, their ionization efficiencies are identical, and differences in the measured relative abundances of isotopically labeled versus naturally abundant labeled peptides allow relative quantitation of the peptide and thus the protein (Oda et al., 1999).

In a variation of this approach, Veenstra et al. used isotopically labeled leucine (Leu-D_{10}) to label biosynthesized proteins of *E. coli* (a multi-auxotrophic K12 strain) grown on minimal media (Veenstra et al., 2000). A parallel experiment was conducted on *E. coli* grown on minimal media without the labeled leucine. Labeled and unlabeled bacterial cells were harvested, combined, lysed, and chromatographically separated by *c*apillary *i*soelectric *f*ocusing (CIEF) and analyzed by FT-ICR-MS. Co-elution of labeled and unlabeled proteins by CIEF allowed simultaneous high-resolution MS measurement of the molecular weight of labeled and unlabeled proteins. The mass difference between labeled and unlabeled proteins allowed calculation of the number of leucines incorporated into the protein.

This information and the very accurate MW of the protein obtained by FT-ICR-MS allowed identification of the protein (Veenstra et al., 2000). This approach was extended and confirmed using other labeled amino acids (arginine, histidine, isoleucine, etc.) and applied to the identification of proteins of *E. coli* as well as *Saccharomyces cerevisiae* (Martinović et al., 2002).

Isotopically labeled amino acids have also been utilized in quantitative proteomics. Stable *i*sotope *l*abeling by *a*mino acids in *c*ell culture (or SILAC) involves the use of isotopically labeled *essential* amino acid (e.g., Leu-^{2}H$_3$ also known as Leu-D$_3$) that is included in cell culture media that is deficient in that amino acid in its naturally abundant isotopic form (Ong et al., 2002). An essential amino acid is not metabolically synthesized by an organism, and thus for normal cell growth and division, these amino acids must be provided from the organism's environment. An isotopically labeled essential amino acid is incorporated into proteins during cell growth and division as there is no metabolic source for this amino acid. In a typical two-state SILAC experiment, cells are cultured under "normal" conditions with media supplemented with the "unlabeled" naturally isotopic abundant essential amino acid. In parallel, cells are cultured under stressed or perturbated conditions, specific to the experimental problem being investigated, with media supplemented with isotopically labeled essential amino acid. Equal amounts of cells from both cultures are harvested, combined, and lysed and the proteins extracted. The proteins may be digested with trypsin directly or separated (e.g., 1-D or 2-D PAGE) and then digested. The difference in *m/z* of a labeled versus unlabeled peptide will be ($n \times$ 3 Th)/z (where n is the number of leucines in the peptide and z is the charge of the peptide). The relative abundance of labeled versus unlabeled peptides when analyzed by mass spectrometry correlates to the relative abundance of the protein from which it came. PMF or MS/MS and database searching are used for peptide/protein identification. This technique has proven successful for identifying the upregulation of specific proteins during differentiation of mammalian cells (Ong et al., 2002). SILAC appears to be applicable to many other cell types (Romijn et al., 2005). A recent review discusses the use of SILAC for quantitative mapping of phosphorylation sites in bacteria (Soufi et al., 2008). SILAC has also been extended to the quantitation of intact proteins (Waanders et al., 2007).

iTRAQ is another strategy involving isotopically labeled chemical reagents for quantitative "shotgun" proteomics (Ross et al., 2004). The iTRAQ approach involves the use of multiple isotopically labeled, but *isobaric*, chemical reagents that covalently bond to primary or secondary amines. For peptides and proteins, this means the label attaches to the N-terminus and/or at the ϵ-amines of lysine side chains. The labeling process does not significantly affect the ionization efficiency of the labeled peptide by ESI or MALDI. An iTRAQ-labeled tryptic peptide that is subjected to fragmentation by CAD/MS/MS will generate a low *m/z* iTRAQ fragment ion (reporter ion) that is isotopically distinguishable from the other iTRAQ reagents used in the experiment. This reporter fragment ion is then used to quantitate the abundance of the peptide to which it was previously bound before fragmentation. A typical sample protocol follows. Equal amounts of protein from multiple samples are each subjected to disulfide reduction, alkylation, and tryptic digestion. Each digested sample is then labeled with an isotopically unique iTRAQ reagent. After labeling, all samples are combined, chromatographically separated, and analyzed by LC/LC/MS/MS. Because each iTRAQ reagent is isotopically unique (but isobaric in total mass to the other iTRAQ reagents), iTRAQ labeling of the same peptide from different samples will result, upon sample combination, in detection of an isobaric m/z for the labeled peptides. However, upon fragmentation, the labeled peptides generate isotopically distinguishable iTRAQ reporter fragment ions. Each isotopically distinguishable iTRAQ fragment ion is then used to quantitate the abundance of that peptide from each particular sample. In addition, iTRAQ labeling and subsequent fragmentation does not affect the efficiency of peptide fragmentation used for peptide identification. Finally, absolute quantitation can also be accomplished by including an iTRAQ-labeled peptide of known concentration to the sample (Ross et al., 2004). An excellent example of the use of iTRAQ reagents was demonstrated in a proteomic study of the extremophile *Sulfolobus solfataricus* (Chong et al., 2007).

11.2.4.5 Label-Free Quantitative Mass Spectrometry-Based Proteomics

Early on, proteomic researchers using mass spectrometry-based techniques noted that the peptides of high-copy proteins were detected more frequently than peptides from lower copy proteins. It was hypothesized that it might be possible to correlate peptide detection to protein abundance, i.e., protein quantitation, without the necessity of using isotopic labeling or other labeling techniques. A number of label-free approaches have been investigated, and some have become highly developed. These techniques have used various criteria for correlating to protein abundance, e.g., protein sequence coverage, number of peptides hits, the number of peptide hits normalized to protein size, spectral counting, and the integrated peak area of an extracted ion chromatogram peptide precursor ion in LC/MS mode. Sequence coverage was used by Florens and coworkers as an approximate measure of relative protein abundance (Florens et al., 2002). Liu et al. have relied on statistical analysis of sampling in a shotgun proteomic experiment to argue that spectrum counting, i.e., the number of MS/MS of a peptide, is a better criteria for estimating relative protein abundance (Liu et al., 2004). Other researchers have suggested that the number of peptide "hits" is a more useful criteria (Gao et al., 2003; Pang et al., 2002). Mann and coworkers have estimated relative protein abundance from the number of sequenced peptides of a protein normalized to the size of the protein, "size" being determined by the number of theoretically possible tryptic peptides, and a criteria/formula was referred to as the *protein abundance index* or PAI (Rappsilber et al., 2002). A further elaboration of this approach, *exponentially modified* PAI or emPAI, was demonstrated for absolute quantitation (Ishihama et al., 2005). Bondarenko and Chelius used the integrated peak area of an extracted ion chromatogram of a peptide precursor ion (in LC/MS mode) to determine the relative abundance of proteins in a protein mixture (Bondarenko et al., 2002; Chelius and Bondarenko, 2002). This approach was further tested on complex biological samples (Chelius et al., 2003) and further validated by Shen and coworkers on data collected on QIT and FT-ICR instruments applying a rigorous statistical analysis to the data (Wang et al., 2006). Guina and collaborators used a similar label-free shotgun proteomic approach to study the regulation of virulence proteins in *Francisella tularensis* subsp. *novicida*, the pathogen responsible for the disease tularemia (Guina et al., 2007). Using their *accurate mass and time* or AMT tag strategy (to be discussed in the next section), Smith and coworkers at Pacific Northwest National Laboratory (PNNL) used integrated peak areas of peptide ions generated by nano-LC-ESI-FT-ICR to globally quantitate protein expression in *Shewanella oneidensis* under aerobic and anaerobic growth conditions (Fang et al., 2006).

11.2.5 Software for Data Analysis of Mass Spectrometry-Based Proteomics

11.2.5.1 Database Search Engines for Bottom-Up Proteomics

The four most commonly used database search engines for analysis of bottom-up proteomic MS/MS data are Sequest (Eng et al., 1994), MASCOT (Perkins et al., 1999), X! Tandem (Craig and Beavis, 2004), and OMSSA (Geer et al., 2004). Protein identifications are accomplished by comparing a peptide MS/MS spectrum to in silico MS/MS spectra in a database derived from the putative proteins from open reading frames in genomic databases. Each software uses its own algorithm for comparing, scoring, and ranking in silico MS/MS spectra to an MS/MS spectrum. Scoring algorithms typically utilize a p-value calculation, i.e., a statistical calculation of the probability that an identification occurred randomly. A very low p-value indicates a high probability that the identification is not random and therefore correct. Software search engines allow the researcher latitude adjust search parameters, such as (1) enzyme used for protein digestion; (2) the number of allowable missed enzymatic cleavage sites; (3) the charge state(s) of the precursor ion (if that information is available); (4) fixed and variable modifications of residues (e.g., methionine oxidation); (5) mass tolerances of the precursor ion and fragment ions; and (6) the database(s) to be searched. It is not unusual for proteomic researchers to process their data using multiple search engines as the results can vary to some degree. Protein identification using multiple search engines gives a researcher greater confidence in the identification, especially in cases where the number of peptides used for identification is relatively few, e.g., "one hit wonders."

11.2.5.2 De Novo Sequencing for Bottom-Up Proteomics

The foregoing database search engines rely for success on the sequence of the protein being present in the database. If the sequence (or a protein sequence of high homology) is not present in the database, then the database search will produce an identification of low confidence (high p-value). If the experimental MS/MS data are of high quality, it is possible to manually determine the sequence of all (or part) of a peptide sequence. This is referred to as de novo sequencing.

There are software programs (both open-access and commercial versions) that can analyze MS/MS data automatically to produce a de novo sequencing without comparison to protein sequences in a database. One such program that is well known is Lutefisk (Taylor and Johnson, 1997, 2001; Johnson and Taylor, 2002). Lutefisk uses a "graph theory approach" to de novo sequencing (Taylor and Johnson, 1997). PEAKS is an example of a commercially available de novo sequencing software program (Bioinformatic Solutions Inc., Waterloo, ON, Canada). PEAKS uses a dynamic programming mathematical algorithm for de novo sequencing different from that used by Lutefisk and other programs that utilize the graphical theory approach (Ma et al., 2003). PEAKS developers have also created a program for utilizing de novo sequence tags generated by PEAKS to conduct an error-tolerant database search. The error tolerance feature allows searching of protein databases for sequences that may be similar to but not *exactly* the same as the de novo sequence. This feature is particularly useful when the gene of the unknown protein may not have been sequenced, but the protein may have some homology to a protein in the database assisting in its identification. This program called *software protein identifier* or SPIDER (Han et al., 2005) which was initially stand-alone but has now been integrated into the latest version of PEAKS. A comparative study of different de novo sequencing algorithms was conducted and reported recently (Pevtsov et al., 2006).

11.2.5.3 Predictive Software for PTMs (e.g., Signal Peptides) and Protein Subcellular Localization

The most common posttranslational modification (PTM) of bacterial proteins is N-terminal methionine cleavage. The presence (or absence) of this PTM has been correlated to the protein half-life $(t_{1/2})$ and thus related to the cycle of protein synthesis and degradation in an organism. There is a rule-of-thumb that predicts the probability of N-terminal methionine removal on the basis of the penultimate N-terminal residue, i.e., the residue immediately adjacent to the N-terminal methionine (Hirel et al., 1989; Gonzales and Robert-Baudouy, 1996; Solbiati et al., 1999). Exceptions to this rule have also been noted (Frottin et al., 2006). The presence or absence of N-terminal methionine is relatively easy to confirm from top-down proteomic data as the measured MW of the protein will differ from the predicted MW by -131 Da. N-terminal methionine can also be confirmed by bottom-up proteomic data if the MS/MS data include peptides that cover the N-terminal region of the protein.

Signal peptides play a critical role in the export of proteins to specific subcellular locations. Once a protein has been exported to its proper location based on its signal peptide, the signal peptide is removed from the polypeptide chain as part of post-translational processing to create a mature, functional protein. Signal peptides are present at the N-terminus of the polypeptide chain and are typically 20–40 residues in length. Signal peptide removal is the most common PTM among bacterial proteins after N-terminal methionine cleavage. Signal peptides are very important to cellular function in both eukaryotes and prokaryotes. In consequence, software programs have been developed for their identification in silico. One of the most widely used is SignalP (and it variants) which combines *neural network* (or NN) and *hidden Markov model* (or HMM) algorithms to predict secreted proteins on the basis of their signal peptides (Nielsen et al., 1996; Bendtsen et al., 2004). Software developers also created a program, LipoP, to specifically identify the signal peptides of lipoproteins in gram-negative bacteria (Juncker et al., 2003).

Prediction of secreted proteins by identification of signal peptides is only one example of a more general bioinformatics goal of classifying proteins based on their subcellular localization. The importance of determining the subcellular localization of a protein is that localization is suggestive of protein function. In addition to signal peptides, recognition of secondary structures such as transmembrane α-helices (membrane proteins) and β-barrel strands in proteins is also used to predict subcellular localization. Software widely used for prediction of subcellular localization is PSORT (Nakai and Kanehisa, 1991) and its later versions PSORTb (Gardy et al., 2005) and WoLF PSORT (Horton et al., 2007). A very recent addition of software for subcellular location of bacterial proteins is LocateP (Zhou et al., 2008). Gardy and Brinkman have provided an excellent review of the various software programs available for determining subcellular location of bacterial proteins (Gardy and Brinkman, 2006). Rodríguez-Ortega et al. used PSORT to predict surface-exposed proteins of group A *Streptococcus* cells. In that study, proteins partially exposed at the bacterial surface were enzymatically "shaved" from the surface without cell lysis. The tryptic peptides released were then identified by mass spectrometry-based proteomic techniques and compared to predictions by PSORT (Rodríguez-Ortega et al., 2006).

11.2.5.4 Software for Deconvolution of LC-ESI-MS of Proteins

Mass spectrometry instrumentation with ESI capability often includes data analysis software with the ability to deconvolute ESI-MS data. The multiply charged ions generated by ESI of proteins may result in spectral congestion due to multiple charge state envelopes produced as a result of co-eluting proteins. In order to identify the proteins eluting at a specific LC retention time, it is necessary to sum MS spectra over a LC retention time range and deconvolute the data. A list of protein molecular weights and their corresponding ion intensities is generated with the option to highlight ion charge states in the MS spectrum. ProTrawler™ is a commercially available software (Bioanalyte, Portland, ME) designed to automatically deconvolute the charge state envelopes of LC/ESI/MS analysis of protein mixtures. As an LC run may last for minutes or even hours automated deconvolution reduces the labor-intensive nature of data analysis. Such software was used successfully by Williams and Musser at FDA/CFSAN (in collaboration with Bioanalyte) to "fingerprint" foodborne pathogens from their profile of protein molecular weights generated from deconvolution LC/MS data (Williams et al., 2002). This work will be discussed at length in a subsequent section.

11.2.5.5 Top-Down Mass Spectrometry-Based Proteomic Software

There are also software programs designed specifically for top-down proteomic analysis. Foremost among these is ProSight PTM 2.0 from the Kelleher group at University of Illinois, Urbana (Zamdborg et al., 2007).

11.2.6 Protein Microarrays: A Non-mass Spectrometry-Based Proteomic Technique

Protein microarrays have gradually made in-roads into the proteomic field. In contrast with DNA microarray, protein microarrays present significant technical challenges. Currently there are three different types of protein microarrays: quantitative, functional, and the recently developed reverse-phase microarray. Griffiths has provided an excellent overview of the current state of the art of this rapidly changing field (Griffiths, 2007). The quantitative array is probably the most well known and involves the immobilization on a prefabricated chip of a series of antibodies. The variable regions of each antibody are designed to bind to a specific epitope of a target molecule, e.g., a protein. For instance, if one is attempting to determine whether a specific protein is being synthesized by a cell, a collection of cells are lysed and their contents are brought into contact with the microarray chip. After incubation, the chip is washed repeatedly in order to remove nonspecifically bound material. A second antibody, which is fluorescently tagged, is then applied to the chip. This second antibody binds to a different region of the target molecule than the capture antibody. Once again an incubation period allows the second antibody to bind to any target molecules captured by the first antibody. The chip is then washed to remove nonspecifically bound fluorescently tagged antibody. The chip is then placed into a reader or a confocal microscope which "lights up" the fluoresce tag for visualization. Spots that fluoresce indicate that the second antibody is bound to the target molecule which itself is bound to capture antibody. This approach is referred to as a "sandwich" assay because the target molecule is bound by the capture antibody on one side of the target molecule and the fluorescently tagged antibody binds on the other "side." Obviously, the success of this approach relies heavily on antibody specificity. The possibility of nonspecific binding is probably the most significant caveat of a protein microarray, although a sandwich assay relies on the specificity of *two* antibodies reducing the likelihood of false positives. There continues to be a need for robust and validated antibodies with high affinity and specificity.

In a very interesting study, Anjum et al. used a protein microarray to capture and detect 17 different O-serotypes of *E. coli* using a cell-bound "sandwich" assay (Anjum et al., 2006). The capture antibody was generated from the antisera produced by different *E. coli*. Capture antibodies for each specific serotype were spotted at four different locations onto a 13 × 13 glass chip array measuring 3 × 3 mm. The capture antibodies are bound to the glass surface by an epoxy. The microarray chip is then inserted and secured to the base of a microtube (1.5 mL). An overnight culture of *E. coli* is boiled briefly (1 h), and an aliquot is added to the microtube. The capture antibody binds to a whole *E. coli* cell by way of its surface molecules. As part of the "sandwich" assay, the second antibody is an anti-*E. coli* core lipopolysaccharide (LPS) antibody which binds at a different location of the bound bacterial cell. Instead of the second antibody having a fluorescent tag, horseradish peroxidase binds to a biotynlated anti-mouse antibody which in turns binds to the second antibody. Addition of a chemical substrate reacts with the peroxidase generating a blue precipitate, the amount of which is measured by the amount of absorption under exposure to a red light (ArrayTube™; CLONDIAG Chip Technologies GmbH, Jena, Germany) (Anjum et al., 2006). When the array was tested against pure cultures of the original 17 *E. coli*, there was a 100% correlation between the array and the conventional serotyping methods. They also reported 88% success rate of identification in a blind trial study of 100 clinical samples of *E. coli* that had previously been serotyped using conventional methods. A failure to identify appeared to be due to the organism becoming untypeable by microarray or conventional technique after very long storage.

11.3 Applications

11.3.1 Bacterial Identification of Foodborne Pathogens by Proteomic Analysis

This section addresses the use of proteomics (specifically, mass spectrometry-based proteomics) for the identification and characterization of pathogens and, in particular, foodborne pathogens. A number of top-down and bottom-up proteomic techniques have been utilized for identification and characterization of pathogens.

11.3.2 Bacterial Identification by Top-Down Proteomic Analysis

11.3.2.1 MALDI-TOF-MS and MALDI-TOF-TOF-MS/MS

In a series of papers, Demirev and coworkers demonstrated identification of microorganisms by combining MALDI-TOF-MS analysis of intact proteins (or cell lysates) with protein database searches (Demirev et al., 1999). This approach involves comparison of the *m/z* of intact proteins observed in the MALDI-TOF-MS spectrum against the molecular weights (MWs) of proteins in a database derived from the open reading frames (ORFs) of genomic sequencing of microorganisms. This bioinformatic-based approach, initiated by Demirev et al. for identification of microorganisms by analysis of MALDI-TOF-MS data, represented an alternative to the bacterial identification by analysis of MALDI-TOF-MS data by pattern recognition analysis of the MS "fingerprint" (Jarmon et al., 2000). Implicit in the microorganism identification by bioinformatic analysis is that both the biomarker proteins and the microorganism are identified, although protein "identification" relies exclusively on a comparison of a peak *m/z* and a protein MW. A *p*-value calculation is employed that estimates the probability that an identification occurred randomly (Pineda et al., 2000). As only protein mass is the only measurable, if PTMs are present in a protein, the peak m/z will not match the database protein MW. Further development of the bioinformatic algorithm approach incorporated the possibility of the simplest PTM: N-terminal methionine cleavage (Demirev et al., 2001a). High-resolution, high mass accuracy MALDI-FT-ICR-MS confirmed many of the protein/microorganism identifications found by MALDI-TOF-MS (Jones et al., 2003). There is a general rule that predicts N-terminal methionine cleavage for bacterial proteins on the basis of the identity of the residue immediately adjacent to the N-terminal methionine. The rule is fairly reliable, although exceptions to the rule have been reported (Frottin et al., 2006; Fagerquist et al., 2005). More complex PTMs, for instance, N-terminal cleavage of signal peptides, have not been incorporated into such bioinformatic algorithms due to the complexity of accurately predicting the presence of signal peptides.

Demirev et al. demonstrated fragmentation of protein biomarkers of *Bacillus cereus* T spores ionized by ESI and fragmented by SORI-CAD and ECD using an FT-ICR mass spectrometer (Demirev et al., 2001b). They demonstrated that it was possible to obtain sequence tags from the fragment ions generated and search against genomic/proteomic databases to identity the protein and thus the microorganism. More recently, Demirev demonstrated sequence-specific fragmentation of protein biomarkers of *Bacillus atrophaeus* and *B. cereus* spores ionized by MALDI and fragmented by MS/MS using TOF-TOF mass spectrometer (Demirev et al., 2005). Singly charged protein ions were fragmented by PSD or *l*aser-*i*nduced *d*issociation (LID) from energy deposited into the protein during the ionization/desorption process. As these protein ions are singly charged (protonated), fragmentation of the polypeptide backbone is more likely to occur at aspartic acid (D) or glutamic acid (E) residues because of transfer of the acidic hydrogen of the side chain of these residues to the polypeptide backbone facilitating its cleavage. Using in-house developed software, MS/MS fragment ions were compared against a database of in silico fragment ions derived from bacterial proteins whose MWs were the same as the MW of the precursor protein ion. A *p*-value algorithm was used to calculate the probability that an identification occurred randomly and the identifications were ranked accordingly. If the top identification had a *p*-value that was orders of magnitude lower than the runner-up identification, then the top identification was considered correct (Demirev et al., 2005). This top-down proteomic approach relies on the assumption that the microorganism being analyzed has been genomically sequenced or has significant homology to a closely related genomically sequenced strain.

Very recently, Fagerquist et al. reported using web-based software (also developed in-house) to rapidly identify protein biomarkers (and their source microorganisms) from sequence-specific fragmentation of intact protein ions by MALDI-TOF-TOF-MS/MS and top-down proteomic analysis (Fagerquist et al., 2009). A simple peak-matching algorithm was used to score/rank identifications, and for purposes of comparison, identifications were independently scored/ranked using Demirev's *p*-value algorithm. Both algorithms gave quite similar rankings. The accuracy of the web-based software was tested against protein biomarkers of *Campylobacter* that had been previously identified by bottom-up proteomic techniques. The software was able to correctly identify the protein as well as the species of the *Campylobacter* from which it originated. Another feature of this software is that it allows MS/MS fragment ions to be compared to in silico fragment ions from polypeptide cleavage adjacent to specific amino acid residues, e.g., D, E, or P specific. A D/E/P-specific or D-specific comparison resulted in a relative enhancement of the score of the correct identification compared to its score from a non-residue-specific comparison (Fagerquist et al., 2009). In a subsequent report, Fagerquist et al. reported identification of protein biomarkers from pathogenic (O157:H7) and nonpathogenic *E. coli* (Fagerquist et al., 2010). It was shown that a single amino acid substitution (D ↔ N) in the YahO protein could be used to distinguish between pathogenic and nonpathogenic *E. coli* strains. Significantly, the D ↔ N substitution, which is not easily detected by MALDI-TOF-MS because it results in a protein MW change of only 1 Da, is easily detected by MS/MS due primarily to the effect of D residues on protein fragmentation efficiency (Fagerquist et al., 2010).

11.3.2.2 LC/ESI-MS

In a series of papers, Williams and coworkers at the FDA, Center for Food Safety and Applied Nutrition, reported development of an LC/MS approach for identifying and discriminating between closely related strains of bacteria (Williams et al., 2002). Their approach involves separating intact proteins from bacterial cell lysates by *high-performance* reverse-phase *liquid chromatography* (HPLC). The HPLC was interfaced to a quadrupole/TOF-MS with an ESI source. ESI of the HPLC eluent resulted in multiply charged protein ions. In order to obtain protein molecular weights, the LC/MS data are deconvoluted in 30 s LC intervals. As the LC run is typically 1–1.5 h, the deconvoluted data are combined into a single data file containing protein MW versus intensity versus LC retention time. Commercially available software has been developed to perform these data processing steps in a more automated fashion. Advantages of this approach are that a greater number of proteins are detected than that observed in a MALDI-TOF-MS spectra. In addition, the accuracy of the protein MW measurement is typically higher than that obtained by MALDI-TOF-MS. In consequence, this approach is more likely to "resolve" more closely related bacterial strains because a greater number of proteins are detected. A combined/deconvoluted LC/MS-generated spectrum of a bacterial strain can then be compared to another closely related (or more distant) bacterial strain. For instance, pathogenic and nonpathogenic strains of a bacterial species could be compared. A unique peak (or peaks) in the spectrum of the pathogenic strain that is absent from the spectrum of the nonpathogenic strain could be used as a strain-specific discriminating biomarker ion. In addition, the proteins responsible for generating unique peaks in a spectrum can be isolated from their retention time and identified by bottom-up proteomic techniques. The protein in question may be involved in (or correlated to) the virulence of the pathogenic strain (Williams et al., 2002). The only significant disadvantage of this approach is the time required for LC analysis as well as the number of LC replicates required to obtain confident detection of lower intensity unique protein biomarkers.

A few examples of this experimental approach are highlighted. *Vibrio parahaemolyticus* is a gram-negative bacterium found in fish and shellfish that can cause serious illness. Using their LC/ESI/MS deconvolution technique, Williams et al. were able to identify a uniquely discriminating protein biomarker (HU-α) from the pandemic strain O3:K6 of *V. parahaemolyticus* (Williams et al., 2004). Having identified the protein by bottom-up proteomics, they were then able to reverse engineer the primer gene that encodes HU-α in the O3:K6 strain and identify a 16-kbp insertion into the gene that could be used as a PCR-based discriminating test for this strain (Williams et al., 2004). In another study, Williams et al. used their technique to identify a unique protein biomarker in thermally tolerant strains of *Enterobacter sakazakii* (Williams et al., 2005a). *E. sakazakii* is a bacteria found occasionally in milk-based products that can cause neonatal meningitis. A unique protein biomarker was identified that was homologous to a hypothetical protein in the thermally tolerant: *Methylobacillus flagellatus* KT. Identification of this protein in only the thermally tolerant strains of *E. sakazakii* suggested a way to discriminate between thermally tolerant and intolerant strains. They went on to reverse engineer primers for the gene of this protein and successfully demonstrated that PCR amplified this gene in thermally tolerant strains only (Williams et al., 2005a). Finally, Williams et al. demonstrated the top-down LC/ESI/MS approach to distinguish between several strains of pathogenic and nonpathogenic *E. coli*. In one particular case, the LC/MS approach was used to distinguish between two O157:H7 strains having "identical phenotypic, serologic and genetic traits" (Williams et al., 2005b). The only difference between the two strains was that one did not have shiga toxin 1 and 2 genes whereas the other did. A comparison of their deconvoluted LC/ESI/MS protein profiles of these two strains revealed a protein of MW = 9,594.4 Da that was present in the strain with the toxin genes but was absent in the strain that did not. Williams et al. also reported detecting a protein with a mass of 19,107 Da that was present in *E. coli* strain O157:H$^-$ only and thus could be used as a discriminating biomarker ion. This protein was identified as cytolethal distending toxin C (*Ccdt*) with an open reading frame (ORF) MW = 19,990 Da (Williams et al., 2005b). The bottom-up proteomic identification was apparently not sufficient to identify the cause of the discrepancy between observed MW and ORF MW of this protein.

11.3.2.3 LC/LC Separations

In addition to its use in separating complex mixtures of peptides from digested proteins, multidimensional liquid chromatography (LC/LC) has also been applied to the separation of complex mixtures of intact proteins. Lubman and coworkers describe using LC/LC to separate the cell lysates of pathogenic and nonpathogenic *E. coli* by combining chromatofocusing and reverse-phase chromatography (Zheng et al.,

2003). Chromatofocusing is a technique wherein proteins are chromatographically separated by their isoelectric point (pI) by establishing a pH gradient across an ion-exchange column. Proteins are eluted from an ion-exchange column as a result of a linear pH gradient in the mobile-phase buffer. The pH is closely monitored during protein elution such that the pI of the protein can be determined based on its elution time from the column. Protein fractions collected on the basis of their pI are then subsequently separated by reverse-phase chromatography using a nonporous reverse-phase HPLC. The final protein fractions are thus identified on the basis of their pI and hydrophobicity. A 2-D protein expression map, similar to the image of a 2-D gel, allows easy comparison of LC/LC separations. Protein fractions of interest can be analyzed by top-down or bottom-up proteomic techniques. Zheng et al. used this approach to analyze cell lysates from two strains of *E. coli* O157:H7 and one nonpathogenic strain of *E. coli* to distinguish any differences in protein expression between the strains. They reported no detectable differences in protein expression between the two O157:H7 strains at a pH range of 4.0–7.0. However, they observed significant differences in protein expression between the O157:H7 strains and the nonpathogenic *E. coli* strain. Proteins that were differentially expressed between the pathogenic and the nonpathogenic strains were identified by peptide mass "fingerprinting" using MALDI-TOF-MS. The MWs of intact proteins were also measured by ESI-TOF-MS (Zheng et al., 2003).

Gunther et al. utilized a commercialized version of chromatofocusing/reverse-phase LC/LC approach in the analysis of the nonpathogenic bacteria *Pseudomonas chloroaphis* (Gunther et al., 2006). In an earlier communication, Gunther et al. had reported that *P. chloroaphis* generated rhamnolipids, an industrially useful biosurfactant molecule, at room temperature (Gunther et al., 2005). Most of the other bacteria that produced rhamnolipids were found to be pathogenic and required elevated temperatures for production of this molecule. In order to better understand the biology of *P. chloroaphis* with respect to rhamnolipid production, Gunther et al. used the LC/LC approach to study differences in protein expression under growth conditions that favored (static growth) and inhibited (shaking growth) rhamnolipid production. Proteins that were differentially expressed between the two culture conditions (and rhamnolipid productions) were identified by a combination of peptide mass "fingerprinting" and MS/MS of tryptic peptides using a MALDI-TOF-TOF-MS. A number of proteins were found to be up- or down-regulated with respect to the different growth conditions. In addition, four proteins that were identified as being differentially expressed for rhamnolipid production were identified as having an affect on biofilm production (Gunther et al., 2006).

Finally, Yates and coworkers reported using 1-D LC chromatofocusing to reduce sample complexity of cell lysates of a breast cancer cell line prior to analysis by MudPIT. After chromatofocusing fractionation, each of the 11 samples generated was subjected to digestion, and the peptides generated were analyzed by nano-LC/LC MudPIT interfaced to a linear ion trap mass spectrometer (Chen et al., 2006). Prefractionation of the proteins prior to digestion assisted in the detection of less abundant proteins (Chen et al., 2006). Thus, it is important to consider multiple proteomic approaches when deciding which technique is most likely to provide useful information about the problem under investigation.

11.3.3 Bacterial Identification by Bottom-Up Proteomic Analysis

An innovative approach for bacterial identification and characterization pioneered by Smith and colleagues at Pacific Northwest National Laboratory (PNNL) involves a hybrid bottom-up proteomic analysis. Initially used for identification of pure cultures (Lipton et al., 2002), it is now used to identify genus/species/strains within a complex bacterial community (Norbeck et al., 2006). The objective is to identify a microorganism present in a mixture of microorganisms (such as in a bacterial community) by identifying a set of peptides that are unique to a particular microorganism. Bacterial cell lysates are digested as part of a typical "shotgun" proteomic experiment. The tryptic peptides are separated by high-performance liquid chromatography (HPLC) followed by CAD/MS/MS on a medium- to low-resolution mass spectrometer (e.g., quadrupole ion trap). The tryptic peptides (and thus their proteins) are identified by database searching. The peptide/protein/microorganism identification along with the peptide LC retention time is stored in a database for subsequent use. The bacterial digest sample is also separated and analyzed by LC-FT-ICR-MS. The high resolution (>100,000 FWHM) and sub-ppm mass accuracy of FT-ICR-MS allow very precise mass measurement of the peptide molecular weight. When combined with the LC retention time of the peptide, this approach is referred to as *a*ccurate *m*ass and elution *t*ime tag strategy or AMT (Norbeck et al., 2006; Strittmatter et al., 2003). LC retention times are normalized to account for variations from sample to sample. The AMT data can then be compared against the previously acquired peptide identifications and LC retention times acquired on the lower resolution mass spectrometer. By this combined bottom-up strategy one is able to identify peptides unique to a microorganism and then use these uniquely identifying peptides to identify the microorganism in a community of microorganisms (Norbeck et al., 2006).

11.3.4 Bacterial Identification by Top-Down and Bottom-Up Proteomic Analysis

Increasingly, proteomic studies often combine both a top-down and a bottom-up analysis. The reason for this approach is that, in addition to a bottom-up identification, it may be useful (or necessary) to know the molecular weight of the mature, intact protein. It is rare that a bottom-up analysis results in 100% sequence coverage of a protein. This is usually not a problem for protein identification as only a few good MS/MS spectra of two or three peptides are required for definitive identification of the protein. However, if a protein has undergone posttranslational modifications (PTMs) or if there is some uncertainty about the correctness of the open reading frame (ORF) sequence or its start codon, then measurement of the

MW of the expressed protein is the first step in a top-down analysis that could provide further information about the protein not obtainable by bottom-up analysis. In addition, the advantage of a combined "top-down/bottom-up" analysis is that the two analyses are orthogonal to one another and thus independent. Each analysis can be used to confirm correctness of the other analysis.

VerBerkmoes et al. conducted an in-depth proteomic study of *S. oneidensis*, a gram-negative bacterium with possible applications in bioremediation because of its ability to reduce metal ions (VerBerkmoes et al., 2002). The objective of this study was to identify as many expressed proteins of *S. oneidensis* as possible using a combination of "shotgun" proteomics (bottom-up) and a protein MW measurement using FT-ICR-MS (top-down). *S. oneidensis* was grown under mid-log growth phase conditions. Cells lysates were fractionated into a number of components in order to maximize the number of proteins detected. In particular, ribosomal proteins (because of their high abundance) were selectively partitioned in order to detect less abundant proteins. Protein fractionation of crude extract was accomplished using a strong anion-exchange column using a fast protein liquid chromatography (FPLC) platform. The fractionated protein samples were then subjected to tryptic digestion. The digested samples were separated using reverse-phase LC and/or LC/LC (MudPIT with salt steps) interfaced to a quadrupole ion trap mass spectrometer. The ion trap was operated in a number of different scan modes and mass ranges to maximize the number proteins detected. Of the 5,177 ORFs identified in the genome of *S. oneidensi*, 868 proteins were detected and identified (VerBerkmoes et al., 2002). Fractionated protein samples were also analyzed by ESI-FT-ICR-MS (VerBerkmoes et al., 2002). Because of the very high resolution and mass accuracy obtainable using an FT-ICR, it is possible to determine the MW of protein with ppm (or sub-ppm) mass accuracy. In addition, the charge state of an ion is readily determined from the isotopomer peaks that are fully resolved at very high resolution (>100,000 FWHM). A single FPLC fraction often contained multiple proteins each producing its own charge state envelope. This complex MS spectral data were deconvoluted providing highly accurate MW determination of the multiple proteins in a sample. This information was then compared to the hypothetical protein MWs derived from the ORFs of the genome. In addition, this top-down analysis was also compared to the bottom-up analysis of the same FPLC fraction. VerBerkmoes reported several proteins with N-terminal methionine cleavage as well as N-terminal cleavage of signal peptides (VerBerkmoes et al., 2002). SignalP software was also used to confirm hypothetical signal peptides to that observed experimentally. It is difficult to underestimate the power of high-resolution, high mass accuracy mass spectrometry in confirming the identity of a particular protein and/or the presence or absence of a specific PTM(s).

In another example of combining bottom-up and top-down proteomics, Fagerquist et al. reported definitive identification of protein biomarkers observed in MALDI-TOF-MS analysis of cell lysates of *Campylobacter* species/strains (Fagerquist et al., 2005). *Campylobacter* is an important foodborne pathogen that causes an estimated 2.4 million cases of gastrointestinal illness every year in the USA (www.cdc.gov/ncidod/dbmd/diseaseinfo/campylobacter). As mentioned previously, MALDI-TOF-MS has become an increasingly powerful tool in the identification/classification of microorganisms including foodborne pathogens (Mandrell et al., 2005). Proteins from microorganisms ionized and detected by MALDI-TOF-MS are, in most cases, never definitively identified. Use of MALDI generated protein ions to generate a unique spectral "fingerprint" of a microorganism has been validated on the basis of analysis of hundreds (and possibly thousands) of bacterial strains. The apparent robustness of this technique is surprising given the limited number of proteins that are typically ionized and detected by MALDI-TOF-MS, i.e., 20–50. This is a relatively small fraction of proteins compared to the total number of ORFs in most bacterial genomes, e.g., 2,000–5,000. In order to examine more fully the proteomic (and ultimately the genomic) underpinnings of the MALDI-TOF approach, a systematic investigation was conducted to identify definitively protein biomarkers of *Campylobacter* observed in MALDI-TOF-MS analysis of cell lysates (Fagerquist et al., 2005). A combined "top-down/bottom-up" proteomic approach was pursued as the most effective strategy for definitive identification of (1) the origin of "shifts" of protein biomarkers' *m/z* observed in MS spectra across different species and strains of *Campylobacter*; (2) protein biomarkers with posttranslational modifications (PTMs), and if such PTMs varied across species and strains of *Campylobacter*; (3) incorrect start codons in the genes of protein biomarkers.

Fagerquist et al. reported a proteomic identification of a single protein biomarker at ∼10 kDa prominently observed in MALDI-TOF-MS spectra of all *Campylobacter* species and strains (Fagerquist et al., 2005). The species/stains analyzed were *Campylobacter jejuni*, *Campylobacter coli*, *Campylobacter upsaliensis*, *Campylobacter lari*, *Campylobacter helveticus*, and *Campylobacter concisus*. *Campylobacter* cell lysates were analyzed by MALDI-TOF-MS and also fractionated by reverse-phase HPLC. The HPLC fractions were analyzed by MALDI-TOF-MS, and the ∼10 kDa protein biomarker observed in the MALDI-TOF spectrum of cell lysates could thus be identified to a specific HPLC fraction. The specific HPLC fraction was then analyzed by high–resolution ESI-MS using a hybrid quadrupole/TOF instrument to obtain a more accurate MW of the protein than that possible using MALDI-TOF-MS. The eluent of the HPLC fraction was evaporated to 10 μL and then analyzed by 1-D PAGE. Protein bands were excised, in-gel digested with trypsin, and the resulting peptides analyzed by nano-HPLC-ESI-MS/MS using a quadrupole/TOF mass spectrometer. Database searching identified the protein biomarker at ∼10 kDa as the DNA-binding protein HU. In addition to genomically sequenced *Campylobacter* strains, several of the strains analyzed were not genomically sequenced. In consequence, the HU gene (*hup*) for these nongenomically sequenced strains was genetically sequenced using primers of the *hup* gene from genomically sequenced strains. These amino acid sequences derived from these *hup* gene sequences were then added to the proteomic databases. This combined "bottom-up/top-down" approach identified definitively that "shifts" in *m/z* of this protein biomarker in MALDI-TOF-MS spectra across species and strains were caused by amino acid variations due to nonsynonymous mutations in its gene. The biomarker mass "shifts" were not due to variations in PTMs. In fact, the HU protein was found not to conform to the rule predicting N-terminal methionine cleavage in bacterial proteins on the basis of the penultimate residue. The HU protein had no PTMs.

In a subsequent report, Fagerquist et al. analyzed the protein biomarkers of three different strains of *C. jejuni*, two of which gave nearly identical MALDI-TOF-MS spectra whereas the third strain appeared to have a greater number of protein biomarkers whose masses were "shifted" compared to the other two strains (Fagerquist et al., 2006). The "top-down/bottom-up" approach once again confirmed that amino acid variations were responsible for the mass "shifts" observed. The cumulative effect of these mass "shifts" observed in multiple protein biomarkers give the MS spectrum its unique protein "signature" or "fingerprint." In addition, it was noted that the *number* of biomarkers that were shifted in mass between spectra appeared to suggest the phylogenetic distance between strains. This was confirmed by a nitrate reduction test that confirmed that the third *C. jejuni* strain was of the sub-species *C. j. doylei* whereas the other two strains were of the sub-species *C. j. jejuni*. It was also noted that the magnitude of the mass difference of a "shift" was not, in itself, indicative of the phylogenetic distance between strains because an amino acid variation may cause the protein mass to increase *or* decrease. In consequence, the net change of a biomarker mass caused by multiple amino acid variations may not differ in magnitude significantly from the mass "shift" caused by a *single* amino acid substitution. However, it was also found, not surprisingly, that the *number* of protein amino acid variations between strains appeared to correspond to the phylogenetic distance between strains.

The most significant effect of phylogenetic distance on protein biomarker detection is caused by a nonsynonymous mutation that results in the premature termination of protein translation, i.e., a nonsense mutation. The transthyretin-like periplasmic protein was detected by MALDI-TOF-MS and identified by "top-down/bottom-up" proteomics in both *C. j. jejuni* strains but was absent in the *C. j. doylei* strain. Genetic analysis determined that its absence in the *C. j. doylei* strain was due to several nonsense mutations (stop codons) in the gene rendering the protein untranslatable, i.e., a pseudogene (Fagerquist et al., 2006). The "top-down/bottom-up" proteomic also detected the presence of a 20-residue N-terminal signal peptide for this protein.

Another example of the utility of a combined top-down/bottom-up proteomic approach is the development of *composite sequence proteomics analysis* or CSPA (Fagerquist, 2007). CSPA involves the identification of the full amino acid sequence of a protein of a non-genomically sequenced bacterial strain by combining multiple proteomic identifications of homologous protein sequence regions from genomically sequenced bacteria. The genomically sequenced bacterial strains may (or may not) be of the same species as the non-genomically sequenced strain. The combined homologous sequence regions form a composite sequence whose MW can be compared to the measured MW of the protein. If there is agreement between the measured and the composite sequence MW, then the composite sequence is entered into the database and compared against the MS/MS data. If there is a discrepancy between the measured and the composite sequence MW, then de novo analysis is performed on the MS/MS data with special attention focused on unconfirmed "gaps" in the composite sequence, i.e., sequence regions lacking MS/MS confirmation from homologous sequence regions of genomically sequenced strains. The MW of the combined composite/de novo sequence is compared to the measured MW of the protein. If agreement is obtained, then the composite/de novo sequence is entered into the database and compared against the MS/MS data (Fagerquist, 2007). CSPA was successfully used to determine the full amino acid sequence of several protein biomarkers of several non-genomically sequenced species/strains of *Campylobacter* (Fagerquist et al., 2007). It is clear that as the number of bacterial genomes in public databases increases there will be an increase in the ability to proteomically sequence proteins from non-genomically sequenced bacterial strains from sequence homologies to genomically sequenced strains. It is interesting to speculate as to the cause of this sequence homology, whether it is an artifact of bacterial evolution or evidence of lateral gene transfer between microorganisms that share the same environmental niche. Interspecies homology may suggest lateral gene transfer (Fagerquist et al., 2007).

11.3.5 Selected Proteomic Studies of Foodborne Pathogens

Foodborne pathogens have been the subject of numerous proteomic studies. However, the distribution of studies is uneven across the various pathogens. For instance, there have been numerous proteomic studies of *Staphylococcus aureus* whereas *Clostridium perfringens* has received considerably less attention, and to date there have been no reported proteomic studies of *Clostridium botulinum*. What follows is a brief discussion of a few representative proteomic studies drawn from the current literature.

11.3.5.1 *Bacillus* spp.

Of the various species of *Bacillus*, *B. cereus* is most associated with foodborne illness. An inhabitant of soil and an occasional contaminant found in milk, there have only been a few proteomic studies reported of *B. cereus*. Very recently, Luo et al. reported a differential proteomic study of *B. cereus* (ATCC14579) grown on a liquid extract of forest soil (soil extracted solubilized organic matter or SESOM and on standard Luria–Bertani (LB) broth) (Luo et al., 2007). Cells were harvested in mid-exponential phase growth. After washing with potassium phosphate buffer, cells were solubilized in an isoelectric buffer of urea, thiourea, CHAPS, DTT, betaine, etc. Cells were lysed by freeze thaw and sonication, and insoluble cellular debris was removed by high-speed centrifugation. The amount of protein extracted was quantified by the Bradford assay. Proteins were separated by 2-D gel electrophoresis: first dimension by isoelectric focusing (IEF) and the second dimension by SDS-PAGE. Gels were silver stained for the purpose of visual comparison because of its greater sensitivity. Triplicate silver-stained gels were performed for each growth media, and triplicate gel images were combined visually to provide a single composite image. These composite gel images were then compared between the two growth media to detect differences in protein expression. They reported that 234 gel spots appeared upregulated in the soil extract compared to LB broth, and 201 proteins were reported to be downregulated (Luo et al., 2007). A protein was considered to be upregulated protein if its SESOM gel volume was greater than the LB gel volume by ≥ 2. A protein was considered downregulated if its SESOM gel volume was less than the LB gel volume by ≤ 0.5. A much smaller fraction of upregulated (35) or downregulated (8) proteins were identified by proteomic techniques. One gel stained with Coomassie blue was used for the purpose of protein identification. The 43 selected gel spots were excised, subjected to in-gel digestion with trypsin and peptide mass fingerprinting and MS/MS using a MALDI-TOF-TOF instrument. Proteins were identified by database searching (Voigt et al., 2006). Significant changes in protein expression between growth in LB and nutrient-poor SESOM media were observed. Differentially expressed proteins included proteins involved in catabolism, biosynthesis, substrate uptake, and the shape of the microorganism. All of these results suggest a significant response of the microorganism to environmental stressors.

A very interesting proteomic study was conducted by Oosthuizen et al. of growth of *B. cereus* in the presence of glass wool to stimulate biofilm formation (Oosthuizen et al., 2002). Biofilm formation is an important, cell density-dependent function of microorganisms that can impair the effectiveness of the use of antibiotic and disinfectants. *B. cereus* has been reported to attach to stainless steel surfaces via the formation of biofilms (Oosthuizen et al., 2001). Oosthuizen et al. conducted a differential proteomic analysis comparing cell growth and biofilm formation in the presence of glass wool for 2- and 18-h periods (Oosthuizen et al., 2002). Glass wool had previously been shown to provide an excellent surface for biofilm formation due to the nature of its surface as well as its high surface-to-volume ratio (Oosthuizen et al., 2001). Proteomic experiments were also conducted on planktonic *B. cereus* cells in the absence and presence of glass wool for 2- and 18-h periods. Proteins from planktonic and biofilm-associated *B. cereus* cells were extracted and separated by 2-D gel electrophoresis. The overall experiment was repeated on three different occasions, and gels were run in triplicate for each experimental condition on each occasion. The "control" gel, i.e., the gel to which all other gels were compared for relative changes in protein expression, was of planktonic cells after 2-h growth with no glass wool. Proteins were identified by N-terminal sequencing by Edman degradation and searching of the *B. cereus* database. These differential proteomic studies revealed a number of unique proteins that were upregulated in the presence of glass wool for both planktonic and surfaced-attached cells for the two different growth periods. Of particular interest was the identification of a protein (YhbH) that was upregulated in both planktonic and surface-attached cells after only 2 h of growth in the presence of glass wool. The authors speculated that YhbH is involved in biofilm regulation (Oosthuizen et al., 2002). This study is also of interest from the point of view of proteomic technique, i.e., mass spectrometry was not used for protein identification but N-terminal Edman sequencing. Although commercially available Edman sequencers were widely used before the development of mass spectrometry-based proteomic techniques, Edman sequencing still retains a niche in the proteomic field.

11.3.5.2 *Campylobacter* spp.

Recently, Seal et al. conducted a comparative proteomic study of two isolates of *C. jejuni* (Seal et al., 2007). One isolate was an excellent colonizer of GI tract of chickens whereas the other isolate was a poor colonizer. Proteins that were differentially expressed between the two isolates were analyzed and identified by 2-D gel electrophoresis, in-gel digestion, and a combination of MS peptide mass fingerprinting and MS/MS of peptides from digested proteins using a MALDI-TOF-TOF mass spectrometer. Differences in gel spot density (from triplicate gels) were used to identify qualitative differences in protein expression between the two isolates. Several proteins, e.g., fibronectin-binding protein and serine protease, were found to be differentially expressed between these two isolates. It was concluded that proteomic analysis was very useful for identifying phenotypic variations between the two isolates studied (Seal et al., 2007).

In another very recent study, Cordwell et al. conducted a comprehensive proteomic analysis of membrane-associated proteins of two strains of *C. jejuni*: JHH1 and ATCC 700297 (Cordwell et al., 2008). JHH1 was a recent clinical isolate obtained from a patient with gastrointestinal campylobacteriosis. The ATTC strain was isolated from a patient diagnosed with Guillain–Barré syndrome. Membrane proteins were isolated by sodium carbonate precipitation.

This study also compared results from two proteomic identifications. First, 2-D gel electrophoresis, followed by peptide mass mapping and/or MS/MS of peptides using a MALDI-TOF-TOF-MS. Second, an LC/LC approach where the digested proteins are fractionated off-line by SCX followed by separation "online" by reverse-phase chromatography interfaced to a quadrupole/TOF mass spectrometer. Proteins identified by only a single peptide ("one-hit wonders") were subjected to MS/MS analysis to confirm the identity of the peptide.

Significant differences in detection of proteins were obtained by the two proteomic approaches. For instance, proteins of high basicity or high hydrophobicity were more easily identified by LC/LC/MS/MS approach than by 2-DE and peptide mass "fingerprinting" (Cordwell et al., 2008).

In a subsequent report from the same group, a number of variants of *Campylobacter ad*herence *f*actor (CadF), a membrane-associated protein, were detected from the JHH1 and NCTC 11168 strains of *C. jejuni* (Scott et al., 2010). The binding of CadF to the extracellular host protein fibronectin (Fn) is an important determinant of *C. jejuni* virulence. Mass and p*I* variants of CadF were analyzed and identified by 2-DE and peptide mass mapping using a MALDI-TOF-MS and by MALDI-MS/MS using a quadrupole-TOF instrument. Although some of the variants were identified as gel artifacts (i.e., refolding of CadF resulting in multiple spots on the 2-D gel), other variants were found to be the result of posttranslational cleavage of CadF from the C-terminus. Certain of these membrane-associated variants would still bind to Fn but when probed against patient sera were found to be nonreactive. Only the full CadF sequence was reactive to patient sera (Scott et al., 2010).

11.3.5.3 *Clostridium* spp.

C. perfringens and *C. botulinum* are the species of *Clostridium* most associated with foodborne illness. However, there have been only a small number of proteomic studies of *C. perfringens* and no reported studies of *C. botulinum* based on a PubMed search. In 2002, Shimizu et al. sequenced and annotated the full genome of *C. perfringens* (Shimizu et al., 2002a). In a subsequent study, the same group studied differences in secreted protein expression, which are regulated by the two-component system VirR/VirS, of a wild-type strain of *C. perfringens* and a *virR* mutant *C. perfringens* strain (Shimizu et al., 2002b). The VirR/VirS system had been known to be involved in control of virulence factors. Proteins were extracted from the cell culture supernatant of each strain and separated by 2-D gel electrophoresis with silver staining. Relative differences in protein expression between the two strains were determined by measurement and analysis of gel spot densities. Gel spots revealing significant differential densities between the two strains were excised, in-gel digested, and peptide mass fingerprinted by MALDI-TOF-MS and database searching. The function of some of the proteins was determined from homology to protein sequences of other genomically sequenced/annotated microorganisms. Fifteen proteins were found to be differentially expressed between the two strains, and Northern hybridization was performed on the genes of these proteins in order to determine if protein expression was regulated at the transcriptional or translational level (Shimizu et al., 2002b).

Although *C. difficile* is not considered a foodborne pathogen, it is an anaerobic pathogen that can lead to infection of the large intestine of the human GI tract when there is a reduction of natural microflora due to antibiotic treatment. The multidrug-resistant *C. difficile* has been genomically sequenced and annotated (Sebaihia et al., 2006). In order to identify virulence factors of *C. difficile* (beyond those previously identified), Wright et al. conducted a qualitative differential proteomic study to identify proteins associated with the peptidoglycan cell wall and a protein surface layer (S-layer) surrounding the cell wall of this gram-positive bacterium (Wright et al., 2005). Proteins extracted from two different methods were subjected to proteomic analysis. A low pH glycine treatment was used to extract proteins primarily from the S-layer, whereas lysozyme treatment was used to extract all proteins external to and/or associated with the cell wall. Qualitative bottom-up proteomic experiments were performed in triplicate using 2-D gel electrophoresis and in-gel digestion and analysis of the tryptic peptides by nano-LC/MS/MS using a quadrupole-TOF instrument. For most of the proteomic identifications, multiple tryptic peptides were obtained for protein identification. A greater number of proteins were extracted and identified by the lysozyme treatment than by the glycine treatment. SignalP software was used to predict which proteins had signal peptides (indicating secretion by a dedicated mechanism). There was no indication whether the predicted signal peptides were confirmed by proteomic analysis. Predicted signal peptides were more prevalent in proteins extracted and identified by glycine extraction (S-layer) than by lysozyme treatment (cell wall) (Wright et al., 2005). A number of proteins that are paralogues of the two most abundant proteins (45 and 36 kDa) of the S-layer were also identified.

11.3.5.4 *Escherichia coli* and *Shigella* spp.

There have been only a limited number of proteomic studies of pathogenic *E. coli*. A few examples are summarized. Li et al. conducted a comparative proteomic study of extracellular proteins of *E. coli* enterohemorrhagic (strain EDL-933) and *E. coli* enteropathogenic (strain E2348/69) as well as mutant strains missing *ifh* and *ler* genes that regulate extracellular virulence factors (Li et al., 2004). Strains were grown in overnight broth cultures. Cells were pelleted by centrifugation, and the supernatant was filtered. Extracellular proteins were precipitated from the supernatant with trichloroacetic acid (TCA), washed multiple times with acetone, and stored at −20°C for subsequent analysis. Protein samples were resolubilized in appropriate buffers for separation by 1-D and 2-D gel electrophoresis. Very few non-extracellular proteins were detected suggesting very little cell lysis. Three separate cultures were analyzed for each strain, and the gels were run in triplicate. Silver-stained spots were excised, in-gel digested, and analyzed by peptide mass fingerprinting using MALDI-TOF-MS and protein database searching. At least four peptides and 20% sequence coverage was considered adequate for protein identification. Fifty-nine extracellular proteins were identified with 26 of those being common to both the enterohemorrhagic and the enteropathogenic wild-type strains. In addition, the mutant strains lacking *ifh* and *ler* genes affected the expression of extracellular proteins related to virulence: EspA, EspB, EspD, and Tir. In addition, using 1-D gel electrophoresis and peptide mass fingerprinting a 110 kDa protein, TagA, was identified as being present in the wild-type strains but absent in the mutants indicating that it is also regulated by both *ifh* and *ler* genes (Li et al., 2004).

In another study, Asakura et al. conducted a comparative proteomic analysis of *E. coli* O157:H7 strain F2 (an oxidative-resistant strain) and MP37: a variant strain of F2 that is oxidative stress sensitive and can be induced into a viable but not culturable state (VBNC) by oxidative stress (Asakura et al., 2007). Oxidative stress in MP37 was simulated by exposure to 0.05% hydrogen peroxide in PBS. The VBNC state was determined from the maintenance of membrane integrity as measured using a LIVE/DEAD BacLight kit as well as the absence of plate counts during culturing. Extracted proteins were separated by 2-D gel electrophoresis in duplicate. Gels were silver stained, and relative spot densities were compared between the F2 strain and the VBNC MP37 strain. Spots showing significant differences in density between the two strains were excised, in-gel digested, peptide mass fingerprinted using MALDI-TOF-TOF-MS, and identified by database searching. Thirty-one proteins were found to have a twofold difference in spot intensity between the two strains, and 17 of those proteins were proteomically identified. The differential protein expression observed between the two strains suggested that ribosomal proteins may be involved in the transition of the MP37 strain to a VBNC state (Asakura et al., 2007).

In a very recent study, Burt et al. studied the antimicrobial effect of the essential oils, carvacrol and *p*-cymene, on protein expression of *E. coli* O157:H7 (ATCC43895) (Burt et al., 2007). Cells were grown overnight at 37°C in Mueller–Hinton broth with a 1.0 mM concentration of carvacrol or *p*-cymene. In another experiment, cells were grown to mid-exponential phase growth at which point different concentrations of carvacrol and *p*-cymene were added and cell growth was continued for an additional 3 h before harvesting. Proteins were separated by 1-D SDS-PAGE, identified by N-terminal Edman sequencing and/or peptide mass fingerprinting by MALDI-TOF-MS and database searching. Carvacrol had a strong inhibitory effect on the formation of flagellin for cells grown overnight at a carvacrol concentration of 1.0 mM. HSP60 (*groEL*) was also significantly upregulated consistent with a stress response by the microorganism. In contrast, carvacrol added at mid-exponential phase growth had little effect on existing flagellin formation. Flagellin formation (or lack thereof) was determined by differential interference contrast microscopy. Cellular motility was measured using the hanging-drop technique (Burt et al., 2007).

Maillet et al. conducted a global proteomic analysis of the nonpathogenic *E. coli* strain K12 (JM109) (Maillet et al., 2007). The study had several objectives. One objective was to demonstrate that state-of-the-art global proteomic techniques can be used to identify sequencing and annotation errors in genomic databases even of a microorganism as highly studied as *E. coli* strain K12. The second purpose was to show the continued importance of 2-D gel electrophoresis (2-DE) as a proteomic technique compared to other techniques, e.g., MudPIT. This study utilized 2-DE for protein separation using three separate gels for the p*I* ranges: 3.0–10.0, 4.0–7.0, and 5.0–6.0. Protein identification was performed using peptide mass fingerprinting (PMF) and MALDI-TOF-MS or N-terminal Edman sequencing. In addition, protein MW and p*I* were experimentally determined from curves generated from regression analysis of plots of theoretical MW and p*I* versus gel migration distance. Proteins identified at multiple spots in a gel were used as internal standards.

Plots of theoretical versus experimental MW and theoretical versus experimental p*I* allowed identification of discrepancies which led to identification of possible PTMs from PMF or N-terminal sequencing data or identification of possible sequencing or annotation errors in the genome database. By this approach, they reported identification of 1,151 proteins as well as numerous instances of errors in the annotation of genome ORFs (Maillet et al., 2007).

The three species of *Shigella* most often linked to foodborne illness are *Shigella flexneri* (Reller et al., 2006), *Shigella sonnei* (Gaynor et al., 2008), and *Shigella dysenteriae* which is also classified as a Class B biological terrorism agent (Centers for Disease Control, 2006). There have been no reported proteomic studies of *S. sonnei* or *S. dysenteriae*, but there have been a handful of proteomic studies of *S. flexneri*. Liao et al. generated a proteome map of *S. flexneri* 2a strain 2457T (Liao et al., 2003). The study was an early attempt to identify as many as possible intracellular and extracellular proteins. This effort involved the use 2-D gel electrophoresis, PMF, and database searching. Isoelectric focusing in the first dimension was accomplished with immobilized pH gradient strips. Multiple and overlapping p*I* gels were used to maximize the number of spots/proteins visualized. Intracellular proteins were extracted from harvesting and lysis of cells in early stationary phase of growth. Extracellular proteins were extracted from the media supernatant also in early stationary phase growth using the TCA precipitation method. Analytical gels were silver stained for visualization and enumeration of spots. Coomassie-stained gels were used for protein identification by spot excision, in-gel digestion, PMF, and database searching. The standard for definitive protein identification included at least five peptides and 15% coverage. Searches included the genomes of *E. coli*, *S. flexneri* 2a 301, and *S. flexneri* 2457T. Out of a total of 488 spots, 388 proteins were successfully identified which represented 169 genes. An identification subset was the extracellular proteins with 131 spots being identified representing 116 genes, but only 39 of those were extracellular proteins indicating that there was some contribution due to cell lysis. Theoretical and experimental p*I* and MW were plotted, and in most cases a good correlation was obtained. Discrepancies were ascribed primarily to posttranslational processing. A great deal of homology (95%) was observed between the identified proteins of *S. flexneri* strain 2457T and the *E. coli* genome, confirming the phylogenetic closeness of these two microorganisms (Liao et al., 2003).

In a more recent study, Jennison et al. identified eight proteins from *S. flexneri* 2457T that were immunoreactive to the sera of five *Shigella* patients (Jennison et al., 2006). Six of these proteins had not been previously identified as immunogenic. Soluble and membrane proteins were separately extracted from *S. flexneri* 2457T cells and separated by 2-D gel electrophoresis. For each experiment, two gels were run in parallel with one gel being stained with Coomassie (for later proteomic identification) and the other gel being used for protein transfer to a PVDF membrane for probing with patient sera. In addition to testing five patients with acute shigellosis, sera from five healthy individuals were also tested. Protein spots exhibiting immunoreactivity were matched to protein spots in the Coomassie-stained gel which were then excised, in-gel digested, and identified by PMF and database searching. Eight proteins were found to be immunoreactive: TolC, OmpA, IpaD, AnsB, GroEL, Spa33, Ggt, and TolB. Only IpaD and OmpA had been previously identified as being immunoreactive (Jennison et al., 2006). Immunoproteome experiments such as this show great promise for identifying proteins involved in pathogen virulence as well as providing potential antigenic candidates for development of vaccines.

11.3.5.5 *Listeria* spp.

There have been a number of excellent proteomic studies of pathogenic and non-pathogenic *Listeria*. A few are summarized. Schaumburg et al. conducted an in-depth proteomic analysis of "cell wall-associated proteins" of *Listeria monocytogenes* (Schaumburg et al., 2004). Proteins of this subproteome were extracted sequentially using a series of different high-concentration salt solutions (Tris, KSCN, CHAPS, CTAB, octylglucoside, and SDS). The proteins from each extraction solution were precipitated using the TCA/acetone method. Proteins were resolubilized, separated by 1-D SDS-PAGE, transferred to PVDF membrane (Western blot), stained with Coomassie blue, and identified by N-terminal Edman sequencing. Each extraction was also subjected to an assay for detection of cytoplasmic enzymatic activity (aminopeptidase C) indicating cell lysis. The results of the assay and 1-D gel analysis indicated that extractions with Tris and KSCN showed the least evidence of cellular disruption or lysis. In consequence, the Tris and KSCN extractions were repeated, and the extracted proteins, after precipitation by TCA/methanol, were resolubilized and separated by 2-D gel electrophoresis. Two-dimensional gels were stained with RuBPS, gel spots excised and in-gel digested, and protein spots identified by PMF using MALDI-TOF-MS. A protein identification was considered definitive if it included a sequence coverage of 20% and a significance score twice that of the search engine threshold. Fifty-five proteins were identified as being associated with the cell wall. None of the proteins identified possessed the C-terminal LPXTG motif indicative of proteins that are covalently bound to the cell wall. Software bioinformatic tools were used to predict signal peptides and confirm subcellular localization of proteins, including SignalP, SMART, Sosui Signal beta, PSORT, iPSORT, and AnTheProt. Nine of the 55 proteins identified were cytoplasmic proteins not previously known to be affiliated with the cell wall. It was conjectured that these cytosolic proteins (GroEL, DnaK, EF-Tu, Enolase, GAPDH) may have "moonlighting" functions (Jeffery, 1999) at the cell wall. Surface plasmon resonance (Biacore) was used to confirm and measure the binding of these cytosolic proteins to a fragment of human plasminogen protein (Schaumburg et al., 2004).

In a subsequent study, Trost et al. identified and compared secreted proteins ("the secretome") from pathogenic *L. monocytogenes* (EGD-e wild type) and non-pathogenic *Listeria innocua* (CLIP11262 wild type) (Trost et al., 2005). Bacterial cells were grown overnight in BHI medium and then transferred to minimal medium broth culture and grown for 17 h at 37°C. Growth in minimal medium was considered to increase expression of virulence factors. The cells were pelleted by centrifugation in late exponential phase. The supernatant was filtered, and proteins were precipitated using the TCA/acetone method. Proteins were analyzed by two proteomic techniques. First, resolubilized proteins were separated by 2-D gel electrophoresis. Gel spots were excised, in-gel digested, and analyzed by PMF and post-source dissociation (PSD) using a MALDI-TOF-MS and MASCOT database searching. In a second approach, precipitated protein samples were digested overnight and their peptides analyzed by LC/MS/MS using nano-LC-quadrupole/TOF mass spectrometry and database searching. A number of software tools were used to predict secreted proteins starting with SignalP (version 2.0.b2) but also including SMART, Sosui Signal beta, PSORT I, iPSORT, AnTheProt, and InterPro. The use of multiple bioinformatics tools was to reduce the possibility of false positives. Putative exported proteins were also subcategorized as secreted, cell wall proteins, integral membrane proteins, etc. This bioinformatic analysis resulted in a prediction of 121 secreted proteins. Proteomic techniques identified 105 proteins in the culture supernatant of *L. monocytogenes* which included all previously identified virulence factors. Roughly, half of these proteins were surface-associated or secreted proteins with the other half being primarily cytosolic proteins. Proteomic analysis also confirmed the existence of about 50 secreted proteins that were common to both *L. monocytogenes* and *L. innocua* with roughly the same number unique to each species. Sixteen secreted proteins of *L. monocytogenes* had no corresponding gene in *L. innocua* (Trost et al., 2005).

Lenz et al. conducted a proteomic study of the protein secretion system SecA2 of *L. monocytogenes* (Lenz et al., 2003). Seventeen proteins were identified as being secreted by the SecA2 system to either the extracellular milieu or associated with the cell surface. Bacterial cells were grown in LB broth to exponential phase. Cells were pelleted by centrifugation, and the supernatant was treated with the TCA/acetone method for protein precipitation from the supernatant (Lenz and Portnoy, 2002). Precipitated proteins were resolubilized in SDS-PAGE buffer and separated by 1-D gel electrophoresis. Gel bands were excised, in-gel digested with trypsin, analyzed by PMF using MALDI-TOF-MS, and identified by database searching. Proteomic data were analyzed using MS-FIT of Protein Prospector proteomic software. SignalP software was used to predict proteins with signal peptides. The 2 most abundant of the 17 identified proteins were autolysins which are involved in bacterial virulence (Lenz et al., 2003).

11.3.5.6 *Salmonella* spp.

There have been a number of proteomic studies of *Salmonella*. Two noteworthy studies are summarized. Recently, Hu et al. combined a complementation assay and proteomics to study antibiotic resistance in *Salmonella enterica* serovar Typhimurium (SET) (Hu et al., 2007). A gene (*yjeH*), thought to have an affect on antibiotic resistance, was found to have a transposon insertion in a mutant strain of SET resulting in a fourfold lowered resistance to cephalosporin antibiotics. Hu and colleagues were able to insert a functional *yjeH* gene back into the mutant strain by a plasmid insertion. A proteomic analysis was then conducted of the outer membrane proteins of the wild-type strain, the mutant strain, and the mutant with the plasmid insertion using 2-DE, in-gel digestion of gel spots, and MS/MS using a quadrupole/TOF instrument. Qualitative differences in protein expression were identified by significant changes in spot density between three gel replicates. Proteomic analysis identified several outer membrane proteins that appeared to be up- or downregulated as a result of the functionality of the *yjeH* gene (Hu et al., 2007). This study represents an excellent example of the targeted use of proteomics to address a specific question related to protein expression within the context of single gene function.

In another recent proteomic study, Adkins et al. reported a global bottom-up proteomic analysis of two strains of *S. enterica* serovar Typhimurium (SET) under three different growth conditions (Adkins et al., 2006). The two strains were SET strain LT2 (a less virulent strain because of a defective *rpoS* gene) and the SET strain ATCC 14028 (a more virulent strain). Three different growth conditions were utilized to enhance differences in expression of virulence factors between the two strains: logarithmic phase growth in rich medium, stationary phase growth and growth at low pH in minimal media (magnesium depleted). In addition, a number of protein extraction and digestion protocols were tested to determine which gave the greatest number of protein identifications. Digested protein samples were fractionated off-line using an SCX column. The fractionated peptides were then separated by reverse-phase chromatography followed by ESI/MS/MS using a linear ion trap mass spectrometer. Data-dependent scanning was utilized, and the five most abundant peptide ions detected in an MS survey scan were subjected to MS/MS by CID. Peptides were identified by database searching using SEQUEST™, and appropriate statistics were used to calculate the false discovery rate. Relative protein abundances were determined by the number of peptide identifications for each protein, i.e., a label-free approach. Global proteomic analysis revealed that the relative abundances of most of the proteins of the two strains were the same under different growth conditions. However, a small subset of proteins for each growth condition showed differences in expression between the two strains. For example, proteins involved in propanediol utilization were only observed under low-pH/Mg-depleted minimal media growth conditions, and these proteins were five times more abundant in the more virulent strain (ATCC 14028) than in the less virulent strain (LT2) (Adkins et al., 2006).

11.3.5.7 *Staphylococcus* **spp.**

There have been numerous proteomic studies of *Staphylococcus* and in particular *S. aureus*. What follows is a summary of a few representative studies. Taverna et al. isolated and identified surface/cell wall-associated proteins of a strain of *S. aureus* isolated from a case of bovine mastitis (Taverna et al., 2007). Harvested cells were treated under isotonic conditions with lysostaphin, an endopeptidase that cleaves the chemical cross-links in the peptidoglycan cell wall of *Staphylococcus* (Vytvytska et al., 2002). After lysostaphin treatment, protoplasts (i.e., cells without cell walls) were pelleted by low-speed centrifugation. The supernatant was dialyzed overnight, and proteins were precipitated from the supernatant by TCA treatment. Precipitated proteins were resolubilized and separated by 2-DE. Replicate gels were performed: four analytical gels (silver stained for sensitivity) and four preparative gels (colloidal blue stained for proteomic identification). Preparative gel spots were excised, in-gel digested with trypsin, and analyzed by PMF using MALDI-TOF-MS and database searching with MASCOT. Sequence coverage of proteomic identifications was ~30–60%. Identified proteins were further analyzed by SignalP (to predict/identify putative signal peptides) and by TMPred (to predict/identify transmembrane motifs). A proteomic map of surface-associated proteins was tabulated, and most of the proteins identified were cell wall-associated or membrane-associated proteins and had predicted signal peptides and/or transmembrane motifs. A number of cytosolic proteins were also identified, e.g., GroEL, DnaK, EF-TU (Taverna et al., 2007). Identification of cytosolic proteins in a surface-associated subproteome has been reported in other proteomic analyses of microorganisms, for example, *L. monocytogenes* (Schaumburg et al., 2004).

Gatlin et al. identified cell envelope-associated proteins of two isogenic strains of *S. aureus* having resistance to the antibiotic vancomycin (Gatlin et al., 2006).

Cells were grown to mid-exponential phase growth in broth to enhance cell envelope-associated proteins. For purposes of comparison, proteins were extracted from the bacterial cells under three different extraction conditions: proteins extracted from digestion of the cell wall, proteins extracted without cell wall digestion, and proteins extracted from whole-cell lysates. The bacterial cell wall was digested by incubation with lysostaphin and mutanolysin. Protoplasts were pelleted by centrifugation, and protein extracted from supernatant was separated by 2-DE. Proteins extracted by cell wall digestion were separated by anion-exchange (DEAE) chromatography to remove cell wall carbohydrates prior to 2-DE. Gels were stained with Coomassie over 3 days, and relative quantification of protein expression was calculated by differences in the average spot intensity in four gel replicates. Gel spots were excised, in-gel digested with trypsin, and analyzed using a MALDI-TOF-TOF as well as nanoflow-LC-MS/MS using a LIT mass spectrometer. Proteins extracted by the cell wall digestion method were also analyzed by shotgun proteomics after removing cell wall carbohydrates by column chromatography (anion exchange, DEAE), followed by trypsin digestion of fractions and separation of the tryptic peptides by SCX. Each SCX fraction was then analyzed by nanoflow-LC-MS/MS using a LIT mass spectrometer (an off-line MudPIT approach). Proteins were identified by database searching using MASCOT. Identified proteins were analyzed by a number of protein function/subcellular localization algorithms. Proteins were classified as to function and subcellular localization, e.g., lipoproteins, cell well-anchored proteins, cell wall-associated proteins. Of particular note was identification of nearly 96 proteins detected in the cell envelope which were not predicted to have either an export signal peptide or motifs indicative of cell envelope localization (Gatlin et al., 2006).

Finally, Scherl et al. describe a very interesting combined proteomic/transcriptomic study of three strains of *S. aureus* having varying resistance to glycopeptide antibiotics. Proteins from membrane extracts were analyzed by 2-DE. In addition, differences in protein expression were quantified using a shotgun proteomic approach with iTRAQ reagents for quantitation, off-line LC-MALDI, and MALDI-TOF-TOF MS/MS of tryptic peptides (Scherl et al., 2006).

11.3.5.8 *Vibrio* spp.

There have been only a limited number of proteomic studies of *Vibrio* spp. Kan et al. conducted a global proteome analysis of *Vibrio cholerae* strain N16961 under aerobic and anaerobic growth conditions to simulate conditions in the environment and in the intestine, respectively (Kan et al., 2004). Proteins were extracted from cells grown in aerobic and anaerobic cultures. The extracted proteins were separated by 2-D gel electrophoresis, in-gel digested, and identified by PMF with MALDI-TOF-MS and database searching. Gel spot intensity was measured using software, and reproducible changes in spot intensity (protein abundance) between gels obtained from aerobic and anaerobic culture conditions were identified. Ten proteins were identified as being reproducibly upregulated under aerobic conditions, and 14 proteins were identified as being reproducibly upregulated under anaerobic conditions. The functions of these differentially expressed proteins could be related to the conditions under which the cells were grown (Kan et al., 2004).

Xu and coworkers reported conducting a proteomic analysis of the outer membrane proteins (OMPs) of *V. parahaemolyticus* under two different salt concentrations during culturing to mimic the environments found in the intestine of a mammalian host (Xu et al., 2004). OMPs were extracted and analyzed by 2-D gel electrophoreses. A relatively small number of gel spots were visualized by 2-D. Those gel spots that showed a significant difference in density between the two gels (representing different salt concentrations) were excised, in-gel digested, and identified by PMF using a MALDI-TOF-MS. A small number of OMPs appeared to vary in protein expression as a result of changes in salt concentration, i.e., osmolarity. The only apparent weakness in the proteomic analysis was that gel replicates were seemingly not conducted. This study also reported experiments of cell growth and motility as a function of salt concentration (Xu et al., 2004).

11.3.5.9 Viruses Associated with Foodborne Illness

Several viruses are associated with foodborne illness, for example, norovirus (or Norwalk virus) (Le Guyader et al., 2006) and hepatitis A (Wheeler et al., 2005); however, there are no reports in the scientific literature on the use of proteomics to study these human pathogens, perhaps due to the simplicity and lack of differential expression of viral proteomes.

11.3.6 How Microbial Genomics Impacts Microbial Proteomics

As an analytical technique, proteomics draws upon knowledge from biochemistry, molecular biology, bioanalytical chemistry, protein chemistry, computer science, physical and analytical chemistry, and physics. As a research field, proteomics is an interdisciplinary science involving biologists, microbiologists, molecular biologists, computer scientists, medical researchers, drug designers, and anyone interested in understanding the dynamic processes of protein expression in an organism. Increasingly, proteomics (along with genomics, metabolomics, transcriptomics, etc.) is viewed as only one component of the broader research area of systems biology whose overarching goal is to map and understand the expression, interactions, and degradation of multitudinous molecules that make up an organism from conception to death and how that organism interacts with its environment during the different stages of its development.

Proteomics is inextricably bound to the information provided from genomic sequencing. Without the bedrock of genomic information obtained through genomic sequencing, proteomics would never have achieved (or been spurred to achieve) the rapid advances that have already occurred in such a short space of time. As the putative proteins obtained from the open reading frames (ORFs) of a genome are only the starting point for a more comprehensive understanding of an organism, the increased number of fully sequenced and annotated genomes has spurred the improvement of existing proteomic platforms, e.g., 2-D gel electrophoresis, as well as the development of new platforms, e.g., protein microarrays. Continued advances in the speed of genomic sequencing will increase pressure for development of faster proteomic technologies.

Mass spectrometry-based proteomics will continue to play a central role in this development.

As the most abundant organisms on the planet, microorganisms have been of particular interest to biologists and the genomics community. In consequence, the field of microbial genomics has advanced rapidly. As of the writing of this chapter, nearly 700 microbial genomes have been completed (with another ~1,000 in progress) which represent ~260 different genera (http://www.ncbi.nlm.nih.gov/sites/entrez/). The sequencing of increasing numbers of microbial genomes of the same genera or species allows comparative analysis of genomes across species and strains. For instance, a comparison of the genomes of pathogenic *L. monocytogenes* (EGD-e) and nonpathogenic *L. innocua* (Clip11262) allowed identification of genes possibly associated with virulence (Glaser et al., 2001). Genes present in *L. monocytogenes* but absent from *L. innocua* (especially genes with known extracellular or cell wall-associated protein functions) suggest they may be important to pathogenicity. The information derived from such genomic sequencing and analysis can become the springboard for further investigation, such as the use of mutant strains of *L. monocytogenes* lacking specific putative virulence genes in order to better understand the mechanism of virulence (Lenz et al., 2003). Further investigations may involve proteomic analysis involving extraction and identification of proteins from specific subcellular locations, e.g., extracellular secreted proteins (Trost et al., 2005) or cell wall-associated proteins (Schaumburg et al., 2004). Genomic sequencing may also spur the development of DNA microarrays for the study of gene regulation under different environmental conditions. Such studies, in turn, may lead to quantitative proteomic studies to confirm changes in protein expression. If the primary goal of *qualitative* proteomics is to map all the expressed proteins (and their PTMs) in an organism, the primary goal of *quantitative* proteomics is to measure the dynamics of protein expression (and their PTMs) as a function of environmental stresses on an organism. DNA microarrays measure gene regulation based on the amount of mRNA generated by an organism, the assumption being that protein expression should correlate to changes in mRNA levels. However, there is not always a direct correlation between mRNA levels and protein expression (Gygi et al., 1999b). Protein expression can be controlled at the level of transcription or at the level of translation.

The feasibility of CSPA (described in an earlier section) would not be possible without the increasing numbers of sequenced microbial genomes within a genera and/or species (Fagerquist, 2007). The discovery of homologous protein sequence regions across species/strains of *Campylobacter* from bottom-up proteomic MS/MS and database searching of genomically sequenced *Campylobacter* species/strains allowed not only the construction of composite protein sequences of non-genomically sequenced strains of *Campylobacter* but also the significance of such homology as to its origins, i.e., bacterial evolution or lateral gene transfer. The greater the number of sequenced microbial genomes of closely related as well as more phylogenetically distant strains, the greater the opportunity to study the phenomena such as lateral gene transfer and phylogenetic relationships using proteomic techniques.

An increasingly important area of microbial genomics is comparative genomics, i.e., the comparison of the genomes of multiple, sometimes closely related, microbes which may (or may not) occupy the same environmental niche. For example, Perna and coworkers compared six genomes of *E. coli* and *S. flexneri* strains (Mau et al., 2006). Among the strains analyzed, they found genomic evidence for recombination events (lateral gene transfer) across lineages. Quinones and coworkers conducted a comparative genomic analysis of 32 strains of *C. jejuni* from South Africa obtained from clinical isolates using a combination of multi-locus sequence typing (MLST) and DNA microarray analysis (Quiñones et al., 2008). In that study, they reported identification of regions of isolated genes (*C. jejuni* integrated elements or CJIEs) in a particular heat-stable serotype of *C. jejuni* that could be used to distinguish the serotype isolated in South Africa from the same heat-stable serotype isolated in another country, e.g., Mexico or Canada (Quiñones et al., 2008).

Banfield and collaborators used shotgun genomic sequencing and shotgun MudPIT proteomics to study a microbial community comprised of multiple microorganisms occupying the same environmental niche, i.e., biofilm (Lo et al., 2007). Cells of *Leptospirillum* groups II and III (acidophilic chemoautotrophs) were sampled from natural biofilm present in mine drainage which exists at very low pH and high metal concentration. They reported evidence of recombination (lateral gene transfer) among *Leptospirillum* strains from analysis of single nucleotide polymorphisms (SNPs) and identification of strain-specific peptides by bottom-up proteomics. SNP and proteomic analysis both confirmed the same recombination points in a number of proteins, e.g., a Rubisco-like protein (Lo et al., 2007). This study represents an excellent example of how genomics and proteomics can be used in tandem to study complex biological systems.

11.3.7 How Microbial Proteomics Impacts Microbial Genomics

Microbial proteomics has a number of impacts on microbial genomics. First, the open reading frames (ORFs) of an annotated microbial genome represent only putative proteins. The actual expression of specific proteins can only be determined by their extraction, digestion, and proteomic analysis. Confirmation of protein expression by proteomic analysis not only confirms the phenotypic expression of the gene but also confirms the correctness of the amino acid sequence of the ORF. It is not unusual to find DNA sequencing errors (and specifically start codon errors) in fully annotated and completed genomes (Fagerquist, 2007). This problem is even more prevalent in non-annotated or unassembled or incomplete microbial genomes that are increasingly appearing in public databases. Although a non-annotated fragmentary genome may appear as a hindrance to proteomic research, it actually represents an opportunity. Proteomic identification can prove the existence of specific proteins whose genes may be incomplete, incorrect, or missing from an incompletely sequenced genome. The proteomic identification is possible because fully sequenced annotated genomes in the database may contain proteins whose sequences have sufficient homology to non-genetically sequenced proteins. In consequence, proteomic confirmation may facilitate and motivate the work of genomic researchers to assemble, annotate, correct, and close all "gaps" in an unfinished genome project.

Another major impact of microbial proteomics on microbial genomics is the identification of PTMs. ORFs do not indicate directly the presence (or absence) of PTMs in the mature intact protein. PTMs, such as removal of N-terminal methionine, removal of N-terminal and C-terminal signal peptides, phosphorylation, glycosylation, lipidation, must be determined experimentally. Although ever more sophisticated software has been developed to predict protein PTMs (e.g., signal peptides) as well as protein subcellular localization from analysis of the primary amino acid sequence of ORFs, definitive proof of a PTM or the subcellular location of a protein can only come from bottom-up or top-down proteomic identification. In that regard, top-down proteomics is uniquely positioned to identify not only the type of PTM but also which amino acid residue has the PTM. PTM determination from bottom-up proteomics is often more challenging because PTMs can be lost during peptide backbone fragmentation by CAD. Although the information obtained from proteomic analysis may involve significant labor, definitive confirmation of the protein's primary sequence as well as determination of its mature form (with all PTMs) will ultimately be informative to the annotation of future genomic sequencing efforts, especially of closely related microorganisms.

Large high-throughput shotgun proteomic efforts of microorganisms, such as those at Pacific Northwest National Laboratory, Oak Ridge National Laboratory, and elsewhere in the USA and the world, can often proceed prior to the sequencing of a microbial genome (Norbeck et al., 2006). Proteomic research groups worldwide are increasingly generating (and archiving) vast amounts of MS and MS/MS proteomic data. Such data can be processed de novo to identify high-quality sequence tags for peptide and protein identification or by database searching in the hope that homologous sequence regions are present in genomically sequenced microorganisms. Alternatively, such data can remain archived until completion of genomic sequencing. However, the fact that much of the proteomic data of a microorganism may already exist should be an added incentive for rapid completion of genomic sequencing projects.

Another impact of microbial proteomics on microbial genomics is more indirect insofar as it deals with measuring changes in protein expression. Protein expression is a dynamic process whereas the information contained in a genome is (by comparison) relatively static. DNA microarrays address the issue of dynamics by measuring rates of transcription from changes in mRNA levels and attempting to extrapolate such changes to protein expression. As mentioned previously, the linkage between changes in mRNA levels and changes in protein expression (as measured by quantitative proteomics) is not always as tight as one might have expected (Gygi et al., 1999b). However, the fact that this linkage (or correlation) is not perfect only points to the mutually confirmatory as well as complementary nature of transcriptomics and proteomics. If the ultimate objective is a systems biology understanding of a microorganism and how it responds to, and interacts with, its host or environment, then one needs genomics, transcriptomics, proteomics, and metabolomics.

11.4 Conclusions and Future Directions

The proteomics of microorganisms, and in particular foodborne pathogens, is a relatively new area of research. Although proteomics is only one component of systems biology, it plays a critical role in the understanding of microorganisms and their response to their environment through changes in protein expression and posttranslational processing. Genomics, transcriptomics, and metabolomics are not appropriate to address questions regarding protein expression. The rapid development of mass spectrometry-based proteomics and the parallel development of protein microarrays suggest that this field will continue to evolve with the emergence of new proteomic technologies and improvements in existing proteomic platforms. As future outbreaks of microbially linked foodborne illness occur, proteomics will undoubtedly continue to focus on issues such as antibiotic resistance, biofilm formation, virulence and pathogenicity, microbial forensics, microbial persistence in the environment, pathogen/host interactions, lateral gene transfer, bacterial communities, quorum sensing, and vaccine development.

Disclaimer The views expressed in this chapter are those of the author and are not intended to represent, explicitly or implicitly, the views of the US Department of Agriculture. Mention of a brand or firm name does not constitute an endorsement by the US Department of Agriculture over others of a similar nature not mentioned.

Acknowledgment I wish to acknowledge Dr. William H. Vensel for useful discussions during the preparation of this chapter.

References

Adkins JN, Mottaz HM, Norbeck AD, Gustin JK, Rue J, Clauss TRW, Purvine SO, Rodlandt KD, Heffron F, Smith RD (2006) Analysis of the *Salmonella typhimurium* proteome through environmental response toward infectious conditions. Mol Cell Proteomics 5:1450–1461

Anjum MF, Tucker JD, Sprigings KA, Woodward MJ, Ehricht R (2006) Use of miniaturized protein arrays for *Escherichia coli* O serotyping. Clin Vaccine Immunol 3:561–567

Asakura H, Panutdaporn N, Kawamoto K, Igimi S, Yamamoto S, Makino S (2007) Proteomic characterization of enterohemorrhagic *Escherichia coli* O157:H7 in the oxidation-induced viable but non-culturable state. Microbiol Immunol 51:875–881

Bakhoum SFW, Agnes GR (2005) Study of chemistry in droplets with net charge before and after Coulomb explosion: ion-induced nucleation in solution and implications for ion production in an electrospray. Anal Chem 77:3189–3197

Bendtsen JD, Nielsen H, von Heijne G, Brunak S (2004) Improved prediction of signal peptides: SignalP 3.0. J Mol Biol 340:783–795

Bondarenko PV, Chelius D, Shaler TA (2002) Identification and relative quantitation of protein mixtures by enzymatic digestion followed by capillary reversed-phase liquid chromatography-tandem mass spectrometry. Anal Chem 74:4741–4749

Brown RS, Lennon JJ (1995) Mass resolution improvement by incorporation of pulsed ion extraction in a matrix-assisted laser desorption/ionization linear time-of-flight mass spectrometer. Anal Chem 67:1998

Burt SA, van der Zee R, Koets AP, de Graaff AM, van Knapen F, Gaastra. W, Haagsman HP, Veldhuizen EJ (2007) Carvacrol induces heat shock protein 60 and inhibits synthesis of flagellin in *Escherichia coli* O157:H7. Appl Environ Microbiol 73:4484–4490

Centers for Disease Control and Prevention (CDC) (2006) Emergency preparedness & response: bioterrorism agents/diseases. CDC, Atlanta, GA. http://www.bt.cdc.gov/agent/agentlistcategory.asp. Accessed 31 July 2006

Chelius D, Bondarenko PV (2002) Quantitative profiling of proteins in complex mixtures using liquid chromatography and mass spectrometry. J Proteome Res 1:317–323

Chelius D, Zhang T, Wang G, Shen RF (2003) Global protein identification and quantification technology using two-dimensional liquid chromatography nanospray mass spectrometry. Anal Chem 75:6658–6665

Chen EI, Hewel J, Felding-Habermann B, Yates JR 3rd (2006) Large scale protein profiling by combination of protein fractionation and multidimensional protein identification technology (MudPIT). Mol Cell Proteomics 5:53–56

Chernushevich IV, Loboda AV, Thomson BA (2001) An introduction to quadrupole-time-of-flight mass spectrometry. J Mass Spectrom 36:849–865

Chong PK, Burja AM, Radianingtyas H, Fazeli A, Wright PC (2007) Proteome analysis of *Sulfolobus solfataricus* P2 propanol metabolism. J Proteome Res 6:1430–1439

Comisarow MB, Marshall AG (1974) Fourier transform ion cyclotron resonance spectroscopy. Chem Phys Lett 25:282–283

Coon JJ, Ueberheide B, Syka JE, Dryhurst DD, Ausio J, Shabanowitz J, Hunt DF (2005) Protein identification using sequential ion/ion reactions and tandem mass spectrometry. Proc Natl Acad Sci USA 102:9463–9468

Cordwell SJ, Len ACL, Touma RG, Scott NE, Falconer L, Jones D, Connolly A, Crossett B, Djordjevic SP (2008) Identification of membrane-associated proteins from *Campylobacter jejuni* strains using complementary proteomic technologies. Proteomics 8:122–139

Cornish TJ, Cotter RJ (1993) A curved-field reflectron for improved energy focusing of product ions in time-of-flight mass spectrometry. Rapid Commun Mass Spectrom 7:1037–1040

Cotter RJ, Gardner BD, Iltchenko S, English RD (2004) Tandem time-of-flight mass spectrometry with a curved field reflectron. Anal Chem 76:1976–1981

Craig R, Beavis RC (2004) TANDEM: Matching proteins with tandem mass spectra. Bioinformatics 20:1466–1467

Demirev PA, Feldman AB, Kowalski P, Lin JS (2005) Top-down proteomics for rapid identification intact microorganisms. Anal Chem 77:7455–7461

Demirev PA, Ho Y-P, Ryzhov V, Fenselau C (1999) Microorganism identification by mass spectrometry and protein database searches. Anal Chem 71:2732–2738

Demirev PA, Lin JS, Pineda FJ, Fenselau C (2001a) Bioinformatics and mass spectrometry for microorganism identification: proteome-wide post-translational modifications and database search algorithms for characterization of intact *H. pylori*. Anal Chem 73:4566–4573

Demirev PA, Ramirez J, Fenselau C (2001b) Tandem mass spectrometry of intact proteins for characterization of biomarkers from *Bacillus cereus* T spores. Anal Chem 73:5725–5731

Douglas DJ, Frank AJ, Mao D (2005) Linear ion traps in mass spectrometry. Mass Spectrom Rev 24:1–29

Edman P (1956) On the mechanism of the phenyl isothiocyanate degradation of peptides. Acta Chem Scand 10:761

Eng JK, McCormack AL, Yates JR 3rd (1994) An approach to correlate tandem mass spectral data of peptides with amino acid sequences in a protein database. J Am Soc Mass Spectrom 5:976–989

Fagerquist CK (2007) Amino acid sequence determination of protein biomarkers of *Campylobacter upsaliensis* and *C. helveticus* by "composite" sequence proteomic analysis. J Proteome Res 6:2539–2549

Fagerquist CK, Bates AH, Heath S, King BC, Garbus BR, Harden LA, Miller WG (2006) Sub-
speciating *Campylobacter jejuni* by proteomic analysis of its protein biomarkers and their post-
translational modifications. J Proteome Res 5:2527–2538

Fagerquist CK, Garbus BR, Miller WG, Williams KE, Yee E, Bates AH, Boyle S, Harden LA,
Cooley MB, Mandrell RE (2010) Rapid identification of protein biomarkers of *Escherichia
coli* O157:H7 by matrix-assisted laser desorption ionization-time-of-flight-time-of-flight mass
spectrometry and top-down proteomics. Anal Chem 82:2717–2725

Fagerquist CK, Garbus BR, Williams KE, Bates AH, Boyle S, Harden LA (2009) Web-based
software for rapid top-down proteomic identification of protein biomarkers, with implications
for bacterial identification. Appl Environ Microbiol 75:4341–4353

Fagerquist CK, Miller WG, Harden LA, Bates AH, Vensel WH, Wang G, Mandrell RE (2005)
Genomic and proteomic identification of a DNA-binding protein used in the "fingerprinting"
of *Campylobacter* species and strains by MALDI-TOF-MS protein biomarker analysis. Anal
Chem 77:4897–4907

Fagerquist CK, Yee E, Miller WG (2007) Composite sequence proteomic analysis of protein
biomarkers of *Campylobacter coli*, *C. lari* and *C. concisus* for bacterial identification. Analyst
132:1010–1023

Fang R, Elias DA, Monroe ME, Shen Y, McIntosh M, Wang P, Goddard CD, Callister SJ, Moore
RJ, Gorby YA, Adkins JN, Fredrickson JK, Lipton MS, Smith RD (2006) Differential label-free
quantitative proteomic analysis of *Shewanella oneidensis* cultured under aerobic and suboxic
conditions by accurate mass and time tag approach. Mol Cell Proteomics 5:714–725

Fenn JB, Mann M, Meng CK, Wong SF, Whitehouse CM (1989) Electrospray ionization for mass
spectrometry of large biomolecules. Science 246:64–71

Florens L, Washburn MP, Raine JD, Anthony RM, Grainger M, Haynes JD, Moch JK, Muster N,
Sacci JB, Tabb DL, Witney AA, Wolters D, Wu Y, Gardner MJ, Holder AA, Sinden RE, Yates
JR, Carucci DJ (2002) A proteomic view of the *Plasmodium falciparum* life cycle. Nature
419:520–526

Frottin F, Martinez A, Peynot P, Mitra S, Holz RC, Giglione C, Meinnel T (2006) The proteomics
of N-terminal methionine cleavage. Mol Cell Proteomics 5:2336–2349

Gao J, Opiteck GJ, Friedrichs MS, Dongre AR, Hefta SA (2003) Changes in the protein expression
of yeast as a function of carbon source. J Proteome Res 2:643–649

Gardy JL, Brinkman FSL (2006) Methods for predicting bacterial protein subcellular localization.
Nat Rev Microbiol 4:741–751

Gardy JL, Laird MR, Chenm F, Rey S, Walsh CJ, Ester M, Brinkman FS (2005) PSORTb v.2.0:
expanded prediction of bacterial protein subcellular localization and insights gained from
comparative proteome analysis. Bioinformatics 21:617–623

Gatlin CL, Pieper R, Huang S-T, Mongodin E, Gebreorgis E, Parmar PP, Clark DJ, Alami H,
Papazisi L, Fleischmann RD, Gill SR, Peterson SN (2006) Proteomic profiling of cell envelope-
associated proteins from *Staphylococcus aureus*. Proteomics 6:1530–1549

Gaynor K, Park SY, Kanenaka R, Colindres R, Mintz E, Ram PK, Kitsutani P, Nakata M, Wedel
S, Boxrud D, Jennings D, Yoshida H, Tosaka N, He H, Ching-Lee M, Effler PV (Jan 2008)
International foodborne outbreak of *Shigella sonnei* infection in airline passengers. Epidemiol
Infect 4:1–7 [Epub ahead of print]

Geer LY, Markey SP, Kowalak JA, Wagner L, Xu M, Maynard DM, Yang X, Shi W, Bryant SH
(2004) Open mass spectrometry search algorithm. J Proteome Res 3:958–964

Gerber SA, Rush J, Stemman O, Kirschner MW, Gygi SP (2003) Absolute quantification of
proteins and phosphoproteins from cell lysates by tandem MS. Proc Natl Acad Sci USA
100:6940–6945

Glaser P, Frangeul L, Buchrieser C, Rusniok C, Amend A, Baquero F, Berche P, Bloecker H,
Brandt P, Chakraborty T, Charbit A, Chetouani F, Couvé E, de Daruvar A, Dehoux P, Domann
E, Domínguez-Bernal G, Duchaud E, Durant L, Dussurget O, Entian KD, Fsihi H, García-del
Portillo F, Garrido P, Gautier L, Goebel W, Gómez-López N, Hain T, Hauf J, Jackson D, Jones
LM, Kaerst U, Kreft J, Kuhn M, Kunst F, Kurapkat G, Madueno E, Maitournam A, Vicente

JM, Ng E, Nedjari H, Nordsiek G, Novella S, de Pablos B, Pérez-Diaz JC, Purcell R, Remmel B, Rose M, Schlueter T, Simoes N, Tierrez A, Vázquez-Boland JA, Voss H, Wehland J, Cossart P (2001) Comparative genomics of *Listeria* species. Science 294:849–852

Gonzales T, Robert-Baudouy J (1996) Bacterial aminopeptidases: properties and functions. FEMS Microbiol Rev 18:319–344

Griffiths J (2007) The way of the array. Anal Chem 35:8833–8837

Guina T, Radulovic D, Bahrami AJ, Bolton DL, Rohmer L, Jones-Isaac KA, Chen J, Gallagher LA, Gallis B, Ryu S, Taylor GK, Brittnacher MJ, Manoil C, Goodlett DR (2007) MglA regulates *Francisella tularensis* subsp. *novicida* (*Francisella novicida*) response to starvation and oxidative stress. J Bacteriol 189:6580–6586

Gunther NW IV, Nunez A, Fett W, Solaiman DK (2005) Production of rhamnolipids by *Pseudomonas chlororaphis*, a nonpathogenic bacterium. Appl Environ Microbiol 71: 2288–2293

Gunther NW IV, Nunez A, Fortis L, Solaiman DKY (2006) Proteomic based investigation of rhamnolipid production by *Pseudomonas chlororaphis* strain NRRL B-3061. J Ind Microbio Biotech 33:914–920

Gygi SP, Rist B, Gerber SA, Turecek F, Gelb MH, Aebersold R (1999a) Quantitative analysis of complex protein mixtures using isotope-coded affinity tags. Nature Biotechnol 17:994–999

Gygi SP, Rochon Y, Franza BR, Aebersold R (1999b) Correlation between protein and mRNA abundance in yeast. Mol Cell Biol 19:1720–1730

Han Y, Ma B, Zhang K (2005) SPIDER: software for protein identification from sequence tags with de novo sequencing error. J Bioinform Comput Biol 3:697–716

Hardwidge PR, Donohoe S, Aebersold R, Finlay BB (2006) Proteomic analysis of the binding partners to enteropathogenic *Escherichia coli* virulence proteins expressed in *Saccharomyces cerevisiae*. Proteomics 6:2174–2179

Henzel WJ, Watanabe C, Stults JT (2003) Protein identification: the origins of peptide mass fingerprinting. J Am Soc Mass Spectrom 14:931–942

Hewick RM, Hunkapiller MW, Hood LE, Dreyer WJ (1981) A gas-liquid solid phase peptide and protein sequenator. J Biol Chem 256:7990–7997

Hirel PH, Schmitter MJ, Dessen P, Fayat G, Blanquet S (1989) Extent of N-terminal methionine excision from *Escherichia coli* proteins is governed by the side-chain length of the penultimate amino acid. Proc Natl Acad Sci USA 86:8247–8251

Hoaglund CS, Valentine SJ, Sporleder CR, Reilly JP, Clemmer DE (1998) Three-dimensional ion mobility/TOFMS analysis of electrosprayed biomolecules. Anal Chem 70:2236–2242

Hoaglund-Hyzer CS, Lee YJ, Counterman AE, Clemmer DE (2002) Coupling ion mobility separations, collisional activation techniques, and multiple stages of MS for analysis of complex peptide mixtures. Anal Chem 74:992–1006

Hoaglund-Hyzer CS, Li J, Clemmer DE (2000) Mobility labeling for parallel CID of ion mixtures. Anal Chem 72:2737–2740

Horton P, Park KJ, Obayashi T, Fujita N, Harada H, Adams-Collier CJ, Nakai K (2007) WoLF PSORT: protein localization predictor. Nucleic Acids Res 35:W585–W587

Hu WS, Lin Y-H, Shih C-C (2007) A proteomic approach to study *Salmonella enterica* serovar Typhimurium putative transporter YjeH associated with ceftriaxone resistance. Biochem Biophys Res Commun 361:694–699

Hu Q, Noll RJ, Li H, Makarov A, Hardman M, Cooks GR (2005) The Orbitrap: a new mass spectrometer. J Mass Spectrom 40:430–443

Ishihama Y, Oda Y, Tabata T, Sato T, Nagasu T, Rappsilber J, Mann M (2005) Exponentially modified protein abundance index (em-PAI) for estimation of absolute protein amount in proteomics by the number of sequenced peptides per protein. Mol Cell Proteomics 4:1265–1272

Jarmon KH, Cebula ST, Saenz AJ, Petersen CE, Valentine NB, Kingsley MT, Wahl KL (2000) An algorithm for automated bacterial identification using matrix-assisted laser desorption/ionization mass spectrometry. Anal Chem 72:1217–1223

Jeffery CJ (1999) Moonlighting proteins. Trends Biochem Sci 24:8–11

Jennison AV, Raqib R, Verma NK (2006) Immunoproteome analysis of soluble and membrane proteins of *Shigella flexneri* 2457T. World J Gastroenterol 12:6683–6688

Johnson RS, Taylor JA (2002) Searching sequence databases via de novo peptide sequencing by tandem mass spectrometry. Mol Biotechnol 22:301–315

Jones JJ, Stump MJ, Fleming RC, Lay JO Jr, Wilkins CL (2003) Investigation of MALDI-TOF and FT-MS techniques for analysis of *Escherichia coli* whole cells. Anal Chem 75:1340–1347

Juncker AS, Willenbrock H, Von Heijne G, Brunak S, Nielsen H, Krogh A (2003) Prediction of lipoprotein signal peptides in Gram-negative bacteria. Protein Sci 12:1652–1662

Kan B, Habbi H, Schmid M, Liang W, Wang R, Wang D, Jungblut PR (2004) Proteome comparison of *Vibrio cholerae* cultured in aerobic and anaerobic conditions. Proteomics 4:3061–3067

Karas M, Bachmann D, Bahr U, Hillenkamp F (1987) Matrix-assisted ultraviolet laser desorption of non-volatile compounds. Int J Mass Spectrom Ion Process 78:53–68

Karas M, Hillenkamp F (1988) Laser desorption ionization of proteins with molecular masses exceeding 10,000 daltons. Anal Chem 60:2299–2301

Lancaster KS, An HJ, Li B, Lebrilla CB (2006) Interrogation of N-linked oligosaccharides using infrared multiphoton dissociation in FT-ICR mass spectrometry. Anal Chem 78:4990–4997

Le Guyader FS, Bon F, DeMedici D, Parnaudeau S, Bertone A, Crudeli S, Doyle A, Zidane M, Suffredini E, Kohli E, Maddalo F, Monini M, Gallay A, Pommepuy M, Pothier P, Ruggeri FM (2006) Detection of multiple noroviruses associated with an international gastroenteritis outbreak linked to oyster consumption. J Clin Microbiol 44:3878–3882

Lee YJ, Hoaglund-Hyzera CS, Srebalus Barnes CA, Hilderbrand AE, Valentine SJ, Clemmer DE (2002) Development of high-throughput liquid chromatography injected ion mobility quadrupole time-of-flight techniques for analysis of complex peptide mixtures. J Chromatogr B Analyt Technol Biomed Life Sci 782:343–351

Lenz LL, Mohammadi S, Geissler A, Portnoy DA (2003) SecA2-dependent secretion of autolytic enzymes promotes *Listeria monocytogenes* pathogenesis. Proc Indian Natl Sci Acad B Biol Sci 100:12432–12437

Lenz LL, Portnoy DA (2002) Identification of a second *Listeria secA* gene associated with protein secretion and the rough phenotype. Mol Microbiol 45:1043–1056

Li M, Rosenshine I, Tung SL, Wang XH, Freidberg D, Hew CL, Leung KY (2004) Comparative proteomic analysis of extracellular proteins of enterohemorrhagic and enteropathogenic *Escherichia coli* strains and their *ihf* and *ler* mutants. Appl Environ Microbiol 70:5274–5282

Liao X, Ying T, Wang H, Wang J, Shi Z, Feng E, Wei K, Wang Y, Zhang X, Huang L, Su G, Huang P (2003) A two-dimensional proteome map of *Shigella flexneri*. Electrophoresis 24:2864–2882

Lipton MS, Pasa-Tolic L, Anderson GA, Anderson DJ, Auberry DL, Battista JR, Daly MJ, Fredrickson J, Hixson KK, Kostandarithes H, Masselon C, Markillie LM, Moore RJ, Romine MF, Shen Y, Stritmatter E, Tolic N, Udseth HR, Venkateswaran A, Wong KK, Zhao R, Smith RD (2002) Global analysis of the *Deinococcus radiodurans* proteome by using accurate mass tags. Proc Natl Acad Sci USA 99:11049–11054

Liu H, Sadygov RG, Yates JR III (2004) A model for random sampling and estimation of relative protein abundance in shotgun proteomics. Anal Chem 76:4193–4201

Lo I, Denef VJ, Verberkmoes NC, Shah MB, Goltsman D, DiBartolo G, Tyson GW, Allen EE, Ram RJ, Detter JC, Richardson P, Thelen MP, Hettich RL, Banfield JF (2007) Strain-resolved community proteomics reveals recombining genomes of acidophilic bacteria. Nature 446: 537–541

Loboda AV, Ackloo S, Chernushevich IV (2003) A high-performance matrix-assisted laser desorption/ionization orthogonal time-of-flight mass spectrometer with collisional cooling. Rapid Commun Mass Spectrom 17:2508–2516

Luo Y, Vilain S, Voigt B, Albrecht D, Hecker M, Brözel VS (2007) Proteomic analysis of *Bacillus cereus* growing in liquid soil organic matter. FEMS Microbiol Lett 271:40–47

Ma B, Zhang K, Hendrie C, Liang C, Li M, Doherty-Kirby A, Lajoie G (2003) PEAKS: powerful software for peptide de novo sequencing by tandem mass spectrometry. Rapid Commun Mass Spectrom 17:2337–2342

MacGillivray AJ, Rickwood D (1974) The heterogeneity of mouse-chromatin nonhistone proteins as evidenced by two-dimensional polyacrylamide-gel electrophoresis and ion-exchange chromatography. Eur J Biochem 41:181–190

Maillet I, Berndt P, Malo C, Rodriguez S, Brunisholz RA, Pragai Z, Arnold S, Langen H, Wyss M (2007) From the genome sequence to the proteome and back: evaluation of *E. coli* genome annotation with a 2-D gel-based proteomics approach. Proteomics 7:1097–1106

Makarov AA (1999) US Patent 5,886,346

Makarov A (2000) Electrostatic axially harmonic orbital trapping: a high-performance technique of mass analysis. Anal Chem 72:1156–1162

Makarov A, Denisov E, Kholomeev A, Balschun W, Lange O, Strupat K, Horning S (2006) Performance evaluation of a hybrid linear ion trap/orbitrap mass spectrometer. Anal Chem 78:2113–2120

Mamyrin BA, Karataev VI, Shmikk DV, Zagulin VA (1973) The mass-reflectron, a new nonmagnetic time-of-flight mass spectrometer with high resolution. *Sov. Phys.* JETP 37:45

Mandrell RE, Harden LA, Bates A, Miller WG, Haddon WF, Fagerquist CK (2005) Speciation of *Campylobacter coli, C. jejuni, C. helveticus, C. lari, C. sputorum*, and *C. upsaliensis* by matrix-assisted laser desorption ionization-time of flight mass spectrometry. Appl Environ Microbiol 71:6292–6307

March RE (1997) An introduction to quadrupole ion trap mass spectrometry. J Mass Spectrom 32:351–369

Marshall AG, Wang T-CL, Ricca TL (1985) Tailored excitation of Fourier transform ion cyclotron resonance mass spectrometry. J Am Chem Soc 107:7893–7897

Martinović S, Veenstra TD, Anderson GA, Pasa-Tolić L, Smith RD (2002) Selective incorporation of isotopically labeled amino acids for identification of intact proteins on a proteome-wide level. J Mass Spectrom 37:99–107

Mau B, Glasner JD, Darling AE, Perna NT (2006) Genome-wide detection and analysis of homologous recombination among sequenced strains of *Escherichia coli*. Genome Biol 7:R44

McCormack AL, Schieltz DM, Goode B, Yang S, Barnes G, Drubin D, Yates JR 3rd (1997) Direct analysis and identification of proteins in mixtures by LC/MS/MS and database searching at the low-femtomole level. Anal Chem 69:767–776

Medzihradszky KF, Campbell JM, Baldwin MA, Falick AM, Juhasz P, Vestal ML, Burlingame AL (2000) The characteristics of peptide collision-induced dissociation using a high-performance MALDI-TOF/TOF tandem mass spectrometer. Anal Chem 72:552–558

Merenbloom SI, Koeniger SL, Valentine SJ, Plasencia MD, Clemmer DE (2006) IMS-IMS and IMS-IMS-IMS/MS for separating peptide and protein fragment ions. Anal Chem 78: 2802–2809

Morris HR, Paxton T, Dell A, Langhorne J, Berg M, Bordoli RS, Hoyes J, Bateman. RH (1996) High sensitivity collisionally-activated decomposition tandem mass spectrometry on a novel quadrupole/orthogonal-acceleration time-of-flight mass spectrometer. Rapid Commun Mass Spectrom 10:889–896

Nakai K, Kanehisa M (1991) Expert system for predicting protein localization sites in Gram-negative bacteria. Proteins 11:95–110

Nielsen H, Engelbrecht J, von Heijne G, Brunak S (1996) Defining a similarity threshold for a functional protein sequence pattern: the signal peptide cleavage site. Proteins 24:165–177

Norbeck AD, Callister SJ, Monroe ME, Jaitly N, Elias DA, Lipton MS, Smith RD (2006) Proteomic approaches to bacterial differentiation. J Microbio Methods 67:473–486

Oda Y, Huang K, Cross FR, Cowburn D, Chait BT (1999) Accurate quantitation of protein expression and site-specific phosphorylation. Proc Natl Acad Sci USA 96:6591–6596

O'Farrell PH (1975) High resolution two-dimensional electrophoresis of proteins. J Biol Chem 250:4007–4021

Ong S-E, Blagoev B, Kratchmarova I, Kristensen DB, Steen H, Pandey A, Mann M (2002) Stable isotope labeling by amino acids in cell culture, SILAC, as a simple and accurate approach to expression proteomics. Mol Cell Proteomics 246:376–386

Oosthuizen MC, Steyn B, Lindsay D, Brözel VS, von Holy A (2001) Novel method for the proteomic investigation of a dairy-associated *Bacillus cereus* biofilm. FEMS Microbiol Lett 194:47–51

Oosthuizen MC, Steyn B, Theron J, Cosette P, Lindsay D, Von Holy A, Brözel VS (2002) Proteomic analysis reveals differential protein expression by Bacillus cereus during biofilm formation. Appl Environ Microbiol 68:2770–2780

Opiteck GJ, Lewis KC, Jorgenson JW, Anderegg RJ (1997) Comprehensive on-line LC/LC/MS of proteins. Anal Chem 69:1518–1524

Pan S, Rush J, Peskind ER, Galasko D, Chung K, Quinn J, Jankovic J, Leverenz JB, Zabetian C, Pan C, Wang Y, Oh. JH, Gao J, Zhang J, Montine T, Zhang J (2008) Application of targeted quantitative proteomics analysis in human cerebrospinal fluid using a liquid chromatography matrix-assisted laser desorption/ionization time-of-flight tandem mass spectrometer (LC MALDI TOF/TOF) platform. J Proteome Res 7:720–730

Pang JX, Ginanni N, Dongre AR, Hefta SA, Opitek GJ (2002) Biomarker discovery in urine by proteomics. J Proteome Res 1:161–169

Perkins DN, Pappin DJ, Creasy DM, Cottrell JS (1999) Probability-based protein identification by searching sequence databases using mass spectrometry data. Electrophoresis 20:3551–3567

Pevtsov S, Fedulova I, Mirzaei H, Buck C, Zhang X (2006) Performance evaluation of existing de novo sequencing algorithms. J Proteome Res 5:3018–3028

Pineda FJ, Lin JS, Fenselau C, Demirev PA (2000) Testing the significance of microorganism identification by mass spectrometry and proteome database search. Anal Chem 72:3739–3744

Price WD, Schnier PD, Williams ER (1996) Tandem mass spectrometry of large biomolecule ions by blackbody infrared radiative dissociation. Anal Chem 68:859–866

Qian MG, Lubman DM (1995) Analysis of tryptic digests using microbore HPLC with an ion trap storage/reflectron time-of-flight detector. Anal Chem 67:2870–2877

Qian MG, Zhang Y, Lubman DM (1995) Collision-induced dissociation of multiply charged peptides in an ion-trap storage/reflectron time-of-flight mass spectrometer. Rapid Commun Mass Spectrom 9:1275–1282

Quiñones B, Guilhabert MR, Millerm WG, Mandrell RE, Lastovica AJ, Parker CT (2008) Comparative genomic analysis of clinical strains of *Campylobacter jejuni* from South Africa. PLoS ONE 3:e2015

Rabilloud T (2002) Two-dimensional gel electrophoresis in proteomics: old, old fashioned, but it still climbs up the mountains. Proteomics 2:3–10

Rappsilber J, Ryder U, Lamond AI, Mann M (2002) Large-scale proteomic analysis of the human spliceosome. Genome Res 12:1231–1245

Reid GE, McLuckey SA (2002) 'Top down' protein characterization via tandem mass spectrometry. J Mass Spectrom 37:663–675

Reller ME, Nelson JM, Mølbak K, Ackman DM, Schoonmaker-Bopp DJ, Root TP, Mintz ED (2006) A large, multiple-restaurant outbreak of infection with *Shigella flexneri* serotype 2a traced to tomatoes. Clin Infect Dis 42:163–169

Reyzer ML, Caprioli RM (2005) MALDI mass spectrometry for direct tissue analysis: a new tool for biomarker discovery. J Proteome Res 4:1138–1142

Rodríguez-Ortega MJ, Norais N, Bensi G, Liberatori S, Capo S, Mora M, Scarselli M, Doro F, Ferrari G, Garaguso I, Maggi T, Neumann A, Covre A, Telford JL, Grandi G (2006) Characterization and identification of vaccine candidate proteins through analysis of the group A *Streptococcus* surface proteome. Nat Biotechnol 24:191–197

Romijn EP, Christis C, Wieffer M, Gouw JW, Fullaondo A, van der Sluijs P, Braakman I, Heck AJ (2005) Expression clustering reveals detailed co-expression patterns of functionally related proteins during B cell differentiation: a proteomic study using a combination of one-dimensional gel electrophoresis, LC-MS/MS, and stable isotope labeling by amino acids in cell culture (SILAC). Mol Cell Proteomics 4:1297–1310

Ross PL, Huang YN, Marchese JN, Williamson B, Parker K, Hattan S, Khainovski N, Pillai S, Dey S, Daniels S, Purkayastha S, Juhasz P, Martin S, Bartlet-Jones M, He F, Jacobson A, Pappin

DJ (2004) Multiplexed protein quantitation in *Saccharomyces cerevisiae* using amine-reactive isobaric tagging reagents. Mol Cell Proteomics 3:1154–1169

Ruotolo BT, Hyung SJ, Robinson PM, Giles K, Bateman RH, Robinson CV (2007) Ion mobility-mass spectrometry reveals long-lived, unfolded intermediates in the dissociation of protein complexes. Angew Chem Int Ed Engl 46:8001–8004

Schaumburg J, Diekmann O, Hagendorff P, Bergmann S, Rohde M, Hammerschmidt S, Jansch L, Wehland J, Karst U (2004) The cell wall subproteome of *Literia monocytogenes*. Proteomics 4:2991–3006

Scherl A, Francois P, Charbonnier Y, Deshusses JM, Koessler T, Huyghe A, Bento M, Stahl-Zeng J, Fischer A, Masselot A, Vaezzadeh A, Galle F, Renzoni A, Vaudaux P, Lew D, Zimmermann-Ivol CG, Binz P-A, Sanchez J-C, Hochstrasser DF, Schrenzel J (2006) Exploring glycopeptide-resistance in *Staphylococcus aureus*: a combined proteomics and transcriptomics approach for the identification of resistance-related markers. BMC Genomics 7:296

Schnaible V, Wefing S, Resemann A, Suckau D, Bucker A, Wolf-Kummeth S, Hoffmann D (2002) Screening for disulfide bonds in proteins by MALDI in-source decay and LIFT-TOF/TOF-MS. Anal Chem 74:4980–4988

Schwartz JC, Senko MW (2004) US Patent 6,797,950

Scott NE, Marzook NB, Deutscher A, Falconer L, Crossett B, Djordjevic SP, Cordwell SJ (2010) Mass spectrometric characterization of the *Campylobacter jejuni* adherence factor CadF reveals post-translational processing that removes immunogenicity while retaining fibronectin binding. Proteomics 10:277–288

Seal BS, Hiett KL, Kuntz RL, Woolsey R, Schegg KM, Ard M, Stintzi A (2007) Proteomic analyses of a robust versus a poor chicken gastrointestinal colonizing isolate of *Campylobacter jejuni*. J Proteome Res 6:4582–4591

Sebaihia M, Wren BW, Mullany P, Fairweather NF, Minton N, Stabler R, Thomson NR, Roberts AP, Cerdeño-Tárraga. AM, Wang H, Holden MT, Wright A, Churcher C, Quail MA, Baker S, Bason N, Brooks K, Chillingworth T, Cronin A, Davis P, Dowd L, Fraser A, Feltwell T, Hance Z, Holroyd S, Jagels K, Moule S, Mungall K, Price C, Rabbinowitsch E, Sharp S, Simmonds M, Stevens K, Unwin L, Whithead S, Dupuy B, Dougan G, Barrell B, Parkhill J (2006) The multidrug-resistant human pathogen *Clostridium difficile* has a highly mobile, mosaic genome. Nat Genet 38:779–786

Senko MW, Speir JP, McLafferty FW (1994) Collisional activation of large multiply charged ions using Fourier transform mass spectrometry. Anal Chem 66:2801–2808

Shevchenko A, Chernushevich I, Ens W, Standing KG, Thomson B, Wilm M, Mann M (1997) Rapid 'de Novo' peptide sequencing by a combination of nanoelectrospray, isotopic labeling and a quadrupole/time-of-flight mass spectrometer. Rapid Commun Mass Spectrom 11:1015–1024

Shimizu T, Ohtani K, Hirakawa H, Ohshima K, Yamashita A, Shiba T, Ogasawara N, Hattori M, Kuhara S, Hayashi H (2002a) Complete genome sequence of *Clostridium perfringens*, an anaerobic flesh-eater. Proc Natl Acad Sci USA 99:996–1001

Shimizu T, Shima K, Yoshino K, Yonezawa K, Shimizu T, Hayashi H (2002b) Proteome and transcriptome analysis of the virulence genes regulated by the VirR/VirS system in *Clostridium perfringens*. J Bacteriol 184:2587–2594

Sinha P, Poland J, Schnölzer M, Rabilloud T (2001) A new silver staining apparatus and procedure for matrix-assisted laser desorption/ionization-time of flight analysis of proteins after two-dimensional electrophoresis. Proteomics 1:835–840

Solbiati J, Chapman-Smith A, Miller JL, Miller CG, Cronan JE Jr (1999) Processing of the N termini of nascent polypeptide chains requires deformylation prior to methionine removal. J Mol Biol 290:607–614

Soufi B, Jers C, Hansen ME, Petranovic D, Mijakovic I (2008) Insights from site-specific phosphoproteomics in bacteria. Biochim Biophys Acta 1784:186–192

Stephens WE (1946) A pulsed mass spectrometer with time dispersion. Phys Rev 69:691

Strittmatter EF, Ferguson PL, Tang K, Smith RD (2003) Proteome analyses using accurate mass and elution time peptide tags with capillary LC time-of-flight mass spectrometry. J Am Soc Mass Spectrom 14:980–991

Syka JEP, Coon JJ, Schroeder MJ, Shabanowitz J, Hunt DF (2004) Peptide and protein sequence analysis by electron transfer dissociation mass spectrometry. Proc Natl Acad Sci 101: 9528–9533

Tanaka K, Ido Y, Akita S, Yoshida Y, Yoshida T (1987) Second Japan-China joint symposium on mass spectrometry (Abstract), Osaka, Japan, 15–18 Sept

Taverna F, Negri A, Piccinini R, Zecconi A, Nonnis S, Ronchi S, Tedeschi G (2007) Characterization of cell wall associated proteins of a *Staphylococcus aureus* isolated from bovine mastitis case by a proteomic approach. Vet Microbiol 119:240–247

Taylor JA, Johnson RS (1997) Sequence database searches via de novo peptide sequencing by tandem mass spectrometry. Rapid Commun Mass Spectrom 11:1067–1075

Taylor JA, Johnson RS (2001) Implementation and uses of automated de novo peptide sequencing by tandem mass spectrometry. Anal Chem 73:2594–2604

Taylor D, Schwartz J, Zhou J, James M, Bier M, Korsak A, Stafford G (1995). Application of tailored waveform generation to the quadrupole ion trap. Proceedings of the 43rd ASMS conference on mass spectrometry and allied topics, Atlanta, Georgia, May 21–26, 1995, 1103

Tonge R, Shaw J, Middleton B, Rowlinson R, Rayner S, Young J, Pognan F, Hawkins E, Currie I, Davison M (2001) Validation and development of fluorescence two-dimensional differential gel electrophoresis proteomics technology. Proteomics 1:377–396

Trost M, Wehmhoner D, Karst U, Dieterich G, Wehland J, Jansch L (2005) Comparative proteome analysis of secretory proteins from pathogenic and nonpathogenic *Listeria* species. Proteomics 5:1544–1557

Unlü M, Morgan ME, Minden JS (1997) Difference gel electrophoresis: a single gel method for detecting changes in protein extracts. Electrophoresis 18:2071–2077

Valentine SJ, Counterman AE, Hoaglund CS, Reilly JP, Clemmer DE (1998) Gas-phase separations of protease digests. J Am Soc Mass Spectrom 9:1213–1216

Veenstra TD, Martinović S, Anderson GA, Pasa-Tolić L, Smith RD (2000) Proteome analysis using selective incorporation of isotopically labeled amino acids. J Am Soc Mass Spectrom 11:78–82

VerBerkmoes NC, Bundy JL, Hauser L, Asano KG, Razumovskaya J, Larimer F, Hettich RL, Stephenson JL Jr (2002) Integrating top-down and bottom-up mass spectrometric approaches for proteomic analysis of *Shewanella oneidensis*. J Proteome Res 1:239–252

Voigt B, Schweder T, Sibbald MJ, Albrecht D, Ehrenreich A, Bernhardt J, Feesche J, Maurer KH, Gottschalk G, van Dijl JM, Hecker M (2006) The extracellular proteome of *Bacillus licheniformis* grown in different media and under different nutrient starvation conditions. Proteomics 6:268–281

Vytvytska O, Nagy E, Blüggel M, Meyer HE, Kurzbauer R, Huber LA, Klade CS (2002) Identification of vaccine candidate antigens of *Staphylococcus aureus* by serological proteome analysis. Proteomics 2:580–590

Waanders LF, Hanke S, Mann M (2007) Top-down quantitation and characterization of SILAC-labeled proteins. J Am Soc Mass Spectrom 18:2058–2064

Wang G, Wu WW, Zeng W, Chou CL, Shen RF (2006) Label-free protein quantification using LC-coupled ion trap or FT mass spectrometry: reproducibility, linearity, and application with complex proteomes. J Proteome Res 5:1214–1223

Washburn MP, Wolters D, Yates JR III (2001) Large-scale analysis of the yeast proteome by multidimensional protein identification technology. Nat Biotechnol 19:242–247

Wheeler C, Vogt TM, Armstrong GL, Vaughan G, Weltman A, Nainan OV, Dato V, Xia G, Waller K, Amon J, Lee TM, Highbaugh-Battle A, Hembreem C, Evenson S, Ruta MA, Williams IT, Fiore AE, Bell BP (2005) An outbreak of hepatitis A associated with green onions. N Engl J Med 353:890–897

Wiley WC, MacLaren IH (1955) Time-of-flight spectrometer with improved resolution. Rev Sci Instr 26:1150

Williams TL, Leopold P, Musser S (2002) Automated postprocessing of electrospray LC/MS data for profiling protein expression in bacteria. Anal Chem 74:5807–5813

Williams TL, Monday SR, Edelson-Mammel S, Buchanan R, Musser SM (2005a) A top-down proteomics approach for differentiating thermal resistant strains of *Enterobacter sakazakii*. Proteomics 5:4161–4169

Williams TL, Monday SR, Feng PCH, Musser SM (2005b) Identifying new PCR targets for pathogenic bacteria using top-down LC/MS protein discovery. J Biomol Tech 16:134–142

Williams TL, Musser SM, Nordstrom JL, DePaola A, Monday SR (2004) Identification of a protein biomarker unique to the pandemic O3:K6 clone of *Vibrio parahaemolyticus*. J Clin Microbio 42:1657–1665

Wolters DA, Washburn MP, Yates JR 3rd (2001) An automated multidimensional protein identification technology for shotgun proteomics. Anal Chem 73:5683–5690

Wright A, Wait R, Begum S, Crossett B, Nagy J, Brown K, Fairweather N (2005) Proteomic analysis of cell surface proteins from *Clostridium difficile*. Proteomics 5:2443–2452

Wu JT, Qian MG, Li MX, Liu L, Lubman DM (1996) Use of an ion trap storage/reflectron time-of-flight mass spectrometer as a rapid and sensitive detector for capillary electrophoresis in protein digest analysis. Anal Chem 68:3388–3396

Xu C, Ren H, Wang S, Peng X (2004) Proteomic analysis of salt-sensitive outer membrane proteins of *Vibrio parahaemolyticus*. Research Microbio 155:835–842

Yost RA, Enke CG (1978) Selected ion fragmentation with a tandem quadrupole mass spectrometer. J Amer Chem Soc 100:2274–2275

Zamdborg L, LeDuc RD, Glowacz KJ, Kim YB, Viswanathan V, Spaulding IT, Early BP, Bluhm EJ, Babai S, Kelleher NL (2007) ProSight PTM 2.0: improved protein identification and characterization for top down mass spectrometry. Nucleic Acids Res 35(2):W701–W706

Zheng S, Schneider KA, Barder TJ, Lubman DM (2003) Two-dimensional liquid chromatography protein expression mapping for differential proteomic analysis of normal and O157:H7 *Escherichia coli*. BioTechniques 35:1202–1212

Zhou M, Boekhorst J, Francke C, Siezen RJ (2008) LocateP: genome-scale subcellular-location predictor for bacterial proteins. BMC Bioinformatics 9:173

Zubarev RA, Kelleher NL, McLafferty FW (1998) Electron capture dissociation of multiply charged protein cations. A nonergodic process. J Am Chem Soc 120:3265–3266

Chapter 12
Molecular Epidemiology of Foodborne Pathogens

Yi Chen, Eric Brown, and Stephen J. Knabel

12.1 Introduction

The purpose of this chapter is to describe the basic principles and advancements in the molecular epidemiology of foodborne pathogens. Epidemiology is the study of the distribution and determinants of infectious diseases and/or the dynamics of disease transmission. The goals of epidemiology include the identification of physical sources, routes of transmission of infectious agents, and distribution and relationships of different subgroups. Molecular epidemiology is the study of epidemiology at the molecular level. It has been defined as "a science that focuses on the contribution of potential genetic and environmental risk factors, identified at the molecular level, to the etiology, distribution and prevention of diseases within families and across populations" (http://www.pitt.edu/~rlaporte/who.html). The European Study Group of Epidemiologic Markers (ESGEM) and the Molecular Typing Working Group proposed some important definitions in molecular epidemiology (Struelens, 1996) and some of these definitions are included in Table 12.1.

Recent major foodborne outbreaks have highlighted the importance of developing subtyping methods that can lead to rapid and accurate identification of outbreaks and eventual trace back to the source of contamination. Thus, the primary goal of subtyping methods is to differentiate the source strain from all other background strains in the context of an outbreak. With the development of rapid distribution systems, food products can now be shipped to every corner of the country and even the world in a short period of time. The rapid development of molecular subtyping techniques has greatly enhanced our ability to characterize individual strains. In addition, a pathogen often evolves in response to host and environmental antimicrobial mechanisms. Therefore, identification of the genetic determinants of these evolutionary events is also important for studying the molecular epidemiology of foodborne pathogens.

Y. Chen (✉)
Center for Food Safety and Applied Nutrition, US Food and Drug Administration, College Park, MD, USA
e-mail: yi.chen@fda.hhs.gov

M. Wiedmann, W. Zhang (eds.), *Genomics of Foodborne Bacterial Pathogens*,
Food Microbiology and Food Safety, DOI 10.1007/978-1-4419-7686-4_12,
© Springer Science+Business Media, LLC 2011

Table 12.1 Important definitions in epidemiology

Term	Definition	References
Isolate	A collection of cells derived from a primary colony growing on a solid media on which the source of the isolates was inoculated	Struelens (1996)
Strain	Isolates that have distinct phenotypic and/or genotypic characteristics compared to other isolates from the same species	Struelens (1996)
Type	A common attribute that can define a group of closely related strains	Chen et al. (Chapter 12, this volume)
Subtype	Type at the subspecies level	Chen et al. (Chapter 12, this volume)
Source strain	The original strain that is spread and causes an outbreak or an epidemic	Chen et al. (Chapter 12, this volume)
Clone	A strain or a group of strains descended asexually from a single ancestral cell (source strain) which have identical or similar phenotypes or genotypes as identified by a specific strain typing method	Struelens (1996)
Outbreak	An acute appearance of a cluster of an illness caused by an outbreak clone that occurs in numbers in excess of what is expected for that time and place. A foodborne outbreak is defined as two or more illnesses resulting from the ingestion of a common food	Riley (2004)
Outbreak clone	A strain or a group of strains descended asexually from a single ancestral cell (source strain) that are associated with one outbreak	Chen et al. (Chapter 12, this volume)
Epidemic	One or more outbreaks caused by an epidemic clone that survives and spreads over a long period of time and space	Riley (2004)
Epidemic clone	A strain or a group of strains descended asexually from a single ancestral cell (source strain) that are involved in one epidemic and can include one or more outbreak clones	Riley (2004)
Pandemic	An epidemic that spreads globally and may last for years	Riley (2004)

Modern subtyping techniques have recently undergone extensive improvements with more understanding of genomics, and many new methods have led to enhanced performance. Our understanding of epidemiology and the evolution of foodborne pathogens has been greatly enhanced by the development and application of these advanced molecular subtyping techniques. Understanding the basic epidemiologic principles that govern the evolution and spread of bacterial pathogen populations is critical for correctly interpreting the results from these advanced molecular subtyping methods.

12.1.1 Important Definitions and Basic Principles of Molecular Epidemiology

Some definitions were proposed by Riley (2004) and the Molecular Typing Working Group (Struelens, 1996), and some are proposed in this chapter.

12.1.2 *Applications of Strain Typing Techniques*

Three major applications of strain typing techniques are in taxonomy, epidemiology, and phylogeny (evolutionary genetics). Taxonomy, also known as (bio)systematics, is the practice and science of classification of organisms based on their common characteristics (van Belkum et al., 2001). One definition for species, a controversial subject in the case of asexual microorganisms, is a group of isolates with a common origin or ancestry as demonstrated by a discrete typing unit (van Belkum et al., 2001). Strain typing at the subspecies level is often referred to as subtyping, and analysis of subtyping data can be used for taxonomy. Ribosomal RNA (rRNA) genes have been widely used as markers for determining taxonomy because these genes are present in all cells, show little evidence of horizontal gene transfer, and possess both conserved and variable domains (Lane et al., 1985). Recently, many new strain subtyping methods have provided further insight into the many already assigned species, and numerous species have been reassigned based on new strain subtyping data. Therefore, selection of proper targets, quality of the typing data, and analysis of these data are critical for correct classification of strains within microbial species. One of the most recent changes in the nomenclature for a foodborne pathogen was that of *Enterobacter sakazakii*, which was isolated from many powdered infant formula samples and associated with severe and fatal neonatal meningitis (Iversen et al., 2008). New evidence obtained from various subtyping methods such as biotyping, DNA–DNA hybridization, amplified fragment length polymorphism, and ribotyping have all contributed to the reclassification of this organism as *Cronobacter*.

In evolutionary genetics, subtyping data are used to infer phylogeny (evolutionary history) at the subspecies level (Gurtler and Mayall, 2001). Population geneticists utilize subtyping data and various phylogenetic computational algorithms to study the relationships of various subgroups within a population. However, strain typing techniques have been mostly used to study the epidemiology of pathogens, and most literature has focused on utilizing strain typing techniques for this purpose. Understanding the basic principles of epidemiology is important for scientists who are developing subtyping strategies. Unlike evolutionary genetics, which focuses on long-term evolution, molecular epidemiology must focus on short-term evolution of microorganisms in order to track the transmission of specific clonal groups during weeks, months, or over a few years. Another big difference between molecular epidemiology and evolutionary genetics is that molecular epidemiology can be empirical. For example, it is often assumed that isolates involved in an outbreak are genetically related and they are unrelated to concurrent sporadic isolates. Riley (2004) gave a detailed review of the principles of molecular epidemiology of infectious diseases. Many of the basic principles of molecular epidemiology discussed in that book have been incorporated into and expanded on in this chapter.

Early recognition of foodborne outbreaks. The terms epidemic and outbreak are often used interchangeably in the literature. In this chapter the two concepts are distinguished to differentiate between long-term and short-term spread of infectious agents, respectively. The United States Centers for Disease Control and Prevention (CDC) has defined all isolates belonging to the same outbreak as the outbreak strain

(Graves et al., 2005). However, evidence has shown that isolates involved in the same outbreak can exhibit slightly different genetic characteristics (i.e., subtypes). Therefore, by the definition of strain given by Struelens et al. (1996) these different subtypes should be classified as different strains (Graves et al., 2005). Given the existence of these slightly different genetic subtypes within epidemic clones and the assumption that all isolates associated with an outbreak are clonally related Orskov and Orskov (1983), we propose "outbreak clone" (Table 12.1) as a more appropriate term than "outbreak strain" when referring to these genetically different strains within an outbreak.

Early recognition of foodborne outbreaks is vital for controlling the spread of disease. Before subtyping techniques were developed and widely applied to the field of epidemiology, epidemiologists relied solely on conventional epidemiologic data to track outbreaks. When an increasing number of cases of a specific type of foodborne disease is reported, epidemiologists typically conduct patient interviews and case–control studies to identify the food vehicle responsible. If the cases are found to be due to the consumption of a common food vehicle, they are defined as an outbreak. However, conventional epidemiologic surveillance systems have certain limitations. For example, case–control studies can be time consuming and, sometimes inaccurate because patients often cannot remember what they ate and/or when they ate the food. This is especially problematic for diseases that are widely spread and have a relatively long incubation period. Humans that are infected with *Listeria monocytogenes* demonstrate a wide range of incubation period (7–60 days). Therefore timely detection of listeriosis outbreaks is often problematic using conventional epidemiologic approaches (Wiedmann, 2002). With the development of rapid, high-volume distribution systems in the food industry, it is now critical to recognize outbreaks in a timely manner. Molecular subtyping techniques provide an excellent tool for doing this. Use of molecular subtyping techniques to identify an acute increase of a cluster of isolates with the same or similar subtype indicates that an outbreak may be ongoing. This is especially helpful for detecting widely scattered multistate outbreaks. Comparison of subtyping data generated from different states now allows for the timely detection of these previously "hidden" outbreaks. Therefore, molecular subtyping techniques have greatly enhanced the sensitivity of outbreak identification.

Identifying sources of contamination. A basic assumption in the field of molecular epidemiology is that isolates that are part of the same chain of transmission are the descendents of the source strain and thus can be referred to as a clone Orskov and Orskov (1983). Epidemic strains within a clonal group always have identical or very similar subtypes as shown in many studies, and this is used as a guiding principle in outbreak investigations. Molecular subtyping techniques have overcome many limitations of traditional epidemiology. For example, Bender et al. (2001) incorporated pulsed-field gel electrophoresis (PFGE) into a surveillance system for *Salmonella enterica* serotype Typhimurium in the State of Minnesota from 1994 to1998. PFGE patterns were able to aid them in assigning priorities for treating patients, focusing investigations, and avoiding unnecessary investigation of concurrent increases in unrelated patterns. In contrast, conventional surveillance caused unnecessary investigations of epidemiologically unrelated cases that were temporally, but not epidemiologically, related. Routine PFGE typing allowed the detection of four out of six community-based outbreaks that would not have been detected by traditional surveillance methods, largely because PFGE efficiently separated outbreak-related isolates from sporadic-case isolates. On the other hand, conventional surveillance systems did not have enough sensitivity to detect outbreaks involving common serotypes, especially for those outbreaks with a few cases spanning a period of several weeks. This was confirmed by the fact that conventional surveillance systems failed to detect any outbreaks of this serotype from 1990 to 1993. PFGE typing allowed investigators to intervene to stop transmission from a variety of sources, and PFGE data were also instrumental in the recall of the contaminated food products (Bender et al., 2001).

Strain typing techniques have greatly helped us to understand the epidemiology of various pathogens. For example, before various typing techniques were used, it was unclear whether or not *L. monocytogenes* was a foodborne pathogen because many animals also carried this pathogen. Loessner and Busse (1990) used phage typing to demonstrate that *L. monocytogenes* in human patients came from foods. Litrup et al. (2007) used multilocus sequence typing to demonstrate that *Campylobacter coli* isolates from pigs contained sequence types that were quite different from those of human isolates, and therefore, pigs were likely not a significant source of human *C. coli* infections.

Subtyping data can also be used to detect the routes by which foodborne pathogens are transmitted throughout the food system. For example, various strain typing techniques have been developed to detect routes of transmission of *L. monocytogenes* in shrimp (Destro et al., 1996a), meat (Senczek et al., 2000), salmon (Dauphin et al., 2001), and vegetable (Bansal et al., 1996) processing plants as well as fish smoking and slaughter houses (Wulff et al., 2006). Identifying the routes by which foodborne pathogens are transmitted to finished products in food processing plants facilitates establishment of effective intervention strategies to prevent contamination. These types of studies can also improve our understanding of the ecology of various foodborne pathogens in food processing plants, which can also lead to more effective intervention strategies.

Bacterial reproduction and transmission are not always totally clonal because recombination constantly alters their clonal population structure (Feil et al., 2001). During the spread of the source strain or even during the isolation, passage and storage of isolates in laboratories, various changes, mostly due to recombination, can lead to genetic variations in the isolates which cannot be explained by clonal theory. Therefore, a good epidemiologic subtyping scheme must be able to demonstrate the close relatedness of these clonal subtypes, even though they possess minor genetic variations. Thus, it is critical that epidemiologic typing methods target markers that are not subject to recombination and are epidemiologically relevant. Also, the speed at which these molecular markers undergo genetic changes should match the scope and purpose of the study. As a result, long-term epidemiologic studies and short-term outbreak investigations may require use of distinctly different markers that evolve at different rates. To properly select stable and epidemiologically relevant markers, an understanding of the physiology, evolution, and genomic structure of the target pathogen is essential. Evolutionary relationships between different bacterial isolates may not necessarily be concordant with epidemiologic relatedness in the case of outbreaks due to two or more independent sources. However, in the case of the more common single-source foodborne disease outbreaks, the evolutionary relationship is usually consistent with the spread of a single outbreak clone.

Differentiation of virulent and non-virulent strains. Subtyping methods are frequently used to identify virulent clonal groups that have the ability to cause human disease. Identification, characterization, and tracking of these virulent epidemic clones would facilitate the prevention and control of potential outbreaks. For example, among all *Escherichia coli* serotypes, specific clones of *E. coli* O157:H7 have been recognized as potentially more virulent. Manning et al. (2008) developed a whole genome-based SNP typing strategy and used it to identify different clades of *E. coli* O157:H7, one of which (clade 8) appeared to be more virulent. Nightingale et al. (2008) observed reduced virulence in *L. monocytogenes* strains with premature stop codons in *inlA* and Van Stelten and Nightingale (2008) subsequently developed a multiplex SNP typing strategy to differentiate virulence-attenuated strains from other strains of *L. monocytogenes*.

Identification of strains with different host specificity. Strains of the same bacterial species are not always specific to the same host. Molecular subtyping techniques can be used to differentiate strains with different host ranges. Mohapatra et al. (2007) developed a rep-PCR subtyping technique to group *E. coli* strains that were specific to humans, poultry, and wild birds. The technique is useful for quickly identifying sources of *E. coli* contamination in the environment. Miller et al. (2006) identified host-associated multilocus sequence types of *Campylobacter coli* strains from different food animal sources. These findings can be used for efficient source tracking of *C. coli*, especially for tracing human clinical isolates back to their animal sources (Miller et al., 2006).

12.1.3 Performance Criteria of Strain Typing Techniques

Typeability. Strains in the targeted population should be typeable using a specific subtyping method, which means the markers targeted by a subtyping method need to be present in as many strains as possible. Some examples of low typeability include (i) some strains that do not have H antigens and therefore cannot be assigned a proper serotype and (ii) strains that do not have plasmids cannot be subtyped by plasmid profiles.

Reproducibility. A subtyping method should have both inter-laboratory and intra-laboratory reproducibility. This criterion is related to the stability of the genetic markers, which is affected by the type of genetic changes involved. For example, single base mutations tend to be relatively rare and thus are more stable, while recombination events due to mobile genetic elements such as plasmids, bacteriophages, and transposons occur more frequently, making these kinds of changes relatively less stable. These latter recombination events are known to occur widely over time in food processing plants and during regular laboratory handling of bacterial cultures.

Discriminatory power (D). Discriminatory power describes the ability of a sub-typing system to generate distinct and discrete units of information from different strains (Struelens, 1996). In order to track the routes of transmission of outbreaks, high discriminatory power is needed to separate outbreak-related isolates from non-related isolates. Therefore, discriminatory power is one of the most important criteria for molecular subtyping methods. In addition, in many cases not all strains of one species of a bacterial pathogen are equally virulent. Subtyping schemes are needed to differentiate pathogenic from non-pathogenic strains, track virulent clones and identify subpopulations that are specific to certain hosts or environments (Riley, 2004). Early subtyping methods generally lacked enough discriminatory power to separate epidemiologically unrelated isolates, and therefore, discriminatory power was the main criterion of various subtyping studies during the past three decades. Hunter and Gaston (1988) proposed a numerical index (Simpson's index – see below) for discriminatory power which is based on the probability that two unre-lated strains sampled from the test population will be placed into different typing groups. This index originated in the ecological study of intraspecies populations and has subsequently been used in various microbiological studies to compare the relative discriminatory power of different strain typing techniques.

$$\text{Simpson's index } D = 1 - \frac{1}{N(N-1)} \sum_{j=1}^{S} n_j(n_j - 1)$$

where N is the total number of test strains; S is the total number of subtypes; and n_j is the number of strains belonging to the jth subtype.

Epidemiologic concordance (E). Epidemiologic concordance, sometimes referred to as epidemiologic relevance, describes the ability of a subtyping system to correctly classify into the same clone all epidemiologically related isolates from a well-described outbreak (Struelens, 1996). Therefore, an epidemiologically concor-dant subtyping method should be able to (i) cluster strains that are epidemiologically related with a particular epidemic/outbreak and (ii) separate these strains from those that are not related to the same epidemic/outbreak (Chen et al., 2005). Many strain typing studies only evaluated discriminatory power without validating the epidemi-ologic concordance of their methods. However, high epidemiologic concordance is very important for the detection of a cluster of strains with closely related subtypes for early recognition of outbreaks. A subtyping method with too high discriminatory power may fail to detect such clusters (Sauders et al., 2003). Therefore, too much or too little discriminatory power is not desirable. It should always be remembered that a valid typing technique is ultimately judged by its ability to generate results that are in agreement with epidemiologic observations made in the real world (Foxman et al., 2005).

Portability. With the development of international and national food manufac-turing and distribution systems, many foodborne outbreaks appear as multistate outbreaks. Some epidemics are even multinational. Therefore, the combined efforts of health agencies from different geographic areas are needed to control and investigate these outbreaks/epidemics. This requires that subtyping data be highly portable and easily exchanged electronically throughout the world.

Practical concerns (ease of use, cost, and high throughput). Practical concerns are also important issues that need to be addressed for all subtyping methods. A subtyping strategy that utilizes complicated and expensive technologies may have excellent discriminatory power and epidemiologic concordance; however, if it is too expensive and/or not easy to perform, then this subtyping scheme will not be accessible, especially by developing countries, small food companies, local health agencies, and community microbiology laboratories (van Belkum, 2003). In this case, many local outbreaks may fail to be identified rapidly, and this could be a concern since subsequent multistate outbreaks and pandemics may originate from these local outbreaks. Therefore, simple and inexpensive molecular subtyping methods have important roles in epidemiologic investigations. High throughput is another important criterion for molecular subtyping methods. High-throughput methods can analyze a large number of samples at the same time, and the cost per sample is usually much lower than with low-throughput methods. High-throughput methods can be used by state and federal health agencies to analyze large sample sets. Other types of high-throughput methods, such as whole genome microarray analysis, can measure variations in large portions of the whole genome, but the number of isolates that can be tested is currently limited due to the high costs associated with microarrays (van Belkum, 2003).

To evaluate the performance criteria of subtyping schemes, the first important step is to select the proper strains for validation. The test strains need to include not only isolates from a few well-characterized outbreaks but also isolates that are not related to these outbreaks. These latter strains should be geographically and temporally distinct from those that are associated with the outbreaks. It is also preferable that these strains come from various sources and lineages from the whole population and not be closely related. A common mistake in developing subtyping methods is to attempt to achieve maximum discriminatory power using all isolates in a collection even though some isolates may be closely related or even the same strain. This is especially important when developing subtyping methods for epidemiologic investigations, because it is critical that these methods group those isolates that are epidemiologically related. However, it is often not clear whether or not isolates are closely related (Riley, 2004). Therefore, isolates are often presumed to be related or unrelated based on their origin and source information. When selecting well-characterized outbreaks for evaluation of new molecular subtyping methods, it is recommended that several outbreaks caused by a single epidemic clone be selected. Evidence has shown that some epidemic clones of certain microorganisms have caused multiple outbreaks that were separated by several years (Kathariou, 2002). Successful identification of these epidemic clones will greatly contribute to studies on the distribution, risk factors, and molecular determinants of infectious diseases. A sufficient number of diverse strains representing the entire population are needed in order to be confident in the conclusions reached in these studies. There is no universal rule regarding the total number of strains used in a given study; however, many recent subtyping studies analyzed more than 100 strains (Chen et al., 2007; Ducey et al., 2007; Ward et al., 2004).

12.1.4 Selection of Markers for Molecular Epidemiology

The performance of different strain typing techniques is ultimately decided by the biomarkers that are selected. Therefore, the criteria for selecting the subtyping methods also apply to the selection of markers. Strain typing methods have been developed and improved over the last three decades, with a general shift from phenotype-based to genotype-based markers.

Early subtyping strategies targeted phenotypic markers. Common phenotypes used for subtyping foodborne pathogens included, but were not limited to, surface antigen structures, virulence potential, plasmid profiles, phage profiles, biochemical profiles, and fatty acid profiles. These methods are generally useful only for species identification because their discriminatory power within the same species is limited (Riley, 2004). Subtyping methods based on biochemical characteristics can be costly and time consuming depending on the biochemical tests selected (i.e., sugar fermentation patterns) and their typeability is also limited. Also, the metabolic activities of microorganisms are greatly affected by growth conditions. Many confounding variables associated with biochemical characteristics may provide false discrimination. For example, antimicrobial susceptibility typing has not been widely applied to *L. monocytogenes* because it does not always provide consistent results (Romanova et al., 2002). Overall, phenotypic methods are often not reproducible, because phenotypes are due to gene expressions which are influenced by growth conditions and can suffer from phenotype switching of bacteria (Tardif et al., 1989). Nonetheless, phenotypic subtyping methods have provided some degree of discrimination of various foodborne pathogens at the subspecies level. For example, serotyping is still an important method for subtyping various foodborne pathogens, such as *Salmonella*, *E. coli*, and *L. monocytogenes*. When different phenotypic methods were combined, epidemiologists were able to obtain a relatively reliable and accurate estimation of strain relatedness and this enhanced our understanding of the epidemiology of foodborne outbreaks in the 1980s and 1990s. Early phenotypic typing methods were also used to construct basic population structures of various foodborne pathogens and establish the genetic relationship of many important subgroups of each pathogen. Even though many phenotypic typing methods are no longer in use, some phenotypic characteristics are still used widely and considered as references for the evaluation of newly developed genetic-based subtyping methods. Therefore, it is still important to understand some basic phenotypic typing methods as many underpin current advances in more refined genotypic approaches.

Replacing phenotypic methods with 16S rDNA sequencing and DNA hybridization was a milestone for defining bacterial species. With the development of molecular biology techniques, scientists were able to target genetic markers of microorganisms which are not affected by laboratory handling and culture conditions. Also, these genotypic methods inherently contain more information about a given strain. For example, a gel-based method can yield dozens of amplified or digested fragments, while a sequence-based method can identify tens of hundreds of polymorphic nucleotide sites within the same genome. Genotypic methods can further enhance discriminatory power by detecting DNA polymorphisms that do not change the phenotype. All DNA polymorphisms between different isolates are determined by differences in their genomic sequences. However, when genotypic methods were first developed, direct measurement of DNA sequences was not practical due to the high costs of equipment and reagents. Therefore, scientists utilized technologies such as arbitrary primers, restriction enzymes, hybridization, and PCR to convert DNA sequence polymorphisms into electrophoretic gel patterns or hybridization patterns. With the development of DNA sequencing technologies, subtyping methods emerged which directly targeted partial genomic DNA sequences. Now, subtyping methods have evolved to target a substantial portion or all of the whole genome, and their performance has been enhanced greatly by various advanced genomic technologies.

12.2 Epidemiologic Typing Methods

12.2.1 Phenotypic Methods

Serotyping. Serotyping is one of the most commonly used methods for subspecies differentiation of gram-negative bacterial pathogens such as *S. enterica* and *E. coli.* Serotyping is also an important tool for epidemiologic studies of *L. monocytogenes.* Serotyping targets antigenic variations on cell surfaces, including somatic (O), capsular (K), and flagellar (H) antigens. Different tertiary structures of these antigens react with specific polyclonal and monoclonal antibodies from the bloodstream of an animal host. Different strains can then be assigned serotypes based on their reaction patterns with a panel of antibodies. Serotyping methods for *E. coli,* *Salmonella* spp., and *L. monocytogenes* are well established, and serotype information is often used to validate newly developed subtyping methods. Vogel et al. (2000) evaluated the epidemiologic relevance of serotyping of *E. coli* using a set of epidemiologically unrelated and related strains and found that serotyping provided preliminary assessment of strain relatedness. Molecular subtyping methods which provide data concordant with serotyping are generally believed to be epidemiologically concordant and phylogenetically meaningful (Chen et al., 2007; Nadon et al., 2001; Nightingale et al., 2005). A limitation of serotyping is that it is usually expensive and time consuming due to the need to maintain and handle polyclonal or monoclonal antisera, and therefore it is mostly accessible to large reference laboratories which perform routine serotyping. For example, serotyping of *Salmonella* requires more than 250 different typing sera and 350 different antigens and it typically takes 3 days to finish the serotyping. In addition, the quality of commercial antisera can greatly affect the accuracy of serotyping. Another problem associated with conventional serotyping is that antigens from different strains may cross-react with different antibodies and some strains may not express typing antigens on their surface (Riley, 2004). More than 700 serotypes of *E. coli* have been identified based on combinations of O, H, and K antigens. Certain serotypes of *E. coli*, such as O157:H7, O26, O111, and O103, cause most human diseases, while others are non-pathogenic. Serotype information is always heavily relied upon in epidemiologic investigations of *Salmonella* outbreaks. The serotyping schemes for *L. monocytogenes*, *Campylobacter*, *Shigella*, and *Vibrio* are also well developed. Even though serotyping is always used to identify the relationship between different strains, it possesses relatively low discriminatory power. For example, many *E. coli* outbreaks are caused by the same serotype, O157:H7. Therefore, in order to track the contamination sources of different outbreaks and study the global epidemiology of this pathogen, new subtyping schemes that can differentiate strains within a serotype are necessary.

With the development of molecular biology techniques, scientists have started to explore molecular-based serotyping strategies to overcome the limitations of traditional serotyping. This requires that the molecular determinants of serotypes be well understood. For example, McQuiston et al. (2004) sequenced 280 alleles of *fliC*, *fljB*, and *flpA* that are known to encode 67 flagellar antigens in *Salmonella* and found multiple single-nucleotide polymorphisms between different H antigen types. Yoshida et al. (2007) later designed a DNA microarray assay for rapid determination of *Salmonella* serotypes. Fitzgerald et al. (2007) developed and evaluated a *Salmonella* O-group-specific Bio-Plex assay to detect the six most common serogroups in the United States (B, C_1, C_2, D, E, and O13). Their approach could be completed in 45 min post-PCR and it correctly and specifically identified 362 of 384 (94.3%) isolates tested in comparison to traditional serotyping. Leader et al. (2009) developed an assay for serotyping using multiplex PCR and capillary electrophoresis that could identify the top 50 serovars of *S. enterica*.

Phage typing. Phage typing has been an important epidemiologic typing tool for many foodborne pathogens over the past few decades. This approach separates bacterial pathogens into subspecies or strains based on their susceptibility to lysis by a panel of different bacteriophages (Loessner and Busse, 1990). In 1983, Ralovich et al. (1986) demonstrated the usefulness of phage typing for epidemiologic investigations of *L. monocytogenes* and showed that phage typing and serotyping were concordant epidemiologic approaches for that pathogen. Although phage typing was still found to have limited discriminatory power, it improved subtyping by providing further discrimination between strains within the same serotype, especially serotype 4b, which includes most listeriosis outbreak-associated strains. Frost et al. (1999) used a phage-typing scheme to enhance discriminatory power beyond serotyping of *Campylobacter jejuni*. Phage typing of *Salmonella* spp. is generally used for subgrouping some common *S. enterica* serotypes such as *Salmonella* Typhimurium and *Salmonella* Enteritidis. Different schemes have been developed for these serovars in different countries. Some examples include the Felix/Callow (England) typing system, Lilleengen typing system (Sweden), and Anderson typing system (England) (Bansal et al., 1996). The Anderson typing system can distinguish more than 300 definitive phage types (DTs) of *Salmonella* and has been used worldwide (Bansal et al., 1996). Similarly, Khakhria et al. (1990) developed a phage-typing method for subtyping *E. coli* O157: H7; however, only 13 phage types were identified among 152 outbreak strains of *E. coli* O157: H7. Frost et al. (1999) reported the utility of phage typing for *C. jejuni* and *C. coli*. That study concluded that it was difficult to compare phage-typing results between different laboratories and substantial standardization was required.

Recently, prophages have been identified as possible markers for subtyping foodborne pathogens at the DNA levels. Cooke et al. (2007) found that prophage sequences of *Salmonella* Typhimurium are hot spots of genomic variation in *S. enterica* and were thus used to successfully discriminate between different strains. Chen and Knabel (2008) identified SNPs that could be used to differentiate closely related *L. monocytogenes* outbreak clones within epidemic clones II, III, and IV. Orsi et al. (2008) later reported that SNPs in prophage regions were the only ones that could differentiate strains from a sporadic case and an epidemic clone III outbreak that occurred 12 years apart in the same food processing environment. The hypervariable nature of prophages make them suitable for investigating short-term epidemiology. This makes prophage regions suitable for investigating short-term epidemiology.

Plasmid profiling. Different strains of the same bacterial species may contain plasmids of different sizes and numbers. Therefore, plasmid profiles can be used to differentiate different strains and aid in epidemiologic typing in some studies. However, it has been of limited use because not all bacterial strains contain plasmids which reduce typeability. Sørum et al. (1990) used plasmid profiling to study the epidemiology of cold-water vibrios and found that the discriminatory power of plasmid profiling is sometimes insufficient in certain species. Radu et al. (2001) observed little variation among different isolates of *E. coli* O157:H7.

Multilocus enzyme electrophoresis (MLEE). Subtyping methods underwent a major breakthrough with the invention of multilocus enzyme electrophoresis (MLEE) typing. MLEE differentiates different strains by the electrophoretic mobility of major metabolic enzymes (Bibb et al., 1990). Briefly, water-soluble enzymes are obtained from cell culture, separated by electrophoresis and stained. The amino acid sequences determine the electrostatic charges of the proteins, and therefore, variations in amino acid sequences are reflected by differences in the molecular weight and net charge of the enzymes. The different mobilities of different enzymes produce a protein banding pattern (subtype) which is unique to each strain. MLEE was applied to a variety of bacterial pathogens, such as *E. coli*, *Salmonella* spp., *Haemophilus influenzae*, *Neisseria meningitidis*, *Streptococcus* spp., and yielded very good discriminatory power, typeability, and reproducibility compared to earlier phenotypic typing methods.

MLEE has been used to study the population genetics of *L. monocytogenes*, *E. coli*, *S. enterica*, *C. jejuni*, and *Vibrio cholerae* and provided valuable information for population studies of these pathogens. The population structure of *L. monocytogenes* determined by MLEE has been widely accepted and is consistent with that determined by many recent molecular subtyping methods (Caugant et al., 1996). Sails et al. (2003) demonstrated that clonal complexes of *C. jejuni* determined by MLEE also correlated well with the clonal complexes recently identified by multilocus sequence typing.

Other phenotypic methods. Many other phenotypic methods have been applied to subtype-specific foodborne pathogens, such as biotyping (Notermans et al., 1989), antibiotic-resistance typing (Harvey and Gilmour, 2001), and fatty acid profiling (Harvey and Gilmour, 2001; Wilhelms and Sandow, 1989). However, they have not been widely applied.

12.2.2 PCR-Based Subtyping

Randomly amplified polymorphic DNA (RAPD). RAPD is a PCR technique used widely for subtyping various bacterial pathogens. Unlike conventional PCR, the target PCR products of RAPD are unknown and so arbitrary PCR primers are used. The primers are usually 10-mer long and are chosen arbitrarily by the researcher or can be generated randomly by computers. The arbitrary primers can simultaneously anneal to multiple sites across the whole genome and generate multiple PCR amplicons. These products can then be separated by gel electrophoresis and the banding patterns of different strains compared. RAPD-PCR can differentiate strains with only a few nucleotide differences. Because of this marked sensitivity, RAPD has been used successfully for differentiating both gram-positive and gram-negative bacteria, especially closely related species or epidemiologically related strains. Nilsson et al. (1994) developed and optimized an RAPD subtyping method for *Bacillus cereus* that showed excellent reproducibility. Mazurier and Wernars (1992) evaluated the epidemiologic relevance of RAPD using well-characterized outbreak isolates of *L. monocytogenes* and found that RAPD correctly classified 92 out of 102 isolates into corresponding epidemic groups. Vogel et al. (2000) demonstrated that RAPD provided higher discriminatory power than did ribotyping and serotyping for epidemiologic typing of *E. coli*.

Figure 12.1 illustrates RAPD analysis. To enhance priming with short primers, many primers are designed with a GC content between 70 and 80% and low annealing temperatures are often used. Reproducibility is one of the biggest concerns with RAPD since factors such as DNA purity and concentration, the type

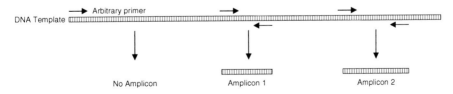

Fig. 12.1 Randomly amplified polymorphic DNA schematic showing amplification using arbitrary primers. Arbitrarily designed short primers (8–12 nucleotides) anneal to many locations on the template genomic DNA. When two primers anneal in the opposite direction to two genomic locations that are reasonably distant from each other, a fragment is amplified. These randomly amplified fragments are then analyzed by conventional gel electrophoresis

of Taq polymerase employed, and PCR reaction conditions used can all affect the reproducibility of RAPD-PCR. Therefore, it is critical to maintain consistency in DNA template quality, reagents, and experimental conditions for successful and reproducible RAPD-PCR.

Amplified fragment length polymorphism (AFLP). AFLP is a highly discriminatory subtyping method which was first described by Zabeau and Vos (1993). With AFLP, genomic DNA is purified and treated with two different restriction enzymes and then two different restriction-specific adaptors are ligated to ends of the restriction fragments. PCR primers, which are complementary to the adaptors, are designed to selectively amplify the ligated restriction fragments. The PCR amplicons are then analyzed by gel electrophoresis and gel patterns (polymorphisms between and within restriction sites) are used to assign subtypes. Hundreds of fragments can be amplified this way. Therefore a random nucleotide is added to the 3′-end of the primer so that only the few fragments that perfectly match the primers are amplified, thus limiting the number of bands that appear on a gel (Fig. 12.2).

AFLP has been used to differentiate strains of many foodborne pathogens. Melles et al. (2007) analyzed 994 *Staphylococcus aureus* strains using multilocus variable number tandem repeat analysis (MLVA) and high-throughput AFLP and found these two methods had similar discriminatory power. Herrera et al. (2002) reported a discrepancy between *Shigella flexneri* strain relatedness identified by plasmid profiling, serotyping, and AFLP analysis, and thus no definitive conclusions could be drawn about the epidemiologic concordance of AFLP. Both Ripabelli et al. (2000) and Guerra et al. (2002) developed AFLP schemes for subtyping *L. monocytogenes* and found that although not sufficiently discriminatory, AFLP results were congruent

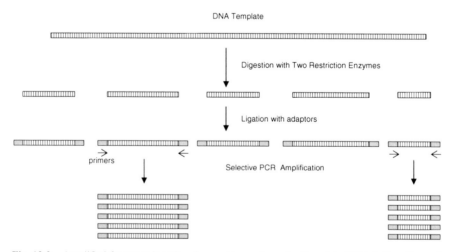

Fig. 12.2 Amplified fragment length polymorphism schematic. Template DNA is first digested with two restriction enzymes, preferably a hexa-cutter and a tetra-cutter, and then the restriction fragments are ligated to the adaptors. Primers are designed to be complementary to the adapters, and their 3′-ends contain a random nucleotide for selective amplification. The amplicons of selective amplification are visualized by gel electrophoresis

with serotyping, phage typing, and other subtyping methods, thereby confirming the three known genetic lineages of *L. monocytogenes*. Keto-Timonen et al. (2003) described the improvement of AFLP by careful selection of restriction enzymes. The discriminatory power of their AFLP scheme was over 0.99 and the results were congruent with PFGE.

A major disadvantage of AFLP is that it requires DNA purification, restriction, and ligation of adaptors, all of which are technically demanding. Internal variabilities due to incomplete digestion and/or ligation are also known to affect the final banding patterns (Riley, 2004).

Repetitive PCR. Repetitive PCR (Rep-PCR) typing is another PCR-based subtyping technique (Fig. 12.3). In bacteria, dispersed chromosomal repetitive elements are randomly distributed throughout their genomes. These sequences differ in size and often do not encode proteins. With Rep-PCR, primers that are complementary to these repeats are designed and used to amplify differently sized DNA fragments lying between the repeats. Several families of repetitive elements have been identified in some eubacterial species. One type of repeat sequence is repetitive extragenic palindromes (REPs), which may be regulatory sequences within untranslated regions of bacterial operons (Craven et al., 1993). Another type of repetitive sequence, enterobacterial repetitive intergenic consensus (ERIC), has been identified in *E. coli* and *Salmonella* spp. (Hulton et al., 1991). Another class of repeats, BOX, was identified in *Streptococcus pneumoniae* (Koeuth et al., 1995). There are three subunits of BOX: BOX-A, BOX-B, and BOX-C. Even though the evolutionary origin and functions of these BOX regions remain unclear, they have proven useful for the differentiation of enteric species and the development of strain-specific subtyping methods. Both REP and ERIC sequences contain conserved regions for primer targeting and variable regions for polymorphism detection. For example, REP primers usually target the left and right sides of a conserved palindromic sequence and are oriented in opposite directions so that the primer extends outwardly in a 3′-direction away from the palindrome. The regions between the repetitive palindromic islands are thus amplified. These regions range in size from 0.2 kb up to 4 kb, which results in a total PCR-based chromosomal fingerprint for a given strain.

Jersek et al. (1999) developed a rep-PCR scheme targeting short repetitive extragenic palindromic (REP) elements and enterobacterial repetitive intergenic

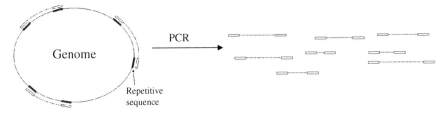

Fig. 12.3 Schematic of repetitive PCR. Primers are designed to bind to the repetitive elements, and regions between these repeats are amplified. These fragments are then analyzed by conventional gel electrophoresis or capillary-based electrophoretic technologies

consensus (ERIC) sequences in *L. monocytogenes* and found that its discriminatory power was high (D = 0.98). Hahm et al. (2003) compared REP and BOX-based rep-PCR, AFLP, PFGE, and ribotyping for subtyping *E. coli*. The authors concluded that PFGE had the highest discriminatory power, while AFLP, REP-PCR, BOX-PCR, and ribotyping can serve as first-step screening methods. The authors also found that although there were differences between the strain relationships identified by each method, the subtype profiles of the *E. coli* O157:H7 isolates were virtually identical with REP-PCR and Box-PCR.

One of the limitations associated with Rep-PCR technologies is their lack of specificity due to the relatively short primers used. Gillings and Holley (1997) demonstrated that ERIC-PCR generated complex gel patterns from eukaryotes, even though ERIC elements have not been found in those species. This indicated that the short primers might bind to non-specific regions.

Multiple-locus variable number tandem repeat analysis (MLVA). MLVA is one of the CDC's candidates for second-generation molecular subtyping (Hyytia-Trees et al., 2006). MLVA targets variable number tandem repeat (VNTR) polymorphisms in the genomes of different bacterial pathogens. PCR primers are designed to amplify tandem repeats based on whole genome sequences. The size and the number of repeats at each locus are then analyzed by computer, and combinations of these repeats define MLVA types (MTs). Tandem repeats are recognized as containing phylogenetic signals because the repeats sometimes are targets of evolutionary events such as mutation and recombination. These evolutionary events may change the size and number of the repeats, and as a result, the number of these repeats at a specific locus is similar among strains that are closely related and variable between unrelated strains. VNTRs also correlate with many genomic changes essential for bacterial survival under stress conditions. Such changes include deletions, insertions, and mutations that affect gene regulation, antigenic shifts, and inactivation of mismatch repair systems (Hyytia-Trees et al., 2006). VNTRs actually play an important role in the adaptation of bacteria, especially those with small genomes (Hyytia-Trees et al., 2006). Therefore, MLVA is expected to provide relatively accurate information about the genetic relatedness between different bacterial strains. Unlike PFGE, the targets of MLVA are specific VNTRs that can be PCR amplified using primers designed based on whole genome sequences. Thus, MLVA is easier to interpret than PFGE. In addition, the essential steps in MLVA are multiplex PCR and capillary electrophoresis, which are easy to perform, standardize, and automate, making MLVA a potential high-throughput subtyping strategy. The final results in MLVA are discreet sizes of each VNTR loci, and thus, the results are easier to compare than gel banding patterns generated by other fragment-based methods. The key to the development of reliable and accurate MLVA schemes is the identification of VNTRs. One of the problems associated with the development of MLVA for *Salmonella* is that different serovars differ slightly in their genomic organizations and thus some well-characterized VNTRs may not be present in all serovars. In this case, different MLVA typing schemes may be needed for each specific serotype of *Salmonella*. To develop a good MLVA scheme, large amounts of complete genome sequence data are essential, and in this aspect, PFGE is superior since no prior knowledge of the whole genome sequence data is required to develop PFGE schemes.

Like other subtyping schemes, stability of the targets is important for development of reproducible and epidemiologically relevant MLVA schemes. Some VNTRs can be very unstable and potentially can separate strains within the same outbreak clone, thus confounding the true epidemiology. Some extremely unstable VNTRs may even change during regular laboratory sub-culturing and thus affect the reproducibility of the MLVA approach. Another potential drawback of MLVA is that since the primers for amplifying the VNTRs are designed based on currently available whole genome sequences, it may not be possible to successfully amplify all VNTRs from all strains in the same species, thereby making typeability an issue. For example, an insertion within a VNTR would confound the analysis of the size of the VNTR. Therefore, selection of VNTRs and design of PCR primers are critical to developing an epidemiologically relevant MLVA scheme. Intensive evaluation and validation are also needed for each MLVA scheme. MLVA has been applied to many foodborne pathogens such as *E. coli*, *Salmonella* spp., and *L. monocytogenes* and has been proven to yield very high discriminatory power. Hyytia-Trees et al. (2006) evaluated the epidemiologic relevance of an MLVA scheme for subtyping *E. coli* O157:H7 and claimed MLVA possessed promising epidemiologic relevance by correctly clustering isolates belonging to eight well-characterized outbreaks. In 2006, an MLVA scheme for subtyping *L. monocytogenes* was described by Murphy et al. (2006). In that study, MLVA was shown to be able to discriminate strains of the same serotype and correlated well with PFGE data from the same set of isolates. Kawamori et al. (2008) compared MLVA and PFGE for subtyping *E. coli* O157: H7 and found a good correlation between MLVA and PFGE. Although MLVA is a fragment-based method, the utilization of meaningful molecular markers, PCR, and capillary electrophoresis generate a more phylogenetically meaningful and less-ambiguous output, providing a major advantage over other fragment-based subtyping methods. Hall et al. (2009) evaluated MLVA and PFGE for subtyping *Salmonella* Enteritidis, and the discriminatory indexes were 0.968 and 0.873, respectively. These studies demonstrated the potential of MLVA for reliable and rapid subtyping of this particular foodborne pathogen.

12.2.3 Restriction Pattern-Based Typing

Restriction fragment length polymorphism (RFLP). RFLP is a subtyping technique which targets the polymorphisms within and between restriction sites. Briefly, genomic DNA is purified from cell cultures and cut into fragments using restriction enzymes, and the fragments separated using gel electrophoresis. Different strains can differ in the distances between restriction sites or in the sequences within the restriction sites, and thus yield different gel patterns. The number of restriction sites in the whole genome varies from 10 to 1,000 depending on the type of restriction enzyme used. Some "frequent cutter" restriction enzymes can produce more than

1,000 fragments with different sizes. Some "rare cutters" can produce around 10 fragments with sizes ranging from 500 to 800,000 bp.

There are three categories of RFLP analysis. First, PCR can be used to amplify a specific region of the whole genome, and this region can then be analyzed by RFLP using frequent-cutting restriction enzymes. Second, the whole genome can be analyzed using frequent-cutting restriction enzymes followed by gel electrophoresis and Southern blotting using probes specific to certain genes. When the probes target rDNA genes, the method is known as ribotyping (Jacquet et al., 1995). Third, the whole genome can be analyzed using rare-cutting restriction enzymes, yielding fragments up to 800 kb (macrorestriction). Traditional gel electrophoresis is not able to analyze these larger fragments, therefore, a special technique, called pulsed-field gel electrophoresis (PFGE) is required to accurately separate these large fragments. This approach is discussed below.

PCR-RFLP. PCR-RFLP was used for molecular subtyping of foodborne pathogens in some early studies (Paillard et al., 2003; Wiedmann et al., 1997). The performance of PCR-RFLP is determined by the genes and enzymes used. Genes that are selected for this analysis include housekeeping genes, virulence genes, and those genes encoding important surface proteins (Vines et al., 1992; Wiedmann et al., 1997). PCR-RFLP analysis using a single gene and enzyme usually provides limited discriminatory power and is often used for species identification. Kärenlampi et al. (2004) used *Alu*I to digest partial *groEL* PCR products and found that the PCR-RFLP assay performed better than 16S rDNA sequencing for the identification of *Campylobacter* spp. By targeting *ial*, Kingombe et al. (2005) developed a PCR-RFLP assay to differentiate enteroinvasive *E. coli* and *Shigella* spp. Multiple restriction enzymes are used to enhance the discriminatory power of PCR-RFLP and generate data containing phylogenetic signals. For example, PCR-RFLP analysis of four virulence genes classified *L. monocytogenes* into two divisions with division I containing serotypes 1/2a, 1/2c, and 3c and division II containing serotypes 1/2b, 3b, and 4b (Vines et al., 1992). This classification expanded previous findings concerning the population structure of *L. monocytogenes* by MLEE (Bibb et al., 1989) and was subsequently confirmed by many other molecular subtyping methods.

Ribotyping. In ribotyping, chromosomal DNA is digested by a frequent-cutting restriction enzyme and small DNA fragments are produced. DNA is then analyzed by Southern blotting with rRNA probes to generate unique banding patterns. An automated ribotyping system, the DuPont Qualicon RiboPrinter®, is used to generate and analyze these banding patterns. Automated ribotyping has been used to subtype a variety of foodborne pathogens including *E. coli*, *Salmonella* spp., *Vibrio* spp., and *L. monocytogenes* (De Cesare et al., 2001; Destro et al., 1996b; Nadon et al., 2001; Wiedmann et al., 1997). A Web-based database (www.pathogentracker.net) has been developed to allow exchange of ribotyping data and other subtyping data from various foodborne pathogens.

Pulsed-field gel electrophoresis (PFGE). PCR-RFLP and ribotyping have enhanced discriminatory power, epidemiologic concordance, reproducibility, and typeability compared to phenotypic methods. These methods have aided the epidemiologic investigations of foodborne outbreaks for many years and have also clarified the population structure of various pathogens. However, sometimes, their discriminatory power can be insufficient for some situations and certain pathogens. This is most likely because they target only restriction enzyme polymorphisms in several genes within the whole genome. In contrast, PFGE targets macrorestriction enzyme polymorphisms spanning the whole genome, thereby ameliorating the overall performance of fragment-based subtyping methods, especially in terms of discriminatory power. As a result, PFGE is currently used by the CDC as a gold standard subtyping method for investigating the molecular epidemiology of foodborne pathogens, including *E. coli* O157, *L. monocytogenes*, *S. enterica*, *Shigella* spp., and *Campylobacter* (Swaminathan et al., 2001). The method underpins the PulseNet molecular epidemiologic typing network both nationally and internationally. Schwartz and Cantor (1984) first described the use of PFGE as a method to separate larger DNA fragments, which could not be properly resolved by conventional gel electrophoresis. Under conventional gel electrophoresis conditions, DNA fragments above 30–50 kb migrate with similar mobility and thus cannot be accurately separated. During PFGE the electric field is periodically reoriented, which forces DNA fragments to change directions and thus large fragments can be separated from each other based on their molecular weight. When performing PFGE analysis, it is vital to obtain pure and intact chromosomal DNA prior to digestion. Using DNA of low purity or partially degraded chromosomal DNA may lead to loss of large-size fragments and the generation of smaller "artifactual" fragments that may confound subsequent gel pattern interpretations. To prevent DNA degradation, chromosomal DNA used for PFGE analysis must be prepared using special and standardized procedures. To accomplish this, a bacterial suspension is mixed with melted agarose and immobilized as the agarose solidifies creating an agarose plug. Bacterial cells within the agarose plug are lysed initially and then the bacterial genomic DNA is digested using a rare-cutting restriction enzyme (macrorestriction). Plugs containing the purified and digested DNA are then placed into wells in an agarose gel and separated by PFGE. As evident, the above-mentioned protocol is labor intensive and time consuming compared to many other subtyping methods.

A number of enzymes have shown promise in the macrorestriction of the genomes of foodborne pathogens. These enzymes have relatively long recognition sites and are rare cutters that only cleave the bacterial genome in a few locations. Some rare-cutting restriction enzymes that have been used successfully for PFGE are *Bln*I, *Spe*I, *Swa*I, *Xba*I, *Xho*I, *Not*I, *Sma*I, *Asc*I, and *Apa*I. Not all enzymes yield macrofragments on species analyzed. Partial digestions, diffuse banding patterns, and small fragments were observed for certain enzyme-template combinations. Therefore, PFGE protocols for each foodborne pathogen have to be individually optimized and validated.

Starting from the early 1990s, PFGE has been used for epidemiologic investigation of foodborne outbreaks (Swaminathan et al., 2001). Saidijam et al. (2003) used a two-enzyme PFGE (*Sma*I and *Sal*I) to differentiate sporadic-case strains of *C. jejuni*. Khambaty et al. (1994) used PFGE to successfully track the first documented foodborne outbreak caused by *Staphylococcus intermedius*. In 1996, CDC established PulseNet, a national subtyping and surveillance system involving public health and food regulatory laboratories for the detection and tracking of foodborne outbreaks in the USA (Swaminathan et al., 2001). In this CDC-coordinated network, participating laboratories utilize standardized PFGE protocols to subtype foodborne pathogens. PulseNet USA currently has standardized PFGE protocols for STEC O157, *S. enterica*, *Shigella* spp., *L. monocytogenes* (Figs. 12.4 and 12.5), thermo-tolerant *Campylobacter* spp., *Clostridium perfringens*, and *V. cholerae*. For each pathogen, the standardized protocol uses two or more macrorestriction enzymes with the primary enzyme used to screen for relatedness and the second enzyme used for confirmation or further discrimination. PFGE patterns are submitted electroni-cally to a shared database at CDC, which allows for rapid exchange and comparison of nationwide subtyping data. After PulseNet was introduced, many more outbreaks were detected and sources of contamination were successfully identified. All 50 US states participated in the PulseNet system in 2001 and the number of PFGE patterns submitted to the database reached 270,000 in 2005 (Gerner-Smidt et al., 2006). The PulseNet system has now expanded to include Canada, Europe, Asia Pacific, and Latin America (http://www.cdc.gov/pulsenet/).

Although PFGE provides excellent discriminatory power for subtyping many foodborne pathogens, sometimes the banding patterns are ambiguous and difficult to interpret. In 1995, Tenover et al. (1995) proposed the following criteria for interpre-tation of PFGE banding patterns: (i) isolates that are indistinguishable from source strain are part of the outbreak; (ii) isolates that differ from the outbreak strain by 2–3 band differences in PFGE banding patterns are probably part of the outbreak; (iii) isolates that differ from the outbreak strain by 4–6 band differences in PFGE banding patterns are possibly part of the outbreak; and (iv) isolates that differ from the outbreak strain by more than seven band differences in PFGE banding patterns are not part of the outbreak. The above "Tenover criteria" was subsequently used in many studies for over 10 years following the publication of the paper. However, some recent studies have suggested that the Tenover criteria do not apply in many situations. For example, Barrett et al. (2006) found that different PFGE patterns were present in a single chain of *E. coli* O157 transmission and suggested that the source of this outbreak would likely not have been identified if epidemiologists relied solely on PFGE data. In this case, PFGE was too discriminatory and sepa-rated isolates in the same outbreak clone. Hanninen et al. (1999) showed that two human isolates of *C. jejuni* changed their PFGE patterns after intestinal passage. Insertion sequences, plasmids, and genome rearrangements are known to contribute to the instability of PFGE banding patterns. In contrast, an identical pulsotype is sometimes detected in two samples that are not epidemiologically linked (Maslanka et al., 1999). For example, Gerner-Smidt et al. (2006) found that PFGE did not pro-vide satisfactory discriminatory power when studying certain groups of *S. enterica* and it was often not possible to separate potential outbreak-related strains from spo-radic strains. Barrett et al. (2006) concluded that PFGE data alone cannot prove an epidemiologic connection, and therefore PFGE results must be combined with con-ventional epidemiologic data to confirm the true epidemiology. PFGE utilizes gel banding patterns to infer genetic relatedness of different bacterial strains. Because of these limitations, scientists have recommended that the Tenover criteria be applied

Cleavage Position	Length of sequence (bp)	Length of sequence (bp) (sorted)	In silico PFGE	Actual PFGE
460161	460164	778227	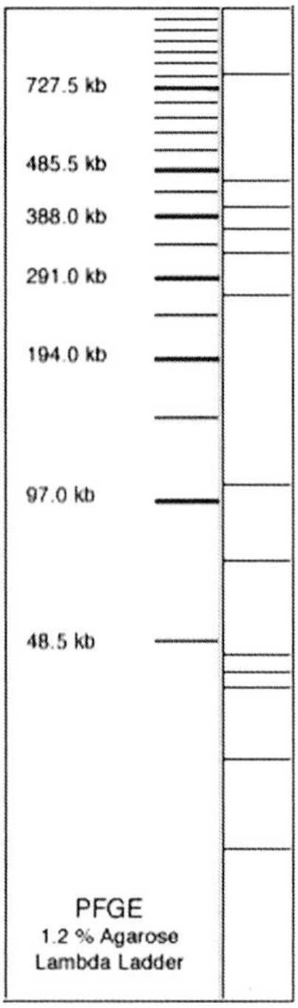	
477522	17361	460164		
522603	45081	405980	727.5 kb	
560798	38195	362888		
632104	71306	326918	485.5 kb	
896922	264818	264818		
1675149	778227	104516	388.0 kb	
1779665	104516	71306		
2185645	405980	45081	291.0 kb	
2186257	612	41361		
2513175	326918	38195	194.0 kb	
2540276	27101	27101		
2581637	41361	17361		
2944525	362888	612	97.0 kb	
			48.5 kb	
			PFGE 1.2 % Agarose Lambda Ladder	

Fig. 12.4 *In silico* restriction map and actual PFGE pattern of *Asc*I based on the whole genome sequence of *L. monocytogenes* EGDe (generated from www.in-silico.com) and Neves et al. (2008)

only to short-term outbreak investigations (1 year) when there is already an implied epidemiologic association (Barrett et al., 2006). Tenover criteria may also be applied to some species, like methicillin-resistant *S. aureus* (MRSA), because MRSA strains generally have a recent origin and are expected to have minimal genetic divergence.

PFGE schemes based on two or more enzymes have been studied to overcome the limitations of two-enzyme PFGE. For example, Davis et al. (2003) found that similarity coefficients between two gel patterns were not good indicators of genetic relatedness, because matching bands do not always represent homologous genetic materials and there are limitations of the power of PFGE to resolve bands of nearly identical size. The authors suggested a combination of analyses using six or more restriction enzymes might provide a reliable estimate of genetic relatedness without reference to epidemiologic data. Singer et al. (2004) explored possible reasons for the discrepancies between PFGE banding patterns and genetic relationships using computer-simulated populations of *E. coli* strains with a known genetic relationship and found that PFGE could not accurately infer the true genetic relationship

Cleavage Position	Length of sequence (bp)	Length of sequence (bp) (sorted)	In silico PFGE	Actual PFGE
688	694	466785		
2475	1787	386644		
6011	3536	262875	727.5 kb	
6709	698	168867		
8496	1787	168180	485.5 kb	
69562	61066	140746		
238429	168867	137706	388.0 kb	
328658	90229	125422		
385325	56667	111151	291.0 kb	
390497	5172	93458		
464910	74413	90229	194.0 kb	
633090	168180	88245		
1019734	386644	74413		
1084723	64989	64989		
1138780	54057	61066	97.0 kb	
1150330	11550	56667		
1159036	8706	54057		
1270187	111151	48598		
1310570	40383	48145		
1352558	41988	45208	48.5 kb	
1446016	93458	42281		
1488297	42281	41988		
1498860	10563	40383		
1507374	8514	37677		
1509161	1787	15191		
1509859	698	11550		
1598104	88245	10563		
1613295	15191	10108	PFGE	
1615082	1787	8706	1.2 % Agarose	
1615780	698	8514	Lambda Ladder	
1660988	45208	5172		
1798694	137706	3536		
2061569	262875	2038		
2202315	140746	2038		
2204353	2038	1787		
2205051	698	1787		
2253649	48598	1787		
2301794	48145	1787		
2427216	125422	698		
2437324	10108	698		
2439362	2038	698		
2440060	698	698		
2477737	37677	698		
2944522	466785			

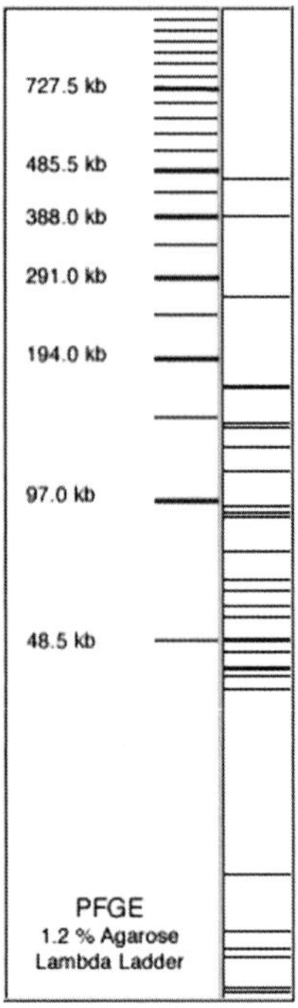

Fig. 12.5 *In silico* restriction map and actual PFGE pattern of *Apa*I based on the whole genome sequence of *L. monocytogenes* EGDe (generated from www.in-silico.com) and Neves et al. (2008)

among the simulated populations. This study demonstrated that the use of multiple enzymes significantly improved the correlation between PFGE banding patterns and phylogeny. Zheng et al. (2007) evaluated the PFGE schemes based on different combinations of up to six enzymes (*Sfi*I/*Pac*I/*Not*I/Xba I/*Bln*I/*Pac*I) to improve PFGE for *Salmonella* Enteritidis and found that the protocol using a three-enzyme set, *Sfi*I/*Pac*I/*Not*I, was highly discriminatory and enhanced PFGE typing for this pathogen. The authors concluded that the selection of correct combinations of enzymes is equally important as the number of enzymes. The same research group also found that the PFGE scheme using the total six enzymes was epidemiologically concordant and correctly clustered *Salmonella* Enteritidis according to their geographic and food sources (Zheng and Brown, personal communication).

Direct genome restriction enzyme analysis (DGREA). A new restriction pattern-based subtyping methodology using an endonuclease with a high cutting frequency is also available. DGREA, in contrast to PFGE, produces small and discrete DNA fragments that can be visualized by nondenaturing polyacrylamide gel electrophoresis. DGREA has been used to analyze *V. parahaemolyticus* strains associated with an outbreak in Chile and was demonstrated to have a discriminatory power similar to PFGE (Fuenzalida et al., 2006). Gonzalez-Escalona et al. (2007) compared the discriminatory power and epidemiologic concordance of DGREA with PFGE and MLST and demonstrated that DGREA was consistent with MLST in identifying the strain relatedness of *V. vulnificus* strains and provided a better picture of strain relatedness than PFGE. DGREA had better typeability than PFGE for subtyping *V. vulnificus*. DGREA uses frequent cutters that generate fragments with sizes ranging between 500 and 3,000 bp, which explains why DGREA was able to detect more genomic variations than PFGE. In addition, DGREA does not require expensive PFGE equipment and time-consuming PFGE gel electrophoresis protocols. The whole procedure can take less than 6 h post-genomic DNA extraction. Therefore, DGREA may represent a good alternative to PFGE for rapid differentiation of strains at low cost.

Analysis of gel banding patterns. Fragment-based subtyping methods that utilize PCR and restriction technologies sometimes suffer from poor reproducibility due to the internal variability of PCR and restriction digestions. The same PCR assay applied to the same cultures in different operations may generate slightly different patterns due to the variability of primers, polymerases, buffers, thermocyclers, template DNA and reaction conditions. Gel electrophoresis can result in uneven lane-to-lane migration of DNA fragments and variation in intensity of bands in separate runs (Riley, 2004). The presence of multiple bands of similar sizes can also confound gel pattern analysis. Therefore, many software packages have been developed to aid gel banding pattern recognition and analysis. However, analysis of gel banding patterns using commercial software is not always reliable and the usefulness and reliability of currently available software is still being debated. Cardinali et al. (2002) compared three analytical systems for DNA banding patterns of *Cryptococcus neoformans* and found that different algorithms provided slightly different topologies using the same set of isolates. However, after evaluating two commercial software packages, Gerner-Smidt et al. (1998) concluded that the computer software were robust and performed satisfactorily. In 2002, Rementeria et al. (2001) conducted a thorough comparison and evaluation of three commercial software packages for analysis of gel banding patterns for RAPD and PFGE. The authors found general agreement between different software and visual observation, but slight discrepancies still existed. The authors finally concluded that computerized analyses based on gel banding patterns "do not provide an indisputably correct analysis in genotype definition" (Rementeria et al., 2001). The computerized analysis of different gel images must go through a normalization process which needs to be supervised by operators, and all programs require operators to make decisions at some steps. Thus, final results could be subjective. Singer et al. (2004) found that subjectivity can influence divergence between gel banding patterns and true genetic relationship of strains. A commonly used algorithm for analysis of banding pattern data is unweighted pair group method with arithmetic mean analysis (UPGMA), which is based on the number of different and common bands. However, UPGMA itself is not a good algorithm for inferring the genetic relationship between different bacterial strains, therefore, it is difficult to accurately infer relatedness of isolates (Nei and Kumar, 2000). In 2004, Duck et al. (2003) showed that parameters of computer software need to be optimized for each species to compensate for the various intra- and inter-gel variations in PFGE libraries and that algorithms used for gel analysis still require improvement. Therefore, van Belkum et al. (2001) suggested that a binary output (numbers or characters) is preferred over gel banding patterns for molecular subtyping strategies.

Another limitation of fragment-based methods such as AFLP, RAPD, and PFGE is that genetic variations that cause banding pattern differences detected by these methods are not well understood, which makes it difficult to infer phylogeny from gel banding patterns. To solve this problem, small fragments from AFLP and RAPD gels can be extracted and sequenced by DNA cloning. Recently, Chen and Knabel (2008) developed a ligation-mediated PCR approach to obtain the nucleotide sequence at ends of large fragments (> 300 kbp) from PFGE gels. This new method provides a valuable tool for understanding PFGE banding patterns and was recently used to identify critical SNPs that are suitable for epidemiologic subtyping of *L. monocytogenes* (Chen and Knabel, 2008).

Despite the disadvantages discussed above, fragment-based subtyping methods, especially PFGE, have proven useful in many circumstances and have greatly facilitated epidemiologic investigations of foodborne outbreaks. Another advantage of PFGE is that it has been used to subtype tens of thousands of isolates worldwide in the past few decades with the implementation of PulseNet. Therefore, the very large database of clinical, food, and environmental isolates will greatly aid any ongoing or future epidemiologic investigations and also help us understand long-term transmission of important infectious agents.

While sequence-based subtyping methods are widely used to determine the genetic relationship of different strains, cluster analysis using gel banding patterns has also been used for this purpose. A variety of formulas have been developed to calculate indices that measure genetic distance or similarities between different gel banding patterns or hybridization patterns. These calculations are based on the presence and absence of positive signals (bands or hybridization) between two strains (*i* and *j*). The following is a list of formulas that have been proposed for calculating similarity index: (Carrico et al., 2005; Riley, 2004).

$$\text{Dice index or coefficient: } S_D = \frac{n_i + n_j}{2n_{ij} + n_i + n_j}$$

$$\text{Jaccard index: } S_J = \frac{n_i}{n_{ij} + n_i + n_j}$$

$$\text{Ochiai index: } S_O = \frac{n_{ij}}{\sqrt{(n_{ij} + n_i)(n_{ij} + n_j)}}$$

$$\text{Sneath and Sokal index: } S_{SS} = \frac{2(n_{ij} + n_0)}{2(n_{ij} + n_0) + n_i + n_j}$$

$$\text{Simple matching index: } S = \frac{n_{ij} + n_0}{n_{ij} + n_i + n_j + n_0}$$

where

n_{ij} is the number of common positive signals present in both patterns [the positive signals (bands or hybridization) are listed according to their positions and the presence of a positive signal in the same position of two strains is called a common positive signal);

n_i is the number of positive signals present only in strain i;

n_j is the number of positive signals present only in strain j;

n_0 is the number of common negative signals in both strains (knowledge of all possible positions is needed to calculate this value).

It is generally believed that the number of bands absent from both strains (n_0) does not provide meaningful results and therefore the Sneath and Sokal index and the simple matching index are used less frequently (Carrico et al., 2005; Riley, 2004). In contrast, the number of common negative signals is used when calculating indices based on hybridization patterns. As discussed above, the similarity indices calculated by computer software need to be visually cross-examined by investigators, and the strain relatedness determined by similarity indices is used only as a reference for inferring the epidemiologic relationships of different bacterial strains (Carrico et al., 2005; Riley, 2004). Again, scientists need to combine conventional epidemiologic data with similarity indices to determine the epidemiologic relationship between different strains. A frequently asked question in cluster analysis (using gel banding patterns) is the threshold value of similarity that should be used to define an outbreak clone or epidemic clone. The answer depends on the discriminatory power of the subtyping method being employed. For example, when ribotyping is used for *L. monocytogenes*, the threshold value should probably be 100% for identifying outbreak clones, while PFGE based on different restriction enzymes may have different threshold values.

12.2.4 Hybridization-Based Subtyping

Another type of molecular subtyping is based on hybridization technologies using subtype-specific DNA probes. Different strains have their own specific genomic regions which may react with a certain probe panel to generate different patterns. Probes can be PCR products amplified from reference strains that are then deposited onto surfaces by spotting or short oligonucleotides synthesized in situ by photolithographic and non-contact photo-activated methods (Hughes et al., 2001). Different surface materials can allow for different numbers of probes to be deposited: nitrocellulose or nylon can hold hundreds of probes and the array they form is called a macroarray or low-density array; glass slides can hold tens of thousands of high-density probes and are called microarrays. The performance of hybridization-based typing methods is determined by the number and selection of appropriate and specific probes. When the number of probes is high, the technology is referred to as DNA array technology.

Chip-based arrays. DNA microarrays are an extension of hybridization-based array technology and are powerful tools for molecular subtyping of foodborne pathogens. Thousands of oligonucleotide probes are designed based on whole genome sequences and spotted onto small solid surfaces, such as glass, plastic, and silicon chips. The genomic DNA of "unknown" strains are extracted, labeled, and hybridized to the oligonucleotide array. The array then detects the presence or the absence of genomic regions that are complementary to the oligonucleotide probes, which allows identification of sequences unique to each "unknown" strain. The length of oligonucleotide probes can vary depending on the markers targeted. For example, when PCR products are used as probes, presence or absence of certain genes can be detected. When lengths of the oligonucleotides are miniscule, small deletions, sequence variations, or even single-nucleotide mutations can be detected. Short oligo arrays can detect many more genomic variations, but at a higher cost than arrays that rely on PCR amplicons.

DNA array technologies have been widely used for subtyping foodborne pathogens. Fitzgerald et al. (2007) developed a liquid suspension array to determine common *Salmonella* serogroups. Doumith et al. (2006) developed an array typing scheme and demonstrated its usefulness in epidemiologic investigation of listeriosis outbreaks. The authors concluded that a big advantage of the array typing scheme over PFGE is that the genetic basis for strain variations can be inferred from the hybridization patterns.

DNA array-based technologies have both advantages and limitations. The biggest advantage of microarrays over other methods is that they can simultaneously detect variations (i.e., chromosomal rearrangements and insertion/deletions) throughout genomes of many strains. They can be used as a high-throughput subtyping tool. Another advantage of microarray technology is that it does not require prior knowledge of genome sequences of test strains. It can reveal a large number of previously unidentified genomic regions that are specific to various species and subspecies and, therefore, may be functionally important. However, microarrays also have several limitations or disadvantages, the biggest one being that it is not cost effective. Additionally, data analysis is also a major challenge of microarray technology. When genomic DNA hybridizes to the array chip, not all regions generate absolute positive and negative signals. Random partial hybridization creates noises which can pose a major problem for interpretation of microarray data. Therefore, finding out how to filter out the noise and target only true signals is critical to developing accurate and epidemiologically relevant microarray analysis. Different materials and labeling effects of different dyes can generate artifacts, and therefore, microarray procedures must be standardized and data normalized. A program "Minimum Information about a Microarray Experiment" (MIAME) was proposed to facilitate reproducible and unambiguous interpretation of microarray data (Brazma et al., 2001). The proposed principles used to ameliorate signal/noise ratio include experimental design, samples used, extract preparation and labeling, hybridization procedures and parameters, measurement data, specifications, and array design (Brazma et al., 2001).

Microarrays have been widely used for molecular classification and genomic mutation studies. Koreen et al. (2004) performed microarray analysis to determine genomic variations among 14 environmental, veterinary, and clinical *S. enterica* serovar Dublin, Agona, and Typhimurium strains isolated in Ireland and Canada between 2000 and 2003. The authors used PCR amplicons as probes and identified ORF regions and prophage regions that were unique to each strain, but the approach allowed limited differentiation among the strains within the same serovar. Fukiya et al. (2004) developed a whole genome microarray approach to analyze the genomic diversity in pathogenic *E. coli* and *Shigella* strains and demonstrated that this microarray technique can be used for phylogenetic analysis of certain foodborne pathogens. Borucki et al. (2004) developed a mixed-genome microarray containing 629 probes for subtyping *L. monocytogenes*. Their mixed-genome array provided high epidemiologic relevance by correctly grouping isolates from well-identified outbreaks and was more discriminatory than were ribotyping and MLST targeting six housekeeping genes. Data generated by this microarray were consistent with PFGE, ribotyping, and serotyping. Results also demonstrated acceptable reproducibility.

Suspension-based microarrays. A new microarray technology was recently developed by Luminex (Luminex, Inc.) based on specific target capture and flow cytometry Dunbar (2006). This method allows simultaneous identification of genomic variations by mixing different sets of beads that are conjugated with specific capture probes derived from target sequences. With this technology, up to 100 biomarkers can be targeted simultaneously in a single reaction. Beads are dyed with two spectral fluorochromes with different intensities and therefore contain their unique spectral emissions that are detected by the laser. Up to 100 colors can be coated to beads, and target DNAs that are attached to each bead can be individually recognized. High multiplex capacity and relatively fast reaction time make this technology very promising for rapid detection and subtyping of foodborne pathogens (Fig. 12.6).

12.2.5 DNA Sequence-Based Subtyping

From the mid 1990s, DNA sequence-based subtyping methods have gained popularity due to increased availability of whole genome sequences, large and unambiguous

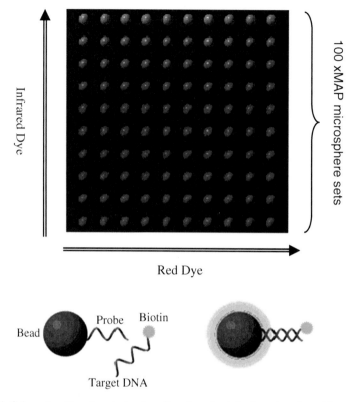

Fig. 12.6 Schematic of Luminex technology (Luminex, Inc.) for detection of specific target DNAs (adapted from www.luminexcorp.com). Combination of *infrared dye* and *red dye* can create 100 different *colors* to be coated onto *tiny beads*. *Colored beads* are then coupled with genotype-specific probes, which subsequently hybridize to target DNAs that are detected by a spectral laser during flow cytometry

information content of sequence data, better cost effectiveness, speed of DNA sequencing, and the ability to analyze and share sequence data via the internet (Chan et al., 2001). In general, DNA sequence data (A, T, G. C) are inherently more specific, discreet and informative than fragment-based data. On the other hand, fragment-based methods are more ambiguous and less informative. However, there are thousands of genes in a bacterial genome, and it is difficult to find a gene that can accurately represent the evolutionary history of a bacterial species. Genes that undergo intensive recombination are unsuitable for inferring genetic related-ness among different strains. Sequence typing strategies targeting multiple genes that are not subjected to recombination were subsequently developed to prevent this problem (Lemee et al., 2005).

Multilocus sequence typing (MLST). MLST, which evolved from MLEE, was developed by Maiden et al. (1998). Unlike MLEE, which targets the electrophoretic mobility of multiple enzymes, MLST, when originally developed, targeted the sequences of housekeeping genes, which code for proteins essential for cell survival

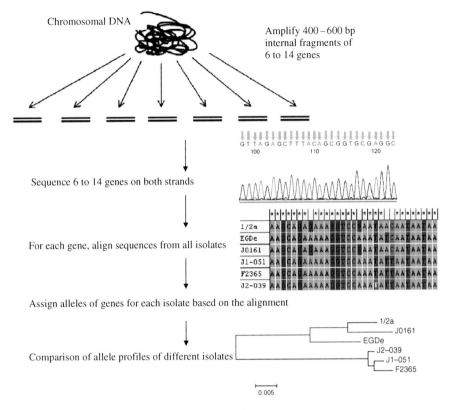

Fig. 12.7 Flow diagram of MLST (adapted from Feil and Spratt, 2001)

and reproduction. In brief, 6–14 genes are selected from the whole genome of a bacterial species, internal fragments (400–600 bp) of each gene are sequenced, and the concatenated sequences from different strains are aligned to differentiate the strains. To avoid problems with horizontal gene transfer, generally the 6–14 genes are selected so that they are evenly distributed throughout the genome. Fragments with a size of 400–600 bp are selected because they can be sequenced easily by rapid automated Sanger sequencing (Fig. 12.7). One of the major advantages of sequence-based methods over fragment-based methods is that DNA sequences can be used for direct and reliable phylogenetic analysis and are therefore expected to provide more accurate information on the relatedness of different strains. Sequence data are also more portable and unambiguous than fragment-based data and thus are easier to compare and interpret both within and across laboratories. MLST has been applied to various pathogens including *S. aureus*, *Streptococcus pyogenes*, *E. coli* O157:H7, *S. enterica*, and *L. monocytogenes*, and has greatly facilitated our understanding of the epidemiology and population genetics of these pathogens. For example, Salcedo et al. (2003) developed an MLST scheme for *L. monocytogenes* using nine house-keeping genes and identified 29 sequence types from 62 strains. They found that the

MLST results were congruent with PFGE data. Housekeeping genes evolve relatively slowly and thus are highly conserved. Therefore, while they are good markers for studying population structure of bacterial pathogens, they do lack the ability to discriminate among closely related strains. To enhance the discriminatory power of MLST, Zhang et al. (2004) subsequently developed an MLST-based scheme termed multi-virulence-locus sequence typing (MVLST), which was based solely on virulence gene sequences. MVLST was able to separate 28 diverse *L. monocytogenes* strains into 28 sequence types and was more discriminatory than *Apa*I-PFGE. Virulence genes are believed to be generally more susceptible to recombination than are housekeeping genes and that might be why virulence genes were not at first incorporated into MLST schemes to study the molecular epidemiology of foodborne pathogens (Cooper and Feil, 2004). Chen et al. (2007) evaluated the epidemiologic relevance of MVLST using well-identified outbreak and non-outbreak isolates of *L. monocytogenes* and demonstrated that MVLST had excellent discriminatory power and epidemiologic concordance and correctly identified known genetic lineages and epidemic clones of *L. monocytogenes*. These findings were later confirmed by Lomonaco et al. (2008) using additional virulence genes and regions in *L. monocytogenes*. Similarly, Tankouo-Sandjong et al. (2007) developed an MLST scheme that targeted only virulence genes for subtyping *S. enterica* subsp. *enterica* serovars, and their scheme yielded a clear differentiation of all strains analyzed. Another study demonstrated that sequence variations of three virulence-associated loci of methicillin-resistant *S. aureus* (MRSA) were able to identify epidemic clones and thus could be used to study the epidemiology of this pathogen (Gomes et al., 2005).

However, there are also some confounding variables associated with MLST schemes. Some genes often undergo constant recombination, and thus the clusters identified using these gene sequences are not good indicators of true strain relatedness. For example, Brown et al. (2003) revealed that only three of six housekeeping genes used to construct *S. enterica* subspecies evolution were actually phylogenetically congruent, while the remaining had been substantially affected by lateral gene transfer between distinct *Salmonella* groups.

To solve this problem, computational biologists designed various algorithms to estimate the rates of recombination based on sequence data. Another method to analyze MLST data is to use only the gene allelic profiles instead of the actual gene sequences. eBURST is a software package that is designed for analyzing allelic profiles of MLST data (Fig. 12.8). It first identifies mutually exclusive groups of related subtypes in the population and then the founding subtype of each group. It then predicts the descent of the founding subtype and uses a radial diagram to display strain relatedness. This approach may be suitable for analyzing evolutionary relationships that are difficult to represent using a tree topology as a result of frequent recombination. eBURST allows for identification of clonal complexes and provides a snapshot of the whole population and does not attempt to infer true phylogenetic relationship among different strains. More specifically, "it simply produces an hypothesis about the way each clonal complex may have emerged and diversified – and any additional phenotypic, genotypic, or epidemiologic data that are available should be

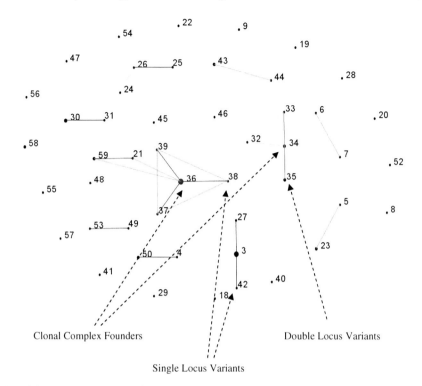

Fig. 12.8 eBURST analysis of sample multilocus sequence typing data (Courtesy of Narjol González-Escalona). eBURST diagrams display the model of evolution from the predicted founding sequence type and shows the radial links from the founder to each of its single locus variants (SLVs) and double locus variants (DLVs)

used to explore the plausibility of the proposed ancestry and patterns of descent" (http://eburst.mlst.net/v3/instructions/1.asp).

To achieve an accurate analysis of strain relationships using MLST sequencing data, various methods have been developed to detect recombination and phylogenetically congruent genetic regions. These genetic regions can thus be excluded from sequence analysis. Alternatively, depending on the extent of recombination identified, proper approaches (discussed below) can be chosen for accurate strain relationship determination. One of the classic methods for the detection of recombination is the "Sawyer's test." Sawyer's test detects recombination intervals by identifying shared patterns of polymorphisms in a sequence alignment. It can also infer possible gene recombination events from outside the alignment. This method is now implemented in a computer program, "GENECONV" (http://www.math.wustl.edu/~sawyer/geneconv/). Another method to detect the phylogenetically congruent genetic regions is the incongruence length difference (ILD) test. The ILD test uses phylogenetic incongruence to trace sequences that may have been acquired promiscuously by exchange of DNA during chromosome evolution. Brown et al. (2002) used the ILD test to detect recombination in several housekeeping genes of

Salmonella. Their ILD test results were supported using additional independent tests such as split decomposition analysis and compatibility of sites. In contrast to classic tree building methods, split decomposition does not force bifurcate trees when inferring phylogeny from a given set of alignments. It detects recombination among different strains and constructs a mesh-like network to visualize phylogenetic information among different strains. Similar to the concept of constructing a mesh-like phylogenetic network, a computer algorithm, RETICULATE, was developed to analyze the reticulate evolutionary relationship among different strains. RETICULATE generates a compatibility matrix of all binary nucleotide sites in a sequence alignment. Two nucleotide sites are considered compatible if they can be explained by the same phylogeny. Incompatible sites may be the result of recombination or redundant mutations at a single site.

There are three major approaches for analyzing strain relatedness: phenetic, cladistic, and statistical approaches. Phenetic approaches do not make assumptions about the overall evolutionary process and isolates are clustered based on overall similarities which are given numerical values using mathematical formulas. Two major phenetic approaches are the unweighted pair group method using arithmetic averages (UPGMA) and the neighbor-joining (NJ) method. Before discussing these two tree building methods, it is necessary to discuss the algorithms used to transfer molecular data (DNA sequence or gel banding pattern) to numerical similarities.

Phenetic approaches first convert DNA sequence data or gel banding patterns to numerical values of genetic distance. Various mathematical formulas for calculating evolutionary distance have been explored. Although there are still controversies regarding which methods are most accurate and reliable, some methods are generally believed to be useful for calculating genetic distances from DNA sequencing data. The simplest method of calculating divergence between two microorganisms is using the number of different nucleotides between homologous sequences. A ratio of the number of different nucleotides between two sequences to the total number of nucleotides is calculated and called p distance. The Jukes and Cantor's model (1969) modifies the equation of p distance and is expected to be more accurate. However, the above-mentioned methods assume that substitutions are random and the nucleotide frequency is 0.25. Each substitution is independent of the preceding nucleotide, and substitution rates are equivalent which ignore the difference between transition and transversion mutations. Some of these assumptions are easily violated, especially when evolutionary distances between highly divergent taxa are calculated. For example, transition rates are often higher than transversion rates. Kimura's two-parameter model (Kimura, 1980) was developed to take the difference between transition and transversion into account. However, even transitional mutations (or transversional mutations) may occur at different rates. In summary, substitution rates of different nucleotide pairs (i.e., A to T, G to C, A to G, T to C) can vary and can also be influenced by nucleotide frequencies of A, T, G, and C in specific sequences. More recent models such as Tajima and Nei's model (Nei and Tajima, 1985), Tamura's model (Tamura, 1992), and Tamura and Nei's model (Tamura and Nei, 1993) combine parameters for differences in both substitution rates and nucleotide frequencies to estimate substitution rates. These models generally perform well; however, they all ignore the fact that rates of substitution also vary with genomic locations of nucleotide sites. For this reason, gamma distances for all models were developed for estimating rates of substitution (Nei and Kumar, 2000). In recent years, more complicated models, such as logDet (Zharkikh, 1994) and paralinear models (Lake, 1994), have also been proposed. Interestingly, these complicated models do not necessarily perform better than much simpler models, even though they significantly increase the computational load (Nei and Kumar, 2000).

UPGMA is one of the first and simplest algorithms for cluster analysis of different strains based on numerical values of genetic distance (Nei and Kumar, 2000). UPGMA is a sequential clustering algorithm and relationships between taxonomic units are identified in order of similarity, and the resultant phylogenetic tree is built stepwise. UPGMA calculates the average genetic distance between different clusters, and different taxa in the same cluster are considered to have identical genetic distances to a taxon outside of the cluster (which is called ultrametric). This may not hold true in real-world situations. As stated above, UPGMA is commonly used to analyze gel banding patterns obtained by ribotyping, RAPD, and PFGE. However, it does not always construct reliable trees, especially when different taxa are not closely related and/or when there is recombination between two sequences. Nevertheless, if a validated phenogram containing a collection of well-characterized strains is available, UPGMA is useful for assigning any new taxa in the phenogram.

Neighbor joining (NJ) is one of the most commonly used methods for tree construction (Nei and Kumar, 2000). It is a bottom-up clustering method used for stepwise construction of distance-based phylogenetic trees. Neighbor joining has been shown to construct trees that are statistically consistent under many models of evolution, and therefore, it is believed to construct more accurate phenetic trees with high probability, given the presence of sufficient data. The neighbor-joining algorithm is very efficient, and very large data sets can be analyzed in a relatively short period of time. These data sets are computationally prohibitive for other phylogenetic analysis algorithms such as maximum parsimony and maximum likelihood. Other phenetic algorithms include but are not limited to minimum evolution (ME) and least square (LE). Readers are referred to Kumar and Nei (2000) for detailed descriptions of these algorithms.

The second approach for constructing phylogenetic trees is the cladistic approach by which the relationships of different taxa are determined by a parsimony-based evolutionary model instead of distance matrix-based values. A typical approach, maximum parsimony (MP), is widely used as a tree-building method based on the tenets of Occam's razor which states that the explanation retaining the least number of evolutionary assumptions is likely the most accurate solution (Persing et al., 2003). This method constructs a tree which requires the smallest number of character state transitions (i.e., nucleotide substitutions or changes in antibiotic resistance) to explain the whole evolutionary process. While MP is useful in many circumstances (i.e. morphological data or binary molecular data), it tends to yield numerous equally parsimonious but different topologies if the number of sequences is large and the number of nucleotides is small. Also, MP may not always construct statistically consistent trees for short-branch or long-branch attractions. Additionally, MP is not as computationally efficient as NJ, although the grid-computing systems available from Apple, Inc. are now ameliorating this solution.

The third approach for constructing phylogenetic trees is the statistical approach, with maximum likelihood (ML) method as an example. ML approach makes inferences about parameters of the underlying probability distribution from a given data set (Persing et al., 2003). This approach considers rates of nucleotide substitutions and calculates the probability that the proposed topology and the evolutionary process generate the observed data set. The topology and the evolutionary process with a higher probability of producing the observed data set are preferred and the method searches for the tree with the highest probability or likelihood. The biggest advantage of ML over other tree-building methods is that it is statistically meaningful and it is very robust to many violations of the assumptions in the evolutionary model. However, it is computationally intensive and impractical for large data sets. ML is generally used to infer phylogenetic relationships using nucleotide sequences and is not suitable for analyzing gel banding patterns or hybridization patterns. However, a variant of the ML method, Bayesian analysis, has recently been developed for relatedness analysis based on gel patterns or hybridization patterns (Yang and Rannala, 1997).

Because it is impossible to know true phylogenetic topologies, it is impossible to determine which tree-building method produces the most accurate tree. In many instances, different tree-building algorithms will generate different dendrograms even when based on same set of molecular data. If possible, multiple tree construction methods should be evaluated on the same data set, and if they all produce similar topologies, then this "consensus" tree is expected to be reliable and a reasonable hypothesis of strain evolution.

When recombination is frequent in the population, the genetic relationship between closely related strains can still be reliably estimated because they are descendents of a recent ancestor and are expected to have less recombination. However, the genetic relationship of those distantly related strains can be obscured by recombination over the long term. Fortunately, epidemiologists do not always need information on relatedness between major lineages of the entire population. The ability to cluster identical or closely related strains in order to understand the spread of a few epidemic clones is very important in any tree analysis.

12.2.6 *Single-Nucleotide Polymorphism Typing*

Sequence-based typing is a newly developed subtyping method which targets polymorphisms in DNA sequences. Single-nucleotide polymorphism (SNP) is an important example of this subtyping methodology. While whole genome sequencing is still relatively expensive and impractical for molecular epidemiology, direct detection of SNPs has evolved as a next-generation subtyping strategy. Unlike direct sequencing of a fragment in the genome, SNP typing directly targets single-nucleotide polymorphisms (SNPs) in the genome and thus has the potential to be more rapid and cost effective than do MLST-based schemes. In contrast to traditional subtyping methods, such as PFGE and MLST, SNP typing lends itself to high-throughput formats. There are two major types of SNP typing techniques: allele-specific hybridization and primer extension. The following section discusses various technologies to detect single-nucleotide polymorphisms (SNPs) within genomes. Readers are also referred to Sobrino et al. (2005) and Dearlove (2002) for a more detailed review of current SNP typing techniques.

Hybridization-based SNP typing. Allele-specific hybridization is based on hybridization to genomic regions that differ at the SNP site using probes specific to each allele. Taqman® SNP assay (Applied Biosystems, CA) is a representative of this type of SNP techniques (Livak et al., 1995). The Taqman assay targets binary SNP sites and is based on the 5′ nuclease activity of Taq polymerase. Two probes that differ at the SNP site are designed with one probe matching the test allele of the targeted SNP and the other probe matching the control allele. The two probes are labeled by different fluorescent dyes attached to their 5′-ends and a quencher attached to their 3′-ends. The quencher inhibits the fluorescence by the fluorescence resonance energy transfer (FRET) probe. A primer binds upstream of the targeted SNP and Taq polymerase initiates DNA extension. If the probe matches the test strain, then DNA synthesis extends to the SNP site and Taq polymerase cleaves and releases the 5′ fluorescent end of the FRET probe, which results in fluorescence. In the case of a mismatch SNP, the 5′ fluorescent end of the FRET probe is not cleaved resulting in no fluorescence. The Taqman assay is fast, accurate, and permits high throughput. It can detect up to 100 SNPs within hours. In comparison with other assays, this assay is advantageous because it requires only a one-step enzymatic reaction and all reactions can use the same PCR conditions. Workflow is very simple and can be easily automated. However, the Taqman assay cannot easily be multiplexed and therefore the cost is high when large numbers of SNPs need to be detected. Some other SNP typing methods based on allele-specific hybridization include, but are not limited to, molecular beacons and dynamic allele-specific hybridization (DASH). Molecular beacon technology is similar to Taqman technology because they utilize florescent probes. A molecular beacon probe is designed by attaching two complimentary arms to a short target sequence. One arm is conjugated with a fluorophore and the other arm is conjugated with a quencher. The two complimentary arms are initially bound to each other and therefore no fluorescence can be detected. Once the probe binds to template DNA, the beacon probe undergoes a spontaneous conformational reorganization that causes binding arms to dissociate and thus the fluorophore and the quencher are moved away from each other, emitting fluorescence (Fig. 12.9).

An extension of allele-specific hybridization is microarray-based hybridization which is also called microarray resequencing (Sobrino and Carracedo, 2005). In this strategy, hundreds to thousands of allele-specific oligonucleotides are designed and attached to a solid support and are then hybridized with fluorescent genomic fragments containing SNP sites. This method can detect many SNPs from a large portion of the whole genome. Zhang et al. (2006) developed an SNP typing scheme

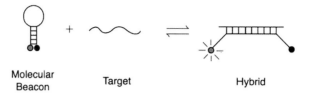

Molecular Beacon   Target   Hybrid

Fig. 12.9 Molecular beacon analysis (adapted from www.molecularbeacons.org). In the Taqman SNP typing assay, a pair of Taqman probes are designed to target each SNP. One probe has perfect match with template DNA and binds to it. The reporter fluorophores (FAM or VIC) are cleaved during the PCR reactions. The other probe has a mismatch with the template DNA and thus does not bind to the template DNA. In the molecular beacon assay, two complementary arms are attached to the probe. When the probe perfectly matches the template, their hairpin is pulled apart and the two complementary arms are separated, and the reporter dye is cleaved during the PCR reactions

using microarray resequencing to discover 906 SNPs in *E. coli* O157. This typing scheme allowed discrimination of 11 outbreak-associated isolates and provided insights into the genomic diversity and evolution of *E. coli* O157. One disadvantage of this method is its technical complexity and lack of cost effectiveness. The GeneChip® (Affymetrix) system improves the performance of this technology by utilizing a tiling strategy (Wang et al., 1998) and has been applied to genotyping various organisms.

There are two variants of hybridization-based SNP typing methods. One is allele-specific oligonucleotide ligation and another is invasive cleavage. In allele-specific oligonucleotide ligation, a probe common to both alleles is designed and binds to the immediate downstream regions of the SNP and two allele-specific probes with the 3'-end nucleotide complementary to each allele of the SNP compete to anneal to the DNA regions immediately upstream of the common probe. The allele-specific probe matching the SNP will ligate to the common probe in the presence of ligase. The ligation products are then amplified by PCR. SNPlex™ genotyping system (Applied Biosystems, CA) is a representative of this technology. Allele-specific ligation products are detected by PCR using ZipChute probes which are fluorescently labeled and hybridize to the complementary sequences that are part of allele-specific amplicons. These probes are eluted and detected by electrophoresis using Applied Biosystem's 3730 or 3730*xl* DNA Analyzers. The major advantage of the SNPlex system is that it can detect up to 48 SNPs per reaction and is therefore suitable for large-scale SNP analysis. However, design and optimization of the multiplex system is time consuming. Vega et al. (2005) reported a comparison between Taqman and SNPlex typing and they found that the concordance between these two methods was 99%. The overall cost of SNPlex typing is higher than pyrosequencing and Taqman typing, but the cost per SNP is lower. Luminex suspension arrays can also be used to develop allele-specific extension schemes for SNP typing. First, PCR is performed to amplify the target regions. Then allelic-specific primer (with an anti-TAG tail) binds to the immediate upstream region of the SNP site. The extension product is biotin labeled. Finally, the extension product with the anti-TAG tail is hybridized to xTAG beads and analyzed by the Luminex system. Other allele-specific ligation-based SNP strategies include but are not limited to Illumina genotyping (Illumina, Inc., www.illumina.com), genotyping using Padlock probes (Nilsson et al., 1994), and sequence-coded separation (Grossman et al., 1994). Another variant of hybridization-based SNP typing methods is based on invasive cleavage. The Invader® assay (Third Wave™ Technology, www.twt.com) is a representative of this SNP typing strategy currently used in many clinical diagnostics. An invading probe complementary to upstream sequence of SNP site binds to the PCR-amplified DNA template. Another allele-specific probe is designed so that the 3′ region is complementary to the downstream sequence of the SNP site (including the SNP site) and the 5′ region (arm) is not complementary to DNA template. When the allele-specific probe binds to DNA template, it will overlap with invading probe and the structure can be recognized and cleaved by the Flap endonuclease, releasing the 5′ arm of the allele-specific probe. The arm sequence then binds to a complementary FRET probe and forms an invasive cleavage structure. A cleavase enzyme recognizes the structure and cuts the FRET probe releasing the fluorescent dye. One advantage of invader assays is that the PCR amplification of template DNA and subsequent SNP detection occurs in the same tube, thus it is labor efficient; however, invader assays cannot be multiplexed and it appears that its assay optimization is time consuming.

Primer extension-based SNP typing. Primer extension is based on the principle of PCR technology. A primer binds to the DNA region upstream of the SNP site and is extended by a single dideoxynucleotide (ddNTP) specific to the SNP site using DNA polymerase. This technology is also called minisequencing. SNaPshot™ multiplex SNP typing (Applied Biosystems, CA) is one of the most common commercial technologies based on minisequencing (Fig. 12.10). In this typing system, several fragments of genomic DNA are targeted and amplified by multiplex PCR. The SNaPshot primers are designed to anneal immediately upstream of the SNP site and are extended by one fluorescently labeled ddNTP. Different ddNTPs are labeled

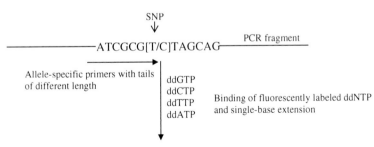

Fig. 12.10 SNaPshot SNP typing assay. Primers are designed to bind immediately upstream of the SNP site. Fluorescently labeled ddNTP then binds to the SNP site during primer extension. Fluorescence is subsequently detected to determine the nucleotide at the SNP site. To allow separation and detection in a capillary sequencer, polyT tails of varying sizes are attached to the 5′-end of genotyping primers

by different dyes and emit light with different wavelengths. The reaction is performed using a DNA sequencer and subtypes are determined by color and size of fluorescent fragments. Several primers can be multiplexed into one SNaPshot reaction for simultaneous detection of several SNPs. Hommais et al. (2005) developed a SNaPshot typing system targeting 13 SNPs for subtyping *E. coli* and showed that the results of SNaPshot typing were consistent with results obtained by MLEE and ribotyping. Dalmasso et al. (2009) developed a SNaPshot assay for the identification of 6 major human pathogenic *Vibrio* species. The SNaPshot multiplex system has a relatively low cost, but it requires prior PCR amplifications and purification of genomic regions containing SNP sites. Other minisequencing-based SNP detection methods include, but are not limited to, matrix-assisted laser desorption/ionization time-of-flight mass spectrometry (MALDI-TOF MS) (Haff and Smirnov, 1997), Sequenom MassARRAY (Sauer et al., 2000), and arrayed primer extension (APEX) (Shumaker et al., 1996).

Pyrosequencing is another popular minisequencing-based SNP detection method. The technology utilizes an enzyme cascade system and generates a luciferase-based light signal whenever a dNTP binds to the DNA template (Fig. 12.11). In brief, sequencing primer is mixed with DNA template, DNA polymerase, ATP sulfurylase, luciferase and apyrase, adenosine-5′-phosphosulfate (APS), and luciferin. Deoxyribonucleotide triphosphates (dNTPs) are then added one base solution at a time (A, T, G, or C) with washing in between. If a specific

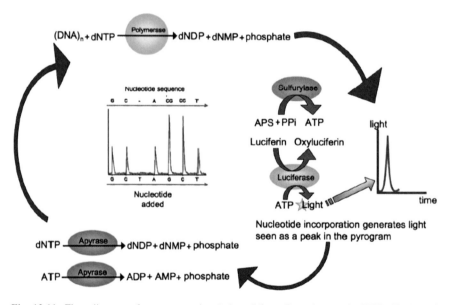

Fig. 12.11 Flow diagram of pyrosequencing (adapted from Petrosino et al., 2009). Each peak in the pyrograms represents a pulse of light detected in the instrument. dNTP, deoxynucleoside triphosphate; dNDP, deoxynucleoside diphosphate; dNMP, deoxynucleoside monophosphate; PPi, pyrophosphate; APS, adenosine-5′-phosphosulfate

nucleotide base can bind to the DNA template, equimolar pyrophosphate (PPi) will be released and subsequently converted to ATP by ATP sulfurylase in the presence of APS. ATP then helps convert luciferin to oxyluciferin which generates visible light. The intensity of the light (represented by the height of the peak in light analysis software) is proportional to the number of dNTPs incorporated. Pyrosequencing allows for cheap, fast, and accurate sequencing of very short genomic regions and is a good alternative to Sanger sequencing. It is especially suitable for SNPs which are close together.

Selection of the appropriate SNP typing strategy. Various SNP typing technologies have been developed over the past few years. Different SNP typing techniques exhibit both advantages and disadvantages. Practical concerns are critical factors when choosing an SNP typing assay, because the cost of SNP typing is still relatively high compared to conventional subtyping techniques such as PFGE and MLST. Common considerations include cost (overall cost and cost per SNP), labor and time efficiency, multiplexing, level of throughput, and instrument requirements. Allele-specific hybridization-based techniques like the Taqman assay are easy to perform because PCR and hybridization occur in the same reaction and samples do not need prior and post-treatment. However, their ability to be multiplexed is limited and thus the cost can be high when a large number of SNPs need to be analyzed. Both pyrosequencing and invader assays are easy to perform but cannot be easily multiplexed. However, their per-test cost is lower than that of allele-specific hybridization-based technologies. Minisequencing assays like SNaPshot have a relatively low cost and can be multiplexed to further reduce cost, thus making it useful for screening large numbers of SNPs. However, the optimization of multiplex systems can be time consuming and multiplexing reduces the flexibility of combining different SNPs. Also, DNA templates need to be pre-amplified and purified before minisequencing reactions are carried out. Pati et al. (2004) performed a direct comparison between invader assay, SNaPshot typing, and pyrosequencing and found that invader typing was the most accurate along with the lowest cost. SNaPshot typing and pyrosequencing had similar performance, but the cost of SNaPshot typing was higher than that of the other two methods. SNaPshot uses standard DNA sequencers; Taqman assays use standard real-time thermocyclers and invader assays use a standard fluorescent spectrophotometer. These instruments are readily available in many research and public health laboratories. In contrast, pyrosequencing-, microarray-, and flow cytometry-based assays require special expensive instruments which are not widely available. Luminex SNP typing system has high capacity for multiplexing to detect multiple SNPs in one reaction. However, the optimization of multiplex PCR prior to the array analysis poses a major challenge. Other major disadvantages of new microarray-based SNP assays are problems associated with reproducibility and data analysis as discussed in the microarray section. It is expected that with the development of genomic technologies, throughput of SNP typing assays will rapidly increase. The cost of setting up equipment and reagents such as probes and chips is also expected to decrease dramatically in the future (Dearlove, 2002).

Careful selection of highly informative SNPs and design of SNP typing primers are key to accurate and cost-efficient SNP typing schemes. Although whole genome sequence data and MLST data are increasingly available for SNP identification, SNPs that are identified based on only a few whole genome sequences may not be representative of the entire population. Similar to choosing the right genetic markers for MLST, choosing the right SNPs in the right genomic locations is also critical for developing a subtyping method with high discriminatory power and epidemiologic concordance. In addition, SNPs harbored in repeat regions or highly recombinogenic regions would probably not be stable, and thus useful for epidemiology or evolutionary genetics. MLST databases are another useful source for SNP identification. DNA alignments in MLST databases generally contain sequences from more than 50 strains, therefore SNPs identified based on these alignments would be expected to provide high discriminatory power. However, an internal problem with any SNP typing strategy is that new SNPs continually evolve and therefore SNP typing schemes based on currently available SNPs would not detect these new SNPs. Thus, it is important to realize that SNP typing still depends on DNA sequencing of newly evolved strains, especially when investigating the molecular epidemiology of new outbreaks.

The above-mentioned new high-throughput subtyping technologies provide the advantage of targeting many more markers than do earlier molecular subtyping methods and therefore are expected to be more discriminatory. They can also overcome the loci bias that MLST may have. However, their overall high cost may compromise their wide application in nationwide surveillance systems such as PulseNet. For epidemiologic typing purposes, if a confined region of a whole genome can provide epidemiologically relevant data, then targeting whole genome variations may be unnecessary and worse, confounding. For example, whole genome variations may be random and may not correlate with the epidemic properties of bacterial species. Incorporation of these variations introduces risk of increasing "noise" and thus confounding data interpretation (Wassenaar, 2003). Therefore, it is important to confirm the epidemiologic concordance of any new subtyping markers, including SNPs.

12.2.7 Genotyping with Mass Spectrometry

Mass spectrometry has been applied to detect SNPs in bacterial genomes. This technology distinguishes itself from other SNP typing techniques because it does not directly detect nucleotide sequences or SNPs. Rather, subtle differences in base composition drive detectable differences between strains. Briefly, short-length PCR products are purified and injected into a mass spectrometer. For small DNA molecules, an exact mass measurement can be used to back-calculate a list of base compositions consistent with the measured molecular weight since we know the exact masses of the four possible bases. Given the mass of a single strand of DNA, many base compositions are possible. However by comparing molecular weights of complementary DNA strands from multiple sequences and algorithmically comparing this information to the reference gene sequences, mass spectrometry is able to determine the exact base composition of small-length PCR amplicons. The IBIS

T5000 system (Abbott Biosciences, Inc.) was developed based on this technology. Basically, this technology correlates changes in sequences with changes in base compositions and uses base compositions as signatures to differentiate bacterial pathogens. Hall et al. (2009) modified an MLST approach for *S. aureus* and used the IBIS system and was able to detect sequence variations in seven housekeeping genes. The IBIS system has the capacity for automation and rapid genotyping. However, a limiting factor is that mass spectrometry equipment is very expensive and therefore not practical to many laboratories.

12.3 Conclusions and Perspectives

This chapter attempted to first discuss the basic principles and criteria associated with molecular epidemiology. It then discussed the disadvantages and limitations of previous technologies and methods and how they have been overcome by novel genomics-based approaches. As a result, molecular epidemiology has been greatly advanced by subtyping methods that specifically target genetic markers that have been identified using more recent genomic and computer technologies. These advancements have dramatically increased discriminatory power, epidemiologic concordance, reproducibility, and typeability of various novel strain typing methods. Genetic markers for molecular epidemiology have evolved from expressed proteins and gene fragments to specific gene sequences to whole genome SNPs.

Understanding strengths and weaknesses associated with each subtyping method is critical to the selection of appropriate subtyping methods, study design, and data interpretation. Traditional phenotypic methods such as serotyping, phage typing, and plasmid profiling generally provide limited discriminatory power. In addition, phage typing and plasmid profiling suffer from poor reproducibility and typeability and thus have been rarely used in recent epidemiologic studies. However, serotyping is still used as a reference in many epidemiologic investigations. It can often be used synergistically with other advanced molecular subtyping techniques. Traditional serotyping is usually slow and expensive; however, rapid molecular serotyping methods are now being developed for subtyping various foodborne pathogens. Even though there are arguments that many current serotyping methods could provide incorrect information about strain relatedness, serotyping will likely continue to contribute to our understanding of epidemiology and evolution of many foodborne pathogens.

Traditional gel-based methods, such as RAPD, AFLP, ribotyping, and PCR-RFLP, generally provide limited resolution for short-term epidemiologic purposes. RAPD and AFLP also have reproducibility and typeability concerns. However, ribotyping and PCR-RFLP may provide valuable information regarding the population structure and long-term transmission of various pathogens. In addition, previous publications using these techniques have been instrumental in designing future molecular subtyping studies. Therefore, these methods can be combined with more recently developed methods to provide a complete picture of both short- and long-term epidemiology of numerous pathogens. PFGE, the current gold standard for molecular epidemiology, has successfully aided the investigation of numerous foodborne outbreaks in the past two decades and will continue being instrumental for investigating the molecular epidemiology of foodborne pathogens in the near future. PFGE is especially useful for the differentiation of closely related strains isolated in a short period of time. However, PFGE sometimes provides too much or too little discriminatory power and is less suitable for inferring phylogeny and constructing the routes of long-term transmission of bacterial pathogens.

Sequence-based subtyping methodologies (especially multilocus sequence typing) have demonstrated excellent discriminatory power and epidemiologic concordance with a variety of pathogens. However, performance of MLST is dependent on the specific pathogen under investigation and specific loci selected. MLST schemes targeting housekeeping genes generally provide satisfactory discriminatory power for long-term epidemiologic and evolutionary analysis; however, many of them have lower discriminatory power compared to PFGE and thus are not suitable for short-term epidemiologic investigations. Also, certain genes that are under frequent recombinational pressures may not be able to generate phylogenetically meaningful subtyping data. In contrast, the inclusion of virulence genes in sequence-based typing schemes has been found to increase both discriminatory power and epidemiologic concordance. Most recently, inclusion of prophage and CRISPR loci have provided additional discriminatory power for accurately differentiating outbreak strains of *L. monocytogenes* (Chen and Knabel, 2008) and the major serovars of *Salmonella enterica* (Liu et al., unpublished observations), respectively. Understanding the population structure and evolutionary history of various pathogens is key to developing good MLST schemes. Caution is needed to select genes that are not under frequent recombination. For example, in some clonal groups, such as *E. coli* O157:H7, a gene may be acquired by an ancestor via a recent recombination event. It therefore may not have enough time to diversify, resulting in little or no sequence variability. Additionally, recombination of one or only a few preferred alleles due to stabilizing selection pressures also precludes sufficient genetic diversity for a meaningful analysis.

SNP typing strategies usually provide higher throughput by targeting more genomic regions and thus allow more samples to be analyzed in the same run. This could potentially provide increased discriminatory power and epidemiologic concordance if appropriate SNPs are selected. However, there are also some concerns with this approach. Discovery of epidemiologically concordant SNPs is the first and most critical step in the development of any SNP typing strategy. Rapid whole genome sequencing and microarray resequencing can be used to discover genomewide informative SNPs. However, these technologies may not be accurate and may be too expensive for sequencing a large number of strains. Therefore, SNPs identified by these methods need to be confirmed using PCR sequencing. These SNPs also need to be screened and validated for discriminatory power and epidemiologic concordance using a large number of diverse, but relevant strains. Bacterial pathogens constantly evolve and new strain or genotypes keep emerging, and those recently evolved strains may contain new SNPs that are not included in the current panel of SNPs. Therefore, SNP panels should be constantly updated by whole genome sequencing of newly generated outbreak isolates.

With rapid development of novel molecular biology technologies and our enhanced understanding of the genomic structures and virulence mechanisms of bacterial pathogens, more advanced subtyping strategies that target highly relevant molecular makers are now possible. Therefore, understanding basic principles of proper marker selection and appropriate interpretation of molecular subtyping data are of critical importance to molecular epidemiology. This chapter has provided insights into those performance criteria that are critical for accurate subtyping of foodborne pathogens. Combining multiple subtyping techniques is often beneficial since none of them are perfect for all applications. Choice of specific subtyping techniques also depends on the availability of laboratory facilities and equipment. Some subtyping methods have special and expensive instrument requirements. In laboratories that lack these facilities, a combination of several simple fragment-based subtyping methods such as multiplex PCR, ribotyping, and PCR-RFLP may be a valuable alternative for accurate strain typing. As evident, combining subtyping schemes typically increases discriminatory power and provides more accurate clustering of strains. In large federal or state laboratories where routine testing of large numbers of samples is performed, high-throughput SNP typing may be more cost efficient.

It is notable that the introduction of molecular subtyping methods has greatly changed the scope of epidemiology and facilitated our understanding of both short- and long-term evolution and transmission of various foodborne pathogens. The choice of specific subtyping methods depends on scope and purpose of the investigation. The ultimate goal of molecular epidemiology is to identify the epidemiologic relationships between bacterial strains. Different computational algorithms and computer software packages have been developed to identify strain relatedness based on output data generated by various subtyping methods. However, strain relatedness revealed by specific strain typing methods is not necessarily the actual epidemiologic relationship of those strains. In many cases, strain typing data are used as a basis to complement conventional epidemiologic findings during outbreak investigations (Riley, 2004). On the other hand, strain typing data often reveal epidemiologic relationships that would not be detected by conventional epidemiology. However, whenever possible, molecular epidemiology and conventional epidemiology should complement and support one another in order to ensure the most accurate epidemiology of foodborne pathogens.

References

Bansal NS, McDonell FH, Smith A, Arnold G, Ibrahim GF (1996) Multiplex PCR assay for the routine detection of Listeria in food. Int J Food Microbiol 33:293–300

Barrett TJ, Gerner-Smidt P, Swaminathan B (2006) Interpretation of pulsed-field gel electrophoresis patterns in foodborne disease investigations and surveillance. Foodborne Pathog Dis 3:20–31

Bender JB, Hedberg CW, Boxrud DJ, Besser JM, Wicklund JH, Smith KE, Osterholm MT (2001) Use of molecular subtyping in surveillance for *Salmonella enterica* serotype Typhimurium. N Engl J Med 344:189–195

Bibb WF, Gellin BG, Weaver R, Schwartz B, Plikaytis BD, Reeves MW, Pinner RW, Broome CV (1990) Analysis of clinical and food-borne isolates of *Listeria monocytogenes* in the United

States by multilocus enzyme electrophoresis and application of the method to epidemiologic investigations. Appl Environ Microbiol 56:2133–2141

Bibb WF, Schwartz B, Gellin BG, Plikaytis BD, Weaver RE (1989) Analysis of *Listeria monocytogenes* by multilocus enzyme electrophoresis and application of the method to epidemiologic investigations. Int J Food Microbiol 8:233–239

Borucki MK, Kim SH, Call DR, Smole SC, Pagotto F (2004) Selective discrimination of *Listeria monocytogenes* epidemic strains by a mixed-genome DNA microarray compared to discrimination by pulsed-field gel electrophoresis, ribotyping, and multilocus sequence typing. J Clin Microbiol 42:5270–5276

Brazma A, Hingamp P, Quackenbush J, Sherlock G, Spellman P, Stoeckert C, Aach J, Ansorge W, Ball CA, Causton HC, Gaasterland T, Glenisson P, Holstege FC, Kim IF, Markowitz V, Matese JC, Parkinson H, Robinson A, Sarkans U, Schulze-Kremer S, Stewart J, Taylor R, Vilo J, Vingron M (2001) Minimum information about a microarray experiment (MIAME)-toward standards for microarray data. Nat Genet 29:365–371

Brown EW, Kotewicz ML, Cebula TA (2002) Detection of recombination among *Salmonella enterica* strains using the incongruence length difference test. Mol Phylogenet Evol 24: 102–120

Brown EW, Mammel MK, LeClerc JE, Cebula TA (2003) Limited boundaries for extensive horizontal gene transfer among *Salmonella* pathogens. Proc Natl Acad Sci USA 100:15676–15681

Cardinali G, Martini A, Preziosi R, Bistoni F, Baldelli F (2002) Multicenter comparison of three different analytical systems for evaluation of DNA banding patterns from *Cryptococcus neoformans*. J Clin Microbiol 40:2095–2100

Carrico JA, Pinto FR, Simas C, Nunes S, Sousa NG, Frazao N, de Lencastre H, Almeida JS (2005) Assessment of band-based similarity coefficients for automatic type and subtype classification of microbial isolates analyzed by pulsed-field gel electrophoresis. J Clin Microbiol 43: 5483–5490

Caugant DA, Ashton FE, Bibb WF, Boerlin P, Donachie W, Low C, Gilmour A, Harvey J, Norrung B (1996) Multilocus enzyme electrophoresis for characterization of *Listeria monocytogenes* isolates: results of an international comparative study. Int J Food Microbiol 32:301–311

Chan MS, Maiden MC, Spratt BG (2001) Database-driven multi locus sequence typing (MLST) of bacterial pathogens. Bioinformatics 17:1077–1083

Chen Y, Knabel SJ (2008) Prophages in *Listeria monocytogenes* contain single-nucleotide polymorphisms that differentiate outbreak clones within epidemic clones. J Clin Microbiol 46:1478–1484

Chen Y, Zhang W, Knabel SJ (2005) Multi-virulence-locus sequence typing clarifies epidemiology of recent listeriosis outbreaks in the United States. J Clin Microbiol 43:5291–5294

Chen Y, Zhang W, Knabel SJ (2007) Multi-virulence-locus sequence typing identifies single nucleotide polymorphisms which differentiate epidemic clones and outbreak strains of *Listeria monocytogenes*. J Clin Microbiol 45:835–846

Cooke FJ, Wain J, Fookes M, Ivens A, Thomson N, Brown DJ, Threlfall EJ, Gunn G, Foster G, Dougan G (2007) Prophage sequences defining hot spots of genome variation in *Salmonella enterica* serovar Typhimurium can be used to discriminate between field isolates. J Clin Microbiol 45:2590–2598

Cooper JE, Feil EJ (2004) Multilocus sequence typing – what is resolved? Trends Microbiol 12:373–377

Craven SE, Cox NA, Bailey JS, Stern NJ, Meinersmann RJ, Blankenship LC (1993) Characterization of *S. california* and *S. typhimurium* strains with reduced ability to colonize the intestinal tract of broiler chicks. Avian Dis 37:339–348

Dalmasso A, Civera T, Bottero MT (2009) Multiplex primer-extension assay for identification of six pathogenic vibrios. Int J Food Microbiol 129:21–25

Dauphin G, Ragimbeau C, Malle P (2001) Use of PFGE typing for tracing contamination with *Listeria monocytogenes* in three cold-smoked salmon processing plants. Int J Food Microbiol 64:51–61

Davis MA, Hancock DD, Besser TE, Call DR (2003) Evaluation of pulsed-field gel electrophoresis as a tool for determining the degree of genetic relatedness between strains of *Escherichia coli* O157:H7. J Clin Microbiol 41:1843–1849

De Cesare A, Bruce JL, Dambaugh TR, Guerzoni ME, Wiedmann M (2001) Automated ribotyping using different enzymes to improve discrimination of *Listeria monocytogenes* isolates, with a particular focus on serotype 4b strains. J Clin Microbiol 39:3002–3005

De la Vega FM, Lazaruk KD, Rhodes MD, Wenz MH (2005) Assessment of two flexible and compatible SNP genotyping platforms: TaqMan SNP genotyping assays and the SNPlex genotyping system. Mutat Res 573:111–135

Dearlove AM (2002) High throughput genotyping technologies. Brief Funct Genomic Proteomic 1:139–150

Destro MT, Leitao M, Farber JM (1996a) Use of molecular typing methods to trace the dissemination of *Listeria monocytogenes* in a shrimp processing plant. Appl Environ Microbiol 62:1852–1853

Destro MT, Leitao MF, Farber JM (1996b) Use of molecular typing methods to trace the dissemination of *Listeria monocytogenes* in a shrimp processing plant. Appl Environ Microbiol 62:705–711

Doumith M, Jacquet C, Goulet V, Oggioni C, Van Loock F, Buchrieser C, Martin P (2006) Use of DNA arrays for the analysis of outbreak-related strains of *Listeria monocytogenes*. Int J Med Microbiol 296:559–562

Ducey TF, Page B, Usgaard T, Borucki MK, Pupedis K, Ward TJ (2007) A single-nucleotide-polymorphism-based multilocus genotyping assay for subtyping lineage I isolates of *Listeria monocytogenes*. Appl Environ Microbiol 73:133–147

Duck WM, Steward CD, Banerjee SN, McGowan JE Jr, Tenover FC (2003) Optimization of computer software settings improves accuracy of pulsed-field gel electrophoresis macrorestriction fragment pattern analysis. J Clin Microbiol 41:3035–3042

Dunbar SA (2006) Applications of Luminex xMAP technology for rapid, high-throughput multiplexed nucleic acid detection. Clin Chim Acta 363:71–82

Feil EJ, Holmes EC, Bessen DE, Chan MS, Day NP, Enright MC, Goldstein R, Hood DW, Kalia A, Moore CE, Zhou J, Spratt BG (2001) Recombination within natural populations of pathogenic bacteria: short-term empirical estimates and long-term phylogenetic consequences. Proc Natl Acad Sci USA 98:182–187

Feil EJ, Spratt BG (2001) Recombination and the population structures of bacterial pathogens. Annu Rev Microbiol 55:561–590

Fitzgerald C, Collins M, van Duyne S, Mikoleit M, Brown T, Fields P (2007) Multiplex, bead-based suspension array for molecular determination of common *Salmonella* serogroups. J Clin Microbiol 45:3323–3334

Foxman B, Zhang L, Koopman JS, Manning SD, Marrs CF (2005) Choosing an appropriate bacterial typing technique for epidemiologic studies. Epidemiol Perspect Innov 2:10

Frost JA, Kramer JM, Gillanders SA (1999) Phage typing of *Campylobacter jejuni* and *Campylobacter coli* and its use as an adjunct to serotyping. Epidemiol Infect 123:47–55

Fuenzalida L, Hernandez C, Toro J, Rioseco ML, Romero J, Espejo RT (2006) *Vibrio parahaemolyticus* in shellfish and clinical samples during two large epidemics of diarrhoea in southern Chile. Environ Microbiol 8:675–683

Fukiya S, Mizoguchi H, Tobe T, Mori H (2004) Extensive genomic diversity in pathogenic *Escherichia coli* and *Shigella* strains revealed by comparative genomic hybridization microarray. J Bacteriol 186:3911–3921

Gerner-Smidt P, Graves LM, Hunter S, Swaminathan B (1998) Computerized analysis of restriction fragment length polymorphism patterns: comparative evaluation of two commercial software packages. J Clin Microbiol 36:1318–1323

Gerner-Smidt P, Hise K, Kincaid J, Hunter S, Rolando S, Hyytia-Trees E, Ribot EM, Swaminathan B (2006) PulseNet USA: a five-year update. Foodborne Pathog Dis 3:9–19

Gillings M, Holley M (1997) Repetitive element PCR fingerprinting (rep-PCR) using enter-obacterial repetitive intergenic consensus (ERIC) primers is not necessarily directed at ERIC elements. Lett Appl Microbiol 25:17–21

Gomes AR, Vinga S, Zavolan M, de Lencastre H (2005) Analysis of the genetic variability of virulence-related loci in epidemic clones of methicillin-resistant *Staphylococcus aureus*. Antimicrob Agents Chemother 49:366–379

Gonzalez-Escalona N, Whitney B, Jaykus LA, DePaola A (2007) Comparison of direct genome restriction enzyme analysis and pulsed-field gel electrophoresis for typing of *Vibrio vulnificus* and their correspondence with multilocus sequence typing data. Appl Environ Microbiol 73:7494–7500

Graves LM, Hunter SB, Ong AR, Schoonmaker-Bopp D, Hise K, Kornstein L, DeWitt WE, Hayes PS, Dunne E, Mead P, Swaminathan B (2005) Microbiological aspects of the investigation that traced the 1998 outbreak of listeriosis in the United States to contaminated hot dogs and estab-lishment of molecular subtyping-based surveillance for *Listeria monocytogenes* in the PulseNet network. J Clin Microbiol 43:2350–2355

Grossman PD, Bloch W, Brinson E, Chang CC, Eggerding FA, Fung S, Iovannisci DM, Woo S, Winn-Deen ES (1994) High-density multiplex detection of nucleic acid sequences: oligonucleotide ligation assay and sequence-coded separation. Nucleic Acids Res 22: 4527–4534

Guerra MM, Bernardo F, McLauchlin J (2002) Amplified fragment length polymorphism (AFLP) analysis of *Listeria monocytogenes*. Syst Appl Microbiol 25:456–461

Gurtler V, Mayall BC (2001) Genomic approaches to typing, taxonomy and evolution of bacterial isolates. Int J Syst Evol Microbiol 51:3–16

Haff LA, Smirnov IP (1997) Single-nucleotide polymorphism identification assays using a ther-mostable DNA polymerase and delayed extraction MALDI-TOF mass spectrometry. Genome Res 7:378–388

Hahm BK, Maldonado Y, Schreiber E, Bhunia AK, Nakatsu CH (2003) Subtyping of foodborne and environmental isolates of *Escherichia coli* by multiplex-PCR, rep-PCR, PFGE, ribotyping and AFLP. J Microbiol Methods 53:387–399

Hall TA, Sampath R, Blyn LB, Ranken R, Ivy C, Melton R, Matthews H, White N, Li F, Harpin V, Ecker DJ, McDougal LK, Limbago B, Ross T, Wolk DM, Wysocki V, Carroll KC (2009) Rapid molecular genotyping and clonal complex assignment of *S. aureus* isolates by PCR/ESI-MS. J Clin Microbiol 47:1733–1741

Hanninen ML, Hakkinen M, Rautelin H (1999) Stability of related human and chicken *Campylobacter jejuni* genotypes after passage through chick intestine studied by pulsed-field gel electrophoresis. Appl Environ Microbiol 65:2272–2275

Harvey J, Gilmour A (2001) Characterization of recurrent and sporadic *Listeria monocytogenes* isolates from raw milk and nondairy foods by pulsed-field gel electrophoresis, monocin typ-ing, plasmid profiling, and cadmium and antibiotic resistance determination. Appl Environ Microbiol 67:840–847

Herrera S, Cabrera R, Ramirez MM, Usera MA, Echeita MA (2002) Use of AFLP, plasmid typing and phenotyping in a comparative study to assess genetic diversity of *Shigella flexneri* strains. Epidemiol Infect 129:445–450

Hommais F, Pereira S, Acquaviva C, Escobar-Paramo P, Denamur E (2005) Single-nucleotide polymorphism phylotyping of *Escherichia coli*. Appl Environ Microbiol 71:4784–4792

Hughes TR, Mao M, Jones AR, Burchard J, Marton MJ, Shannon KW, Lefkowitz SM, Ziman M, Schelter JM, Meyer MR, Kobayashi S, Davis C, Dai H, He YD, Stephaniants SB, Cavet G, Walker WL, West A, Coffey E, Shoemaker DD, Stoughton R, Blanchard AP, Friend SH, Linsley PS (2001) Expression profiling using microarrays fabricated by an ink-jet oligonucleotide synthesizer. Nat Biotechnol 19:342–347

Hulton CS, Higgins CF, Sharp PM (1991) ERIC sequences: a novel family of repetitive elements in the genomes of *Escherichia coli*, *Salmonella typhimurium* and other enterobacteria. Mol Microbiol 5:825–834

Hunter PR, Gaston MA (1988) Numerical index of the discriminatory ability of typing systems: an application of Simpson's index of diversity. J Clin Microbiol 26:2465–2466

Hyytia-Trees E, Smole SC, Fields PA, Swaminathan B, Ribot EM (2006) Second generation subtyping: a proposed PulseNet protocol for multiple-locus variable-number tandem repeat analysis of Shiga toxin-producing *Escherichia coli* O157 (STEC O157). Foodborne Pathog Dis 3:118–131

Iversen C, Mullane N, McCardell B, Tall BD, Lehner A, Fanning S, Stephan R, Joosten H (2008) *Cronobacter* gen. nov., a new genus to accommodate the biogroups of *Enterobacter sakazakii*, and proposal of *Cronobacter sakazakii* gen. nov., comb. nov., *Cronobacter malonaticus* sp. nov., *Cronobacter turicensis* sp. nov., *Cronobacter muytjensii* sp. nov., *Cronobacter dublinensis* sp. nov., *Cronobacter genomospecies* 1, and of three subspecies, *Cronobacter dublinensis* subsp. *dublinensis* subsp. nov., *Cronobacter dublinensis* subsp. *lausannensis* subsp. nov. and *Cronobacter dublinensis* subsp. *lactaridi* subsp. nov. Int J Syst Evol Microbiol 58: 1442–1447

Jacquet C, Catimel B, Brosch R, Buchrieser C, Dehaumont P, Goulet V, Lepoutre A, Veit P, Rocourt J (1995) Investigations related to the epidemic strain involved in the French listeriosis outbreak in 1992. Appl Environ Microbiol 61:2242–2246

Jersek B, Gilot P, Gubina M, Klun N, Mehle J, Tcherneva E, Rijpens N, Herman L (1999) Typing of *Listeria monocytogenes* strains by repetitive element sequence-based PCR. J Clin Microbiol 37:103–109

Jukes TH, Cantor C (1969) Evolution of protein molecules. In: Munro HN (ed) Mammalian protein metabolism. Academic Press, New York, NY

Karenlampi RI, Tolvanen TP, Hanninen ML (2004) Phylogenetic analysis and PCR-restriction fragment length polymorphism identification of *Campylobacter* species based on partial groEL gene sequences. J Clin Microbiol 42:5731–5738

Kathariou S (2002) *Listeria monocytogenes* virulence and pathogenicity, a food safety perspective. J Food Prot 65:1811–1829

Kawamori F, Hiroi M, Harada T, Ohata K, Sugiyama K, Masuda T, Ohashi N (2008) Molecular typing of Japanese *Escherichia coli* O157: H7 isolates from clinical specimens by multilocus variable-number tandem repeat analysis and PFGE. J Med Microbiol 57:58–63

Keto-Timonen RO, Autio TJ, Korkeala HJ (2003) An improved amplified fragment length polymorphism (AFLP) protocol for discrimination of *Listeria* isolates. Syst Appl Microbiol 26:236–244

Khakhria R, Duck D, Lior H (1990) Extended phage-typing scheme for *Escherichia coli* O157:H7. Epidemiol Infect 105:511–520

Khambaty FM, Bennett RW, Shah DB (1994) Application of pulsed-field gel electrophoresis to the epidemiological characterization of *Staphylococcus intermedius* implicated in a food-related outbreak. Epidemiol Infect 113:75–81

Kimura M (1980) A simple method for estimating evolutionary rates of base substitutions through comparative studies of nucleotide sequences. J Mol Evol 16:111–120

Kingombe CI, Cerqueira-Campos ML, Farber JM (2005) Molecular strategies for the detection, identification, and differentiation between enteroinvasive *Escherichia coli* and *Shigella* spp. J Food Prot 68:239–245

Koeuth T, Versalovic J, Lupski JR (1995) Differential subsequence conservation of interspersed repetitive *Streptococcus pneumoniae* BOX elements in diverse bacteria. Genome Res 5: 408–418

Koreen L, Ramaswamy SV, Graviss EA, Naidich S, Musser JM, Kreiswirth BN (2004) spa typing method for discriminating among *Staphylococcus aureus* isolates: implications for use of a single marker to detect genetic micro- and macrovariation. J Clin Microbiol 42:792–799

Lake JA (1994) Reconstructing evolutionary trees from DNA and protein sequences: paralinear distances. Proc Natl Acad Sci USA 91:1455–1459

Lane DJ, Pace B, Olsen GJ, Stahl DA, Sogin ML, Pace NR (1985) Rapid determination of 16S ribosomal RNA sequences for phylogenetic analyses. Proc Natl Acad Sci USA 82:6955–6959

Leader BT, Frye JG, Hu J, Fedorka-Cray PJ, Boyle DS (2009) High-throughput molecular determination of *Salmonella enterica* serovars by use of multiplex PCR and capillary electrophoresis analysis. J Clin Microbiol 47:1290–1299

Lemee L, Bourgeois I, Ruffin E, Collignon A, Lemeland JF, Pons JL (2005) Multilocus sequence analysis and comparative evolution of virulence-associated genes and housekeeping genes of *Clostridium difficile*. Microbiology 151:3171–3180

Litrup E, Torpdahl M, Nielsen EM (2007) Multilocus sequence typing performed on *Campylobacter coli* isolates from humans, broilers, pigs and cattle originating in Denmark. J Appl Microbiol 103:210–218

Livak KJ, Flood SJ, Marmaro J, Giusti W, Deetz K (1995) Oligonucleotides with fluorescent dyes at opposite ends provide a quenched probe system useful for detecting PCR product and nucleic acid hybridization. PCR Methods Appl 4:357–362

Loessner MJ, Busse M (1990) Bacteriophage typing of *Listeria* species. Appl Environ Microbiol 56:1912–1918

Lomonaco S, Chen Y, Knabel SJ (2008) Analysis of additional virulence genes and virulence gene regions in *Listeria monocytogenes* confirms the epidemiologic relevance of multi-virulence-locus sequence typing. J Food Prot 71:2559–2566

Maiden MC, Bygraves JA, Feil E, Morelli G, Russell JE, Urwin R, Zhang Q, Zhou J, Zurth K, Caugant DA, Feavers IM, Achtman M, Spratt BG (1998) Multilocus sequence typing: a portable approach to the identification of clones within populations of pathogenic microorganisms. Proc Natl Acad Sci USA 95:3140–3145

Manning SD, Motiwala AS, Springman AC, Qi W, Lacher DW, Ouellette LM, Mladonicky JM, Somsel P, Rudrik JT, Dietrich SE, Zhang W, Swaminathan B, Alland D, Whittam TS (2008) Variation in virulence among clades of *Escherichia coli* O157:H7 associated with disease outbreaks. Proc Natl Acad Sci USA 105:4868–4873

Maslanka SE, Kerr JG, Williams G, Barbaree JM, Carson LA, Miller JM, Swaminathan B (1999) Molecular subtyping of *Clostridium perfringens* by pulsed-field gel electrophoresis to facilitate food-borne-disease outbreak investigations. J Clin Microbiol 37:2209–2214

Mazurier SI, Wernars K (1992) Typing of *Listeria* strains by random amplification of polymorphic DNA. Res Microbiol 143:499–505

McQuiston JR, Parrenas R, Ortiz-Rivera M, Gheesling L, Brenner F, Fields PI (2004) Sequencing and comparative analysis of flagellin genes fliC, fljB, and flpA from *Salmonella*. J Clin Microbiol 42:1923–1932

Melles DC, van Leeuwen WB, Snijders SV, Horst-Kreft D, Peeters JK, Verbrugh HA, van Belkum A (2007) Comparison of multilocus sequence typing (MLST), pulsed-field gel electrophoresis (PFGE), and amplified fragment length polymorphism (AFLP) for genetic typing of *Staphylococcus aureus*. J Microbiol Methods 69:371–375

Miller WG, Englen MD, Kathariou S, Wesley IV, Wang G, Pittenger-Alley L, Siletz RM, Muraoka W, Fedorka-Cray PJ, Mandrell RE (2006) Identification of host-associated alleles by multilocus sequence typing of *Campylobacter coli* strains from food animals. Microbiology 152: 245–255

Mohapatra BR, Broersma K, Mazumder A (2007) Comparison of five rep-PCR genomic fingerprinting methods for differentiation of fecal *Escherichia coli* from humans, poultry and wild birds. FEMS Microbiol Lett 277:98–106

Murphy M, Corcoran D, Buckley JF, O'Mahony M, Whyte P, Fanning S (2006) Development and application of Multiple-Locus Variable Number of tandem repeat Analysis (MLVA) to subtype a collection of *Listeria monocytogenes*. Int J Food Microbiol 1435–1450

Nadon CA, Woodward DL, Young C, Rodgers FG, Wiedmann M (2001) Correlations between molecular subtyping and serotyping of *Listeria monocytogenes*. J Clin Microbiol 39: 2704–2707

Nei M, Kumar S (2000) Molecular evolution and phylogenetics. Oxford University Press, Oxford

Nei M, Tajima F (1985) Evolutionary change of restriction cleavage sites and phylogenetic inference for man and apes. Mol Biol Evol 2:189–205

Neves E, Lourenco A, Silva AC, Coutinho R, Brito L (2008) Pulsed-field gel electrophoresis (PFGE) analysis of *Listeria monocytogenes* isolates from different sources and geographical origins and representative of the twelve serovars. Syst Appl Microbiol 31:387–392

Nightingale KK, Ivy RA, Ho AJ, Fortes ED, Njaa BL, Peters RM, Wiedmann M (2008) inlA premature stop codons are common among *Listeria monocytogenes* isolates from foods and yield virulence-attenuated strains that confer protection against fully virulent strains. Appl Environ Microbiol 74:6570–6583

Nightingale KK, Windham K, Wiedmann M (2005) Evolution and molecular phylogeny of *Listeria monocytogenes* isolated from human and animal listeriosis cases and foods. J Bacteriol 187:5537–5551

Nilsson M, Malmgren H, Samiotaki M, Kwiatkowski M, Chowdhary BP, Landegren U (1994) Padlock probes: circularizing oligonucleotides for localized DNA detection. Science 265: 2085–2088

Notermans S, Chakraborty T, Leimeister-Wachter M, Dufrenne J, Heuvelman KJ, Maas H, Jansen W, Wernars K, Guinee P (1989) Specific gene probe for detection of biotyped and serotyped *Listeria* strains. Appl Environ Microbiol 55:902–906

Orsi RH, Borowsky ML, Lauer P, Young SK, Nusbaum C, Galagan JE, Birren BW, Ivy RA, Sun Q, Graves LM, Swaminathan B, Wiedmann M (2008) Short-term genome evolution of *Listeria monocytogenes* in a non-controlled environment. BMC Genomics 9:539

Orskov F, Orskov I (1983) From the national institutes of health. Summary of a workshop on the clone concept in the epidemiology, taxonomy, and evolution of the Enterobacteriaceae and other bacteria. J Infect Dis 148:346–357

Paillard D, Dubois V, Duran R, Nathier F, Guittet C, Caumette P, Quentin C (2003) Rapid identification of *Listeria* species by using restriction fragment length polymorphism of PCR-amplified 23S rRNA gene fragments. Appl Environ Microbiol 69:6386–6392

Pati N, Schowinsky V, Kokanovic O, Magnuson V, Ghosh S (2004) A comparison between SNaPshot, pyrosequencing, and biplex invader SNP genotyping methods: accuracy, cost, and throughput. J Biochem Biophys Methods 60:1–12

Persing DH, Tenover FC, Versalovic J, Tang Y-W, Unger ER, David MDR, White TJ 2003. Molecular microbiology: diagnostic principles and practice. ASM Press, Washington, DC

Petrosino JF, Highlander S, Luna RA, Gibbs RA, Versalovic J (2009) Metagenomic pyrosequencing and microbial identification. Clin Chem 55:856–866

Radu S, Ling OW, Rusul G, Karim MI, Nishibuchi M (2001) Detection of *Escherichia coli* O157:H7 by multiplex PCR and their characterization by plasmid profiling, antimicrobial resistance, RAPD and PFGE analyses. J Microbiol Methods 46:131–139

Ralovich B, Ewan EP, Emody L (1986) Alteration of phage- and biotypes of *Listeria* strains. Acta Microbiol Hung 33:19–26

Rementeria A, Gallego L, Quindos G, Garaizar J (2001) Comparative evaluation of three commercial software packages for analysis of DNA polymorphism patterns. Clin Microbiol Infect 7:331–336

Riley LW (2004) Molecular epidemiology of infectious diseases: principles and practices. ASM Press, Washington, DC

Ripabelli G, McLauchin J, Threlfall EJ (2000) Amplified fragment length polymorphism (AFLP) analysis of *Listeria monocytogenes*. Syst Appl Microbiol 23:132–136

Romanova N, Favrin S, Griffiths MW (2002) Sensitivity of *Listeria monocytogenes* to sanitizers used in the meat processing industry. Appl Environ Microbiol 68:6405–6409

Saidijam M, Psakis G, Clough JL, Meuller J, Suzuki S, Hoyle CJ, Palmer SL, Morrison SM, Pos MK, Essenberg RC, Maiden MC, Abu-bakr A, Baumberg SG, Neyfakh AA, Griffith JK, Stark MJ, Ward A, O'Reilly J, Rutherford NG, Phillips-Jones MK, Henderson PJ (2003) Collection and characterisation of bacterial membrane proteins. FEBS Lett 555:170–175

Sails AD, Swaminathan B, Fields PI (2003) Utility of multilocus sequence typing as an epidemiological tool for investigation of outbreaks of gastroenteritis caused by *Campylobacter jejuni*. J Clin Microbiol 41:4733–4739

Salcedo C, Arreaza L, Alcala B, de la Fuente L, Vazquez JA (2003) Development of a multilocus sequence typing method for analysis of *Listeria monocytogenes* clones. J Clin Microbiol 41:757–762

Sauders BD, Fortes ED, Morse DL, Dumas N, Kiehlbauch JA, Schukken Y, Hibbs JR, Wiedmann M (2003) Molecular subtyping to detect human listeriosis clusters. Emerg Infect Dis 9: 672–680

Sauer S, Lechner D, Berlin K, Lehrach H, Escary JL, Fox N, Gut IG (2000) A novel procedure for efficient genotyping of single nucleotide polymorphisms. Nucleic Acids Res 28:E13

Schwartz DC, Cantor CR (1984) Separation of yeast chromosome-sized DNAs by pulsed field gradient gel electrophoresis. Cell 37:67–75

Senczek D, Stephan R, Untermann F (2000) Pulsed-field gel electrophoresis (PFGE) typing of *Listeria* strains isolated from a meat processing plant over a 2-year period. Int J Food Microbiol 62:155–159

Shumaker JM, Metspalu A, Caskey CT (1996) Mutation detection by solid phase primer extension. Hum Mutat 7:346–354

Singer RS, Sischo WM, Carpenter TE (2004) Exploration of biases that affect the interpretation of restriction fragment patterns produced by pulsed-field gel electrophoresis. J Clin Microbiol 42:5502–5511

Sobrino B, Brion M, Carracedo A (2005) SNPs in forensic genetics: a review on SNP typing methodologies. Forensic Sci Int 154:181–194

Sobrino B, Carracedo A (2005) SNP typing in forensic genetics: a review. Methods Mol Biol 297:107–126

Sorum H, Hvaal AB, Heum M, Daae FL, Wiik R (1990) Plasmid profiling of *Vibrio salmonicida* for epidemiological studies of cold-water vibriosis in Atlantic salmon (*Salmo salar*) and cod (*Gadus morhua*). Appl Environ Microbiol 56:1033–1037

Struelens MJ (1996) Consensus guidelines for appropriate use and evaluation of microbial epidemiologic typing systems. Clin Microbiol Infect 2:2–11

Swaminathan B, Barrett TJ, Hunter SB, Tauxe RV (2001) PulseNet: the molecular subtyping network for foodborne bacterial disease surveillance, United States. Emerg Infect Dis 7:382–389

Tamura K (1992) Estimation of the number of nucleotide substitutions when there are strong transition-transversion and G+C-content biases. Mol Biol Evol 9:678–687

Tamura K, Nei M (1993) Estimation of the number of nucleotide substitutions in the control region of mitochondrial DNA in humans and chimpanzees. Mol Biol Evol 10:512–526

Tankouo-Sandjong B, Sessitsch A, Liebana E, Kornschober C, Allerberger F, Hachler H, Bodrossy L (2007) MLST-v, multilocus sequence typing based on virulence genes, for molecular typing of *Salmonella enterica* subsp. *enterica* serovars. J Microbiol Methods 69:23–36

Tardif G, Sulavik MC, Jones GW, Clewell DB (1989) Spontaneous switching of the sucrose-promoted colony phenotype in *Streptococcus sanguis*. Infect Immun 57:3945–3948

Tenover FC, Arbeit RD, Goering RV, Mickelsen PA, Murray BE, Persing DH, Swaminathan B (1995) Interpreting chromosomal DNA restriction patterns produced by pulsed-field gel electrophoresis: criteria for bacterial strain typing. J Clin Microbiol 33:2233–2239

van Belkum A (2003) High-throughput epidemiologic typing in clinical microbiology. Clin Microbiol Infect 9:86–100

van Belkum A, Struelens M, de Visser A, Verbrugh H, Tibayrenc M (2001) Role of genomic typing in taxonomy, evolutionary genetics, and microbial epidemiology. Clin Microbiol Rev 14:547–560

Van Stelten A, Nightingale KK (2008) Development and implementation of a multiplex single-nucleotide polymorphism genotyping assay for detection of virulence-attenuating mutations in the *Listeria monocytogenes* virulence-associated gene inlA. Appl Environ Microbiol 74: 7365–7375

Vines A, Reeves MW, Hunter S, Swaminathan B (1992) Restriction fragment length polymorphism in four virulence-associated genes of *Listeria monocytogenes*. Res Microbiol 143: 281–294

Vogel L, van Oorschot E, Maas HM, Minderhoud B, Dijkshoorn L (2000) Epidemiologic typing of *Escherichia coli* using RAPD analysis, ribotyping and serotyping. Clin Microbiol Infect 6:82–87

Wang DG, Fan JB, Siao CJ, Berno A, Young P, Sapolsky R, Ghandour G, Perkins N, Winchester E, Spencer J, Kruglyak L, Stein L, Hsie L, Topaloglou T, Hubbell E, Robinson E, Mittmann M, Morris MS, Shen N, Kilburn D, Rioux J, Nusbaum C, Rozen S, Hudson TJ, Lipshutz R, Chee M, Lander ES (1998) Large-scale identification, mapping, and genotyping of single-nucleotide polymorphisms in the human genome. Science 280:1077–1082

Ward TJ, Gorski L, Borucki MK, Mandrell RE, Hutchins J, Pupedis K (2004) Intraspecific phylogeny and lineage group identification based on the prfA virulence gene cluster of *Listeria monocytogenes*. J Bacteriol 186:4994–5002

Wassenaar TM (2003) Molecular typing of pathogens. Berl Munch Tierarztl Wochenschr 116: 447–453

Wiedmann M (2002) Molecular subtyping methods for *Listeria monocytogenes*. J AOAC Int 85:524–531

Wiedmann M, Bruce JL, Keating C, Johnson AE, McDonough PL, Batt CA (1997) Ribotypes and virulence gene polymorphisms suggest three distinct *Listeria monocytogenes* lineages with differences in pathogenic potential. Infect Immun 65:2707–2716

Wilhelms D, Sandow D (1989) Preliminary studies on monocine typing of *Listeria monocytogenes* strains. Acta Microbiol Hung 36:235–238

Wulff G, Gram L, Ahrens P, Vogel BF (2006) One group of genetically similar *Listeria monocytogenes* strains frequently dominates and persists in several fish slaughter- and smokehouses. Appl Environ Microbiol 72:4313–4322

Yang Z, Rannala B (1997) Bayesian phylogenetic inference using DNA sequences: a Markov Chain Monte Carlo method. Mol Biol Evol 14:717–724

Yoshida C, Franklin K, Konczy P, McQuiston JR, Fields PI, Nash JH, Taboada EN, Rahn K (2007) Methodologies towards the development of an oligonucleotide microarray for determination of *Salmonella* serotypes. J Microbiol Methods 70:261–271

Zabeau M, Vos P (1993) AFLP: not only for fingerprinting, but for positional cloning. European Patent Application, EP 0534858

Zhang W, Jayarao BM, Knabel SJ (2004) Multi-virulence-locus sequence typing of *Listeria monocytogenes*. Appl Environ Microbiol 70:913–920

Zhang W, Qi W, Albert TJ, Motiwala AS, Alland D, Hyytia-Trees EK, Ribot EM, Fields PI, Whittam TS, Swaminathan B (2006) Probing genomic diversity and evolution of *Escherichia coli* O157 by single nucleotide polymorphisms. Genome Res 16:757–767

Zharkikh A (1994) Estimation of evolutionary distances between nucleotide sequences. J Mol Evol 39:315–329

Zheng J, Keys CE, Zhao S, Meng J, Brown EW (2007) Enhanced subtyping scheme for *Salmonella enteritidis*. Emerg Infect Dis 13:1932–1935

Chapter 13
The Evolution of Foodborne Pathogens

Galeb S. Abu-Ali and Shannon D. Manning

13.1 Introduction

Despite continuous advances in food safety and disease surveillance, control, and prevention, foodborne bacterial infections remain a major public health concern. Because foodborne pathogens are commonly exposed to multiple environmental stressors, such as low pH and antibiotics, most have evolved specific mechanisms to facilitate survival in adverse environments. Consequently, the incidence of foodborne disease in the United States caused by *Campylobacter*, *Escherichia coli* O157:H7, *Listeria*, *Salmonella*, *Shigella*, *Vibrio*, and *Yersinia* has remained unchanged (Centers for Disease Control and Prevention, 2009). Besides emphasizing the need for more efficient methods to combat enteric disease, the steady rates of foodborne infections highlight the ability of these pathogens to rapidly adapt to pressures created by varying food production and animal husbandry practices as well as transport through different food matrices and hosts. As the traditional sources of infection are better recognized and controlled, it is not uncommon for many foodborne pathogens to cause large-scale outbreaks associated with novel food vehicles. *Salmonella* spp., for example, have recently been implicated in outbreaks involving contaminated fruits, vegetables, jalapeno peppers, cantaloupe, cereal, and peanut butter, while enterohemorrhagic *E. coli* (EHEC) O157:H7 has been linked to contaminated spinach, lettuce, cookie dough, and frozen pizza. It is therefore necessary to better understand those bacterial factors and mechanisms important for the evolution of virulence as well as adaptation. The purpose of this chapter is to review recent findings regarding the evolution of foodborne pathogens while emphasizing the contribution of genomics.

S.D. Manning (✉)
Microbiology and Molecular Genetics, Michigan State University, 194 Food Safety and Toxicology Building, E. Lansing, MI 48824, USA
e-mail: shannon.manning@ht.msu.edu

M. Wiedmann, W. Zhang (eds.), *Genomics of Foodborne Bacterial Pathogens*,
Food Microbiology and Food Safety, DOI 10.1007/978-1-4419-7686-4_13,
© Springer Science+Business Media, LLC 2011

13.2 The Evolution of Virulence

Over time, bacteria continuously undergo genomic modification, which promotes a fitness advantage and facilitates colonization of and persistence in new niches. According to Darwin, virulence is defined as a trait that benefits an organism in the struggle for existence (Darwin, 1859). In the evolution of virulence, genomes of pathogenic bacteria are continually reconfigured to develop an ideal set of virulence characteristics that will afford the highest benefit in the current environment. The virulence of a pathogen may increase during adaptation outside or within the host and result in more severe disease, or it may decrease toward a balanced state of mutualism. It has been hypothesized that the ability to cause overt disease is influenced by trade-offs between a pathogen's virulence potential and transmission dynamics among hosts (Ewald, 1987). Ultimately, human morbidity and mortality may be coincidental to infection by bacteria that have evolved by adapting to selective pressures outside humans and have consequently acquired traits that are, by chance, detrimental to the human host (Pallen and Wren, 2007). Although many forces can impact the evolution of virulence, several different mechanisms, such as lateral gene transfer, recombination, mutation, and gene duplication, are important for generating genomic diversity and are likely driving the emergence of more virulent bacterial pathogens.

13.2.1 The Bacterial Pan-Genome

Whole-genome sequencing and subsequent comparative genomics analyses have significantly advanced our understanding of the mechanisms by which bacterial pathogens evolve. For example, a comparison of the nonpathogenic *E. coli* K-12 MG1655 genome (Blattner et al., 1997) to two EHEC O157:H7 genomes from notable outbreaks in Japan (Sakai; Hayashi et al., 2001) and Michigan (EDL-933; Perna et al., 2001) demonstrated that both genomes (~5.2–5.5 Mb) are between 859 kb (Sakai) and 1,387 kb (EDL-933) larger than K-12 MG1655. Among all three genomes, 4.1 Mb was found to be highly conserved and comprises the chromosomal backbone (Hayashi et al., 2001; Perna et al., 2001), which was likely inherited from a common ancestor (Ohnishi et al., 2001). Strain-specific sequences for both Sakai (1,393 kb) (Hayashi et al., 2001) and EDL-933 (1,576 kb) (Perna et al., 2001) are also present and exist mainly on different O157-specific islands (O-islands) and plasmids. These strain-specific sequences have been shown to represent foreign DNA, namely prophages and prophage-like elements, which were acquired via lateral gene transfer (Ohnishi et al., 2001). Multiple lineage-specific differences between the two O157 genomes are also apparent, with EDL-933 having a duplicated 86 kb phage-like element similar to the Sakai prophage-like element 1 (SpLE1) of Sakai, but lacking a Sakai-specific Mu-like phage (Ohnishi et al., 2001), for instance. The frequency and variation of phage elements in these two distinct O157 outbreak strains highlight the role that lateral gene transfer plays in generating genomic diversity among pathogenic bacteria. Indeed, similar findings have been described for other common foodborne pathogens as well, such as *Salmonella enterica* (Brussow et al., 2004) and *Campylobacter jejuni* (Scott et al., 2007), in which prophage involvement markedly mediates intraspecies genome diversification.

Multiple comparative genomic studies have confirmed that the complete bacterial genome, or pan-genome, contains a set of core and dispensable genes. The core genome is comprised of genes found in all strains, whereas the dispensable genome consists of genes found in up to two different strains (Medini et al., 2005). For *E. coli*, a comparison of 1 commensal strain and 16 strains representing six pathotypes, including the five groups of intestinal *E. coli* pathotypes, estimated the core genome to consist of ~2,200 genes among the ~13,000 genes in the *E. coli* pan-genome (Fig. 13.1) (Rasko et al., 2008). Rasko et al. also determined that roughly 300 novel genes will be identified for each new *E. coli* genome sequence, indicating that the *E. coli* pan-genome is not saturated, but 'open' (Rasko et al., 2008). While both strain- and pathotype-specific genes were identified, the latter genes were detected at low frequencies, suggesting that the independent diversification of different pathotypes in newly occupied niches has played a key role in generating genomic variation (Rasko et al., 2008). Another analysis comparing 4 genomes of enterohemorrhagic *E. coli* (EHEC) serogroups O157, O26, O103, and O111 to 21 additional *E. coli* and *Shigella* genomes also demonstrated that the entire *E. coli* pan-genome consists of ~13,000 genes and 12,940 CDS groups (Ogura et al., 2009). In general, EHEC strains were estimated to have a larger genome, with an average of ~500 more genes than an extraintestinal pathogenic *E. coli* (ExPEC) strain and ~1,000 more genes than *Shigella* (Ogura et al., 2009).

Another study using comparative genomic hybridization of 26 intestinal and extraintestinal pathogenic *E. coli* strains representing seven pathotypes and commensal strains demonstrated that 2–10% of nonpathogenic *E. coli* K-12 MG1655 ORFs were absent from each strain regardless of pathotype (Dobrindt et al., 2003). To highlight the genetic diversity, only 3,100 translatable ORFs from K-12 MG1655 are shared among all strains despite the large average genome size (~4.7 Mbp) per strain (Dobrindt et al., 2003). Similarly, an alternative comparative genomic hybridization study of nonpathogenic *E. coli* K-12 W3110 and 22 *E. coli* strains representing six pathotypes demonstrated that ~2,800 ORFs are conserved and 1,424 (35%) ORFs are absent from at least one of the strains examined (Fukiya et al., 2004). Together, these data suggest that the core *E. coli* genome consists of ~2,200–3,100 genes (Dobrindt et al., 2003; Fukiya et al., 2004; Rasko et al., 2008), a number that is dependent on the type and number of *E. coli* strains included in any given analysis. The finding that distinct gene profiles correlate with specific *E. coli* pathotypes (Fukiya et al., 2004) reflects the niche-specific genome adaptation that is characteristic of a successful pathogen. The effect of the habitat on the shaping of bacterial genomic content is perhaps best observed in examples of parallel evolution.

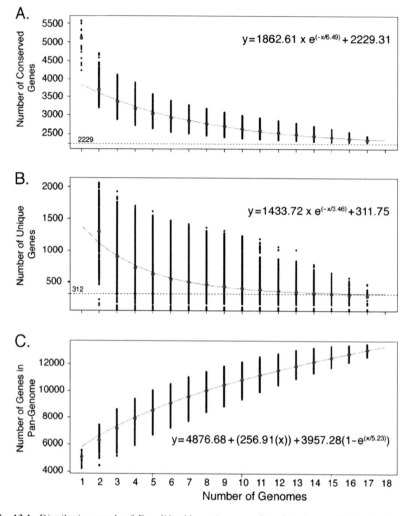

Fig. 13.1 Distribution trends of *E. coli* backbone (conserved) and strain-specific (unique) genes. Note that as more genomes are sequenced, the number of both backbone and strain-specific genes decreases (**a** and **b**), while the pan-genome size increases (**c**). Taken from Rasko et al. (2008)

13.2.2 Parallel Evolution

The process of parallel evolution occurs when similar genetic alterations develop in distinct, but related, bacterial lineages due to comparable environmental pressures and genetic backgrounds. Enteropathogenic *E. coli* (EPEC) and EHEC are noteworthy examples of parallel evolution. By comparing the rate of synonymous substitutions for *E. coli* and *S. enterica*, Reid et al. have shown that the progenitors of EHEC O157:H7 and the nonpathogenic *E. coli* K-12 separated approximately

4.5 million years ago (Reid et al., 2000). Phylogenetic analyses based on multilocus sequence typing (MLST) data for 7 conserved housekeeping genes among 20 pathogenic *E. coli* strains have demonstrated that several *E. coli* clonal groups evolved similar sets of virulence mechanisms in parallel. This process was suggested to have occurred independently through the lateral acquisition of mobile genetic elements (e.g., pathogenicity islands, plasmids, and bacteriophages) that have increased the virulence of evolved lineages (Reid et al., 2000).

The study by Reid et al. confirmed the phylogenetic framework of specific intestinal *E. coli* lineages identified in an earlier study using multilocus enzyme electrophoresis (MLEE), a method for studying genetic variation based on protein polymorphisms caused by amino acid replacements (Whittam et al., 1993). Both the MLEE- and MLST-based phylogenies categorized EPEC and EHEC strains into two distinct clonal groups each and uncovered significant variation in the extant population of intestinal *E. coli* (Reid et al., 2000; Whittam, 1998; Whittam et al., 1993). It was also demonstrated that the intestinal *E. coli* pathotypes are polyphyletic, or are comprised of phylogenetically distinct lineages (clonal groups), and that lineages from different pathotypes can be more closely related than lineages of the same pathotype. For example, the EHEC 2 clonal group comprising non-O157 serotypes is more closely related to the EPEC 2 clonal group than to EHEC 1, or the group containing O157:H7. Another example is EPEC O55:H7, which is similar to other EPEC with regard to clinical presentation but is more closely related to EHEC O157:H7 than other EPEC lineages (Reid et al., 2000; Whittam, 1998). Consequently, Whittam et al. proposed a two-step model of EHEC O157:H7 evolution in which an *E. coli* lineage ancestral to EPEC O55:H7 first acquired the ability to develop attaching and effacing lesions important for colonization that was followed by the acquisition of multiple virulence factors (Whittam et al., 1993).

Although the MLEE- and MLST-based phylogenies supporting the population structure and parallel evolution of EPEC and EHEC clonal groups (Reid et al., 2000; Whittam et al., 1993) were initially based on the nucleotide diversity of only a small fraction of the *E. coli* genome, recent high-throughput genome sequencing data have confirmed the phylogenetic relationships among intestinal *E. coli* pathotypes (Ogura et al., 2009; Rasko et al., 2008). One analysis of 345 non-recombinogenic, orthologous genes in 25 *E. coli* and *Shigella* genomes has confirmed the polyphyletic nature of various pathotypes (Ogura et al., 2009), with EHEC O157:H7, for example, being distantly related to EHEC 2 (Fig. 13.2a). Nevertheless, an examination of complete genome sequences in all 25 strains illustrates that *E. coli* cluster according to pathotype (Fig. 13.2b) (Ogura et al., 2009); therefore, it is the presence of unique DNA segments in the dispensable portion of the genome that distinguishes the different pathotypes.

13.2.2.1 Acquisition of Virulence Traits by Lateral Gene Transfer

It has been suggested that strains with specific genetic backgrounds may be more likely to acquire, integrate, and express DNA from other sources (Escobar-Paramo et al., 2004; Ochman et al., 2000), a suggestion supported by evidence that some

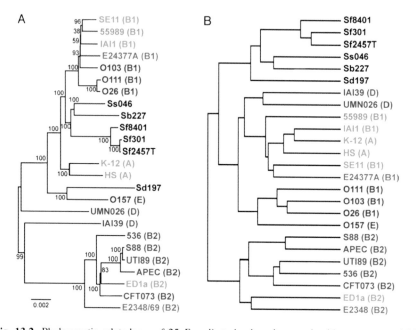

Fig. 13.2 Phylogenetic relatedness of 25 *E. coli* strains based on nucleotide sequences of 345 non-recombinogenic genes (**a**) and whole-genome sequences (**b**). Pathotypes are indicated by different colors: *black, Shigella*; *blue*, extraintestinal *E. coli*; *orange*, enteroaggregative *E. coli*; *green*, commensal; *red*, EHEC; *pink*, EPEC. Taken from Ogura et al. (2009)

bacterial lineages are more pathogenic than others. Indeed, faulty mismatch repair (Karaolis et al., 1999; LeClerc et al., 1996) and high recombination frequencies have been described as mechanisms by which some genotypes can more readily acquire DNA than others (Reid et al., 2000). One classic example is the presence of a chromosomal pathogenicity island carrying the cholera toxin phage receptor in epidemic and pandemic strains of *Vibrio cholerae*, which is absent in non-pathogenic *V. cholerae* strains (Karaolis et al., 1998). For intestinal *E. coli*, it is likely that numerous virulence determinants, which have evolved in divergent lineages through repeated and independent lateral gene transfer events, also provide a selective advantage. EHEC and EPEC, for example, have independently acquired several phages, pathogenicity islands, and plasmids, which work in concert within a chromosomal background that promotes the establishment and dissemination of successful lineages (Reid et al., 2000). These mobile elements vary considerably in genetic composition and integration site (Hayashi et al., 2001; Ogura et al., 2009; Perna et al., 2001) despite having similar functions in disease pathogenesis.

The pathogenicity island carrying the locus of enterocyte effacement (LEE), which encodes intimin (*eae*) and a type 3 secretion system (T3SS) that mediates the formation of attaching and effacing lesions, is an example of a mobile element that is critical for disease pathogenesis (Kaper et al., 2004) and bacterial transmission (Wickham et al., 2007). While the LEE is present in all attaching and effacing

E. coli (AEEC), it was suggested to have been acquired several times in parallel, as its tRNA insertion site and intimin type vary across clonal groups. Specifically, LEE inserts into *pheU* among EHEC 2 and EPEC 2 and *selC* among EHEC 1 and EPEC 1 lineages (Reid et al., 2000). While typical EPEC lineages have acquired the EPEC adherence factor (EAF) plasmid encoding the bundle-forming pilus and other traits important for disease (Goncalves et al., 1997), EHEC lineages have acquired a distinct plasmid. The EHEC plasmid contains enterohemolysin, a type 2 secretion system (T2SS) important for hemorrhagic colitis (Law, 2000), and a number of other genes shown to be important for pathogenesis (Kaper et al., 2004). For EHEC 2, the analysis of three non-O157 genomes identified four other unique plasmids in the O111 genome (Ogura et al., 2009), suggesting additional plasmid acquisition events in this lineage.

Perhaps the most important EHEC virulence characteristic is the ability to produce one or more Shiga toxins, potent two-component cytotoxins also produced by *Shigella dysenteriae* 1 (O'Brien et al., 1983) that contribute to the renal damage associated with hemolytic uremic syndrome (HUS) (Kaper et al., 2004). Shiga toxins are encoded by genes located on lysogenic lambdoid prophages that can insert into multiple locations around the EHEC chromosome (Serra-Moreno et al., 2007). EHEC represent only a small fraction of Shiga toxin-producing *E. coli* (STEC) that possess the LEE island and the ability to elicit the formation of attaching and effacing lesions (Donnenberg, 2002). By contrast, STEC lack the LEE island and only rarely contribute to sporadic cases of human infection (Whittam, 1998), though STEC-associated urinary tract infections have been documented (Donnenberg, 2002). Interestingly, some STEC strains possess a unique pathogenicity island encoding distinct virulence factors, including an adherence gene similar to *iha* of EHEC O157:H7 and a serine protease (EspI), which has been shown to insert into the same *selC* tRNA site as the LEE island (Schmidt et al., 2001).

There is some evidence to suggest that Stx promotes human intestinal colonization (Robinson et al., 2006); however, it is more likely that Stx is maintained to facilitate survival in an environment outside of the human gut. Indeed, when challenged with a grazing protozoa, *Tetrahymena piriformis*, EHEC O157:H7 demonstrates an increased survival rate compared to Stx-negative isogenic mutants (Steinberg and Levin, 2007); K12 transduced with the Stx-converting phage also has increased fitness following *T. piriformis* exposure (Steinberg and Levin, 2007). Similarly, *Legionella pneumophila* has the ability to evade the human immune response, a characteristic suggested to be the result of coevolution with fresh water protozoa (Albert-Weissenberger et al., 2007), while the interaction between *S. enterica* and rumen protozoa has been shown to enhance pathogenicity and invasion (Rasmussen et al., 2005). Therefore, it has been argued that the pathogenicity of EHEC O157:H7 in humans may be coincidental (Pallen and Wren, 2007), resulting from a chance exposure to a bacterium that is highly adapted for survival in a challenging habitat outside the human gut.

In general, the propensity of an *E. coli* pathotype, such as EHEC, to acquire and maintain foreign DNA is not fully understood, but perhaps is a secondary effect of defective methyl-directed mismatch repair (LeClerc et al., 1996). Nevertheless, the retention of similar sets of laterally acquired DNA across divergent lineages suggests that these mobile elements confer a selective advantage, thereby allowing these lineages to persist, diversify, and ultimately give rise to new more fit lineages.

13.2.2.2 Evidence for Recombination and Random Mutation

In addition to the reshuffling of existing genes into new allele combinations, gene recombination events can result in the substitution of gene fragments from different sources. Therefore, varying portions of some genes can have different evolutionary origins. The presence of such genes can potentially obscure ancestral phylogenetic signals, and therefore, their phylogenies cannot be described by a single tree. In order to examine the effect of recombination on the clonal divergence of *E. coli* populations, Reid et al. analyzed the compatibility matrices of polymorphic nucleotides in seven conserved metabolic genes, or MLST loci (Reid et al., 2000). Two sites in a set of aligned sequences are considered compatible if each nucleotide polymorphism can be inferred to have occurred only once so that their phylogeny can be explained by a single tree (Jakobsen and Easteal, 1996). Reticulate analysis of within- and between-gene divergence among conserved MLST loci of EHEC and EPEC strains revealed a high degree of compatibility (Reid et al., 2000), indicating that past recombination has not played a significant role in the evolution of EHEC and EPEC backbone genes.

While point mutations have been documented to be imperative for the evolution of specific genes, they are not as important for the evolution of bacterial virulence. Mutations, for instance, were not found to play a role in the evolution of genes important for the phenotypic characteristics that distinguish intestinal *E. coli* from *S. enterica* (Ochman et al., 2000). In this scenario, lateral gene transfer events appear to be more important (Ochman et al., 2000), a finding also supported by the observation that recombination rates far exceeded mutation rates during the divergence of *E. coli* (Guttman and Dykhuizen, 1994). Similarly, a study of the evolution of genomic content in EHEC O157:H7 strains and close relatives using comparative genomic hybridizations has demonstrated that divergence in genome content due to recombination is ∼140 times greater than divergence at the nucleotide level due to mutation (Wick et al., 2005).

13.2.2.3 Experimental Evidence for Parallel Evolution

A comparison of genome-wide transcription between two non-mutator *E. coli* strains was performed before and after 20,000 generations of evolution in a constant environment to identify parallel changes in gene expression (Cooper et al., 2003). Parallel evolution was evident, as transcriptional profiles of the evolved populations differed more from the respective progenitor populations with regard to gene expression patterns than from each other. Indeed, 59 genes had significant changes in expression between the evolved and ancestral populations, and sequencing of several guanosine tetraphosphate (ppGpp) regulator genes identified a nonsynonymous mutation in *spoT* in one evolved population. The introduction of this *spoT* mutation into the ancestral background resulted in a fitness gain similar to that in the evolved lineage. However, the introduction of the *spoT* mutation into the other evolved strain had no effect on fitness, suggesting that the other evolved *E. coli* had developed a similar beneficial mutation in parallel. Remarkably, distinct *spoT* point mutations were subsequently identified in 8 of the 12 evolved lineages, confirming parallel change at the gene level (Cooper et al., 2003).

13.2.3 Stepwise Evolution

The sudden emergence of clinical illness associated with a new *E. coli* serotype (O157:H7) in 1982, for instance, provided good justification for examining the evolution of virulence in a foodborne pathogen. Data generated in prior studies have engendered an evolutionary model by which EHEC O157:H7 arose from an O55:H7-like ancestor through lateral acquisition of genetic material, as well as mutation of genes that encode various phenotypic traits (Reid et al., 2000; Whittam, 1998; Whittam et al., 1993); such genetic alterations have been suggested to occur in a series of steps or through stepwise evolution (Feng et al., 1998).

Unlike most *E. coli*, EHEC O157:H7 does not ferment sorbitol and lacks β-glucuronidase (GUD) activity, which is encoded by the highly conserved *uidA*. It has been demonstrated that GUD inactivity is due to a single point mutation at the – 10 promoter and at allele +92 in the structural gene of the EHEC 1 lineage relative to other EHEC lineages (Feng et al., 1998). The assessment of genetic polymorphisms in 163 intestinal *E. coli* strains using MLEE revealed that the EHEC 1 clonal complex, which consists of four electrophoretic types (ETs), is distantly related to the EPEC O55:H7 complex (Feng et al., 1998), confirming previous findings (Whittam et al., 1993). The ability to ferment sorbitol (SOR$^+$) and express GUD (GUD$^+$) was found to differ between the O157 and O55 strains in the study, as all O55:H7 strains were SOR$^+$ and GUD$^+$, contained the –10 mutation, lacked the *uidA* +92 mutation, and either possessed the Stx2 gene (*stx2*) or lacked *stx* altogether. Overall, each O157 strain was positive for the *uidA* +92 allele and *stx*, but the phenotypes varied among several distinct O157 populations and included the following: SOR$^+$ and GUD$^+$; an inability to ferment sorbitol (SOR$^-$) and lack of GUD activity (GUD$^-$); and SOR$^-$ and GUD$^+$ (Feng et al., 1998). Although the importance of Stx in disease is well established, the relationships between nonfermenting sorbitol and the lack of GUD expression and virulence are not known; however, it is possible that such changes occurred during adaptation and have allowed EHEC O157 strains to occupy a new niche distinct from EPEC O55.

Based on these findings, which postulate the least amount of evolutionary change, Feng et al. assumed the following: for metabolic housekeeping genes, loss of function is far more likely to occur than gain of function during evolution, and the gain of function most likely occurs through lateral gene transfer (Feng et al., 1998). Accordingly, a SOR$^+$, GUD$^+$ EPEC-like ancestor possessing the LEE island was hypothesized to have given rise to O55:H7 containing the *uidA* –10 mutation, a clone that represents the most recent common ancestor (A1) of all O55:H7 and O157:H7 clones in circulation today (Fig. 13.3). A2 represents an O55:H7 lineage

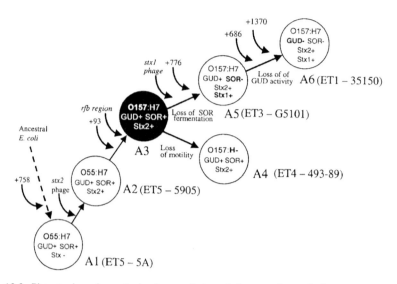

Fig. 13.3 Phenotypic and genotypic changes that mark the stepwise evolutionary emergence of EHEC O157:H7 from an O55:H7 ancestor. Taken from Feng et al. (1998)

that has acquired *stx2*, probably via phage transduction, while A3 is hypothetical as it has yet to be isolated. In the progression from A2 to A3, the *uidA* +92 mutation appeared as well as a change in the somatic antigen from O55 to O157, presumably through lateral transfer of the *rfb* gene cluster. This combination of changes resulted in the SOR⁺, GUD⁺, *stx2*⁺ O157:H7 lineage (A3) (Feng et al., 1998). At this juncture, it has also been postulated that the EHEC plasmid was acquired, as all descendents of A3 possess it, and from A3, two intermediate ancestors arose. A4, which retained the characteristics of A3 but lost its motility, is represented by strains of O157:H⁻, the common lineage associated with HUS outbreaks in Germany (i.e., 'German clone,' Karch et al., 1993), while the A5 O157:H7 lineage is GUD⁺, SOR⁻ and has acquired *stx1* in addition to *stx2* as well as the SpLE1 (Feng et al., 1998; Tarr et al., 2000). The loss of GUD activity (GUD⁻) marks the evolution of A6, which is SOR⁻, GUD⁻, *stx1*⁺, *stx2*⁺, and has both the *uidA* −10 and +92 mutations. A6 remains the most prevalent O157:H7 lineage in circulation in the United States (Fig. 13.3) (Feng et al., 1998).

The genetic basis for the phenotypic switch from GUD⁺ to GUD⁻ was not initially clear, as ancestral lineages A3, A4, and A5 have *uidA* mutations identical to those in the A6 descendent, but can express GUD⁺. A comparison of *uidA*, or *gusA*, nucleotide sequences between the *E. coli* K-12, the O157:H⁻ German clone (A4), and contemporary O157:H7 (A6) lineages identified 19 point mutations in GUD⁺ O157:H⁻ (A4), 3 of which are nonsynonymous (+758, +191, +93; the +92 allele was changed to +93 after the sequence was updated) (Monday et al., 2001). Similar to the +93 mutation, a +758 mutation is also present in the O55:H7 (A1) lineage. GUD⁺ O157:H⁻ (A4) and GUD⁻ O157:H7 (A6) also differ by four mutations in *gusA*, with three representing nonsynonymous substitutions (+686, +776, +1,370) and one representing a G-G dinucleotide insertion (+686) that results in a premature stop codon; the latter mutation is missing from the GUD$^+$ O157:H7 lineage (Monday et al., 2001). Complementation of GUD$^-$ strains with a *gusA* mutant possessing the +776 and +1,370 mutations, but lacking the G-G insertion because of a deletion at +686, restored GUD activity among O157:H7 strains of the A6 lineage (Monday et al., 2001). Consequently, a modified stepwise evolution model was proposed, which hypothesized that the +758 mutation appeared first during the divergence of O55:H7 (A1) from ancestral *E. coli* (Fig. 13.3) (Monday et al., 2001). Although the +93 mutation was likely present in the hypothetical A3 lineage, the +776 mutation was likely present in the O157:H7 (A5) lineage and not its predecessors, including A4; therefore, it was suggested that the +686 G-G insertion gave rise to the GUD$^-$ phenotype of the O157:H7 (A6) lineage (Fig. 13.3) (Monday et al., 2001). Using a similar comparative genomic and reverse genetics approach, Monday et al. have also identified a 12 bp in-frame deletion in *flhC*, which is part of the *flhCD* master flagellar regulator and underlies the loss of motility in the GUD$^+$ and SOR$^+$ O157:H$^-$ German clone (A4) (Monday et al., 2004).

To more completely assess the contribution of lateral gene transfer to the evolution of EHEC O157:H7, the genomic content was compared among O55:H7 and each O157 representative in the stepwise evolution model (Fig. 13.3) (Wick et al., 2005). Using microarrays that probed for ∼6,200 ORFs from three *E. coli* genomes (K-12 MG1655, O157:H7 Sakai, and O157:H7 EDL-933), it was demonstrated that the majority of O157 genes not found in O55:H7 have been introduced via prophage transduction (Wick et al., 2005). Furthermore, it was suggested that the O55:H7-like progenitor retained certain K-12 genes, which were lost as the O157 descendent acquired additional phage DNA from other sources (Wick et al., 2005). Another example illustrating the replacement of DNA fragments is highlighted in the O157:H$^-$ German clone (A4), as it shares several genes with K-12 that are absent in the widespread O157:H7 (A6) lineage (Wick et al., 2005). Together, these findings provide additional support for the stepwise evolution of EHEC O157:H7.

13.2.4 Diversification

Following emergence and establishment, a successful pathogenic population commonly undergoes further independent diversification, which results in a set of closely related, but unique lineages. EHEC O157:H7 represents a good example, as it has continued to diversify following its dissemination throughout human populations. While the ability of MLST to detect genetic variation between EHEC O157:H7 strains is limited (Noller et al., 2003), the combined use of MLST and PFGE data has uncovered as much as 30% genomic divergence across O157 lineages A3–A6 (Fig. 13.3) (Feng et al., 2007). More so, evidence for genomic variation has also been detected in the inferred O157:H7 ancestor, O55:H7 (A2), as these strains typically have the same ST but PFGE patterns can differ by 40% (Feng et al., 2007). Consequently, several additional studies utilizing genomic sequencing data have been conducted to better assess the genetic diversity and population structure of EHEC O157:H7 and its relatives.

The systematic screening of synonymous and nonsynonymous single-nucleotide polymorphisms (SNPs) across the genome has provided informative data that are amenable to phylogenetic and population genetic analyses. In a prior study of 11 EHEC O157:H7 strains isolated from different outbreaks with distinct PFGE patterns, Zhang et al. uncovered 906 SNPs in 523 chromosomal genes via comparative genome sequencing microarrays (Zhang et al., 2006). Considerable diversity in the pO157 plasmid was also observed across strains. By assuming a uniform synonymous substitution rate of 4.7×10^{-9} per site per year for *E. coli* and *S. enterica* (Lawrence and Ochman, 1998), these SNP data implied that the common EHEC O157:H7 lineage evolved ca. 40,000 years ago (Zhang et al., 2006), a more accurate estimate than those derived from MLST or MLEE data. In a follow-up study, SNPs identified by Zhang et al. were combined with SNPs detected in prior studies to establish a set of 96 SNPs in 83 genes for high-throughput genotyping by real-time PCR (Manning et al., 2008). Specifically, the SNP loci included 68 sites found by comparative genome microarrays (Zhang et al., 2006), 15 from housekeeping genes (Hyma et al., 2005), 4 from two complete O157 genomes (Sakai (Hayashi et al., 2001), EDL-933 (Perna et al., 2001)), and 9 from three virulence genes (*eae*, *espA*, and *fimA*). SNP data from >500 EHEC O157:H7 clinical isolates and close relatives (i.e., O55:H7) distinguished 39 SNP genotypes that grouped into 9 distinct clades using the minimum evolution (ME) algorithm (Fig. 13.4) (Manning et al., 2008). The deepest node in the ME phylogeny separates the ancestral O55:H7 lineage and the German clone (clade 9) from contemporary EHEC O157:H7 (clades 1–8), thereby confirming the relationships established in the stepwise evolution model (Feng et al., 1998).

In another follow-up study, (Leopold et al. 2009) resolved the contemporary O157:H7 population into three sequentially evolved clusters by examining backbone SNPs (radial and linear) in seven strains included in the stepwise model of EHEC O157:H7 evolution. These clusters correspond to a subset of clades identified in a prior SNP genotyping study (Manning et al., 2008) and have been noted in the ME phylogeny (Fig. 13.4). To assess the distribution of SNPs, Leopold et al. defined radial SNPs as intra-cluster polymorphic alleles present in one cluster, which implies within-cluster radiation from the founder. By contrast, linear SNPs are defined by their appearance in one cluster founder as well as those evolutionarily derived clusters (Leopold et al., 2009). Through this analysis, the number of SNP differences between clusters 1 and 2 was found to be significantly greater than the number between clusters 2 and 3 (Leopold et al., 2009), confirming that the EHEC O157:H7 subpopulation has diversified considerably over time. Furthermore, examination of the parsimoniously informative SNPs, or polymorphisms present in more than one strain, indicated that there is little evidence for recombination in the EHEC O157:H7 backbone. There are, however, two exceptions including the acquisition of the *rfb* cluster responsible for the O55 to O157 antigen switch as well as the accompanying *gnd* locus (Leopold et al., 2009). In general, these data suggest that recombination has had little to do with the diversification of the

Fig. 13.4 Minimum evolution (ME) tree illustrating the phylogenetic relationships among EHEC O157:H7 and close relatives. *Numbers* at the nodes are the bootstrap confidence values that mark 9 distinct clusters, or clades, which comprise 39 SNP genotypes (SGs). The EHEC O157 strains with ancestral traits (GUD+, SOR+) are found in clade 9, while the remaining clades consist of O157 strains with evolutionarily derived traits (GUD−, SOR−) (Manning et al., 2008). Clusters (numbered 1-3) represent those phylogenetic relationships uncovered by Leopold et al. (2009)

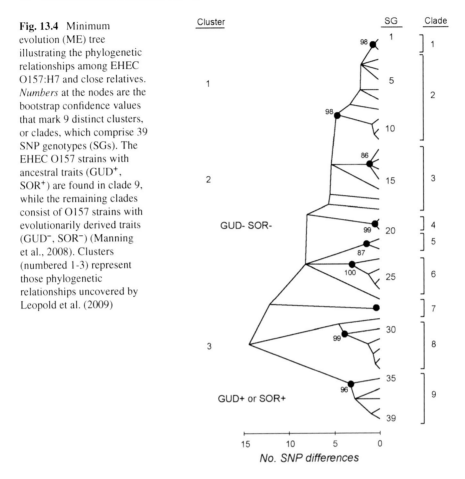

EHEC O157:H7 subpopulation. However, it is important to note that these data slightly contrast results from a prior SNP genotyping study that highlighted significant recombination among a small subset of genotypes (Manning et al., 2008). Additional characterization of those strains with evidence for recombination has detected mixed original cultures, and hence, omitting those strains from the analysis provides no evidence for recombination between clades (unpublished data).

Because the extant EHEC O157:H7 population is more diverse than previously expected, the SNP-based phylogenetic framework (Manning et al., 2008) was used to select divergent strains for complete genome sequencing. Specifically, an EHEC O157:H7 strain (TW14359) implicated in the 2006 North American spinach outbreak was selected, as it is a member of clade 8, an evolutionarily derived lineage most closely related to the ancestral clade 9 GUD+, SOR+ O157, and O55:H7 lineages and most divergent from the EHEC O157:H7 Sakai genome (clade 1) (Hayashi et al., 2001) (Fig. 13.4). The alignment of TW14359 strain sequence contigs to the Sakai genome identified 5,061 (96.3%) significant matches to the 5,253 Sakai strain genes (Manning et al., 2008). Apart from harboring the Stx2c-encoding phage 2851 (Strauch et al., 2008), which encodes the Stx2c toxin variant (Schmitt et al., 1991), TW14359 lacks 192 Sakai genes, 166 of which are encoded by prophage and prophage-like elements (Manning et al., 2008). For example, the Mu-like phage Sp18 that is integrated into the sorbose operon of the Sakai genome (Ohnishi et al., 2002) is absent in TW14359, while the pO157 plasmid was highly conserved (Manning et al., 2008). Among the 4,103 backbone genes shared between Sakai and TW14359, the average sequence identity is 99.8%. Of the 958 shared island genes, the average sequence identity is 97.96%, while the average sequence identity on all shared genes ($n=5,061$) is 99.25%. To determine the degree of similarity between TW14359, Sakai, and the EDL-933 (Perna et al., 2001) genomes, an analysis was conducted to identify highly conserved, shared backbone genes with <0.5% nucleotide divergence and more specifically to identify SNPs with <5 bp nucleotide differences per shared gene. Remarkably, the number of common genes ($n = 2,741$) with at least one SNP ranged from 87 between Sakai and EDL-933 to a distance that was roughly 3× greater ($n \sim 290$ genes) in the TW14359 strain. These data demonstrate that the Sakai (clade 1) and EDL-933 (clade 3) genomes are more similar to each other than both genomes are to the spinach outbreak strain genome, TW14359 (Manning et al., 2008). To support this finding, the complete genome sequence of TW14359 detected $\sim$70 kb of DNA segments that were absent from both Sakai and EDL-933 (Kulasekara et al., 2009).

Optical mapping, a recently developed high-resolution comparative genomic method for generating ordered genome restriction maps from single DNA molecules, identified several additional differences in the TW14359 strain relative to the two other EHEC O157:H7 genomes (Kotewicz et al., 2008). Notably, the optical maps illustrated that the Stx2-encoding phage inserts into *argW* in TW14359 instead of *wrbA*, as it does for both Sakai and EDL-933 (Kotewicz et al., 2008). Although TW14359 does not possess *stx1*, it harbors a phage similar to the Stx1-encoding phage occupying the *yehV* site, the same site that the Stx1-encoding phage occupies in the chromosomes of Sakai and EDL-933 (Kotewicz et al., 2008). The complete TW14359 genome also identified seven putative virulence genes to be exclusive to TW14359 and absent from Sakai and EDL-933 (Kulasekara et al., 2009). These genes share homologous regions with genes from other pathogens that encode factors important for inhibiting host immune responses, sequestering iron, and facilitating protein–protein interactions (Kulasekara et al., 2009). The *norV*, for example, encodes a nitric oxide reductase and has been detected in other *E. coli* pathotypes (Kulasekara et al., 2009). NorV is hypothesized to promote O157:H7 persistence in the large intestine, which is an anaerobic environment with high levels of nitric oxide. Because both Sakai and EDL-933 harbor a 204 bp deletion in *norV*, which likely results in an inactive protein, Kulasekara et al. screened an additional set of 100 O157:H7 strains from multiple sources for the presence of this deletion. Interestingly, an intact *norV* was detected in 42% of the strains and was correlated with the presence of those putative virulence factor genes exclusive to TW14359 as well as the absence of Stx1 (Kulasekara et al., 2009). Because nitric oxide inhibits

Stx production (Vareille et al., 2007), a disruption in *norV* could contribute in part to variable Stx expression, as higher levels of Stx may be produced by strains with an intact gene (Kulasekara et al., 2009).

13.2.5 Microevolution and Clonal Expansion

In some cases, geographic isolation or host restriction leads to either the diversification of specific clones or the dissemination of one common clone. A noteworthy example of the latter possibility is *S. enterica* serovar Typhi, which is considered to lack the genetic diversity routinely observed in other foodborne pathogens. Sequencing of 3,336 bp in seven housekeeping genes, for instance, detected only three polymorphic sites among 26 *S.* Typhi strains isolated from multiple geographic locations (Kidgell et al., 2002). Unlike EHEC O157:H7, which also lacks genetic diversity via sequence analysis of conserved housekeeping genes (Noller et al., 2003), a more thorough sequencing analysis of 19 *S.* Typhi strains has confirmed the lack of genetic diversity and uncovered little evidence of purifying selection, antigenic variation, and recombination between strains (Holt et al., 2008). Instead, among the 19 genome strains, 42 genes were affected by deletion events, 92 new pseudogenes were detected as were 55 nonsense SNPs. The nonsense SNPs, which were absent in the last common ancestor of *S.* Typhi, result in the introduction of a premature stop codon and prevent translation of the affected protein-coding gene (Holt et al., 2008). It was therefore hypothesized that genetic drift, combined with pseudogene formation and loss of gene function, is more important for the evolution of *S.* Typhi, a hypothesis attributable to a small effective population size since *S.* Typhi does not survive outside of humans (Holt et al., 2008). The carrier state of *S.* Typhi has also been suggested to be critical for the establishment and persistence of specific clones in a given population (Roumagnac et al., 2006). Despite the neutral population structure of *S.* Typhi overall (Roumagnac et al., 2006), microevolution, or the accumulation of small-scale genetic mutations, of existing *S.* Typhi clones likely occurs due to various adaptive pressures, including antibiotic use (Holt et al., 2008; Le et al., 2007). These mutations have resulted in the clonal expansion of antibiotic-resistant clones, or haplotypes, in distinct geographic locations (Roumagnac et al., 2006).

13.2.6 Gene Loss and Degradation

Acquisition of pathogenicity islands via lateral transfer is an established paradigm for the evolution of virulence. As indicated in comparative genomic studies of *S.* Typhi (Holt et al., 2008), some bacterial pathogens evolve by gene loss, gene degradation, and loss of gene function. While *S.* Typhi does not necessarily become more virulent due to loss of gene function, other pathogens can develop enhanced pathogenicity via deletions or, in the case of *Shigella* spp. and enteroinvasive *E. coli* (EIEC), via the formation of 'black holes' (Maurelli et al., 1998). Specifically, oligonucleotide probes targeting 14 genes identified from *E. coli* K-12 failed to hybridize to a subset of genes among four species of *Shigella* and EIEC. Of particular interest was the lack of *cadA*, the gene responsible for lysine decarboxylase activity that is absent in all *Shigella* spp. (Silva et al., 1980). Introduction of *cadA* into *Shigella flexneri 2a*, however, resulted in attenuated virulence and inhibition of both plasmid- and chromosomal-encoded enterotoxin activities (Maurelli et al., 1998). It was therefore hypothesized that such large-scale deletions may allow pathogens as well as common commensals to become more virulent.

13.2.7 Virulence Evolution In Vivo

Evolutionary changes in virulence have been suggested to occur because of trade-offs between the effect the pathogen has on the host and its ability to be transmitted to subsequent hosts (Anderson and May, 1979; Ewald, 1987). Alternative factors, including the host immune response and competition with other microbes, are also likely to play a key role in these processes (Alizon et al., 2009). For instance, a study of *S. enterica* serovar Typhimurium in a mouse colitis model demonstrated that the bacterium outcompeted the host microbiota, or collection of microorganisms, by eliciting an inflammatory immune response (Stecher et al., 2007). By contrast, colonization with an avirulent mutant incapable of inducing an inflammatory response was inhibited by the microbiota, a result that was negated in mixed infection models when the wild-type strain was present to initiate an inflammatory response (Stecher et al., 2007). The role that specific host responses play in the evolution of bacterial pathogens has not been fully elucidated, though it is clear that such responses present strong selective pressures that must be overcome in order to colonize the host and initiate the disease process.

13.2.7.1 Selection for Virulence

For the most part, in vivo studies that investigate bacterial pathogenesis have observed increased virulence following serial passage (Ebert, 1998); however, there are important exceptions. In a murine model of human gastroenteritis caused by *C. jejuni*, for example, C57BL/6 interleukin 10 knockout mice were infected with multiple phylogenetically distinct *C. jejuni* strains to assess differences in disease presentation (Bell et al., 2009). While some strains failed to colonize mice or cause disease, other strains caused disease of varying severity. Notably, three of five *C. jejuni* strains became more virulent following serial passage as determined by increased intestinal colonization, decreased time to disease development, and more pronounced pathology. Two of the three strains became more virulent after one passage, while one strain required three passages, illustrating that the rate in which virulence evolves can differ among strains (Bell et al., 2009). Furthermore, microarrays were used to compare the genome content between a virulent strain and a colonizing strain that failed to cause disease; 54 genes that differed between strains or were absent in the colonizing strain were identified. Among these 54 genes, the majority were part of operons encoding surface structures such as capsule, flagella, and lipo-oligosaccharide (LOS), all of which are important for *C. jejuni* disease pathogenesis. One example is the disruption of *gmhA*, a LOS biosynthesis gene (Bell et al., 2009), which was also demonstrated to contribute to enhanced pathogenicity in *S. enterica* (Tenor et al., 2004) and other pathogens in vivo. Therefore, the rate by which a pathogen induces disease increases depending on its adaptive potential, i.e., the potential to restructure its genetic composition so as to secure optimal fitness in a given environment.

13.2.7.2 Transmission Success and Virulence

Although there are several examples of pathogens becoming more virulent following serial passage, few studies have measured how virulence characteristics impact transmission, an important and quantifiable measure of an evolutionarily successful pathogen. One study of *Citrobacter rodentium*, a natural mouse pathogen commonly used to model EPEC- and EHEC-mediated human disease, was performed to examine the effect of disrupting various LEE-encoded type 3 secretion genes (Wickham et al., 2007). The *Δtir* and *ΔescN* mutant strains, for instance, became highly attenuated following colonization of mice and failed to be transmitted to naïve (uninfected) mice. By contrast, moderately attenuated mutant strains lacking *map, espF, espG, espH, sepZ, nleA,* and *nleB* retained the ability to transmit to naïve mice, though this was dependent on exposure timing. Specifically, *ΔespG* and *Δmap* mutant strains required 6 h for 100% transmission to uninfected mice compared to 3 h for the wild-type strain, demonstrating how variation in the combination of virulence determinants influences the rate of transmission (Wickham et al., 2007). A set of iterative competition experiments was also performed whereby mice challenged with *C. rodentium* mutant strains and wild-type strains were co-housed. Those mutant strains with highly attenuated virulence (e.g., *ΔnleA, ΔnleB*) were lost or outcompeted by the wild-type strain following two passages, while moderately attenuated mutant strains (e.g., *ΔespF*) were lost following three to five passages. The mutant *ΔespG* strain, however, was not outcompeted by the wild-type strain (Fig. 13.5), indicating that the retention of certain virulence genes

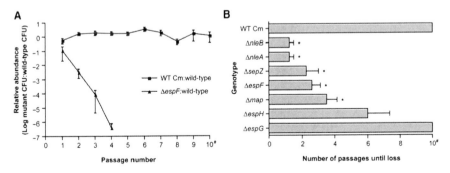

Fig. 13.5 Competitive colonization of mice with wild-type C. rodentium and isogenic mutants containing deletions in LEE genes. Taken from Wickham et al. (2007). (**a**) Mutant:wild-type ratios for the wild type strain versus ΔespF mutants. (**b**) Mean number of passages until loss of competition among multiple LEE gene mutants; all were significantly different ($p < 0.05$*) than wild-type numbers except ΔespH and ΔespG

provides a selective advantage under conditions when pathogen transmission rates are increased (Wickham et al., 2007).

13.2.7.3 Attenuated Virulence

The suppression of virulence properties in pathogenic bacteria during host infection is mediated not only through mechanisms that alter genetic composition, such as mutation and recombination, but also via fine regulation of gene expression. The combination of a genomic approach with in vivo models of disease, for example, has been used to examine virulence attenuation through gene regulation in *S. enterica* serovar Typhimurium (Coombes et al., 2005). Coombes et al. demonstrated that inactivation of *ydgT*, a homologue of nucleoid-associated repressors of virulence in *E. coli* and *Yersinia*, contributed to increased secretion of *S.* Typhimurium pathogenicity (SP) island 2 proteins and a biphasic virulence phenotype in mice (Coombes et al., 2005). Transcriptional profiling by microarrays further illustrated that SP-2 island genes, which mediate intracellular survival and replication, are upregulated in *ydgT* mutant strains relative to wild-type strains (Coombes et al., 2005). It was therefore suggested that in the biphasic phenotype, loss of SP-2 repression by *ydgT* inactivation provides an early advantage to *S.* Typhimurium for survival in the reticuloendothelial system, an ability that is lost later in the course of infection compared to the wild-type strain (Coombes et al., 2005). That is, although the *ydgT* mutant may express increased levels of virulence determinants due to a lack of SP-2 repression, the typhoid clinical illness that is characteristic of *S.* Typhimurium will not develop. This example provides support for the hypothesis that virulence suppression is just as important for pathogenesis as virulence gene activation for some pathogens.

Gene mutation and reassortative recombination are also commonly observed among bacterial pathogens as a means to suppress virulence or evade host immune responses. One study of gnotobiotic mice colonized with *E. coli* K-12 MG1655 reported the isolation of a new, non-motile, and heritable morphotype associated with a considerable fitness gain, which reached 90% prevalence within a week of colonization (Giraud et al., 2008). Transformation of this evolved clone with a DNA plasmid library generated from the ancestor restored the wild-type phenotype when the plasmids carried the *ompB* locus, which codes for the sensor kinase EnvZ and the transcriptional regulator OmpR. This two-component signal transduction system controls the expression of >100 genes. Sequencing of *envZ* and *ompR* revealed parallel missense point mutations in 90% of bacteria following a week of adaptation to the mouse gut, but not in the progenitor. These mutations were shown to indirectly downregulate flagellin expression and reduce membrane permeability. The authors speculate that flagellar repression aids in immune evasion and that decreased permeability promotes growth in the presence of bile salts, a major stressor in the intestinal lumen. More interesting than parallel change, in this case, was the observation that colonization with the wild type did not select for mutations that specifically control either motility or permeability, but for mutations in the *envZ/ompR* system that has multiple effects. This finding lends support to two hypotheses: (i) clonal interference prevents mutations that carry a small fitness gain in favor of those with pleiotropic effects and, therefore, potentially greater adaptive advantage; and (ii) global regulators may have evolved to accelerate adaptation through concerted and synchronized modulation of diverse functions (Giraud et al., 2008).

Gene loss is a common mechanism of virulence attenuation in foodborne pathogens, with the exception of 'black hole' events that increase pathogenicity as in the case of *Shigella* spp. and EIEC (Maurelli et al., 1998). In a classical experiment, EPEC serotype O127:H6 strain E2349/69 was fed to human volunteers to evaluate diarrhea associated with infection. Among over 170 isolates collected from human subjects with diarrhea, 66% had lost the EPEC adherence factor (EAF) plasmid following passage through the host (Levine et al., 1985). The EAF plasmid, in addition to the bundle-forming pili (*bfp*) genes, also encodes the plasmid-encoded regulator (*per*) of the EPEC type 3 secretion system. Loss of the EAF plasmid eliminates the ability of EPEC to develop attaching and effacing lesions and leads to virulence attenuation (Kaper et al., 2004). This was an unexpected finding because the EAF plasmid is highly stable in vitro, and therefore, this in vivo microevolutionary event is possibly the result of host selection pressures (Levine et al., 1985).

A growing body of evidence supports the hypothesis that large-scale genome reduction also occurs in EHEC as it passes through the gastrointestinal tract of humans. In a retrospective study of 787 patients with hemorrhagic colitis or HUS between 1996 and 2006, the examination of stools revealed that 5.5% of patients shed *stx*-negative (*stx⁻*), intimin-positive (*eae⁺*) strains of serogroups O26, O103, O145, and O157 (Bielaszewska et al., 2007). That these *stx⁻*/*eae⁺* strains are EHEC, and not EPEC, was determined by screening for other EHEC virulence factors not present in EPEC as well as MLST for seven housekeeping genes (*purA, adk, icd, fumC, recA, mdh,* and *gyrB*) (Mellmann et al., 2009). Remarkably, a comparison of *stx⁻*/*eae⁺* E. coli to other clinical EHEC strains of corresponding serotypes has uncovered identical virulence gene profiles (Fig. 13.6), while MLST data demonstrated that *stx⁻*/*eae⁺* E. coli are EHEC that have lost the Stx-encoding phage during in vivo passage (Fig. 13.7) (Mellmann et al., 2009)

Excision of the Stx phage leaves the corresponding *tRNA* integration site intact, implying that conversion to *stx⁻*/*eae⁺* by EHEC strains is not unidirectional. Although this is not commonly observed during human infection, experimental evidence shows that EHEC O157 strains that have lost the Stx2 phage during human infection are readily transduced by the Stx2 phage (Mellmann et al., 2008). Indeed, uptake of the Stx2 phage from EHEC O26:H11 and O157:H7 by various *E. coli* strains was shown in several animal models, strongly suggesting that this occurs in humans. Indirect epidemiological support is found in the increased frequency of isolation of *stx1⁻*/*stx2⁺* EHEC O26:H11, which were predominantly *stx1⁺*/*stx2⁻*, over the last 40 years (Mellmann et al., 2009).

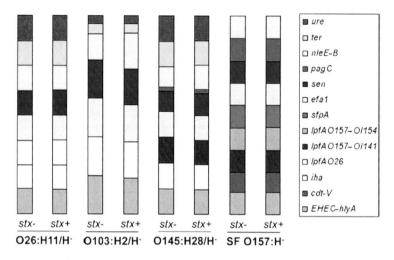

Fig. 13.6 Comparison of virulence gene profiles between *stx⁻/eae⁺ E. coli* isolated from patients with bloody diarrhea or hemolytic uremic syndrome and EHEC strains of the corresponding serotype. Note that virulence profiles between the two groups are virtually identical. Taken from Mellmann et al. (2009)

13.3 Genetic Basis of Adaptation

Adaptive evolution is a process by which natural selection acts to promote the spread of beneficial changes, resulting in increased reproductive success of a particular bacterial type, or clone, in a given environment. While not all evolutionary changes are driven by environmental or adaptive pressures (e.g., genetic drift), all adaptive changes occur by natural selection. For foodborne bacterial pathogens, adaptive evolution can occur in multiple environments and species and can directly affect virulence potential. EHEC O157:H7, for example, can survive in cattle and other ruminants as well as soil, water, and multiple foods. By contrast, more specialized pathogens, such as *S. enterica* serovars Gallinarum and Pullorum (poultry), Choleraesuis (swine), and Dublin (cattle), are restricted to a single host or environment. In both cases, however, human infection occurs following transmission and passage through the gastrointestinal tract, thereby requiring that these microorganisms have the ability to readily adapt to multiple environments. Indeed, some pathogens, including EHEC O157:H7, have acquired the ability to adapt to stressful conditions (e.g., extreme acidity) or have found a specific niche within the host (e.g., intestinal pathogenic *E. coli*) that enhances the likelihood of survival.

Several molecular mechanisms are important for mediating adaptive evolutionary processes such as mutation, recombination, and gene acquisition, loss, and decay. It has long been known that bacteria have evolved parasexual mechanisms of gene exchange that, by and large, explain the presence of strain-specific bacterial DNA fragments, such as pathogenicity islands that carry gene sets important for virulence.

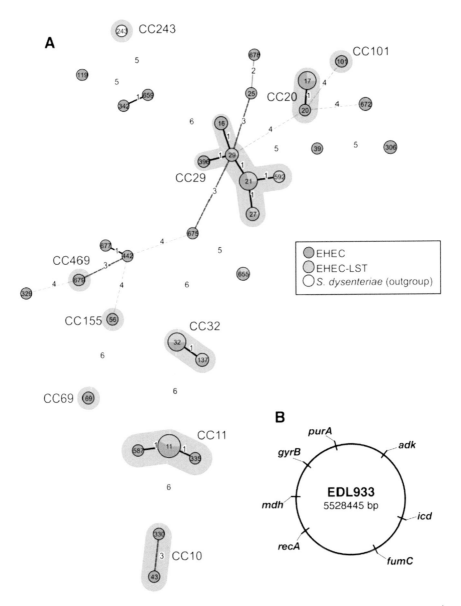

Fig. 13.7 Minimum-spanning tree based on multilocus sequence typing data from *stx⁻/eae⁺* *E. coli* (EHEC-LST), EHEC, and *Shigella dysenteriae*. *Stx⁻/eae⁺ E. coli* strains isolated from patients are members of the same clonal complexes (CCs) as corresponding EHEC strains. Taken from Mellmann et al. (2009)

During these processes, bacterial species have diverged into multiple subspecies, or subpopulations, that contain genetic characteristics unique to each. As illustrated by recent comparative genomic studies, *E. coli* represents an important example of a bacterium with the capacity to acquire and maintain exogenous genetic material.

13.3.1 Lifestyle and Host Adaptation

Several comparative genomic studies have now been conducted to examine trends of gene acquisition, loss, and decay to assess how genomic content is shaped by host specificity and adaptation to specific tissues and environments. For instance, *S. enterica* serovar Enteritidis has a broad host range and slightly larger genome (4.69 Mbp) than its descendent, the chicken-restricted *S. enterica* serovar Gallinarum (4.66 Mbp), with which it shares 99.7% nucleotide sequence homology among shared orthologs (Thomson et al., 2008). Nevertheless, the *S.* Gallinarum strain 287/91 has been shown to have fewer functional genes and significantly more pseudogenes than the *S.* Enteritidis PT4 strain (Thomson et al., 2008). The number of *S.* Gallinarum pseudogenes is comparable to the numbers observed in *S. enterica* serovars Typhi and Paratyphi A, both of which are also restricted to a single host (Parkhill et al., 2001; Thomson et al., 2008). In addition, these three host-restricted serovars are highly invasive, have acquired the potential to induce severe systemic disease, and have lost the ability to colonize the intestinal tract. It has therefore been hypothesized that genome reduction and pseudogene formation, which are a form of reductive evolution, have contributed to the inability of these species to adapt to multiple hosts and tissues (Thomson et al., 2008).

For *S.* Gallinarum 287/91, several mutations have been identified to be important for host and tissue adaptation. Interestingly, some of these mutations have evolved independently and in parallel with other *S. enterica* serovars including both *S.* Typhi CT18 and *S.* Paratyphi (Thomson et al., 2008). A mutation in the cellulose biosynthesis gene (*bcsG*), for example, prevents the production of cellulose, a major ingredient of biofilms (Zogaj et al., 2001). For *Salmonella* spp., biofilm formation has been suggested to facilitate survival in the intestine, as biofilms can protect the bacterium from both chemical and physical stressors (Solano et al., 2002). Therefore, the lack of biofilm production may partly explain why *S.* Gallinarum has a decreased ability to persist in the intestines. Similarly, *S.* Gallinarum 287/91 was found to possess 12 of the 13 fimbrial operons present in the *S.* Enteritidis PT4 chromosome; numerous mutations were identified in key fimbrial genes from *S.* Gallinarum 287/91 as were high levels of pseudogene formation. Nonetheless, both species have also acquired virulence plasmids containing genetically distinct fimbrial operons, a trait shared by *S.* Typhi (Pickard et al., 2003) that demonstrates parallel evolution for the purpose of host adaptation (Thomson et al., 2008). Additional mutations in chemotactic and flagellar genes were also suggested to be important for the dissemination of *S.* Gallinarum to certain tissues and to avoid immune responses, such as pathogen-associated molecular pattern recognition systems (Thomson et al., 2008).

Along with the acquisition of novel virulence factors, a similar loss of gene function has also been observed by comparing genome sequences of *Yersinia enterocolitica* and its descendent, *Y. pestis* (Parkhill et al., 2001). While *Y. enterocolitica* is an enteropathogen, *Y. pestis* is a highly virulent and invasive blood-borne pathogen with limited ability to survive in the environment. There are 149 pseudogenes in *Y. pestis* representing 3.7% of the genome, the majority of which are for surface antigens or secreted proteins; this is not surprising since most of these proteins are required for enteropathogenicity and not the systemic virulence that is characteristic of *Y. pestis* infection (Parkhill et al., 2001). For example, while *Y. pseudotuberculosis* relies on YadA and invasin proteins to colonize the mucosal lining of the intestines, genes encoding these proteins are not functional in *Y. pestis*, as this mechanism of colonization is redundant and unnecessary to the *Y. pestis* lifestyle. Similarly, while motility brings *Y. enterocolitica* into contact with the host cell and indirectly promotes efficient invasion, *Y. pestis* is typically non-motile due to six pseudogenes in flagellar and chemotactic gene clusters (Parkhill et al., 2001). In all, these examples demonstrate that reductive evolution is important for both host specificity and lifestyle adaptation of various foodborne pathogens.

13.3.2 Antibiotic Resistance

Selection pressures created by the long-term use and misuse of antimicrobial agents have resulted in high frequencies of antimicrobial resistance in most bacterial populations. Indeed, antibiotics have been used in disease treatment and animal husbandry practices since the 1940s, which represents a short time span to observe such widespread resistance considering that the evolution of pathogenic bacterial populations has been ongoing for millions of years (Rowe-Magnus and Mazel, 2006). The development of antimicrobial resistance has therefore been suggested to represent an example of evolution in action.

Antibiotic resistance, however, comes with a cost to bacteria, as numerous studies have shown that antibiotic-resistant bacteria have a decreased ability to survive, persist, and compete with their parent antibiotic-sensitive counterparts (Andersson and Levin, 1999). In the case of bacterial pathogens, the burden of antibiotic resistance is also associated with decreased transmission and virulence. Therefore, because resistant pathogens generally exhibit decreased fitness in the absence of antibiotic pressures, it is logical to assume that reduced antibiotic use will lead to a decline in resistance rates. The rationale behind this suggestion is that once the selection pressure acting on a population is lifted, the mutant population will revert back to the wild-type state to avoid extinction. Although reversion to a sensitive state has been observed, resistant bacteria more frequently develop compensatory mutations, which are additional mutations in the same resistance gene or other genes regardless of location, that work in concert to minimize the fitness costs associated with resisting antibiotics (Andersson and Levin, 1999). Compensatory evolution has been suggested to result in the persistence of antibiotic resistance in bacterial populations, and hence, limiting the emergence and spread of resistant strains is critical for combating the problem (Andersson, 2003).

Evidence for compensatory evolution has been observed in multiple studies conducted in vitro and in vivo. The serial passage of avirulent, slow-growing antibiotic-resistant *S. enterica* serovar Typhimurium mutants in mice, for example, has been shown to restore the growth rate and virulence of the mutants via compensatory mutations (Bjorkman et al.,1998). Passage of the same mutant strains in mice and LB medium resulted in different compensatory mutations, indicating that compensatory evolution is also influenced by environmental factors (the habitat that the bacteria occupy) in addition to antibiotic selection. Such findings emphasize the importance of using animals as clinically relevant models of bacterial adaptation to examine the costs associated with antibiotic resistance. In another study, second-site compensatory mutations were also observed in the chromosomally encoded streptomycin-resistant gene (*rpsL*) of *E. coli* following in vitro growth for 180 generations in an antibiotic-free environment (Schrag et al., 1997). To assess the effect of compensatory mutations on antibiotic sensitivity, Schrag et al. replaced the mutated *rpsL* alleles with the wild-type, sensitive *rpsL*$^+$ allele in both the progenitors and evolved clones using phage-mediated transduction. While the resistant progenitors had transduction frequencies comparable to the sensitive clones, the rate of transduction was significantly lower in the evolved, resistant clones, inferring decreased viability. Direct competition assays confirmed the fitness disadvantage of the evolved strains harboring the wild-type *rpsL*$^+$ allele and also suggested that more than one compensatory mutation was generated because the evolved clones originating from the same progenitor had different fitness disadvantages once the sensitive *rpsL*$^+$ allele was re-introduced (Schrag et al., 1997). This study demonstrates that reverting back to the wild-type, sensitive, state may require more than one mutation and, therefore, may partly explain why antibiotic-resistant strains are stabilized in bacterial populations despite having reduced fitness.

There is also evidence to suggest that some antibiotic resistance mutations can confer a fitness advantage without compensatory mutations, even in the absence of antibiotic selection. One study of clonally related *C. jejuni*, for example, compared the fitness of antibiotic-sensitive strains and fluoroquinolone-resistant, spontaneous mutants by direct competition in chickens (Luo et al., 2005). The fluoroquinolone-resistant mutants, which had identical single-base substitutions in *gyrA* (DNA gyrase A subunit), not only retained the ability to efficiently colonize the chicken, but were able to outcompete the wild-type, antibiotic-sensitive strain. Furthermore, eight additional competition experiments, pairing fluoroquinolone-sensitive and -resistant strains isolated on the same day, also revealed that the resistant strains outcompeted the sensitive strains within 3 days post-infection in all but two cases. In the latter two cases, the strain pairs were rearranged with respect to the day of isolation to determine whether possible compensatory mutations developed that are different with respect to the day of isolation. Regardless of the day of isolation, however, one of the two fluoroquinolone-resistant strains failed to outcompete the sensitive strain, excluding a 'later day' compensatory event. Together, these findings demonstrate not only that a single point mutation alone, without compensatory mutations, is responsible for the observed fitness advantage, but also that the genetic constitution of the strain may influence the effect of the antibiotic resistance mutation (Luo et al., 2005).

Similar *gyrA* point mutations have been identified in other pathogens as well and can be both chromosomal or plasmid encoded. For *S. enterica* serovar Typhi, resistance to fluoroquinolones, the first-line agent used to treat typhoid fever, is widespread in many parts of the world, particularly the Indian subcontinent and

Southeast Asia (Wain and Kidgell, 2004). Despite the neutral population structure of *S*. Typhi, as determined by genome sequencing, Holt et al. (2008) detected evidence for positive selection in the chromosomal *gyrA* gene. Interestingly, several SNPs have been uncovered in *gyrA* among 19 *S*. Typhi strains, which is likely due to the strong selective pressure exerted on the *S*. Typhi genome from fluoroquinolone use (Holt et al., 2008). Clonal expansion of several resistant clones has therefore occurred (Roumagnac et al., 2006), which is not surprising given that the increased fitness associated with antibiotic resistance is strong (Holt et al., 2008).

While there are several examples of adaptive mutations that result because of antibiotic pressures, most bacterial pathogens develop resistance via the acquisition of foreign resistance cassettes through lateral gene transfer. The uptake of resistance genes in nature depends on the genetic background of a bacterial population, that is, its propensity to acquire foreign DNA, whereas the rate and extent of fixation of antibiotic resistance in a given population are inversely proportional to the burden it imposes. In a study of *Campylobacter coli* strains from turkeys bred at multiple farms in North Carolina and Virginia, resistance to tetracycline, streptomycin, erythromycin, kanamycin, ciprofloxacin, and nalidixic acid was observed in 59 of 108 (55%) strains (D'Lima et al., 2007). The application of MLST demonstrated that the 59 multidrug-resistant (MDR) strains represented 14 multilocus sequence types (STs), with 3 STs accounting for ∼70% of all MDR strains. Subtyping data generated using different methods (e.g., PFGE) confirmed the *C. coli* population structure derived from MLST data, providing evidence that the 14 MDR STs represent closely related, but distinct subtypes. Furthermore, the three predominant MDR STs were found at multiple farms over a 3-year period and were specific to turkeys; none were isolated from swine farms in the same region. These data imply that specific MDR *C. coli* lineages have disseminated throughout multiple turkey populations and that these lineages have diversified over time (D'Lima et al., 2007). Similar findings have also been observed for MDR *S. enterica* serovar Typhimurium and non-Typhimurium serovars, as MDR due to the diversification of a few successful clones harboring resistance genes present on specific mobile elements, particularly integrons, was found to occur more frequently than the horizontal transfer of integrons across divergent lineages (Krauland et al., 2009).

For the most part, prior studies of antibiotic resistance mechanisms, compensatory mutations, and resistance gene dissemination have focused on the characterization of only a handful of genes or mutations. The use of genomic sequencing, however, has demonstrated that mobile genetic elements as well as chromosomally encoded factors can simultaneously play a role in resistance development, particularly to multiple antibiotics. In addition, full transcriptome analyses have illustrated that multiple genes, including those required for DNA repair, are upregulated following antibiotic exposure, while key virulence genes are downregulated. *S. enterica* serovar Typhimurium exposed to nalidixic acid, for example, had decreased expression of *Salmonella* pathogenicity (SP) islands 1 and 2 in one study (Dowd et al., 2007). Increased expression of genes specific for five families of multidrug-resistant efflux pumps as well as various outer membrane lipoproteins was also observed, indicating that changes in the permeability of the cell membrane were established following nalidixic acid exposure (Dowd et al., 2007). These data indicate that the selective pressures caused by antibiotic use, even at low doses, frequently contribute to multiple changes in the bacterium. Although the ability to rapidly adapt to the presence of harmful antimicrobials initially contributes to lower fitness, bacterial pathogens will become more fit through compensatory evolution and may ultimately develop an enhanced ability to persist and outcompete wild-type sensitive populations. Consequently, it has been suggested that novel drug candidates should be evaluated for the overall effect on the bacterium prior to use. Drugs that cause multiple mutations or mutations by which compensatory evolution will have a pleiotropic negative effect may significantly increase the fitness cost to the bacterium and decrease the likelihood of resistance development (Andersson, 2006). Similarly, the sensible use of drugs and combinations of drugs with different targets or mechanisms could potentially inhibit the stabilization and persistence of resistant clones in the population.

13.3.3 Acid Resistance

The transmission route of foodborne pathogens is such that they are continuously exposed to highly acidic conditions (e.g., food production processes, storage, and ingestion by a broad range of hosts). The ability of E. coli, for instance, to avoid low pH-mediated killing occurs following exposure to mildly acidic environments, which are encountered routinely during food production and passage through the gastrointestinal tract, rather than in environments with a neutral pH (Benjamin and Datta, 1995). Following acid exposure, several foodborne pathogens, such as EHEC O157:H7, S. enterica serovar Typhimurium, and Listeria monocytogenes, have evolved multiple mechanisms of acid resistance, which has also been indirectly linked to increased pathogenicity. Bacterial adaptation to highly acidic conditions contributes to foodborne disease because it facilitates the survival in and contamination of acidic foods (e.g., apple cider and fermented sausage). Prior studies, however, have demonstrated that there is variation in the ability to resist acid among foodborne pathogens; such variation is due to differences in the level of acid encountered, strain source, type of acid resistance system being utilized, and genetic background of the bacterial strain.

One study of survival in a model stomach system using simulated gastric fluid (pH 2.5) has detected differences among divergent EHEC lineages, EHEC O157:H7 and non-O157 EHEC of serogroups O26 and O111. Specifically, strains of EHEC O157:H7 had twice the survival rate of non-O157 EHEC strains. The number of injured cells after passage through the model stomach, and stationary phase transcript levels of the glutamate decarboxylase-dependent (GAD) system genes, gadA and gadB, also varied between EHEC lineages (Bergholz and Whittam, 2007). Although two additional acid resistance systems, the arginine decarboxylase-dependent system and glucose catabolite-repressed system, have been identified among E. coli, the GAD system is thought to be most important for EHEC O157:H7 survival in the gastrointestinal tract (Lin et al., 1996).

A key step in the evolution of acid resistance among *Escherichia* and *Shigella* was the acquisition, via lateral gene transfer, of an acid fitness island (AFI) containing the GAD system genes. The *gadB* gene, located outside the AFI, is hypothesized to have been acquired by duplication of *gadA* instead of through multiple independent acquisition events. This hypothesis is based on the high degree of similarity between *gadA* and *gadB* and the lack of a homologous AFI in closely related *Salmonella* spp. (Bergholz et al., 2007). Duplicated genes can follow different evolutionary trajectories including sequence divergence and the development of new functions or gene conversion (i.e., nonreciprocal recombination) resulting from high sequence similarities (Bergholz et al., 2007). An analysis of *gadA* and *gadB* genes from multiple strains of EHEC and enteropathogenic *E. coli* (EPEC), two phylogenetically and phenotypically distinct pathotypes, identified at least three paralogous gene conversion events that have resulted in the genetic homogenization of both *gad* genes (Bergholz et al., 2007). EHEC O157:H7 strains, members of the EHEC 1 clonal group, have divergent *gadA* and *gadB* genes unlike non-O157 EHEC and EPEC strains, a difference that could have resulted from the different ecological niches utilized by each pathogen (Bergholz et al., 2007).

Recent studies provide support for the hypothesis that different environments can influence the ability of bacteria to adapt to acidic conditions. A comparison of EHEC O157:H7 strains isolated from different sources, including human patients, foods, bovine carcasses and feces, and water, demonstrated significantly different survival rates following exposure to acetic acid. Specifically, strains isolated from patients were less resistant to acetic acid than environmental strains (Oh et al., 2009). Further evidence for variation in acid resistance and survivability among phylogenetically distinct EHEC O157:H7 strains associated with human illness and asymptomatic bovine colonization was recently observed using whole-genome microarrays. Most notably, fitness and stress response genes, including those important for the GAD system, were upregulated in the bovine strains, which also had significantly higher survival rates than human strains after passage through the model stomach (Kailisan Vanaja et al., 2010).

13.4 Concluding Remarks

It is evident that whole-genome sequencing and comparative genomic studies have revolutionized the field of microbial evolution. In general, we now have a better understanding of how (i) genetic variation and evolutionary processes (e.g., gene transfer and mutation) are linked to virulence; (ii) new pathogens emerge in different populations or transmission vehicles; and (iii) different pathogens adapt to hostile environments. While the evolution of bacterial pathogens was traditionally thought to occur through mutations, rearrangements, and lateral gene transfer, it is also evident that many other processes, such as gene loss and degradation, play an important role. The mechanism used by any one pathogen, however, is dependent on multiple factors, including a pathogen's lifestyle, its genetic composition, and the selective pressures it encounters both in the host and in the environment. All three factors as well as those evolutionary processes that guide them are important to determine how to effectively prevent disease caused by foodborne bacterial pathogens.

References

Albert-Weissenberger C, Cazalet C, Buchrieser C (2007) *Legionella pneumophila* – a human pathogen that co-evolved with fresh water protozoa. Cell Mol Life Sci 64:432–448

Alizon S, Hurford A, Mideo N, Van Baalen M (2009) Virulence evolution and the trade-off hypothesis: history, current state of affairs and the future. J Evol Biol 22:245–259

Anderson RM, May RM (1979) Population biology of infectious diseases: Part I. Nature 280: 361–367

Andersson DI (2003) Persistence of antibiotic resistant bacteria. Curr Opin Microbiol 6:452–456

Andersson DI (2006) The biological cost of mutational antibiotic resistance: any practical conclusions? Curr Opin Microbiol 9:461–465

Andersson DI, Levin BR (1999) The biological cost of antibiotic resistance. Curr Opin Microbiol 2:489–493

Bell JA, St Charles JL, Murphy AJ, Rathinam VA, Plovanich-Jones AE, Stanley EL, Wolf JE, Gettings JR, Whittam TS, Mansfield LS (2009) Multiple factors interact to produce responses resembling spectrum of human disease in *Campylobacter jejuni* infected C57BL/6 IL-10$^{-/-}$ mice. BMC Microbiol 9:57

Benjamin MM, Datta AR (1995) Acid tolerance of enterohemorrhagic *Escherichia coli*. Appl Environ Microbiol 61:1669–1672

Bergholz TM, Tarr CL, Christensen LM, Betting DJ, Whittam TS (2007) Recent gene conversions between duplicated glutamate decarboxylase genes (*gadA* and *gadB*) in pathogenic *Escherichia coli*. Mol Biol Evol 24:2323–2333

Bergholz TM, Whittam TS (2007) Variation in acid resistance among enterohaemorrhagic *Escherichia coli* in a simulated gastric environment. J Appl Microbiol 102:352–362

Bielaszewska M, Kock R, Friedrich AW, von Eiff C, Zimmerhackl LB, Karch H, Mellmann A (2007) Shiga toxin-mediated hemolytic uremic syndrome: time to change the diagnostic paradigm? PLoS One 2:e1024

Bjorkman J, Hughes D, Andersson DI (1998) Virulence of antibiotic-resistant *Salmonella* Typhimurium. Proc Natl Acad Sci USA 95:3949–3953

Blattner FR, Plunkett G 3rd, Bloch CA, Perna NT, Burland V, Riley M, Collado-Vides J, Glassner JD, Rode CK, Mayhew GF, Gregor J, Davis NW, Kirkpatrick HA, Goeden MA, Rose DJ, Mau B, Shao Y (1997) The complete genome sequence of *Escherichia coli* K-12. Science 277: 1453–1462

Brussow H, Canchaya C, Hardt WD (2004) Phages and the evolution of bacterial pathogens: from genomic rearrangements to lysogenic conversion. Microbiol Mol Biol Rev 68:560–602

Centers for Disease Control and Prevention (2009) Preliminary FoodNet data on the incidence of infection with pathogens transmitted commonly through food – 10 States, 2008. Morb Mortal Wkly Rep 58:333–337

Coombes BK, Wickham ME, Lowden MJ, Brown NF, Finlay BB (2005) Negative regulation of *Salmonella* pathogenicity island 2 is required for contextual control of virulence during typhoid. Proc Natl Acad Sci USA 102:17460–17465

Cooper TF, Rozen DE, Lenski RE (2003) Parallel changes in gene expression after 20,000 generations of evolution in *Escherichia coli*. Proc Natl Acad Sci USA 100:1072–1077

Darwin C (1859) On the origin of species by means of natural selection, 1st edn. John Murray, London

Dobrindt U, Agerer F, Michaelis K, Janka A, Buchrieser C, Samuelson M, Svanborg C, Gottschalk G, Karch H, Hacker J (2003) Analysis of genome plasticity in pathogenic and commensal *Escherichia coli* isolates by use of DNA arrays. J Bacteriol 185:1831–1840

Donnenberg MS (2002) Escherichia coli virulence mechanisms of a versatile pathogen. Academic Press, San Diego, CA

Dowd SE, Killinger-Mann K, Blanton J, San Francisco M, Brashears M (2007) Positive adaptive state: microarray evaluation of gene expression in *Salmonella enterica* Typhimurium exposed to nalidixic acid. Foodborne Pathog Dis 4:187–200

D'Lima CB, Miller WG, Mandrell RE, Wright SL, Siletzky RM, Carver DK, Kathariou S (2007) Clonal population structure and specific genotypes of multidrug-resistant *Campylobacter coli* from turkeys. Appl Environ Microbiol 73:2156–2164

Ebert D (1998) Experimental evolution of parasites. Science 282:1432–1435

Escobar-Paramo P, Clermont O, Blanc-Potard AB, Bui H, Le Bouguenec C, Denamur E (2004) A specific genetic background is required for acquisition and expression of virulence factors in *Escherichia coli*. Mol Biol Evol 21:1085–1094

Ewald PW (1987) Transmission, modes and evolution of the parasitism-mutualism continuum. Ann NY Acad Sci 503:295–306

Feng P, Lampel KA, Karch H, Whittam TS (1998) Genotypic and phenotypic changes in the emergence of *Escherichia coli* O157:H7. J Infect Dis 177:1750–1753

Feng PC, Monday SR, Lacher DW, Allison L, Siitonen A, Keys C, Eklund M, Nagano H, Karch H, Keen J, Whittam TS (2007) Genetic diversity among clonal lineages within *Escherichia coli* O157:H7 stepwise evolutionary model. Emerg Infect Dis 13:1701–1706

Fukiya S, Mizoguchi H, Tobe T, Mori H (2004) Extensive genomic diversity in pathogenic *Escherichia coli* and *Shigella* Strains revealed by comparative genomic hybridization microarray. J Bacteriol 186:3911–3921

Giraud A, Arous S, De Paepe M, Gaboriau-Routhiau V, Bambou JC, Rakotobe S, Lindner AB, Taddei F, Cerf-Bensussan N (2008) Dissecting the genetic components of adaptation of *Escherichia coli* to the mouse gut. PLoS Genet 4:e2

Goncalves AG, Campos LC, Gomes TA, Rodrigues J, Sperandio V, Whittam TS, Trabulsi LR (1997) Virulence properties and clonal structures of strains of *Escherichia coli* O119 serotypes. Infect Immun 65:2034–2040

Guttman DS, Dykhuizen DE (1994) Clonal divergence in *Escherichia coli* as a result of recombination, not mutation. Science 266:1380–1383

Hayashi T, Makino K, Ohnishi M, Kurokawa K, Ishii K, Yokoyama K, Han CG, Ohtsubo E, Nakayama K, Murata T, Tanaka M, Tobe T, Iida T, Takami H, Honda T, Sasakawa C, Ogasawara N, Yasunaga T, Kuhara S, Shiba T, Hattori M, Shinagawa H (2001) Complete genome sequence of enterohemorrhagic *Escherichia coli* O157:H7 and genomic comparison with a laboratory strain K-12. DNA Res 8:11–22

Holt KE, Parkhill J, Mazzoni CJ, Roumagnac P, Weill FX, Goodhead I, Rance R, Baker S, Maskell DJ, Wain J, Dolecek C, Achtman M, Dougan G (2008) High-throughput sequencing provides insights into genome variation and evolution in *Salmonella* Typhi. Nat Genet 40:987–993

Hyma KE, Lacher DW, Nelson AM, Bumbaugh AC, Janda JM, Strockbine NA, Young VB, Whittam TS (2005) Evolutionary genetics of a new pathogenic *Escherichia* species: *Escherichia albertii* and related *Shigella boydii* strains. J Bacteriol 187:619–628

Jakobsen IB, Easteal S (1996) A program for calculating and displaying compatibility matrices as an aid in determining reticulate evolution in molecular sequences. Comput Appl Biosci 12:291–295

Kailisan Vanaja S, Springman AC, Besser TE, Whittam TS, Manning SD (2010) Differential expression of virulence and stress fitness genes between *Escherichia coli* O157:H7 strains with clinical or bovine-biased genotypes. Appl Environ Microbiol 76:60–68

Kaper JB, Nataro JP, Mobley HL (2004) Pathogenic *Escherichia coli*. Nat Rev Microbiol 2:123–140

Karaolis DK, Johnson JA, Bailey CC, Boedeker EC, Kaper JB, Reeves PR (1998) A *Vibrio cholerae* pathogenicity island associated with epidemic and pandemic strains. Proc Natl Acad Sci USA 95:3134–3139

Karaolis DK, Somara S, Maneval DR Jr, Johnson JA, Kaper JB (1999) A bacteriophage encoding a pathogenicity island, a type-IV pilus and a phage receptor in cholera bacteria. Nature 399: 375–379

Karch H, Bohm H, Schmidt H, Gunzer F, Aleksic S, Heesemann J (1993) Clonal structure and pathogenicity of Shiga-like toxin-producing, sorbitol-fermenting *Escherichia coli* O157:H. J Clin Microbiol 31:1200–1205

Kidgell C, Reichard U, Wain J, Linz B, Torpdahl M, Dougan G, Achtman M (2002) *Salmonella* Typhi, the causative agent of typhoid fever, is approximately 50,000 years old. Infect Genet Evol 2:39–45

Kotewicz ML, Mammel MK, Leclerc JE, Cebula TA (2008) Optical mapping and 454 sequencing of *Escherichia coli* O157:H7 isolates linked to the US 2006 spinach-associated outbreak. Microbiology 154:3518–3528

Krauland MG, Marsh JW, Paterson DL, Harrison LH (2009) Integron-mediated multidrug resistance in a global collection of nontyphoidal *Salmonella enterica* isolates. Emerg Infect Dis 15:388–396

Kulasekara BR, Jacobs M, et al. (2009) Analysis of the genome of the *Escherichia coli* O157:H7 2006 spinach-associated outbreak isolate indicates candidate genes that may enhance virulence. Infect Immun 77:3713–3721

Law D (2000) Virulence factors of *Escherichia coli* O157 and other Shiga toxin-producing *E. coli*. J Appl Microbiol 88:729–745

Lawrence JG, Ochman H (1998) Molecular archaeology of the *Escherichia coli* genome. Proc Natl Acad Sci USA 95:9413–9417

Le TA, Fabre L, Roumagnac P, Grimont PA, Scavizzi MR, Weill FX (2007) Clonal expansion and microevolution of quinolone-resistant *Salmonella enterica* serotype Typhi in Vietnam from 1996 to 2004. J Clin Microbiol 45:3485–3492

LeClerc JE, Li B, Payne WL, Cebula TA (1996) High mutation frequencies among *Escherichia coli* and *Salmonella* pathogens. Science 274:1208–1211

Leopold SR, Magrini V, Holt NJ, Shaikh N, Mardis ER, Cagno J, Ogura Y, Iguchi A, Hayashi T, Mellmann A, Karch H, Besser TE, Sawyer SA, Whittam TS, Tarr PI (2009) A precise reconstruction of the emergence and constrained radiations of *Escherichia coli* O157 portrayed by backbone concatenomic analysis. Proc Natl Acad Sci USA 106:8713–8718

Levine MM, Nataro JP, Karch H, Baldini MM, Kaper JB, Black RE, Clements ML, O'Brien AD (1985) The diarrheal response of humans to some classic serotypes of enteropathogenic *Escherichia coli* is dependent on a plasmid encoding an enteroadhesiveness factor. J Infect Dis 152:550–559

Lin J, Smith MP, Chapin KC, Baik HS, Bennett GN, Foster JW (1996) Mechanisms of acid resistance in enterohemorrhagic *Escherichia coli*. Appl Environ Microbiol 62: 3094–3100

Luo N, Pereira S, Sahin O, Lin J, Huang S, Michel L, Zhang Q (2005) Enhanced in vivo fitness of fluoroquinolone-resistant *Campylobacter jejuni* in the absence of antibiotic selection pressure. Proc Natl Acad Sci USA 102:541–546

Manning SD, Motiwala AS, Springman AC, Qi W, Lacher DW, Ouellette LM, Mladonicky JM, Somsel P, Rudrik JT, Dietrich SE, Zhang W, Swaminathan B, Alland D, Whittam TS (2008) Variation in virulence among clades of *Escherichia coli* O157:H7 associated with disease outbreaks. Proc Natl Acad Sci USA 105:4868–4873

Maurelli AT, Fernandez RE, Bloch CA, Rode CK, Fasano A (1998) 'Black holes' and bacterial pathogenicity: a large genomic deletion that enhances the virulence of *Shigella* spp. and enteroinvasive *Escherichia coli*. Proc Natl Acad Sci USA 95:3943–3948

Medini D, Donati C, Tettelin H, Masignani V, Rappuoli R (2005) The microbial pan-genome. Curr Opin Genet Dev 15:589–594

Mellmann A, Bielaszewska M, Karch H (2009) Intrahost genome alterations in enterohemorrhagic *Escherichia coli*. Gastroenterology 136:1925–1938

Mellmann A, Lu S, Karch H, Xu JG, Harmsen D, Schmidt MA, Bielaszewska M (2008) Recycling of Shiga toxin 2 genes in sorbitol-fermenting enterohemorrhagic *Escherichia coli* O157:NM. Appl Environ Microbiol 74:67–72

Monday SR, Minnich SA, Feng PC (2004) A 12-base-pair deletion in the flagellar master control gene *flhC* causes nonmotility of the pathogenic German sorbitol-fermenting *Escherichia coli* O157:H- strains. J Bacteriol 186:2319–2327

Monday SR, Whittam TS, Feng PC (2001) Genetic and evolutionary analysis of mutations in the *gusA* gene that cause the absence of beta-glucuronidase activity in *Escherichia coli* O157:H7. J Infect Dis 184:918–921

Noller AC, McEllistrem MC, Stine OC, Morris JG Jr, Boxrud DJ, Dixon B, Harrison LH (2003) Multilocus sequence typing reveals a lack of diversity among *Escherichia coli* O157:H7 isolates that are distinct by pulsed-field gel electrophoresis. J Clin Microbiol 41: 675–679

Ochman H, Lawrence JG, Groisman EA (2000) Lateral gene transfer and the nature of bacterial innovation. Nature 405:299–304

Ogura Y, Ooka T, et al. (2009) Comparative genomics reveal the mechanism of the parallel evolution of O157 and non-O157 enterohemorrhagic *Escherichia coli*. Proc Natl Acad Sci USA 106:17939–17944

Oh DH, Pan Y, Berry E, Cooley M, Mandrell R, Breidt F Jr (2009) *Escherichia coli* O157:H7 strains isolated from environmental sources differ significantly in acetic acid resistance compared with human outbreak strains. J Food Prot 72:503–509

Ohnishi M, Kurokawa K, Hayashi T (2001) Diversification of *Escherichia coli* genomes: are bacteriophages the major contributors? Trends Microbiol 9:481–485

Ohnishi M, Terajima J, Kurokawa K, Nakayama K, Murata T, Tamura K, Ogura Y, Watanabe H, Hayashi T (2002) Genomic diversity of enterohemorrhagic *Escherichia coli* O157 revealed by whole genome PCR scanning. Proc Natl Acad Sci USA 99:17043–17048

O'Brien AD, Lively TA, Chang TW, Gorbach SL (1983) Purification of *Shigella dysenteriae 1* (Shiga)-like toxin from *Escherichia coli* O157:H7 strain associated with haemorrhagic colitis. Lancet 2:573

Pallen MJ, Wren BW (2007) Bacterial pathogenomics. Nature 449:835–842

Parkhill J, Dougan G, James KD, Thomson NR, Pickard D, Wain J, Churcher C, Mungall KL, Bentley SD, Holden MT, Sebaihia M, Baker S, Basham D, Brooks K, Chillingworth T, Connerton P, Cronin A, Davis P, Davies RM, Dowd L, White N, Farrar J, Feltwell T, Hamlin N, Haque A, Hien TT, Holroyd S, Jagels K, Krogh A, Larsen TS, Leather S, Moule S, O'Gaora P, Parry C, Quail M, Rutherford K, Simmonds M, Skelton J, Stevens K, Whitehead S, Barrell BG (2001) Complete genome sequence of a multiple drug resistant *Salmonella enterica* serovar Typhi CT18. Nature 413:848–852

Parkhill J, Wren BW, Thomson NR, Titball RW, Holden MT, Prentice MB, Sebaihia M, James KD, Churcher C, Mungall KL, Baker S, Basham D, Bentley SD, Brooks K, Cerdeno-Tarraga AM, Chillingworth T, Cronin A, Davies RM, Davis P, Dougan G, Feltwell T, Hamlin N, Holroyd S, Jagels K, Karlyshev AV, Leather S, Moule S, Oyston PC, Quail M, Rutherford K, Simmonds M, Skelton J, Stevens K, Whitehead S, Barrell BG (2001) Genome sequence of *Yersinia pestis*, the causative agent of plague. Nature 413:523–527

Perna NT, Plunkett G 3rd, Burland V, Mau B, Glasner JD, Rose DJ, Mayhew GF, Evans PS, Gregor J, Kirkpatrick HA, Posfai G, Hackett J, Klink S, Boutin A, Shao Y, Miller L, Grotbeck EJ, Davis NW, Lim A, Dimalanta ET, Potamousis KD, Apodaca J, Anantharaman TS, Lin J, Yen G, Schwartz DC, Welch RA, Blattner FR (2001) Genome sequence of enterohaemorrhagic *Escherichia coli* O157:H7. Nature 409:529–533

Pickard D, Wain J, Baker S, Line A, Chohan S, Fookes M, Barron A, Gaora PO, Chabalgoity JA, Thanky N, Scholes C, Thomson N, Quail M, Parkhill J, Dougan G (2003) Composition, acquisition, and distribution of the Vi exopolysaccharide-encoding *Salmonella enterica* pathogenicity island SPI-7. J Bacteriol 185:5055–5065

Rasko DA, Rosovitz MJ, Myers GS, Mongodin EF, Fricke WF, Gajer P, Crabtree J, Sebaihia M, Thomson NR, Chaudhuri R, Henderson IR, Sperandio V, Ravel J (2008) The pangenome structure of *Escherichia coli*: comparative genomic analysis of *E. coli* commensal and pathogenic isolates. J Bacteriol 190:6881–6893

Rasmussen MA, Carlson SA, Franklin SK, McCuddin ZP, Wu MT, Sharma VK (2005) Exposure to rumen protozoa leads to enhancement of pathogenicity of and invasion by multiple-antibiotic-resistant *Salmonella enterica* bearing SGI1. Infect Immun 73:4668–4675

Reid SD, Herbelin CJ, Bumbaugh AC, Selander RK, Whittam TS (2000) Parallel evolution of virulence in pathogenic *Escherichia coli*. Nature 406:64–67

Robinson CM, Sinclair JF, Smith MJ, O'Brien AD (2006) Shiga toxin of enterohemorrhagic *Escherichia coli* type O157:H7 promotes intestinal colonization. Proc Natl Acad Sci USA 103:9667–9672

Roumagnac P, Weill FX, Dolecek C, Baker S, Brisse S, Chinh NT, Le TA, Acosta CJ, Farrar J, Dougan G, Achtman M (2006) Evolutionary history of *Salmonella* Typhi. Science 314: 1301–1304

Rowe-Magnus D, Mazel D (2006) The evolution of antibiotic resistance.In: Seifert HS, DiRita VJ (eds) Evolution of enteric pathogens. ASM Press, Washington, DC

Schmidt H, Zhang WL, Hemmrich U, Jelacic S, Brunder W, Tarr PI, Dobrindt U, Hacker J, Karch H (2001) Identification and characterization of a novel genomic island integrated at *selC* in locus of enterocyte effacement-negative, Shiga toxin-producing *Escherichia coli*. Infect Immun 69:6863–6873

Schmitt CK, McKee ML, O'Brien AD (1991) Two copies of Shiga-like toxin II-related genes common in enterohemorrhagic *Escherichia coli* strains are responsible for the antigenic heterogeneity of the O157:H- strain E32511. Infect Immun 59:1065–1073

Schrag SJ, Perrot V, Levin BR (1997) Adaptation to the fitness costs of antibiotic resistance in *Escherichia coli*. Proc Biol Sci 264:1287–1291

Scott AE, Timms AR, Connerton PL, El-Shibiny A, Connerton IF (2007) Bacteriophage influence *Campylobacter jejuni* types populating broiler chickens. Environ Microbiol 9:2341–2353

Serra-Moreno R, Jofre J, Muniesa M (2007) Insertion site occupancy by stx2 bacteriophages depends on the locus availability of the host strain chromosome. J Bacteriol 189:6645–6654

Silva RM, Toledo MR, Trabulsi LR (1980) Biochemical and cultural characteristics of invasive *Escherichia coli*. J Clin Microbiol 11:441–444

Solano C, Garcia B, Valle J, Berasain C, Ghigo JM, Gamazo C, Lasa I (2002) Genetic analysis of *Salmonella* Enteritidis biofilm formation: critical role of cellulose. Mol Microbiol 43:793–808

Stecher B, Robbiani R, Walker AW, Westendorf AM, Barthel M, Kremer M, Chaffron S, Macpherson AJ, Buer J, Parkhill J, Dougan G, von Mering C, Hardt WD (2007) *Salmonella enterica* serovar Typhimurium exploits inflammation to compete with the intestinal microbiota. PLoS Biol 5:2177–2189

Steinberg KM, Levin BR (2007) Grazing protozoa and the evolution of the *Escherichia coli* O157:H7 Shiga toxin-encoding prophage. Proc Biol Sci 274:1921–1929

Strauch E, Hammerl JA, Konietzny A, Schneiker-Bekel S, Arnold W, Goesmann A, Puhler A, Beutin L (2008) Bacteriophage 2851 is a prototype phage for dissemination of the Shiga toxin variant gene 2c in *Escherichia coli* O157:H7. Infect Immun 76:5466–5477

Tarr PI, Bilge SS, Vary JC Jr, Jelacic S, Habeeb RL, Ward TR, Baylor MR, Besser TE (2000) Iha: a novel *Escherichia coli* O157:H7 adherence-conferring molecule encoded on a recently acquired chromosomal island of conserved structure. Infect Immun 68:1400–1407

Tenor JL, McCormick BA, Ausubel FM, Aballay A (2004) *Caenorhabditis elegans*-based screen identifies *Salmonella* virulence factors required for conserved host-pathogen interactions. Curr Biol 14:1018–1024

Thomson NR, Clayton DJ, Windhorst D, Vernikos G, Davidson S, Churcher C, Quail MA, Stevens M, Jones MA, Watson M, Barron A, Layton A, Pickard D, Kingsley RA, Bignell A, Clark L, Harris B, Ormond D, Abdellah Z, Brooks K, Cherevach I, Chillingworth T, Woodward J, Norberczak H, Lord A, Arrowsmith C, Jagels K, Moule S, Mungall K, Sanders

M, Whitehead S, Chabalgoity JA, Maskell D, Humphrey T, Roberts M, Barrow PA, Dougan G, Parkhill J (2008) Comparative genome analysis of *Salmonella* Enteritidis PT4 and *Salmonella* Gallinarum 287/91 provides insights into evolutionary and host adaptation pathways. Genome Res 18:1624–1637

Vareille M, de Sablet T, Hindre T, Martin C, Gobert AP (2007) Nitric oxide inhibits Shiga-toxin synthesis by enterohemorrhagic *Escherichia coli*. Proc Natl Acad Sci USA 104:10199–10204

Wain J, Kidgell C (2004) The emergence of multidrug resistance to antimicrobial agents for the treatment of typhoid fever. Trans R Soc Trop Med Hyg 98:423–430

Whittam TS (1998) Evolution of Escherichia coli O157:H7 and other Shiga toxin-producing E. coli strains. In: Kaper JB, O'Brien AD (eds) Escherichia coli O157:H7 and other Shiga toxin-producing E. coli strains. American Society for Microbiology, Washington, DC, pp 195–212

Whittam TS, Wolfe ML, Wachsmuth IK, Orskov F, Orskov I, Wilson RA (1993) Clonal relationships among *Escherichia coli* strains that cause hemorrhagic colitis and infantile diarrhea. Infect Immun 61:1619–1629

Wick LM, Qi W, Lacher DW, Whittam TS (2005) Evolution of genomic content in the stepwise emergence of *Escherichia coli* O157:H7. J Bacteriol 187:1783–1791

Wickham ME, Brown NF, Boyle EC, Coombes BK, Finlay BB (2007) Virulence is positively selected by transmission success between mammalian hosts. Curr Biol 17:783–788

Zhang W, Qi W, Albert TJ, Motiwala AS, Alland D, Hyytia-Trees EK, Ribot EM, Fields PI, Whittam TS, Swaminathan B (2006) Probing genomic diversity and evolution of *Escherichia coli* O157 by single nucleotide polymorphisms. Genome Res 16:757–767

Zogaj X, Nimtz M, Rohde M, Bokranz W, Romling U (2001) The multicellular morphotypes of *Salmonella* Typhimurium and *Escherichia coli* produce cellulose as the second component of the extracellular matrix. Mol Microbiol 39:1452–1463

Index

Note: The letters 'f' and 't' following the locators refer to figures and tables respectively.

M. Wiedmann, W. Zhang (eds.), *Genomics of Foodborne Bacterial Pathogens,*
Food Microbiology and Food Safety, DOI 10.1007/978-1-4419-7686-4,
© Springer Science+Business Media, LLC 2011